Bacteriophages

Bacteriophages: Alternatives to Antibiotics and Beyond

Editors

Pilar García Suárez
Lucía Fernández

MDPI • Basel • Beijing • Wuhan • Barcelona • Belgrade • Manchester • Tokyo • Cluj • Tianjin

Editors
Pilar García Suárez
Instituto de Productos Lácteos de Asturias (IPLA-CSIC)
Spain

Lucía Fernández
Instituto de Productos Lácteos de Asturias (IPLA-CSIC)
Spain

Editorial Office
MDPI
St. Alban-Anlage 66
4052 Basel, Switzerland

This is a reprint of articles from the Special Issue published online in the open access journal *Antibiotics* (ISSN 2079-6382) (available at: https://www.mdpi.com/journal/antibiotics/special_issues/bacteriophages).

For citation purposes, cite each article independently as indicated on the article page online and as indicated below:

LastName, A.A.; LastName, B.B.; LastName, C.C. Article Title. *Journal Name* **Year**, *Article Number*, Page Range.

ISBN 978-3-03943-404-6 (Hbk)
ISBN 978-3-03943-405-3 (PDF)

Contents

About the Editors ix

Preface to "Bacteriophages: Alternatives to Antibiotics and Beyond" xi

Pilar Domingo-Calap and Jennifer Delgado-Martínez
Bacteriophages: Protagonists of a Post-Antibiotic Era
Reprinted from: *Antibiotics* **2018**, *7*, 66, doi:10.3390/antibiotics7030066 1

Sharita Divya Ganeshan and Zeinab Hosseinidoust
Phage Therapy with a Focus on the Human Microbiota
Reprinted from: *Antibiotics* **2019**, *8*, 131, doi:10.3390/antibiotics8030131 17

Lucía Fernández, Susana Escobedo, Diana Gutiérrez, Silvia Portilla, Beatriz Martínez, Pilar García and Ana Rodríguez
Bacteriophages in the Dairy Environment: From Enemies to Allies
Reprinted from: *Antibiotics* **2017**, *6*, 27, doi:10.3390/antibiotics6040027 37

Panos G. Kalatzis, Daniel Castillo, Pantelis Katharios and Mathias Middelboe
Bacteriophage Interactions with Marine Pathogenic Vibrios: Implications for Phage Therapy
Reprinted from: *Antibiotics* **2018**, *7*, 15, doi:10.3390/antibiotics7010015 51

Carlos São-José
Engineering of Phage-Derived Lytic Enzymes: Improving Their Potential as Antimicrobials
Reprinted from: *Antibiotics* **2018**, *7*, 29, doi:10.3390/antibiotics7020029 75

Michael J. Love, Dinesh Bhandari, Renwick C. J. Dobson and Craig Billington
Potential for Bacteriophage Endolysins to Supplement or Replace Antibiotics in Food Production and Clinical Care
Reprinted from: *Antibiotics* **2018**, *7*, 17, doi:10.3390/antibiotics7010017 107

Yen-Te Liao, Alexandra Salvador, Leslie A. Harden, Fang Liu, Valerie M. Lavenburg, Robert W. Li and Vivian C. H. Wu
Characterization of a Lytic Bacteriophage as an Antimicrobial Agent for Biocontrol of Shiga Toxin-Producing *Escherichia coli* O145 Strains
Reprinted from: *Antibiotics* **2019**, *8*, 74, doi:10.3390/antibiotics8020074 133

Jude Ajuebor, Colin Buttimer, Sara Arroyo-Moreno, Nina Chanishvili, Emma M. Gabriel, Jim O'Mahony, Olivia McAuliffe, Horst Neve, Charles Franz and Aidan Coffey
Comparison of *Staphylococcus* Phage K with Close Phage Relatives Commonly Employed in Phage Therapeutics
Reprinted from: *Antibiotics* **2018**, *7*, 37, doi:10.3390/antibiotics7020037 153

Henrike Zschach, Mette V. Larsen, Henrik Hasman, Henrik Westh, Morten Nielsen, Ryszard Miedzybrodzki, Ewa Jończyk-Matysiak, Beata Weber-Dabrowska and Andrzej Górski
Use of a Regression Model to Study Host-Genomic Determinants of Phage Susceptibility in MRSA
Reprinted from: *Antibiotics* **2018**, *7*, 9, doi:10.3390/antibiotics7010009 167

María Cebriá-Mendoza, Rafael Sanjuán and Pilar Domingo-Calap
Directed Evolution of a Mycobacteriophage
Reprinted from: *Antibiotics* **2019**, *8*, 46, doi:10.3390/antibiotics8020046 183

T. Scott Brady, Christopher P. Fajardo, Bryan D. Merrill, Jared A. Hilton, Kiel A. Graves, Dennis L. Eggett and Sandra Hope
Bystander Phage Therapy: Inducing Host-Associated Bacteria to Produce Antimicrobial Toxins against the Pathogen Using Phages
Reprinted from: *Antibiotics* **2018**, *7*, 105, doi:10.3390/antibiotics7040105 . . . . **193**

Ergun Akturk, Hugo Oliveira, Sílvio B. Santos, Susana Costa, Suleyman Kuyumcu, Luís D. R. Melo and Joana Azeredo
Synergistic Action of Phage and Antibiotics: Parameters to Enhance the Killing Efficacy Against Mono and Dual-Species Biofilms
Reprinted from: *Antibiotics* **2019**, *8*, 103, doi:10.3390/antibiotics8030103 . . . . **207**

James J. Bull, Kelly A. Christensen, Carly Scott, Benjamin R. Jack, Cameron J. Crandall and Stephen M. Krone
Phage-Bacterial Dynamics with Spatial Structure: Self Organization around Phage Sinks Can Promote Increased Cell Densities
Reprinted from: *Antibiotics* **2018**, *7*, 8, doi:10.3390/antibiotics7010008 . . . . **227**

Janet Y. Nale, Tamsin A. Redgwell, Andrew Millard and Martha R. J. Clokie
Efficacy of an Optimised Bacteriophage Cocktail to Clear *Clostridium difficile* in a Batch Fermentation Model
Reprinted from: *Antibiotics* **2018**, *7*, 13, doi:10.3390/antibiotics7010013 . . . . **253**

Xiaoran Shang and Daniel C. Nelson
Contributions of Net Charge on the PlyC Endolysin CHAP Domain
Reprinted from: *Antibiotics* **2019**, *8*, 70, doi:10.3390/antibiotics8020070 . . . . **269**

Cristina Howard-Varona, Dean R. Vik, Natalie E. Solonenko, Yueh-Fen Li, M. Consuelo Gazitua, Lauren Chittick, Jennifer K. Samiec, Aubrey E. Jensen, Paige Anderson, Adrian Howard-Varona, Anika A. Kinkhabwala, Stephen T. Abedon and Matthew B. Sullivan
Fighting Fire with Fire: Phage Potential for the Treatment of *E. coli* O157 Infection
Reprinted from: *Antibiotics* **2018**, *7*, 101, doi:10.3390/antibiotics7040101 . . . . **281**

Randolph Fish, Elizabeth Kutter, Daniel Bryan, Gordon Wheat and Sarah Kuhl
Resolving Digital Staphylococcal Osteomyelitis Using Bacteriophage—A Case Report
Reprinted from: *Antibiotics* **2018**, *7*, 87, doi:10.3390/antibiotics7040087 . . . . **295**

Nanna Rørbo, Anita Rønneseth, Panos G. Kalatzis, Bastian Barker Rasmussen, Kirsten Engell-Sørensen, Hans Petter Kleppen, Heidrun Inger Wergeland, Lone Gram and Mathias Middelboe
Exploring the Effect of Phage Therapy in Preventing *Vibrio anguillarum* Infections in Cod and Turbot Larvae
Reprinted from: *Antibiotics* **2018**, *7*, 42, doi:10.3390/antibiotics7020042 . . . . **301**

Tuan Son Le, Thi Hien Nguyen, Hong Phuong Vo, Van Cuong Doan, Hong Loc Nguyen, Minh Trung Tran, Trong Tuan Tran, Paul C. Southgate and D. İpek Kurtböke
Protective Effects of Bacteriophages against *Aeromonas hydrophila* Causing Motile Aeromonas Septicemia (MAS) in Striped Catfish
Reprinted from: *Antibiotics* **2018**, *7*, 16, doi:10.3390/antibiotics7010016 . . . . **317**

Catarina Moreirinha, Nádia Osório, Carla Pereira, Sara Simões, Ivonne Delgadillo and Adelaide Almeida
Protein Expression Modifications in Phage-Resistant Mutants of *Aeromonas salmonicida* after AS-A Phage Treatment
Reprinted from: *Antibiotics* **2018**, *7*, 21, doi:10.3390/antibiotics7010021 . . . . **329**

Expert round table on acceptance and re-implementation of bacteriophage therapy, Wilbert Sybesma, Christine Rohde, Pavol Bardy, Jean-Paul Pirnay, Ian Cooper, Jonathan Caplin, Nina Chanishvili, Aidan Coffey, Daniel De Vos, Amber Hartman Scholz, Shawna McCallin, Hilke Marie Püschner, Roman Pantucek, Rustam Aminov, Jiří Doškař and D. İpek Kurtb́oke
Silk Route to the Acceptance and Re-Implementation of Bacteriophage Therapy—Part II
Reprinted from: *Antibiotics* **2018**, *7*, 35, doi:10.3390/antibiotics7020035 . . . . **343**

Lorena Rodríguez-Rubio, Joan Jofre and Maite Muniesa
Is Genetic Mobilization Considered When Using Bacteriophages in Antimicrobial Therapy?
Reprinted from: *Antibiotics* **2017**, *6*, 32, doi:10.3390/antibiotics6040032 . . . . **367**

About the Editors

Pilar García Suárez, Ph.D., is a Staff Research Scientist of CSIC (Higher Council for Scientific Research National Agency, Spain). Prior to joining the DairySafe team at the Institute of Dairy Products of Asturias (IPLA-CSIC), Dr. García graduated summa cum laude from the University of Oviedo (Spain). She continued her studies in microbiology, earning her PhD at the same university. The majority of her research career has been focused on the exploitation of bacteriophages as genetic tools to be applied in the food industry. In this field, two main research areas can be distinguished: i) bacteriophages infecting lactic acid bacteria as one of the most important problems in the dairy industry, and ii) bacteriophages as antimicrobial agents (human therapy and food safety). Throughout both stages, common subjects have been developed, including: the isolation and morphological characterization of bacteriophages, the sequencing and bioinformatic analysis of bacteriophage's genome, the transcriptomic and proteomic analysis of phage-encoded genes, the expression and purification of relevant phage proteins, the study of phage lytic activities, and the design and analysis of bacteriophage resistant strains. At the moment, the biopreservation of dairy products by phages has become the main topic of her research. The design of phage mixtures and phage-encoded products as biopreservatives against food-borne bacteria led her to apply predictive microbiology and hurdle technology as methods in the improvement of food safety. The removal of biofilms by phage-derived antimicrobials is one of the main interests of her current research.

Lucía Fernández, Ph.D., completed her PhD in Microbiology at the University of Oviedo (Spain). After a postdoctoral stay at the University of British Columbia (Canada), she joined the DairySafe team at the Asturian Institute of Dairy Products (IPLA-CSIC, Spain), where she is currently working as a postdoctoral researcher. Her work focuses on studying the application of bacteriophages and endolysins as new antimicrobials.

Preface to "Bacteriophages: Alternatives to Antibiotics and Beyond"

Bacteriophages, the natural enemies of bacteria, are perpetually under intense investigation in order to exploit their antimicrobial properties. Indeed, so-called phage therapy can have multiple applications in diverse sectors, such as human and veterinary medicine, food production, or aquaculture. Additionally, the use of phages to modulate human microbiota is also a possibility, which shows how much phages can contribute to human health. At the same time, phage lytic proteins (endolysins) are also under the microscope for their many advantages, including the fact that they can be easily engineered and, therefore, adapted to meet specific antimicrobial requirements. However, the design, production and approval of a novel phage-based product requires a long process, starting from the identification of an effective antimicrobial, the subsequent optimization of its application to avoid unwanted effects and, finally, its approval and commercialization. The objective of this Special Issue and book is to publish research papers that add to our understanding of phage–bacteria interactions, which will be useful information in order to improve the lytic ability of these viruses and their derived proteins. Additionally, papers that show the advantages and disadvantages of phage therapy are also welcome.

Pilar García Suárez, Lucía Fernández

Editors

Preface to "Bacteriophages: Alternatives to Antibiotics and Beyond"

Bacteriophages, the natural enemies of bacteria, are perpetually under intense investigation in order to exploit their antimicrobial properties. Indeed, so-called phage therapy can have multiple applications in diverse sectors, such as human and veterinary medicine, food production, or aquaculture. Additionally, the use of phages to modulate human microbiota is also a possibility, which shows how much phages can contribute to human health. At the same time, phage lytic proteins (endolysins) are also under the microscope for their many advantages, including the fact that they can be easily engineered and, therefore, adapted to meet specific antimicrobial requirements. However, the design, production and approval of a novel phage-based product requires a long process, starting from the identification of an effective antimicrobial, the subsequent optimization of its application to avoid unwanted effects and, finally, its approval and commercialization. The objective of this Special Issue and book is to publish research papers that add to our understanding of phage–bacteria interactions, which will be useful information in order to improve the lytic ability of these viruses and their derived proteins. Additionally, papers that show the advantages and disadvantages of phage therapy are also welcome.

Pilar García Suárez
Editor

antibiotics

MDPI

Review

Bacteriophages: Protagonists of a Post-Antibiotic Era

Pilar Domingo-Calap [1,2,*] and Jennifer Delgado-Martínez [1]

[1] Department of Genetics, Universitat de València, 46100 Burjassot, Valencia, Spain; jendel@alumni.uv.es
[2] Institute for Integrative Systems Biology (I2SysBio), Universitat de València-CSIC, 46980 Paterna, Valencia, Spain
* Correspondence: domingocalap@gmail.com; Tel.: +34-963-543-261

Received: 12 July 2018; Accepted: 25 July 2018; Published: 27 July 2018

Abstract: Despite their long success for more than half a century, antibiotics are currently under the spotlight due to the emergence of multidrug-resistant bacteria. The development of new alternative treatments is of particular interest in the fight against bacterial resistance. Bacteriophages (phages) are natural killers of bacteria and are an excellent tool due to their specificity and ecological safety. Here, we highlight some of their advantages and drawbacks as potential therapeutic agents. Interestingly, phages are not only attractive from a clinical point of view, but other areas, such as agriculture, food control, or industry, are also areas for their potential application. Therefore, we propose phages as a real alternative to current antibiotics.

Keywords: bacteriophages; phage therapy; antibiotic resistance; phage display; enzybiotics

1. Introduction

Viruses can infect all types of cells, including bacteria and archaea [1]. Specifically, bacteriophages (phages) are natural killers of bacteria and they were discovered a century ago by Frederick Twort and Félix d'Hérelle, independently [2,3]. In 1915, Twort thought that pathogenic bacteria required an essential substance to grow [4]. By analyzing in detail cultures of *Staphylococcus* sp. from vaccinia virus vaccines, he observed bacteria-free regions in the culture. Although he was unaware of what kind of substance produced those halos, after observing it under the lens, he confirmed that it was bacterial debris and defined it as a bacteriolytic agent [5]. In 1917, Félix d'Hérelle designated as "bacteriophages" some entities that were able to lyse bacterial cells after examining the effect of phages against *Salmonella gallinarum* in the feces of chickens [6–8]. In addition, d'Hérelle was the first to apply phages as a therapy to successfully treat children with severe dysentery.

In 1923, d'Hérelle and his assistant George Eliava established the George Eliava Institute of Bacteriophages, Microbiology, and Virology (Eliava Institute) in Tbilisi, present-day Georgia. During the Second World War, the Eliava Institute's experts provided combinations (cocktails) of different phages to soldiers, especially to treat wounds, gangrene, and diseases, such as cholera. Nowadays, the Eliava Institute carries out clinical trials with patients from all over the world, which often result in high success rates after treatment. Different phage applications in fields, such as human therapy, illness prevention, veterinary, environmental control, and food safety are investigated there [9]. This institute is responsible for the identification of more than 4000 phages, and a large number of phage-related studies have been done there. Because of this, the Eliava Institute is nowadays an important reference center for phages. Due to the discovery of antibiotics and the widespread use of penicillin in the 1940s, the use of phages fell out of favor, and, as a consequence of the Second World War, they were quickly reduced to only being used in Eastern countries, which had no access to antibiotics. Therefore, only Eastern countries (in particular, those of the former USSR), were (and are) using phages to treat bacterial infections, such as *Salmonella* and *Shigella* diseases, among others [10].

In spite of the rapid success of antibiotics, the emergence of multiresistant bacteria is a general concern. Nowadays, some bacterial strains are resistant to almost all available antibiotics. Routinely, surgical interventions can lead to serious complications due to the emergence of resistant bacterial strains that cannot be treated with conventional antibiotics [11]. Regarding the origin of this resistance, horizontal genetic transfer has been thought of as a key factor in the acquisition of antibiotic-resistant genes [7]. Moreover, spontaneous mutations can also occur in some genes under the action of antibiotics, and therefore contribute to its emergence [12]. High mutation and gene flow rates allow for bacteria to evolve quickly under the strong selection pressures that are exerted by antibiotics. In addition, the use of broad-spectrum antibiotics and their misuse promote this problem [13]. As a consequence, bacteria have developed several mechanisms to prevent antibiotic function, such as changes in receptors through enzymes or mutation, removal of the antibiotic by membrane pumps, or antibiotic modification to escape its effect [12–14]. For these reasons, it is important to propose alternative methods to fight against bacterial resistance. Here, we demonstrate how bacteriophages can be useful in this battle, and a wide variety of interesting phage applications are also reviewed.

Bacteriophages can be classified according to their genome, morphology, biological cycle, or the environment where they live [15–17]. Concerning the biological cycles of phages, there are two main types—lytic and lysogenic cycles (Figure 1). The lytic cycle implies bacterial death, which generates virion output. For this, they take advantage of the replication system of the host cell, and when their proteins and viral components are formed, they induce cell lysis [18,19]. In contrast, lysogenic cycles are based in the integration of the genetic material of the phage into the genome of the host cell. At the end of the replication cycle, no new virions are obtained, but bacterial cells with phage genetic material are created, as temperate phages [18]. Due to the variability between phages, it is important to determine which are the most appropriate bacteriophages for each potential application [20].

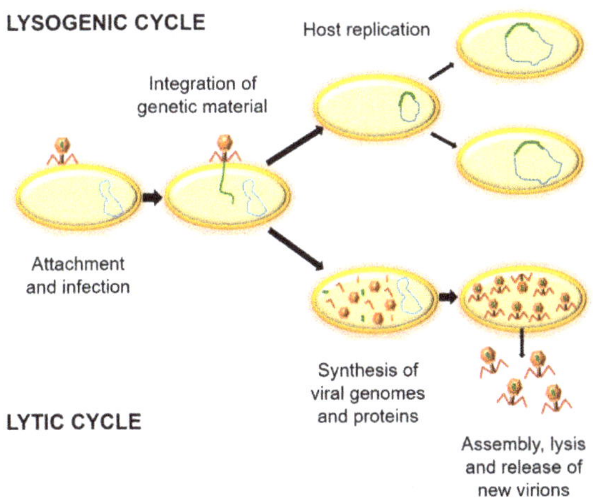

Figure 1. Biological cycles of phages. Firstly, the virus binds to the bacterial cell and injects its genetic material. In the lysogenic cycle, the integration of viral genetic material into the genome of the host occurs, and the bacterial cell replicates without producing virions. In the lytic cycle, viral genetic material is replicated and viral proteins are synthesized. Then, an assembly of virions is achieved, followed by the lysis of the bacteria and the release of new virions.

2. Phages in the Biosphere

Viruses are ubiquitous in the biosphere and can be found in all environments, being the most abundant biological entity [21]. It is believed that there are 10^{31}–10^{32} virions in the biosphere, approximately distributed as 2.6×10^{30} virions in soils, 1.2×10^{30} virions in the ocean, 3.5×10^{30} virions in the oceanic subsurface, and 0.25–2.5 $\times$ 10^{31} virions in the terrestrial subsurface [22]. In addition, significant viral quantities have been found in extreme environments, with between nearly 9.0 $\times$ 10^6 and 1.3 $\times$ 10^8 virions mL^{-1} in sea ice and approximately 5.6–8.7 $\times$ 10^{10} virions cm^{-3} in algal flocks [16]. Other extreme conditions in which viruses have been found include high-temperature environments (thermal waters, geothermal springs, volcanoes, hydrothermal vents, etc.), cold environments (lakes of polar areas, sea ice, etc.), and hypersaline zones [22]. Although few studies have focused on their presence in soils and the rhizosphere, around 1.5 $\times$ 10^8 virions g^{-1} are estimated to be present there [16]. Bacteria can be found in almost any environment, such as seawater, fresh water, and soils [23], so phages are expected to be found in any place where a host is located (Table 1). In addition, some of them may be specifically localized, whilst others can be widely distributed throughout the biosphere [24]. Indeed, only around 6000 different bacteriophage species are known, hiding a great diversity that is still unknown [25]. It is theorized that, in the ocean, there are at least 10^7 phages mL^{-1} [17], and the number of soil phages could be as great as 10^8 virions g^{-1} [26], representing a large proportion of the total amount of viruses in the biosphere.

Phages and bacterial cells have been coevolving for a long time, showing dynamic interactions between them [27]. Experimentally, it has been shown that coevolution between phages and bacteria can increase the rate of molecular evolution. This has been studied in *Pseudomonas fluorescens* SBW25 infected with the phage Φ2, and it has been shown that not all genes evolve equally in the phage. Interestingly, those that evolve quickly are related to the infection of the bacterium, coding for proteins that are related to host attachment [28]. In addition, coevolution leads to the maintenance of bacterial diversity and is responsible for changes in the physiology, abundance, abilities, and virulence of bacteria [29]. Similarly, phages are also influenced by their interaction with bacteria, especially in their defense strategies [30]. Although phages are highly specific, some of them show a wide host range. Moreover, due to phage-bacteria interactions, phages can participate in the biogeochemical cycles of biotic and abiotic environments. When bacterial lysis occurs, bacteria debris remains in the medium, being a nutritional source, and carbon, nitrogen, and phosphorus cycles are enriched or modified [31]. Therefore, phages play an important role in their environment. For example, they participate in nutrient acquisition in marine ecosystems and improve carbon transfer through phage lysis [32].

As with bacteria, we can find phages living in higher organisms, mostly in the digestive tract, vagina, respiratory and oral tract, skin, and mucosal epithelium, forming the so-called "phageome" [33]. Due to the great diversity and quantity of phages in the body, it is possible that they participate in human homeostasis. For example, gut phages are lytic and temperate, and both types are important for avoiding bacterial imbalance. It is suggested that the introduction of viruses in the gut occurs during the first four days of life and that they undergo changes with the development of the body [34].

Thanks to metagenomics, it is possible to determine phage variability and their abundance in each environment. Epifluorescence microscopy and transmission electron microscopy can help to identify new phages [22]. Before these techniques, counts were made in bacterial cultures and by obtaining plaque forming units (PFU), which may lead to difficulties mainly because not all bacteria are cultivable under artificial conditions, and not all phages make lysis plaques [35]. For these reasons, the real abundance and diversity of phages are higher than observed.

Phages can also be found in artificial places or infrastructures that humans inhabit, such as hospitals, showing that natural places are not the only source of phages [36]. Hospital sewage is an especially good reservoir of phages. As explained by numerous articles about multiresistant bacteria, phages that were isolated from the wastewater of medical centres are used in applications against resistant bacteria that cause diseases. Additionally, clinical materials or medical devices can be a source of phages [37,38].

Remarkably, wastewater treatment plants (WWTPs) are the habitat of many types of microorganisms, which makes them interesting considering that many interactions among bacteria and phages take place in them. More than 1000 different types of viruses have been found in WWTPs and a large proportion of them are bacteriophages. The water of WWTPs undergoes several debugging and cleaning processes to obtain potable water, and these physical and chemical methods manage to eliminate many bacterial cells, although they usually fail to remove phages. Therefore, this affords an opportunity to isolate bacteriophages [39].

Accordingly, phages are a ubiquitous ecological solution to develop new treatments against bacteria, although further studies should be done to better understand the relationship between them and bacteria before their application.

Table 1. Summary of the main places where phages can be found: nature, urbanized places, and the human body.

Phages in nature	Soil Terrestrial subsurface Fresh water Ocean Oceanic subsurface Extreme environments: sea ice, algal flocks, hypersaline zones, etc.
Artificial places	Hospital and similar places Wastewater treatment plants Some areas under human impact
Body of animals	Digestive tract Vagina Respiratory and oral tract Skin Mucosal epithelium

3. Potential Application of Phages

Phages should be considered as great potential tools due to their multiple benefits. Since their discovery, phages have been used as models to understand fundamental genetic processes and as great tools in molecular biology. Phage products, such as ligases, polymerases, or recombinant phages, are commonly used in research laboratories. Here, we place an emphasis on the potential application of phages against pathogenic bacteria. The emergence of multidrug-resistant bacteria has led to the need for new treatments. To this end, we assess how phages can help to overcome this critical situation by coming up with potential applications of phages that may be of interest in different areas. Different approaches using phages are proposed, and some of the most relevant ones in the fight against bacterial resistance are described in detail.

3.1. *Phage Therapy*

Phage therapy is based on the therapeutic use of phages to treat pathogenic bacterial infections [40]. Lytic phages are preferably chosen in phage therapy for two main reasons. Firstly, because lytic phages will destroy their host bacteria, whilst temperate ones will not. Secondly, because temperate phages can transfer virulence and resistance genes due to their life cycle, in which the genome of the phage is integrated into and replicates together with the bacterial genetic material [8,19]. The intrinsic characteristics of lytic phages, such as high host-receptor specificity and bacterial cell lysis to release virions, make them highly suitable for clinical applications [7,41]. Some remarkable features of the use of phages include their short replication time and their ability to obtain a large number of viral progeny only in their specific hosts, the specificity to prevent damage in nonpathogenic bacteria (they are ecologically safe and have no known side effects), and their fast and low-cost production. In addition, their short genomes allow for us to understand the molecular mechanisms implicated in controlling resistant cells [7]. Another significant feature of phages is their ability to coevolve with their host, in a hit-and-run response, to counteract possible resistant mutants, with higher mutation rates being described for viruses than for bacteria [30].

3.1.1. Main Applications of Phage Therapy

As previously mentioned, phage therapy has been used since a century ago. However, their use is restricted to Eastern countries, which have different guidelines for clinical trials and research articles are mainly published in Russian or other non-English languages. Because of that, phage therapy is not currently used in European and North American countries. Researchers are now making efforts to follow clinical trials guidelines to use phages in clinics. An interesting European project under human clinical trial is called Phagoburn, which was funded in 2013 by the European Commission. This project is based in the use of phage cocktails for burn injuries infected with *Escherichia coli* or *Pseudomonas aeruginosa* [42].

The main application of phage therapy is its use as a therapeutic agent to eliminate pathogenic bacteria involved in disease or infection as well as those that form biofilm. Another interesting approach is the use of phages as a preventive disinfectant, especially in medical areas or clinical devices. Additionally, current technology or the combination of phages with other techniques can improve these clinical applications. Despite the effectiveness of a single type of bacteriophage against a bacterial strain due to its high specificity, phage cocktails are an interesting strategy to solve issues relating to resistance and a low range of action [43]. They are normally composed of different phages that attack different bacterial strains or species. In this way, phage cocktails can play a decisive role in biofilms by allowing for the phages' effects to last longer by delaying the emergence of resistance to all the phages that are part of the cocktail. Furthermore, it has been proven that these cocktails have other benefits, such as decontaminating food by removing *E. coli*, *Salmonella enterica* or *Listeria* [44].

Outside medicine, phage therapy plays an important role in other fields, such as food production and cattle raising. Phages are useful to ensure food safety because they allow for the removal of bacterial infections in animals and thus prevent the consumption of contaminated food. Some interesting examples are the use of phages to control typical food infections that are caused by *Salmonella* (salmonellosis), *Campylobacter* (campylobacteriosis in poultry), *Listeria monocytogenes*, or *E. coli* [2].

3.1.2. Benefits and Drawbacks of Phage Therapy

Bacteriophages present many benefits that make them excellent tools to treat bacterial diseases and to contribute to the fight against the emergence of bacterial resistance. One of the greatest concerns regarding antibiotics is their side effects, since they can damage the microbiome, which is related to many types of imbalances or diseases. In this regard, the specificity of phages can solve this problem, since they will only replicate inside their specific host (phages cannot infect eukaryotic cells). In contrast to antibiotics, phages can proliferate quickly inside the host (and only if they find their host), can be administered in small doses and with long intervals of time between them, and they are removed once their population is eliminated [2,23]. The action of phages inside the host is very specific, as phage replication only occurs inside bacteria. On the contrary, antibiotics are less precise and they reach more areas without the presence of bacteria in the organism [11]. Another benefit of phages is that they can be used in difficult-to-reach parts of the body, such as treating central nervous system infections, which commonly poses a serious problem [8]. A remarkable feature of phages is that they can evolve, whilst antibiotics are static substances that cannot change even if their environment changes. Another interesting feature of phages is their isolation and production costs, as previously mentioned. The cost of antibiotic production is high, both economically and because antibiotics are not natural and have to be synthesized in a laboratory [23].

It is worth noting that the great specificity of phages is both an advantage and a limitation of phage therapy. Phages avoid damage to the microbiome, but prior to their application, it is necessary to determine in vitro which bacteria are causing the disease. This can be a difficult process because identification must take place quickly in order to apply the treatment to the patient [45,46]. A way to solve this setback is the use of phage cocktails, as this widens the range of action [2]. However, it is possible that in vitro and in vivo phage behavior may be different, and as a result of the lack of in vivo studies, their effectiveness cannot be fully assured [47].

The pharmacology of phages can be very complex, for both the action of phages inside the body (pharmacodynamics) and the body's function on the phages (pharmacokinetics) [33]. In phage therapy, the interactions between phage and bacteria are related to pharmacodynamics. Regarding pharmacokinetics, it is believed that this is linked to the density of phages within the hosts. In the event of a small bacteria population, a large dose of phages must be used in order for them to replicate faster than bacteria. Furthermore, if the bacterial density is small, the phages might not replicate quickly enough and they will not perform the desired action [48]. This can depend on the phage dose, which in turn depends on the bacterial density, the size of phage particles, and the phage virulence, as the more virulent the phage, the better it will attack its host. The resolution of this point is based on a virulent phage with a great burst size (producers of large progenies in a short time) and that is specifically administrated at the infection site.

In addition, it is possible that phages or their products can be recognized by the immune system and induce immune responses. Nevertheless, phage lysis is usually faster than the action of neutralizing antibodies phages. However, some researchers suggest that it is possible that an immune response occurs, owing to the products and enzymes that are released from bacterial lysis. Noteworthy, recent studies showed that phage T4 is highly immunogenic and can be used as potential vaccine candidates [49]. In addition, in many cases immune responses can be avoided by modifying the mode of phage administration [33]. The immune mechanisms that detect phages and subsequently take action are not well understood. Thus, it is necessary to investigate these matters to evaluate phages' effect on the body. On the other hand, different studies are in agreement that the application of phage therapy has no direct consequences on the patient [50]. Despite the fact that phage safety must be confirmed, phages are consumed indirectly by means of fermented foods, breathing, or every time we accidentally drink sea water. For that reason, it seems that bacteriophages do not pose a potential risk [51]. Apart from the route that phages naturally use to arrive in their bacterial host, there are new strategies to improve the lifespan of bacteriophages in an organism. Biomaterials that do not interfere in phage activity should be used, as liposomes or capsules around phages that are made of alginate are accompanied by different ions. One question that needs to be dealt with is the physical limitations of these structures, and the most appropriate way of encapsulating phages must be chosen according to their future function [52].

Above all, the most urgent point that should be solved is the scarcity of basic information related to doses, forms of administration, protocols, and the correct mode of application of this therapy in the specific case of each phage [2,45]. This question, together with the difficulties of patenting phages (since they are natural entities), is an impediment for pharmaceutical companies to accept this therapy [42]. Legal regulations must be established to define limitations and the safe use of phage therapy. Lastly, ethical and social acceptance of phage therapy is a great impediment, since it is difficult to believe that viruses not only are not dangerous for humans, but that they have also the potential to treat diseases.

3.1.3. Emergence of Bacterial Resistance against Phages

Another controversial topic is the emergence of bacterial resistance against phages. Bacteria present natural mechanisms to prevent viral infection (Figure 2). Some of these mechanisms are associated with phages' receptors, e.g., bacteria can hide, change, or even lose phage receptors [8]. Each of these mechanisms is activated in response to a stimulus, for example, receptor loss usually occurs when there is a change in the composition of the bacterium's cell surface, as is seen in *Bordetella* spp. and *Shigella flexneri* [8,41]. If the loss of the receptor occurs, the phage cannot recognize the bacteria, and, subsequently, no new phages will be generated. This occurs, for example, in *E. coli* and *Staphylococcus aureus* as a consequence of membrane protein modifications. Some bacteria even have the ability to secrete extracellular polymeric substances (EPS), glycoconjugates, or alginates in order to prevent the adhesion of the phage to the bacteria. These secretions have been observed in *Pseudomonas* spp., which ejects EPS, and *Enterobacteriaceae*, which secretes glycoconjugates [41].

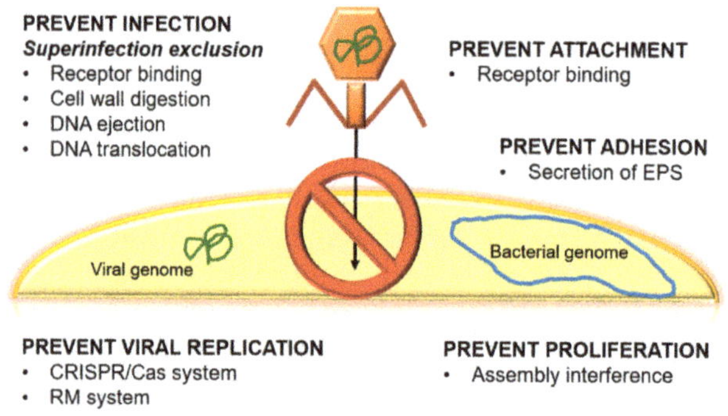

Figure 2. Principal natural mechanisms to prevent viral infection in bacteria. EPS: extracellular polymeric substances, CRISPR/Cas: Clustered Regularly Interspaced Short Palindromic Repeats/Caspase, RM: restriction modification.

There are others systems to escape phages, such as by viral DNA removal by different methods, Clustered Regularly Interspaced Short Palindromic Repeats (CRISPR)-associated proteins, or the superinfection exclusion (Sie) system [9,41]. CRISPR is considered to be the immune system of bacteria to protect the genetic material of the cell against possible attacks of viruses or plasmids [53]. To counteract the CRISPR system, phages have been co-evolving with bacteria thanks to phage-anti-CRISPR [54]. On the other hand, Sie is based on membrane-associated proteins. These proteins interact with proteins that are related to DNA injection when the phage is bound to a membrane receptor. Through this process, it is possible to interrupt the injection of DNA and thus reduce phage virulence, preventing the infection from spreading [55]. Another process is based on the abortive infection (Abi) system, which is responsible for interfering with the replication, transcription, or translation of the phage or in virions formation [41]. This mechanism is activated when the phage has managed to enter the cell because it has evaded the restriction systems and the CRISPR system of the host. The objective of Abi is to destroy the infected cell so that it does not transmit the invasion to the rest of the cells [30]. An interesting way to increase phage fitness against bacteria and to reduce the emergence of resistance is experimental evolution. This method consists of preadapting a phage to its host in vitro for several generations. In addition, experimental evolution produces benefits for phage therapy, since the bacteria–phage coevolution over many generations can make the bacteria diminish their ability to adapt to the environment [56].

The combination of phages with antibiotics is another strategy to apply phages as therapeutic agents. It is possible to reduce resistance because phages can kill antibiotic-resistant variants and vice versa [57]. Owing to these combinations, the phage-antibiotic synergy (PAS) effect usually takes place, which consists of an increase in phage virulence as a result of the administration of sublethal concentrations of antibiotics and it can be convenient to remove pathogenic cells quickly [43]. In addition, additive effects of the components can occur and they can be beneficial. Through animal studies, it has been proven that these treatments have preventive action for bacteria resistance [45], for example, decreasing mutations that confer resistance to bacteria [46]. On the other hand, some of these antibiotics block the cellular cycle of bacteria, and as a result, bacterial cells undergo an increase in volume that facilitates the division of phages and their release at a faster pace [58]. Some studies have confirmed the effectiveness of this method, like the combination of phage SBW25ϕ2 with kanamycin antibiotic against *P. fluorescens* SBW25 [57]. The success of combinations depends on the target cell and phage and antibiotic types. Moreover, it is important that phages and antibiotics detect different regions of union of pathogenic bacteria to ensure the effectiveness of the treatment. These deductions arise from experiments with different phages and antibiotics to decrease *E. coli* in urinary tract infections. Specifically, the best results were obtained from the combination of phages and the antibiotic ciprofloxacin (at a sublethal concentration) against *E. coli* [57]. This synergy has been shown using cefotaxime combined with phage T4 against biofilm formations of *E. coli* [59]. Furthermore, because of the lack of information about the safety of phages, and sometimes the ignorance of phages' effects, it is estimated that the combination of both provides more support than the use of individual bacteriophages only [45]. It is thought that this therapy can be better than phage cocktails because phages and antibiotics act in different ways inside bacterial cells, while all the phages in a phage cocktail may have similar modes of action, which might be a problem for the emergence of resistance [60]. Nevertheless, this association presents some defects. As a consequence of the PAS effect, and mainly due to antibiotics in sublethal concentrations, an SOS (emergency repair) response can arise, which consists of bacterial responses to stress as a consequence of damage to bacterial DNA, causing serious effects by increasing the antibiotic resistance. Other side effects can be the appearance of double resistant mutants, the potentiation of antibiotic resistant bacteria due to the phages preference to infect sensitive variants, or the interference between the action of the phage and the antibiotic resulting in an action that is less effective than the sum of both [45]. It is also important to know when to apply each treatment (phages and antibiotics). It seems that to prevent resistance, an intermediate interval of time between phage and antibiotic applications is better than the simultaneous application of both or their use after a very long period of time between them [60].

3.2. Phage-Derived Enzymes

An alternative to the use of phages against bacterial diseases is the use of phage-derived enzymes. Phages produce several enzyme types, each one being suitable for attacking specific bacteria [61]. Phage enzymes were discovered in 1986 when investigating an enzyme secreted by *Streptococcus* after being infected by a bacteriophage. The activity of this enzyme was the rupture of the cell wall and it was called lysin [62–64]. In nature, these enzymes are found inside phages and help them to penetrate into the host to assure phage replication, among other functions. In general, they degrade peptidoglycan, thereby producing bacterial lysis by osmotic imbalance. Lysins can be used as enzybiotics because, applied exogenously as recombinant proteins, they can remove bacteria. This approach has been used for more than 20 years, especially against Gram-positive bacteria. It was thought that Gram-negative bacteria could physically block the passage of lytic enzymes due to the presence of its outer membrane, but it was later discovered that some phage enzymes can cross this layer. Although it is postulated that these enzymes may cause unwanted immune responses, several studies have shown no serious side effects after their use [65].

Despite their success in killing bacteria, they present some problems of stability and lack of solubility, which may be solved by enzyme engineering. Sometimes, the use of recombinant enzymes is recommended. The great progress in enzyme engineering and synthetic biology, in addition to their low cost of production, makes this technique one of the most effective against antibiotic resistance in its application in clinics. Moreover, lytic enzymes are interesting in other areas, such as agriculture, food industry, diagnostics, environmental control, and bioluminescence [65].

As previously mentioned, one type of phage-derived enzyme is lysins. Within them, we can find endolysins, which are derived from the lytic cycle, and virion-associated lysins (VALs), which are implicated in the entrance to the host cell. Some of them can suffer changes upon the modifications of their host cell and are very specific to species or bacterial serotype [61]. VALs can be part of the phage tail or be inside the capsid. Once the phage is recognized by its receptors in the host, a conformational change in the phage allows entrance into the bacteria thanks to the degradation of peptidoglycan carried out by VALs. The mode of action of VALs against bacteria consists of allowing for the injection of the phage. This fact is possible thanks to the rupture of the peptidoglycan layer of the bacteria through the hydrolysis of chemical bonds. In its clinical application to treat bacterial diseases, this ability is used to induce the osmotic lysis of the bacteria [61]. Differently, endolysins must cross the cell membrane to reach the cell wall and cause lysis, since they are synthesized in the cytoplasm of infected cells at the end of the viral cycle. They are classified into canonical endolysins (the most interesting as enzybiotics) and exported endolysins. Canonical endolysins need other specific phage enzymes, called holins, which will form holes through which the endolysins will leave the cytoplasm [65]. Endolysins are useful as alternatives to antibiotics due to their bactericidal function. This ability has been proven by applying these purified enzymes on bacteria. Endolysins usually have two domains: one is catalytic active (N-terminal), whilst the other attaches to the cell wall (C-terminal). Additionally, some laboratories are creating chimeric endolysins in order to improve bacterial lysis. One example is the chimera Cpl-711, which combines the endolysins Cpl-1 and Cpl-7S from Cp-1 and Cp-7 pneumococcal bacteriophages against *Streptococcus pneumoniae* and their multiresistant strains [66]. An interesting feature of lysins is their high specificity, which reduces the probability of developing bacterial resistance, making resistance an extremely rare event [61].

Holins are proteins derived from phages acting at the end of the lytic process in order to trigger and control the degradation of the cell wall from the bacterial cells. These small membrane proteins control lysis time and are diverse. They are commonly small, have a positive and hydrophobic C terminal domain, and include one to three hydrophobic transmembrane domains by which they can be classified [67]. Membrane channels or pores are created when the concentration of holins exceeds a threshold, and then endolysins will perform their function. The canonical holins form large pores at one side of the host and locally expose the peptidoglycans to cytoplasmic canonical endolysins. Pinholins, another group, form small pores that depolarize the membrane, triggering signal-anchor-release (SAR) endolysin activation and inducing degradation of peptidoglycans in the whole cellular periplasmic space [68]. They can be combined with other enzymes to amplify the host range of endolysins. Since holins have a large strain spectrum, they are interesting against multiresistant strains such as *S. aureus* or *S. suis* [19].

Polysaccharide depolymerases are very useful due to their ability to attack carbohydrates of bacterial membranes [19,69]. Depolymerases present two different forms, as part of the phage particle that is attached to the base plate (in the phage membrane or capsid) or as a protein resulting from cell lysis after phage replication. They also differ in the mode to remove carbohydrate polymers on the membrane of the host bacteria. They are called hydrolases when they are able to hydrolyse the glycosyl–oxygen bond and turn it into a glyosidic bond. In contrast, lyases add a double linkage into uronic acid particles, thanks to the β-elimination system, after the disruption of the glycosidic linkage between a monosaccharide and the C4 of the acid. An example of lyases are the hyaluronan lyases [67]. Lyases would be an indirect way of dealing with bacteria because these enzymes act on bacteria that are encapsulated by weakening the polymer structure that makes up the capsule. This process helps to decrease bacterial virulence by allowing for the immune system to perform its action [70]. One interesting property of depolymerases is their wide bacterial range. In contrast to lysins, resistance against phage depolymerases can emerge due to modifications in polysaccharide composition of capsule, exopolysaccharides, or lipopolysaccharide [61]. New studies are currently emerging to ensure their safety and efficacy against capsulated *E. coli* in mice [69]. Moreover, it seems that some depolymerases eliminate biofilms, such as a depolymerase derived from a phage against *Staphylococcus epidermis*, which is an interesting application due to the difficulties of removing these formations. Depolymerases favor the penetration of the phage into the biofilm and the host cells due to its capacity for degradation of capsular polysaccharides [43]. Another group is the virion-associated peptidoglycan hydrolases (VAPGHs) as lysozymes, lytic transglycosylases, endopeptidases, or glucosaminidases. In contrast to endolysins that act in the final step of viral replication, VAPGHs can favour phage entry to the host by creating a hole through the cell wall of the host cell. With this, the phage can insert its tail and carry out the injection of its genetic material. An interesting example is gp49 from *S. aureus* phage phi11. Under normal conditions, gp49 is not necessary, but when the temperature or density fluctuates, this enzyme can improve infection. These enzymes are located almost anywhere in the phage, as they have been found in the tail, the head, the capsid, and the viral membrane [69,71]. Hydrolases are suitable to carry out their function in Gram-positive and Gram-negative cells because they are encoded by phages that attack both bacterial groups [19]. VAPGHs present several benefits that make them suitable as antimicrobial agents in therapeutic fields and also as food control or food decontaminants, especially in foods that undergo processes at high temperatures during manufacturing, as in dairy products. In addition, it has been shown that they can be useful against multiresistant bacterial strains [71]. Recent research proposed VAPGHs activity to treat plant diseases, such as those provoked by *Agrobacterium tumefaciens* [72]. Finally, to improve their bactericidal capacities, chimeric proteins can be created by combining several enzymes or by exchanging their functional domains. Some of these chimeric proteins have been used to treat diseases that are caused by *S. aureus*. Although few studies have been done to determine bacterial resistance emergence under VAPGHs treatment, bacterial resistance has not been reported so far [71].

Phage-derived enzymes are very useful and an attractive solution against bacterial infections. Synthetic biology allows for creating and modifying phage proteins to improve bacterial range spectrum, reduce bacterial resistance, and reduce immunogenicity. In addition, enzybiotics can be a great tool to be used against intracellular pathogens, where phages have difficulties to reach due to the lack of receptors for eukaryotic cells. Besides, phage-derived enzymes can be easily delivered into specific infection sites, acting locally in the infection and reducing side effects.

3.3. Phage Display

Despite the fact that natural bacteriophages have been studied to diminish bacterial resistance with success, some researchers have gone beyond this to find alternative antibacterial methods based on phages. This leads to phage modifications by gene engineering, the principal advantage of which is greater accuracy capacity [73]. However, there are more benefits of phage engineering, such as obtaining phage elements which are able to detect bacterial hosts, changing phage hosts, or promoting their action [74].

Phage display was developed in 1985 by George P. Smith [75]. It is a technique that is based on the expression of the phage cover of foreign peptides. The phage is exposed to the target (natural or synthetic) of interest until some of the peptides exposed in the phage bind specifically to that target. To carry out these processes, libraries of phage particles are commonly created by means of random peptides. These peptides are bound to components of the phage coat. Random oligonucleotides must be introduced at a specific location within the phage. The function of the library is to facilitate the search for the appropriate peptides through several screenings of ligands for each target. In addition, once the specific peptide has been recognized, it is possible to maximize its specificity and affinity. To check if the proteins that are expressed by the phage bind to the chosen target, a "biopanning" process is performed three to six times. This method consists of the immobilization of the target and its exposure to randomly selected peptide libraries. Then, less to more rigorous washes are made to eliminate phages that are bound nonspecifically to the target. Finally, once the phage has correctly attached to that target, different methods are used to separate it from the target and to sequence it until finding the specific peptide. It is usually said that phage display is a link between phenotype and genotype [76,77].

It is possible to use different types of bacteriophages for phage display, but the most advantageous are the filamentous ones, since they allow for genetic material expansion simply by increasing their filament size. The process of introducing genetic material into the phage does not damage its internal structures. These phages, which are usually temperate, do not kill bacterial cells to carry out their biological cycle and release new virions. Phage M13 is a typical bacteriophage used for this technology. In general, the most interesting gene in phage display is gene VIII, which codes for major structural proteins and is suitable for displaying short peptides and obtaining a high number of desirable molecules. In contrast, gene III codes for minor proteins and is more appropriate for expressing large peptides, although few copies of them will be obtained [78].

Phage display is an excellent tool for vaccine production, the development of new drugs, the study of protein–protein interactions, the selection and modification of substances of interest, the development of monoclonal antibodies with the desired specificity for therapeutic use, the creation of libraries of peptides and other substances, in epitope mapping (as antivenom), or the production of food as biocontrol. In the same way, it is a very useful technique to choose and isolate antibodies against desirable antigens or other targets, and to then create an antibody library [3,77,78]. Related to our subject of study, one application is the use of phage display against bacterial resistance. A typical response of bacteria when they are exposed to antibiotics is the secretion of enzymes. Particularly, they express β-lactamases, which hydrolyses the β-lactam ring with the aim of stopping antibiotic attacks, thus rendering the bacterial cells resistant to them. This resistance is generated as a result of the contact between bacteria and antibiotics presenting a β-lactam ring. β-lactam antibiotics have been widely used since their discovery. As a result, bacteria have been evolving, and, consequently, resistance has emerged. Through the implementation of phage display, this technique can be used to find peptides against β-lactamases enzymes by testing different peptide libraries [79]. Another example is the use of phage display against multiresistant strains of *L. monocytogenes*. These bacteria are usually transmitted by food and cause diseases such as arthritis and infections called listeriosis, which affect the central nervous system. As a result of the increase in antibiotic resistance, alternative methods have been sought to combat these diseases. One of these methods is the use of phage display and peptide libraries. After finding peptides that bind to the bacteria of interest (*L. monocytogenes*), peptides were isolated and it was checked if they presented microbicidal activities [80].

Similarly, the creation of new drugs opens many possibilities for future medicine. This is achieved by following the basic protocol of phage display but against therapeutic targets (e.g., specific points on which to act against pathogenic bacterial diseases). In addition, people under immunosuppression, such as HIV patients, transplant recipients, and pregnant women, are very susceptible to pathogenic infections. It is expected that more than 50% of patients with HIV will develop resistance to many of the current treatments [81]. Therefore, new alternative treatments should be proposed in order to solve this problem.

Moreover, phage display allows for the production of two main types of vaccines: phage display vaccines and phage DNA vaccines. On the one hand, phage display vaccines are based on a virion inside of which is the gene that codes for the antigen that is displayed, and they are more stable than phage DNA vaccines, when considering that the virion protects them. On the other hand, in phage DNA vaccines, inside of the virion there is DNA with the antigen gene that has been cloned in a eukaryotic cassette. As a result of these vaccines, there is a greater immune response than with conventional vaccines. Some of the investigations with these types of vaccines are directed against bacteria, autoimmune diseases, cancer, fungi, parasites, or even contraceptive vaccines [82]. There are also studies to use vaccines with the aim of preventing or reducing antimicrobial resistance. This could be innovative because they produce an immune response if the patient is exposed to a pathogen; as a result, the disease is avoided or weakened. In addition, if vaccination rates increase among the population, it can produce herd immunity, which protects unvaccinated people [13].

Like others methods, phage display presents some drawbacks that should be analysed. In the passage to a soluble medium, the binding capacity of the peptide to its target may be lost. In addition, peptide functions in vitro and in vivo can be different, with the risk of producing side effects in the patient. Peptides are also unstable due to proteolysis, and their ability to develop immune responses can be a problem for their application. Nevertheless, there are some techniques that are being developed to overcome these setbacks. Many of them are related to protein engineering or nanotechnology and are aimed at decreasing immunogenicity and increasing peptide affinity and half-life [83].

4. Conclusions

Multidrug-resistant bacteria are currently emerging for almost all the present-day antibiotics. Antibiotics are stable molecules exerting a selective pressure that allows for bacteria to evolve in order to escape, and new treatments should be proposed. Bacteriophages are a real alternative solution to this problem. Phages have different potential applications, starting from their use as bacterial killers in phage therapy, the use of their derivative enzymes, or their use in phage display. Phage therapy has been used for almost a century and it looks like a safe and effective treatment, although it is necessary to do more research to guarantee its safety in the short and long term. Very interesting and useful variants of phage therapy are emerging to enhance phage functions and to take advantage of them in many different areas. The large amount of possibilities due to the great diversity of phages, the use of phage cocktails, the combination with antibiotics, or promising phage display techniques allows for taking the most convenient approach for each scenario and to open new research areas to determine its advantages and disadvantages in each case. Surprisingly, there are also many new applications derived from phage therapy from which other fields can benefit.

Phages appear to be a great solution not only as an alternative treatment against bacterial diseases (phage therapy or use of phage-derived enzymes) but also as interesting tools in the prevention (phage-delivered enzymes) and diagnosis (bacterial detection and typing). Despite the success that these phage-based treatments are expected to have, we are also facing a big concern: the lack of regulation in developed countries and the acceptance of the general public. Further research in this field will help to create regulatory and safety protocols that will lead to the general use of phages in the clinical and pharmaceutical fields.

Author Contributions: Conceptualization, P.D.-C. Writing-Original Draft Preparation, J.D.-M. and P.D.-C. Writing-Review & Editing, P.D.-C.

Funding: This research received no external funding. P.D.-C. was funded by a Juan de la Cierva Incorporación postdoctoral contract from Spanish MINECO.

Acknowledgments: We thank Rafael Sanjuán for previous discussions.

Conflicts of Interest: The authors declare no conflict of interest.

References

1. Domingo-Calap, P.; Georgel, P.; Bahram, S. Back to the future: Bacteriophages as promising therapeutic tools. *HLA* **2016**, *87*, 133–140. [CrossRef] [PubMed]
2. El-Shibiny, A.; El-Sahhar, S. Bacteriophages: The possible solution to treat infections caused by pathogenic bacteria. *Can. J. Microbiol.* **2017**, *63*, 865–879. [CrossRef] [PubMed]
3. Cisek, A.; Dąbrowska, I.; Gregorczyk, K.; Wyżewski, Z. Phage therapy in bacterial infections treatment: One hundred years after the discovery of bacteriophages. *Curr. Microbiol.* **2016**, *74*, 277–283. [CrossRef] [PubMed]
4. Guzmán, M. El bacteriófago, cien años de hallazgos trascendentales. *Biomédica* **2015**, *35*, 159–161. [PubMed]
5. Trudil, D. Phage lytic enzymes: A history. *Virol. Sin.* **2015**, *30*, 26–32. [CrossRef] [PubMed]
6. Sadava, D.; Heller, G.; Orians, G.; Purves, W.; Hillis, D. *Life: The Science of Biology*, 8th ed.; Médica Panamericana: México, Mexico, 2008; pp. 286–287, ISBN 9789500682695.
7. Gelman, D.; Eisenkraft, A.; Chanishvili, N.; Nachman, D.; Coppenhagem Glazer, S.; Hazan, R. The history and promising future of phage therapy in the military service. *J. Trauma Acute Care Surg.* **2018**, *85*, S18–S26. [CrossRef] [PubMed]
8. Wittebole, X.; De Roock, S.; Opal, S. A historical overview of bacteriophage therapy as an alternative to antibiotics for the treatment of bacterial pathogens. *Virulence* **2013**, *5*, 226–235. [CrossRef] [PubMed]
9. Kutateladze, M. Experience of the Eliava Institute in bacteriophage therapy. *Virol. Sin.* **2015**, *30*, 80–81. [CrossRef] [PubMed]
10. Haddad Kashani, H.; Schmelcher, M.; Sabzalipoor, H.; Seyed Hosseini, E.; Moniri, R. Recombinant endolysins as potential therapeutics against antibiotic-resistant *Staphylococcus aureus*: Current status of research and novel delivery strategies. *Clin. Microbiol. Rev.* **2017**, *31*. [CrossRef] [PubMed]
11. Golkar, Z.; Bagasra, O.; Pace, D. Bacteriophage therapy: A potential solution for the antibiotic resistance crisis. *J. Infect. Dev. Ctries.* **2014**, *8*, 129–136. [CrossRef] [PubMed]
12. Martínez, J. General principles of antibiotic resistance in bacteria. *Drug Discov. Today Technol.* **2014**, *11*, 33–39. [CrossRef] [PubMed]
13. Jansen, K.; Knirsch, C.; Anderson, A. The role of vaccines in preventing bacterial antimicrobial resistance. *Nat. Med.* **2018**, *24*, 10–19. [CrossRef] [PubMed]

14. Bassegoda, A.; Ivanova, K.; Ramón, E.; Tzanov, T. Strategies to prevent the occurrence of resistance against antibiotics by using advanced materials. *Appl. Microbiol. Biotechnol.* **2018**, *102*, 2075–2089. [CrossRef] [PubMed]
15. Ackermann, H. 5500 Phages examined in the electron microscope. *Arch. Virol.* **2006**, *152*, 227–243. [CrossRef] [PubMed]
16. Weinbauer, M. Ecology of prokaryotic viruses. *FEMS Microbiol. Rev.* **2004**, *28*, 127–181. [CrossRef] [PubMed]
17. Ofir, G.; Sorek, R. Contemporary phage biology: From classic models to new insights. *Cell* **2018**, *172*, 1260–1270. [CrossRef] [PubMed]
18. Furfaro, L.; Chang, B.; Payne, M. Applications for bacteriophage therapy during pregnancy and the perinatal period. *Front. Microbiol.* **2018**, *8*. [CrossRef] [PubMed]
19. Criscuolo, E.; Spadini, S.; Lamanna, J.; Ferro, M.; Burioni, R. Bacteriophages and their immunological applications against infectious threats. *J. Immunol. Res.* **2017**, *2017*. [CrossRef] [PubMed]
20. Nilsson, A. Phage therapy—Constraints and possibilities. *Ups. J. Med. Sci.* **2014**, *119*, 192–198. [CrossRef] [PubMed]
21. Viertel, T.; Ritter, K.; Horz, H. Viruses versus bacteria—Novel approaches to phage therapy as a tool against multidrug-resistant pathogens. *J. Antimicrob. Chemother.* **2014**, *69*, 2326–2336. [CrossRef] [PubMed]
22. Parikka, K.; Le Romancer, M.; Wauters, N.; Jacquet, S. Deciphering the virus-to-prokaryote ratio VPR: Insights into virus-host relationships in a variety of ecosystems. *Biol. Rev.* **2016**, *92*, 1081–1100. [CrossRef] [PubMed]
23. Matsuzaki, S.; Uchiyama, J.; Takemura-Uchiyama, I.; Daibata, M. Perspective: The age of the phage. *Nature* **2014**, *509*. [CrossRef] [PubMed]
24. Clokie, M.; Millard, A.; Letarov, A.; Heaphy, S. Phages in nature. *Bacteriophage* **2011**, *1*, 31–45. [CrossRef] [PubMed]
25. De Vos, D.; Pirnay, J. Phage therapy: Could viruses help resolve the worldwide antibiotic crisis? In *AMR Control 2015: Overcoming Global Antibiotic Resistance*; Carlet, J., Upham, G., Eds.; Global Health Dynamics Limited: Ipswich, UK, 2015; pp. 110–114, ISBN 9780957607231.
26. Ashelford, K.; Day, M.; Fry, J. Elevated abundance of bacteriophage infecting bacteria in soil. *Appl. Environ. Microbiol.* **2003**, *69*, 285–289. [CrossRef] [PubMed]
27. Sharma, S.; Chatterjee, S.; Datta, S.; Prasad, R.; Dubey, D.; Prasad, R.K.; Vairale, M.G. Bacteriophages and its applications: An overview. *Folia Microbiol.* **2016**, *62*, 17–55. [CrossRef] [PubMed]
28. Buckling, A.; Rainey, P. Antagonistic coevolution between a bacterium and a bacteriophage. *Proc. R. Soc. B Biol. Sci.* **2002**, *269*, 931–936. [CrossRef] [PubMed]
29. Paterson, S.; Vogwill, T.; Buckling, A.; Benmayor, R.; Spiers, A.J.; Thomson, N.R.; Quail, M.; Smith, F.; Walker, D.; Libberton, B.; et al. Antagonistic coevolution accelerates molecular evolution. *Nature* **2010**, *464*, 275–278. [CrossRef] [PubMed]
30. Stern, A.; Sorek, R. The phage-host arms race: Shaping the evolution of microbes. *BioEssays* **2010**, *33*, 43–51. [CrossRef] [PubMed]
31. Díaz-Muñoz, S.; Koskella, B. Bacteria–phage interactions in natural environments. *Adv. Appl. Microbiol.* **2014**, *89*, 135–183. [PubMed]
32. Srinivasiah, S.; Bhavsar, J.; Thapar, K.; Liles, M.; Schoenfeld, T.; Wommack, K.E. Phages across the biosphere: Contrasts of viruses in soil and aquatic environments. *Res. Microbiol.* **2008**, *159*, 349–357. [CrossRef] [PubMed]
33. Forde, A.; Hill, C. Phages of life-the path to pharma. *Br. J. Pharmacol.* **2018**, *175*, 412–418. [CrossRef] [PubMed]
34. Manrique, P.; Dills, M.; Young, M. The human gut phage community and its implications for health and disease. *Viruses* **2017**, *9*, 141. [CrossRef] [PubMed]
35. Van Geelen, L.; Meier, D.; Rehberg, N.; Kalscheuer, R. Some current concepts in antibacterial drug discovery. *Appl. Microbiol. Biotechnol.* **2018**, *102*, 2949–2963. [CrossRef] [PubMed]
36. Hua, Y.; Luo, T.; Yang, Y.; Dong, D.; Wang, R.; Wang, Y.; Xu, M.; Guo, X.; Hu, F.; He, P. Phage therapy as a promising new treatment for lung infection caused by carbapenem-resistant *Acinetobacter baumannii* in mice. *Front. Microbiol.* **2017**, *8*, 2659. [CrossRef] [PubMed]

37. Jeon, J.; Ryu, C.M.; Lee, J.Y.; Park, J.H.; Yong, D.; Lee, K. In vivo application of bacteriophage as a potential therapeutic agent to control OXA-66-like carbapenemase-producing Acinetobacter baumannii strains belonging to sequence type 357. *Appl. Environ. Microbiol.* **2016**, *82*, 4200–4208. [CrossRef] [PubMed]
38. Peng, F.; Mi, Z.; Huang, Y.; Yuan, X.; Niu, W.; Wang, Y.; Hua, Y.; Fan, H.; Bai, C.; Tong, Y. Characterization, sequencing and comparative genomic analysis of vB_AbaM-IME-AB2, a novel lytic bacteriophage that infects multidrug-resistant *Acinetobacter baumannii* clinical isolates. *BMC Microbiol.* **2014**, *14*, 181. [CrossRef] [PubMed]
39. Lood, R.; Ertürk, G.; Mattiasson, B. Revisiting antibiotic resistance spreading in wastewater treatment plants–bacteriophages as a much neglected potential transmission vehicle. *Front. Microbiol.* **2017**, *8*. [CrossRef] [PubMed]
40. Phage Therapy Center. Available online: http://www.phagetherapycenter.com/pii/PatientServlet?command=static_phagetherapy&secnavpos=1&language=0 (accessed on 29 June 2018).
41. Drulis-Kawa, Z.; Majkowska-Skrobek, G.; Maciejewska, B.; Delattre, A.; Lavigne, R. Learning from bacteriophages-advantages and limitations of phage and phage-encoded protein applications. *Curr. Protein Pept. Sci.* **2012**, *13*, 699–722. [CrossRef] [PubMed]
42. Reardon, S. Phage therapy gets revitalized. *Nature* **2014**, *510*, 15–16. [CrossRef] [PubMed]
43. Pires, D.; Melo, L.; Vilas Boas, D.; Sillankorva, S.; Azeredo, J. Phage therapy as an alternative or complementary strategy to prevent and control biofilm-related infections. *Curr. Opin. Microbiol.* **2017**, *39*, 48–56.
44. Knoll, B.; Mylonakis, E. Antibacterial bioagents based on principles of bacteriophage biology: An overview. *Clin. Infect. Dis.* **2013**, *58*, 528–534. [CrossRef] [PubMed]
45. Torres-Barceló, C.; Hochberg, M. Evolutionary rationale for phages as complements of antibiotics. *Trends Microbiol.* **2016**, *24*, 249–256. [CrossRef] [PubMed]
46. Kutateladze, M.; Adamia, R. Bacteriophages as potential new therapeutics to replace or supplement antibiotics. *Trends Biotechnol.* **2010**, *28*, 591–595. [CrossRef] [PubMed]
47. Ghannad, M.; Mohammadi, A. Bacteriophage: Time to re-evaluate the potential of phage therapy as a promising agent to control multidrug-resistant bacteria. *Iran. J. Basic Med. Sci.* **2012**, *15*, 693–701.
48. Levin, B.; Bull, J. Opinion: Population and evolutionary dynamics of phage therapy. *Nat. Rev. Microbiol.* **2004**, *2*, 166–173. [CrossRef] [PubMed]
49. Tao, P.; Zhu, J.; Mahalingam, M.; Batra, H.; Rao, V.B. Bacteriophage T4 nanoparticles for vaccine delivery against infectious diseases. *Adv. Drug Deliv. Rev.* **2018**, *6*. [CrossRef] [PubMed]
50. Roach, D.; Debarbieux, L. Phage therapy: Awakening a sleeping giant. *Emerg. Top. Life Sci.* **2017**, *1*, 93–103. [CrossRef]
51. Sarker, S.; McCallin, S.; Barretto, C.; Berger, B.; Pittet, A.C.; Sultana, S.; Krause, L.; Huq, S.; Bibiloni, R.; Bruttin, A.; et al. Oral T4-like phage cocktail application to healthy adult volunteers from Bangladesh. *Virology* **2012**, *434*, 222–232. [CrossRef] [PubMed]
52. Cortés, P.; Cano-Sarabia, M.; Colom, J.; Otero, J.; Maspoch, D.; Llagostera, M. Nano/Micro formulations for bacteriophage delivery. *Methods Mol. Biol.* **2018**, *1693*, 271–283.
53. Makarova, K.; Haft, D.H.; Barrangou, R.; Brouns, S.J.; Charpentier, E.; Horvath, P.; Moineau, S.; Mojica, F.J.; Wolf, Y.I.; Yakunin, A.F.; et al. Evolution and classification of the CRISPR–Cas systems. *Nat. Rev. Microbiol.* **2011**, *9*, 467–477. [CrossRef] [PubMed]
54. Yosef, I.; Manor, M.; Kiro, R.; Qimron, U. Temperate and lytic bacteriophages programmed to sensitize and kill antibiotic-resistant bacteria. *Proc. Natl. Acad. Sci. USA* **2015**, *112*, 7267–7272. [CrossRef] [PubMed]
55. Seed, K. Battling phages: How bacteria defend against viral attack. *PLoS Pathog.* **2015**, *11*, e1004847. [CrossRef] [PubMed]
56. Scanlan, P.; Buckling, A.; Hall, A. Experimental evolution and bacterial resistance: Coevolutionary costs and trade-offs as opportunities in phage therapy research. *Bacteriophage* **2015**, *5*. [CrossRef] [PubMed]
57. Valério, N.; Oliveira, C.; Jesus, V.; Branco, T.; Pereira, C.; Moreirinha, C.; Almeida, A. Effects of single and combined use of bacteriophages and antibiotics to inactivate Escherichia coli. *Virus Res.* **2017**, *240*, 8–17. [CrossRef] [PubMed]
58. Comeau, A.; Tétart, F.; Trojet, S.; Prère, M.; Krisch, H. La «synergie phages-antibiotiques». *Med. Sci.* **2008**, *24*, 449–451. [CrossRef] [PubMed]
59. Ryan, E.; Alkawareek, M.; Donnelly, R.; Gilmore, B. Synergistic phage-antibiotic combinations for the control of *Escherichia coli* biofilms in vitro. *FEMS Immunol. Med. Microbiol.* **2012**, *65*, 395–398. [CrossRef] [PubMed]

60. Torres-Barceló, C.; Arias-Sánchez, F.I.; Vasse, M.; Ramsayer, J.; Kaltz, O.; Hochberg, M.E. A window of opportunity to control the bacterial pathogen *Pseudomonas aeruginosa* combining antibiotics and phages. *PLoS ONE* **2014**, *9*, e106628. [CrossRef] [PubMed]
61. Maciejewska, B.; Olszak, T.; Drulis-Kawa, Z. Applications of bacteriophages versus phage enzymes to combat and cure bacterial infections: An ambitious and also a realistic application? *Appl. Microbiol. Biotechnol.* **2018**, *102*, 2563–2581. [CrossRef] [PubMed]
62. Maxted, W. The active agent in nascent phage lysis of streptococci. *J. Gen. Microbiol.* **1957**, *16*, 584–595. [CrossRef] [PubMed]
63. Krause, R. Studies on the bacteriophages of *hemolytic streptococci*: II. Antigens released from the streptococcal cell wall by a phage-associated lysin. *J. Exp. Med.* **1958**, *108*, 803–821. [CrossRef] [PubMed]
64. Fischetti, V. Purification and physical properties of group C streptococcal phage-associated lysin. *J. Exp. Med.* **1971**, *133*, 1105–1117. [CrossRef] [PubMed]
65. São-José, C. Engineering of phage-derived lytic enzymes: Improving their potential as antimicrobials. *Antibiotics* **2018**, *7*. [CrossRef] [PubMed]
66. Diez-Martínez, R.; De Paz, H.D.; García-Fernández, E.; Bustamante, N.; Euler, C.W.; Fischetti, V.A.; Menendez, M.; García, P. A novel chimeric phage lysin with high in vitro and in vivo bactericidal activity against Streptococcus pneumoniae. *J. Antimicrob. Chemother.* **2015**, *70*, 1763–1773. [CrossRef] [PubMed]
67. Drulis-Kawa, Z.; Majkowska-Skrobek, G.; Maciejewska, B. Bacteriophages and phage-derived proteins—Application approaches. *Curr. Med. Chem.* **2015**, *22*, 1757–1773. [CrossRef] [PubMed]
68. Wang, I.N.; Smith, D.L.; Young, R. Holins: The protein clocks of bacteriophage infections. *Annu. Rev. Microbiol.* **2000**, *54*, 799–825. [CrossRef] [PubMed]
69. Lin, H.; Paff, M.; Molineux, I.; Bull, J. Therapeutic application of phage capsule depolymerases against K1.; K5.; and K30 capsulated E. coli in mice. *Front. Microbiol.* **2017**, *8*, 2257. [CrossRef] [PubMed]
70. Pires, D.; Oliveira, H.; Melo, L.; Sillankorva, S.; Azeredo, J. Bacteriophage-encoded depolymerases: Their diversity and biotechnological applications. *Appl. Microbiol. Biotechnol.* **2016**, *100*, 2141–2151. [CrossRef] [PubMed]
71. Rodríguez-Rubio, L.; Martínez, B.; Donovan, D.; Rodríguez, A.; García, P. Bacteriophage virion-associated peptidoglycan hydrolases: Potential new enzybiotics. *Crit. Rev. Microbiol.* **2012**, *39*, 427–434. [CrossRef] [PubMed]
72. Attai, H.; Rimbey, J.; Smith, G.; Brown, P. Expression of a peptidoglycan hydrolase from lytic bacteriophages Atu_ph02 and Atu_ph03 triggers lysis of *Agrobacterium tumefaciens*. *Appl. Environ. Microbiol.* **2017**, *83*. [CrossRef] [PubMed]
73. Olsen, I. Modification of phage for increased antibacterial effect towards dental biofilm. *J. Oral Microbiol.* **2016**, *8*. [CrossRef] [PubMed]
74. Hauser, A.; Mecsas, J.; Moir, D. Beyond antibiotics: New therapeutic approaches for bacterial infections. *Clin. Infect. Dis.* **2016**, *63*, 89–95. [CrossRef] [PubMed]
75. Smith, G.P. Filamentous fusion phage: Novel expression vectors that display cloned antigens on the virion surface. *Science* **1985**, *228*, 1315–1317. [CrossRef] [PubMed]
76. Pande, J.; Szewczyk, M.; Grover, A. Phage display: Concept, innovations, applications and future. *Biotechnol. Adv.* **2010**, *28*, 849–858. [CrossRef] [PubMed]
77. Christensen, D.; Gottlin, E.; Benson, R.; Hamilton, P. Phage display for target-based antibacterial drug discovery. *Drug Discov. Today* **2001**, *6*, 721–727. [CrossRef]
78. Ebrahimizadeh, W.; Rajabibazl, M. Bacteriophage vehicles for phage display: Biology, mechanism and application. *Curr. Microbiol.* **2014**, *69*, 109–120. [CrossRef] [PubMed]
79. Muteeb, G.; Rehman, M.T.; Ali, S.Z.; Al-Shahrani, AM.; Kamal, M.A.; Ashraf, G.M. Phage display technique: A novel medicinal approach to overcome antibiotic resistance by using peptide-based inhibitors against β-lactamases. *Curr. Drug Metab.* **2017**, *18*, 90–95. [CrossRef] [PubMed]
80. Flachbartova, Z.; Pulzova, L.; Bencurova, E.; Potocnakova, L.; Comor, L.; Bednarikova, Z.; Bhide, M. Inhibition of multidrug resistant *Listeria monocytogenes* by peptides isolated from combinatorial phage display libraries. *Microbiol. Res.* **2016**. [CrossRef] [PubMed]
81. Kovacs, J.; Masur, H. Prophylaxis against opportunistic infections in patients with human immunodeficiency virus infection. *N. Engl. J. Med.* **2000**, *342*, 1416–1429. [CrossRef] [PubMed]

82. Bazan, J.; Całkosiński, I.; Gamian, A. Phage display—A powerful technique for immunotherapy. *Hum. Vaccin. Immunother.* **2012**, *8*, 1829–1835. [CrossRef] [PubMed]
83. Omidfar, K.; Daneshpour, M. Advances in phage display technology for drug discovery. *Expert Opin. Drug Discov.* **2015**, *10*, 651–669. [CrossRef] [PubMed]

antibiotics

MDPI

Review

Phage Therapy with a Focus on the Human Microbiota

Sharita Divya Ganeshan [1] and Zeinab Hosseinidoust [1,2,3,4,*]

1 School of Biomedical Engineering, McMaster University, Hamilton, ON L8S 4K1, Canada
2 Department of Chemical Engineering, McMaster University, Hamilton, ON L8S 4L7, Canada
3 Farncombe Family Digestive Health Research Institute, McMaster University, Hamilton, ON L8S 4K1, Canada
4 Michael DeGroote Institute for Infectious Disease Research, McMaster University, Hamilton, ON L8S 4L8, Canada
* Correspondence: doust@mcmaster.ca; Tel.: +1-905-525-9140

Received: 5 July 2019; Accepted: 23 August 2019; Published: 27 August 2019

Abstract: Bacteriophages are viruses that infect bacteria. After their discovery in the early 1900s, bacteriophages were a primary cure against infectious disease for almost 25 years, before being completely overshadowed by antibiotics. With the rise of antibiotic resistance, bacteriophages are being explored again for their antibacterial activity. One of the critical apprehensions regarding bacteriophage therapy, however, is the possibility of genome evolution, development of phage resistance, and subsequent perturbations to our microbiota. Through this review, we set out to explore the principles supporting the use of bacteriophages as a therapeutic agent, discuss the human gut microbiome in relation to the utilization of phage therapy, and the co-evolutionary arms race between host bacteria and phage in the context of the human microbiota.

Keywords: microbiome therapy; phage therapy; evolution; antibiotic resistance

---

1. Introduction

Bacteriophages (phages) are bacteria's natural predators and could be employed for treating infections. Phage therapy was actively utilized immediately after discovery of phages in the early 1900s [1], years prior to the introduction of antibiotics in the 1940s, when infectious diseases were the leading cause of mortality and morbidity within human populations [2,3]. However, a lack of understanding of phage biology in the early days, combined with exaggerated claims, a lack of controlled trials, and poor documentation, among others, led to phage therapy being overshadowed by antibiotics in Western medicine [4]. Antibiotics have since been highly attractive due to their broad-spectrum activity, allowing them to be used against a wide range of infections without necessarily identifying or characterizing the exact infective agent(s).

With increased understanding of the role of human microbiota in our overall well-being, this broad-spectrum killing activity is increasingly presenting itself as a major disadvantage. We now know that a symbiotic relationship is established directly between the human gut microbiome (the population of bacteria, viruses, and fungi that occupy surfaces of the human body) and the physical and mental health of the human host, the disturbance of which could lead to the initiation or progression of many chronic and degenerative diseases [5–9]. The same is true for other niche microbiota such as nasal [10], eye [11], oral [12], and genital tracts [13]. As bacteria's natural predators, bacteriophages offer a powerful advantage over antibiotics, namely that they can be highly specific, targeting only their host bacteria, suggesting a milder therapy towards niche microbiota. However, one of the critical apprehensions concerning lytic phage therapy is the potential for perturbations of the niche microbiota as a result of the strong selective pressure exerted by lytic phages on their host communities [14–16].

In this review, we provide an overview of the general use of bacteriophages as therapeutic agents, then discuss the relationship between bacteriophage therapy and the microbiota in the context of the gut microbiota and possible effects of phage therapy on the human microbiota.

2. Bacteriophages as Therapeutic Agents

2.1. Phage Biology

Bacteriophages (phages) are viruses of bacteria. With an estimated 10^{32} phages, these are the most ubiquitous and diversified biological group residing on Earth [17]. As a result of their obligate requirement of a bacterial host, bacteriophages are abundantly found distributed essentially anywhere their host exists in the biosphere [14–16], with ten to hundreds of millions of phages in every gram of soil [18], water [19], and billions on and inside the human body at any moment [20]. Bacteriophages are diverse in their complexity, structure, genetic material, and are variant in their shape (tailed, filamentous, and icosahedral), and size (Figure 1) [21]. Phage virions generally consist of a protein envelope (sometimes containing lipids) encapsulating a genome consisting of two to hundreds of kilobase pairs of single or double-stranded DNA or RNA [22]. Phage virions could be effectively visualized with electron microscopy; specifically, Transmission Electron Microscopy (TEM) (Figure 2A) [23].

Phages are more genetically diverse than their bacterial hosts (and prey); however, these bacterial viruses only infect a narrow range of bacteria that are closely related due to a combination of various factors. Such limiting factors include the specificity of the virion's host binding proteins, biochemical interactions during infection, the presence of related prophages or particular plasmids in the host, and bacterial resistance mechanisms to phages (its predator) [24–26].

Bacteriophages are classified through a structural and sequence-based taxonomic system; initially into families, and each family is further categorized in accordance to the capsid structure, the structural and chemical composition of the genes and the mechanism of their mRNA production (Table 1) [27,28]. These viruses are further broadly categorized in terms of their propagation cycle as lytic, temperate, and chronic phages [29]. Regardless of the nature of their propagation cycle, bacteriophages first have to bind to specific sites on the host cell surface (Figure 2B). Lytic phages bind and adsorb to specific receptors on the host cell surface and inject their genome into the host cell and undergo propagation, which ultimately results in lysis of the host, further, releasing progeny phages into the surrounding medium (Figure 2C,D) [30]. This process is known as the lytic cycle. As an example, Figure 2E depicts the process of absorption and genome injection of the bacteriophage T7, a well-studied lytic bacteriophage belonging to the *Podoviridae* family [22]. T7, an *Enterobacteria* phage, attaches to outer membrane proteins OmpA and OmpF proteins on the bacterial cell wall, sending multiple internal capsid proteins into the host cell wall to construct an ejectosome from the tail of the phage and induce a pore within the bacterial cell wall. This then permits DNA from the capsid of the phage to translocate into the cell, and in turn, initiates the process to replicate phage DNA within the host [31]. Temperate viruses typically do not immediately "kill" the host bacteria; instead, they integrate their genome into the host chromosome, amplifying with every bacteria reproduction cycle; this embedded genome (known as a prophage) can be expelled from the genome of the host bacteria through the lytic cycle, once induced (Figure 2D) [30]. It is not clear what causes the induction of lytic cycle for temperate phages, but most factors that stress the host cell or cause DNA damage have been shown to induce temperate phages into a lytic cycle [32,33]. Thus, a temperate bacteriophage experiences both lytic and lysogenic cycles. This lytic-lysogeny switch has been a topic of research for decades [34,35]. In the lytic cycle, the phage replicates and lyses the host cell. In the lysogenic cycle, phage DNA is integrated into the host genome, where it is passed on to daughter cells. Phage λ is an example of a heavily studied temperate phage and uses the bacterial maltose pore LamB (λ-receptor) for delivery of its genome into the bacterial cell [36]. There is a third class of phages, known as the chronic phages, which do not lyse the cell or integrate their genome into the host genome, but instead use the host as a continuous phage-making factory—filamentous phage such as M13 belong to this class [29,37]. The common step in all three replication cycles is the specific binding of receptors on the phage virion surface to receptors on the bacterial host cell surface.

Table 1. Classification and Basic Properties of Bacteriophages [21].

Symmetry	Nucleic Acid	Order and Families	Genera	Members	Particulars
Binary (tailed)	DNA, ds, L	*Caudovirales*	15	4950	
		Myoviridae	6	1243	Tail contractile
		Siphoviridae	6	3011	Tail long, noncontractile
		Podoviridae	3	696	Tail short
Cubic	DNA, ss, C	*Microviridae*	4	40	
	ds, C, T	*Corticoviridae*	1	3	Complex capsid, lipids
	ds, L	*Tectiviridae*	1	18	Internal lipoprotein vesicle
	RNA, ss, L	*Leviviridae*	2	39	
	ds, L, S	*Cystoviridae*	1	1	Envelope, lipids
Helical	DNA, ss, C	*Inoviridae*	2	57	Filaments or rods
	ds, L	*Lipothrixviridae*	1	6	Envelope, lipids
	ds, L	*Rudiviridae*	1	2	Resembles TMV
Pleomorphic	DNA, ds, C, T	*Plasmaviridae*	1	6	Envelope, lipids, no capsid
	ds, C, T	*Fuselloviridae*	1	8	Spindle-shaped, no capsid

C: circular; L: linear; S: segmented; T: superhelical; ss: single-stranded; ds: double-stranded.

Phage therapy exploits the natural ability of bacteriophages to specifically target and kill host bacteria with high specificity, for the purpose of treating bacterial infections. For phage therapy, only strictly lytic phages are of immediate interest [38]. Although there are strong arguments to be made supporting the use of temperate phages for therapeutic purposes [39], regulatory agencies have a long way to go before accepting the use of viruses that could transfer their genes directly to bacterial cells in the body [40].

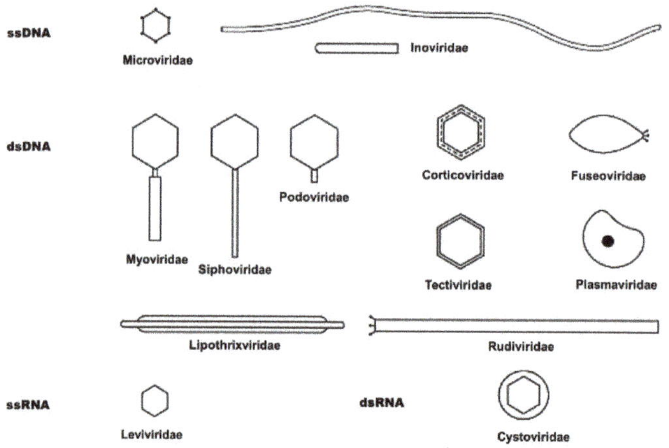

Figure 1. Schematic representation of major groups of bacteriophages. Reproduced with permission from reference [41].

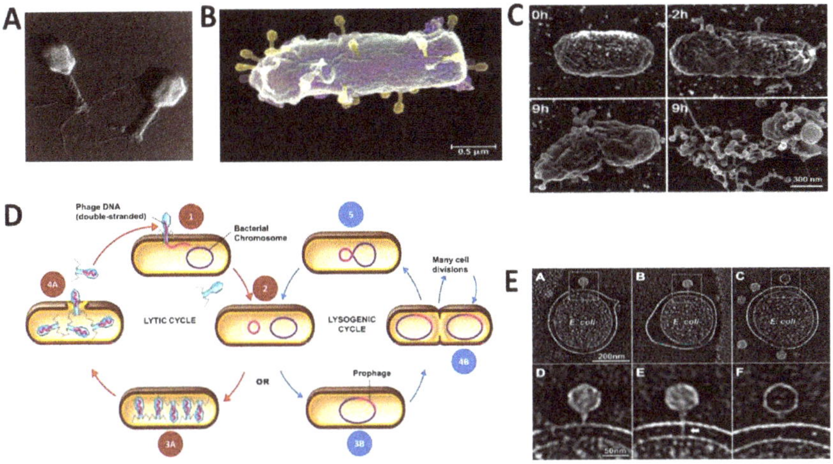

Figure 2. Introduction to lytic phage biology. (**A**) Shadowed Transition Electron Micrograph image of T4 phage (Mag 380,000×). This phage, a member in the *Myoviridae* family of the *Caudovirales* order, is one of the seven *Escherichia coli* phages (T1–T7) in this family. This image shows the icosahedral capsid head containing the genetic material, the contractile tail, and the long tail fibers of the phage. T4 head is approximately 90 nm, wide and the virion is 200 nm in length. This TEM was photographed at Wurtzbiozentrum at the University of Basel, reproduced with permission from references [42,43]. (**B**) Colorized scanning electron micrograph (SEM) images of multiple T4 bacteriophages infecting an *E. coli* cell reproduced with permission from reference [44]. (**C**) SEM images at different stages showing the infection of *Synechococcus* WH8102 by the S-TIM5 phage. 0 h-uninfected cells. 2 h-phage adsorption. 9-h cell lysis. 9-h viral release. These SEM images were collected from Sabehi. G., from the Israel Institute of Technology, reproduced with permission from reference [45]. (**D**) The lytic and lysogenic infection cycles. The first two stages are shared for both the cycles. Step 1- Attachment of the phage tail fibers to a specific receptor site on the bacterial cell wall and injection of the viral genome. Step 2- Phage DNA is then circularized and enters the lytic cycle or the lysogenic cycle. Lytic cycle: Step 3A- Synthesis of new viral proteins within the host. Step 4A- Virions are liberated as mature phages upon cell lysis. Lysogenic cycle: Step 3B- Phage DNA integrates within the bacterial chromosome by recombination, in turn becoming a prophage. Step 4B- Lysogenic bacterium reproduces normally and has the potential to do so over many cell divisions. The prophage may be released from the bacterial chromosome through external triggers, resulting in the initiation of the lytic cycle—reproduced with permission from reference [30]. (**E**) T7 bacteriophage infecting *E. coli* as seen with cryoelectron tomography at ~4 nm resolution. A/D- Adsorption of T7 phage into the outer membrane. B/E- Injection of the extended tail into the cell envelope. C/F- DNA ejection. These images are collected from Bo Hu, University of Texas—reproduced with permission from reference [31].

2.2. History of Phage Therapy

Frederick Twort, an English bacteriologist, first reported evidence of bacteriophages (lysis of bacterial cultures) in 1915 and suggested the effect might be due to the presence of a virus or an enzyme [46]. In 1917, Felix d'Herelle, a French–Canadian, and a self-taught microbiologist working at the Institute Pasteur, independently made similar observations, but he was quick to attribute the effect to a virus and named these microbes "bacteriophages" or bacteria-eaters; he was the first to utilize phages as antimicrobial agents to treat infections [47,48]. The discovery of phages played a crucial role in the development and understanding of further scientific discoveries, including the initiation of understanding the structure of DNA [49], and more recently, in the discovery of new clusters of regularly interspaced short palindromic repeat (CRISPR) systems [50].

Phage therapy was actively utilized, immediately by d'Herelle and later by others [1], for the treatment of infections years prior to the introduction of antibiotics in the 1940s, when infectious diseases were the leading cause of mortality and morbidity within human populations [2,3]. In 1919, d'Herelle treated three brothers, whose sister had just died of dysentery, with phages. The brothers, aged three, seven, and twelve were in a serious condition but began to recover in 24 h [4]. He then moved on to treating Bubonic plague in 1925 by injecting phages directly into the buboes, after which the patient showed signs of recovery in 24 h and exhibited full recovery in less than a month [51]. Data collected from the plague in India (in 1926) using phages isolated by d'Herelle showed a drastic decrease of mortality for the phage-treated group compared to the control group [51]. As the word spread, more trials followed, some of which were inconclusive.

Antibiotics quickly overshadowed the use of phage therapy in Western medicine due to the lack of understanding of both phage biology and microbiology of infectious diseases, exaggerated claims, a lack of controlled trials, poor documentation, and politics [4]. This in turn, sparked scientific controversy about the treatment approach [52,53], rendering phage therapy essentially obsolete in Western medicine for close to four decades. Due to the occurrence of World War II, certain regions of the Soviet Union and Eastern Europe had limited access to antibiotics and thus focused on developing phage therapy [54]. The practice of phage therapy within the Soviet Union was well supported, and this is in fact still heavily practiced in Russia and the Eastern European countries for over 80 years, especially within the Eliava Institute in Tbilisi, Georgia, co-established by Felix d'Herelle himself [55,56].

2.3. *Phage Therapy Compared to Antibiotic Therapy*

As noted above, the 1940s brought upon a golden era for the utilization of antibiotics as antimicrobial agents. Penicillin was not immediately utilized as an antibiotic upon its serendipitous discovery in 1928 [3]. However, it was quickly guided into optimization due to the crucial necessity to treat sick or wounded soldiers in the US and Allies' military forces during the war [57]. Antibiotics were discovered at a time when bacterial infections had a high mortality rate; antibiotics were highly attractive due to their broad-spectrum activity allowing them to be used against a wide range of bacteria without necessarily identifying or characterizing the exact infective bacterium. However, this main advantage was also a great disadvantage. Due to their relative non-specificity towards bacteria, antibiotics also destroy the commensal microflora, especially within the intestine for oral antibiotics. Similar effects on the microbiota are observed for other niches in the body. This is further associated with side-effects such as intestinal problems, or secondary infection, such as *C. difficile* infection [58]. Furthermore, secondary side-effects often occur as antibiotics are also mostly needed in repeated administrations [59]. Using lytic phage, however, could circumvent this issue because bacteriophages are very specific to their host. This means that phage therapeutics could be designed to kill an infection but not harm the microbiota. This specificity could, however, become a challenge because many infections are polymicrobial. This requires phage therapeutics to be cocktails of different phages rather than a pure stock [60].

The non-essential overuse and abuse of antibiotics (in clinics, aquaculture, and agriculture) has fueled the volatile era of antibiotic resistance globally. Specifically, a group of gram-positive and gram-negative species, made up of *Enterococcus faecium*, *Staphylococcus aureus*, *Klebsiella pneumoniae*, *Acinetobacter baumannii*, *Pseudomonas aeruginosa*, and *Enterobacter* species, known as ESKAPE pathogens, most of which are multidrug-resistant isolates, are the leading cause of hospital-acquired infections throughout the world. It is approximated that by the year 2050, we would have 10 million deaths per year, with more people dying from antibiotic resistance than cancer [61]. As a consequence of this current dangerous state of infective antibiotics and antibiotic resistance, interest in phage therapy as an antimicrobial strategy against lethal pathogens has resurfaced. A recent review investigating 30 clinical studies of phage therapy against ESKAPE pathogens, 87% showed efficacy and safety (67%) of the tested phages, but only 35% examined phage resistance [62].

3. Phage Therapy: Where Are We Now?

3.1. Advantages and Challenges of Phage Therapy

Bacteriophages are not a fully established alternative to antibiotics as an antibacterial therapeutic. However, they hold great promise not only for tackling antibiotic-resistant infections, but also for treating at-risk patients that are less tolerant to antibiotics such as infants, pregnant women, and immunocompromised patients. As phage therapy is in vogue right now as a potential alternative/adjuvant, it is imperative to carefully assess all possible characteristics, parameters, and limitations. Some advantages of phage therapy over antibiotics include:

- Specificity: Bacteriophages are generally very specific in their host range, which promises less harm to our microbiota. However, this specificity also means that for most clinically relevant infections, a mixture of different phage will have to be used to address the polymicrobial nature of the infection [60].
- Bactericidal versus bacteriostatic: lytic phages infect their target host bacteria and cause cell death, in comparison to certain bacteriostatic antibiotics [63,64].
- Active on-site propagation: bacteriophages increase in concentration in situ as they propagate in the presence of bacterial host (infectious agents). Unlike antibiotics, which often require frequent doses to kill the bacteria efficiently, only one bacteriophage is theoretically needed to target a single corresponding host bacterium (single-hit kinetics) [65]. It is possible to administer a single low-dose of phage, which will then propagate itself, given the existing bacterial density as an active therapy, resulting in continued bacterial adsorption and killing [66].
- Low inherent toxicity: bacteriophages are primarily composed of nucleic acid and proteins that have been studied to be non-toxic with the use of highly purified phage preparations [67].
- Enormous diversity: Phage is found in diverse abundances across the biosphere and therefore isolating and purifying new phages, necessary to target a known pathogenic bacteria, is achievable even when bacteria evolve resistance [68].
- Formulation and application versatility: Multiple strains of phages can be added together into a "phage cocktail" to target multiple bacteria of interest with a broader killing spectrum [69]. Furthermore, mode of administration (liquid, ointment [70], powder [71], oral tablets [72]) could also vary in accordance with each unique situation.
- Narrow potential for antibiotic cross-resistance: The mechanism of bacterial resistance to antibiotics and phage are entirely different [61,73]. Thus, bacteria that are resistant to antibiotics may be targeted and treated with the use of phage therapy, and vice-versa, presenting combination therapies as an attractive strategy [74].
- Biofilm clearance: phages have been demonstrated to lyse and penetrate through some biofilms that have shown resistance to antibiotics [75]. This is partially attributed to the presence of depolymerases and lysins that can chew through the biofilm extracellular polymeric matrix [76,77].
- Low environmental impact: bacteriophages are natural components of the environment that can be naturally evolved (as opposed to genetic modification), thus easing public acceptance of phage therapy. Furthermore, because this natural product is composed primarily of proteins and nucleic acids, unused phage materials can easily be inactivated and discarded [63].
- Relatively low discovery and production cost: the costs associated with the discovery, isolation, and purification of bacteriophages are significantly decreasing with the progression of screening and sequencing technologies [78,79].

3.2. Outstanding Challenges of Phage Therapy

Regardless of the numerous known advantages of bacteriophages, there remain outstanding challenges and un-addressed limitations to this approach that must be addressed and further investigated. In particular, there are currently four primary concerns:

- Narrow host-range: phages generally have a narrow host-range, which limits exactly which strain(s) of bacteria can be targeted. This is a challenge because most infections in the body are polymicrobial. Thus it is imperative to employ effective and efficient phage cocktails, curated from a combination of multiple selected phages to develop a broader host spectrum [80].
- Even though phage therapy was employed in Western medicine before antibiotics took over, it currently does not have the approval for human administration from the Food and Drug Administration (FDA) or the European Medicines Agency (EMA) due to an increase in regulatory standards, partially as a result of the increased understanding within the scientific community about the potential effect of drugs on the human microbiome and partially due to concerns about phage resistance and the role of phages in bacterial genome evolution. To obtain regulatory approvals, current research is heavily exploring roadblocks through numerous animal studies and clinical trials [81–83]. The FDA has, however, approved the use of phages for food decontamination [84], dietary supplements [85], and environmental prophylaxis. Phage is allowed only on a compassionate care basis for human therapeutic use [86,87].
- Phages as drivers of evolution: unlike common pharmaceuticals, bacteriophages are DNA/RNA-containing protein-based biological agents that have the potential to interact with the body's immune system and other microbial cells in the body and can actively replicate and evolve within the body. This evolution can, in turn, result in the evolution of the commensal host bacterial communities and possibly even impact the composition of the niche microbiome [88].
- Phage selection criteria: The criteria for the selection of therapeutic phage is not well determined. Most reports to date have focused on phage host range, but factors such as the ability of phage to infect stationary phase bacteria [89], phage enzymes, stability to serum inactivation, and mutation rate have been shown to be important and deserve more investigation beyond proof-of-concept reports [68,90].
- Lack of well-curated public phage libraries: there is a serious lack of well-curated, well-characterized libraries for therapeutic phage. Various research labs in academia, industry, and army research centers have small to mid-sized collections that, except for very few cases, are considered proprietary and are not shared with the broader scientific community. This need was acknowledged by the community recently and led the DSMZ (German Collection of Microorganisms and Cell Cultures) to start the Phage Call Project and encourage researchers to deposit characterized phages to their public phage library. This effort, although valuable, is progressing very slowly, with only a few phages per strain currently in the library. The slow pace of progress in this regard can partially be attributed to the vast diversity of phages around the globe, partially to the lack of high throughput methods for rapid isolation and characterization of new phage, and partially to the lack of a mechanism to acknowledge phage as the intellectual property of the discovering lab, leading to the unwillingness of researchers to share so as to maintain competitive advantage.

3.3. Clinical Trials and Current Phage Therapy

The Eliava Institute in Tbilisi Georgia remains a leading expert clinic for active practice of phage therapy, with hundreds of local, as well as international patients that have received lifesaving phage treatment [55]. Moreover, there have been several notable clinical trials of phage therapy within recent time. PhagoBurn, funded by the European Commission, was the first phage therapy clinical trial utilizing good manufacturing practices (GMP) [91]. Their phage cocktail was curated as a treatment for *Escherichia coli* and *Pseudomonas aeruginosa* burn wound infections. The development and validation of PhagoBurn served as an eye-opening experience for the community in terms of pitfalls in designing an efficient manufacturing process with a phage cocktail. The associated clinical trial was also met with significant challenges. Thus, PhagoBurn also provided a major understanding of the challenges in the design of clinical trials for human phage therapy, and other potential limitations that may occur with the manufacturing and administration of a phage treatment [91]. Table 2 lists the more recent phage therapy clinical trials that have been registered in ClinicalTrials.gov.

Table 2. Recent Phage Therapy Clinical Trials.

Trial Title	Condition/Infection	Intervention	Status	Country	Ref.
Standard Treatment Associated with Phage Therapy Versus Placebo for Diabetic Foot Ulcers Infected by *S. aureus*	Diabetic Foot, Staphylococcal Infections	PhagoPied: Topical anti-Staphylococcus bacteriophage therapy	Not Yet Recruiting	France	[92]
Individual Patient Expanded Access for AB-SA01, an Investigational Anti- *S. aureus* Bacteriophage Therapeutic	MDR *Staphylococcus aureus* infections	AB-SA01 (3- phage cocktail)	In Progress	USA	[93]
Individual Patient Expanded Access for AB-PA01, an Investigational Anti-*Pseudomonas aeruginosa* Bacteriophage Therapeutic	*Pseudomonas aeruginosa* infections (incl. MDR stains)	AB-PA01 (4-phage cocktail)	In Progress	USA	[94]
Safety and Efficacy of EcoActive on Intestinal Adherent Invasive *E. coli* in Patients With Inactive Crohn's Disease	Crohn's Disease	EcoActive (collection of bacteriophages)	Recruiting	USA	[95]
Analysis of changes in inflammatory markers in patients treated with bacterial viruses	Wide-range, non-healing postoperative wounds or bone, upper respiratory tract, genital or urinary tract infections whom extensive antibiotic therapy failed	oral, rectal and/or topical bacteriophage lysates/purified phage formulations/phage cocktails	Completed	Poland	[96]
Evaluation of Phage Therapy for the Treatment of Escherichia coli and Pseudomonas aeruginosa Wound infections in Burned Patients	Wound infection	PhagoBurn: *E. coli* phages cocktail (15-phage cocktail), *aeruginosa* Phages cocktail (13-phage cocktail)	Completed	Belgium, France, Switzerland	[91]
Bacteriophage Effects on *Pseudomonas aeruginosa*	Cystic Fibrosis (CF)	Mucophages (10-phage cocktail)	Completed	France	[97]
Therapeutic bacteriophage preparation in chronic otits due to antibiotic-resistant *Pseudomonas aeruginosa*	Antibiotic-resistant *Pseudomonas aeruginosa* in chronic otitis	Biophage-PA	Completed	United Kingdom	[98]
Antibacterial Treatment against Diarrhea in Oral Rehydration Solution	ETEC and EPEC Diarrhea	Oral T4 phage cocktail	Completed	Bangladesh	[81]
Bacteriophages for treating Urinary Tract Infections in Patients Undergoing Transurethral Resection of the Prostate	Urinary Tract Infections (UTI)	Intravesical instillation-PYO phage	Completed	Georgia	[99]
A Prospective, Randomized, Double-Blind Controlled Study of WPP-201 for the Safety and Efficacy of Treatment of Venous Leg Ulcers	Venous Leg Ulcers	WPP-201 (8-phage cocktail)	Completed	USA	[100]

As mentioned above, phage therapeutic products are currently on the market in certain parts of the world. Sextaphage is one such commercial pharmaceutical phage composition from the Russian company Microgen (Figure 3) [101]. This phage therapeutic contains phages against six specific pathogens, with the intent of treating urinary tract infections in pregnant women. Pregnant women are prone to urinary tract infections; however, very limited antibiotics have been proven safe for the developing fetus [102]. Microgen offers multiple other phage products for other infections.

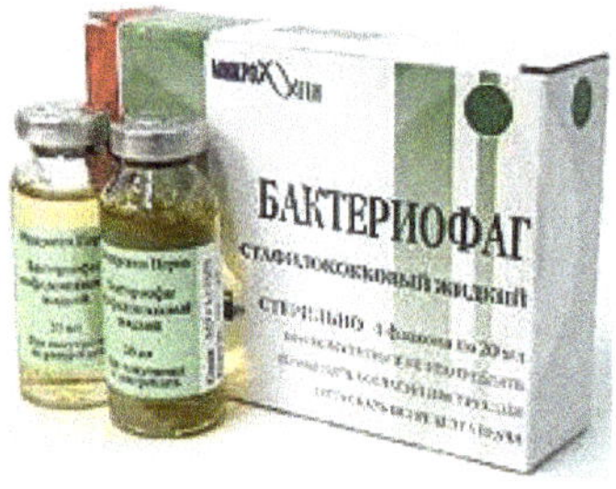

Figure 3. Sextaphage pharmaceutical product from microgen. Image reproduced with permission form reference [103].

A recent notable successful clinical attempt of phage therapy was reported from scientists and physicians at the University of California San Diego School of Medicine, who had worked alongside collaborators from the U.S. Navy Medical Research Center-Biological Defense Research Directorate. As the first attempt in the United States, they successfully utilized intravenous phage therapy to treat a patient with a severe systemic infection caused by multidrug-resistant organisms. Due to the utilization of a specially curated phage therapeutic on a compassionate care level, this patient, a professor in the Department of Psychiatry at UC San Diego School of Medicine, was saved from an end-stage comatose condition [87]. The phages used in that case were isolated from various environmental samples. The UC San Diego case attracted significant publicity leading to the first case of clinical trial authorized in North America for intravenous phage therapy to be carried out at UC San Diego.

In 2016, Paul Turner and colleagues reported isolating a phage that could restore antibiotic sensitivity in multidrug-resistant *Pseudomonas aeruginosa* [104]. This phage was later used to treat a patient with a longstanding aortic graft infection that did not respond to multiple surgical interventions and aggressive antibiotic therapy with a single application of phage [105]. The same team recently reported treating two cystic fibrosis patients with antibiotic-resistant infections. Aside from the wild type phage isolated from natural sources, genetically engineered phages have also been reported for phage therapy. A young patient with cystic fibrosis and bilateral lung transplantation who had developed a *Mycobacterium abscessus* infection was reportedly treated with a phage cocktail of genetically engineered lytic phages that were administered intravenously [106]. Genetically modified phages can circumvent some of the limitations of wild type phage (e.g., narrow host range); however, more indepth investigations are needed to assess the possibility of reversion under the niche selective pressures in the body. Finally, phage-derived proteins have been used as therapeutics [107]. The current review is mainly focused on whole phage virions and the reader is referred to the review article by Roach et al. for more information on using phage-derived biomolecules for therapy [108].

4. Phage-Host Evolutionary Dynamics and Diversification

Bacteria-phage interactions have remained central to the evolution and ecology of microbial communities, as they alter the competition of bacterial species and result in the evolution of new mutants/variants [109]. Phages and their bacterial hosts represent a predator-prey system. Hundreds of new predators (phages) are born each time a prey (bacterial cell) is eaten. Under selective pressure from phage predation, bacteria develop resistance to phages, and consequently, the bacteriophage counter-adapts and evolves to counter this resistance [110–112]. As a result of this co-evolution, phages have acquired counter bacterial defense mechanisms, including anti-CRISPRs [113]. It is partially due to this co-evolution that effective phage infection is not always observed regardless of phage adsorption [114].

Most mechanisms of phage resistance differ from that of antibiotic resistance [73,115]. In addition, as bacterial populations evolve to resist phage predation via random mutations, the predator phage will also shadow the bacterial community and mutate to combat the resistant bacteria, consequently increasing genomic diversity of both host and phage communities [110].

Due to differences in the mutational supply of both the bacteriophage and corresponding bacterial host and levels of resource supply for bacterial resistant mutations and phage infectivity mutations, there is a further difference in the strength of selection, population divergence, trajectory, and adaptation patterns [88,116]. Moreover, lytic phages typically have larger population sizes and shorter generation time in comparison to their corresponding bacterial host, which impacts the evolutionary race between bacteria and phage, leading to a higher diversification of the phage community compared to its host [117]. The strategic exploitation of multiple phages in the form of a therapeutic phage cocktail (and thus multiple selective pressures) has the potential to limit the bacteria's ability to evolve resistance against a single phage species [118].

As the co-evolutionary arms race of phage and their corresponding host bacteria progresses, bacteria mutants resistant to phage predation will evolve. The nature of these mutations is believed to differ in vitro and in vivo and even in different niches in the body and for different phage-host pairs. What is common in all systems is the degree of diversification of the host community in response to phage predation, leading to a range of phenotypes that behave differently from the parental strain in terms of fitness [14,119]. In fact, the difference in behavior between the phage-resistant bacterial mutants and the parental strain was utilized to design phage therapy cocktails that lead an antibiotic-resistant bacterial population to regain antibiotic susceptibility [104].

Competition between the phage-resistant bacterial mutants and the entire niche microbiota for colonization and resources, nutrient availability, as well as the spatial heterogeneity of the niche microenvironment, could all affect the co-evolution of the phage-host system in a phage therapy case, leading to an outcome different from that predicted based on observations in a typical lab culture [120–123]. There are currently very few investigations on phage-host co-evolution in the gut environment. One of the few notable recent attempts is the work from the Institute Pasteur that reports phages evolving to infect different hosts in the gut [124]. It is important to study if the phenotypic and genetic diversity within the gut microbiota, and if the overall community environment structure could be maintained during the co-evolutionary processes underlying the interactions between bacteria and therapeutic lytic phages.

5. Bacteriophages and the Microbiota

5.1. Introduction to the Microbiota

The human body harbors a rich and complex community of microbial cells. This plethora of cells come from all three domains of life (archaea, bacteria, eukarya), with the number of bacteria in the body approximately the same order as that of eukaryotic-human cells, and phage outnumbering the bacterial population by a factor of 10 [125,126]. A symbiotic relationship exists between the human gut microbiome and the human host, the disturbance of which could lead to the initiation or progression of many inflammatory diseases of the gut, including Crohn's disease [127], ulcerative colitis [5], and inflammatory bowel disease [128], among others. In addition, recent research has associated the gut microbiota with non-intestinal diseases such as rheumatoid arthritis [6], diabetes [129], obesity [130], certain forms of cancer [7], various neurological degenerative diseases [131], and various mental disorders [8]. Furthermore, evidence has been found linking the use of a wide range of drugs to alterations in the human gut microbiome, which could, in turn, impact the body's susceptibility towards chronic and degenerative diseases [132–135].

5.2. Bacteriophages as Part of the Human Microbiota

Bacteriophages are naturally present within the human gut, oral and nasal cavity, genital tract, and eye, in collection with other bacteria, fungi, protozoa, and fungi. Most bacteriophages naturally present in the human body are temperate phages and many thus bacteria in our microbiome harbor prophages. Induction of this prophage under different conditions can lead to disturbance of the balance of the microbiome and dysbiosis (imbalance between the organisms present in our natural microflora) [136,137], necessitating further mechanistic investigations on the environmental triggers that induce temperate phage in the gut. Such induction can be a byproduct of diet, lifestyle, or other therapies such as chemotherapy. It can be deduced that controlled induction of the prophages in our commensal bacteria can be used as an approach to control the human microbiome in the case of perturbations of dysbiosis, and could thus be regarded as the next generation of phage therapy. Doing so, however, requires more knowledge about the triggers for prophage induction, a topic on which the current body of knowledge is at best incomplete. Another factor to consider is the interaction of the lytic phages used for therapy with the commensal phages (or prophages) in the microbiota. Specifically, induction of prophages as a direct or indirect result of lytic phage action could result in disturbance to the microbiota and thus further complication of the landscape in which therapeutic phage is expected to function.

5.3. *Microbiome Therapy with Bacteriophages*

Microbiome therapy aims to modulate the human microbiome as a therapeutic strategy to combat many of the chronic or degenerative diseases believed to be associated with dysbiosis in the gut microbiota or other niches in the body [138]. Specifically, due to the natural ability of bacteriophages for highly specific bacteria targeting, the exploitation of lytic or temperate phages as tools for microbiome therapy has the potential to manipulate and engineer the niche microbiota to achieve desirable effects for healthy balanced microflora [139,140]. This could be achieved using lytic phage for selective killing of certain species that have overgrown, leading to dysbiosis, or by selective induction of the temperate phage residing in commensal bacteria. The former, although more close to practice, suffers from the same bottlenecks as standard phage therapy, and the latter, although an attractive notion, is hampered by our very limited understanding of factors than can (selectively) trigger the lytic cycle for a temperate phage. Besides, the latter form of phage therapy requires the selective induction of temperate phage, a topic that has remained relatively unexplored to date.

5.4. *Possible Interactions of Therapeutic Phage and Niche Microbiota*

Despite the presence of commensal bacteriophages, naturally present within the human microbiota, the introduction of therapeutic bacteriophage to the microbiota will still act as an external agent [141]. The effect of lytic phage predation on densely colonized, nutrient-rich environments such as the human oral cavity [142] is different from the gastrointestinal environment, which although densely colonized, is affected strongly by nutrient limitations and thus promotes intense competition and cooperation among species [143].

In a recent report, researchers from the Wyss Institute at Harvard report that lytic phage simultaneously coexists and knocks down its host in the gut in a gnotobiotic mouse model colonized with human microbiota [144]. In addition, they show that phage predation can modulate the gut metabolome and claim that the effect on the metabolome is a result of non-susceptible bacteria co-colonizing species in the gut within a community of commensal bacteria colonizing the gut through cascading effects [144]. This work highlights the importance of the interactions between the commensal bacteria when a colonizing species is under phage predation and the potential impact of gut phages on the mammalian host with implications for both phage and microbiome therapies.

Another investigation that highlights the impact of phage predation on non-targeted species is detailed in a report from Institute Pasteur, which investigates the possibility of phage co-evolution in the gut to evolve the ability to infect non-host bacteria [124]. In this report, Debarbieux et al. study a tripartite network consisting of a virulent bacteriophage, its bacterial host, and a phage-insensitive bacterial strain in the murine gut. They observed that single amino acid substitutions and an unusual homologous intragenomic recombination event within the genome of the bacteriophage enabled it to infect the insensitive strain in the mouse gut [124]. An intermediate bacterial host isolated from the murine microbiota was claimed to mediate bacteriophage adaptation.

It is important to note that the limited reports published on the effect of phage predation on the microbiota are mainly limited to the gut microbiome and do not seem to agree of the actual effect. Where certain groups claim phages induce minimal changes in phylogenetic compositions [145], others showed that phage predation results in compositional changes in the murine gut microbiota [146]. The inconsistency in reports can be partially attributed to the differences in models used (gnotobiotic versus different microbiota models) as well as the phage-host systems under study. Certain phages do not persist for more than a few hours in the gut environment, whereas others can persist for weeks or even months [147]. The ability of phage to persist in the gut environment can be attributed to its ability to infect and produce progeny in a predominately stationary phase host community. Furthermore, external factors may impact the successful administration and persistence of therapeutic phages within the gut microbiome, specifically, the external physical and chemical factors, such as acidity and presence of various ions. The utilization of certain therapeutic phages in acidic gastrointestinal environment may significantly reduce phage stability and phage titer [148–150]. In a fasting human, the median gastric pH is 1.7, in comparison to a pH over 6 in certain regions of the digestive tract [151,152]. To overcome this concern regarding the decreased stability from acidity, an antacid could be simultaneously consumed upon the administration of oral phage therapy, or the therapeutic phage is encapsulated in protective matrices [153–155].

Another possible contributing factor to the indirect effect of lytic phage predation on the niche microbiome is that immune response by the human host could be triggered as bacteriophages do have the potential to cause rapid and massive bacterial lysis, subsequently releasing components from the cell wall and phage proteins into the bloodstream [156]. Hence, the potential of an immune response that may induce the production of antibodies against phage action is of concern [141,157], and calls for more indepth studies on phage-immune interaction as well as effective dosing of phage therapeutics.

Currently, for acquiring regulatory approval it is important to prove the administered therapeutic phage (1) does not have the ability to integrate its genome into the genome of the target bacteria or any of the components of the niche microbiome setting off further indirect effects within the microbiome, (2) will not indirectly affect the microbiome through selective pressure on various non-target commensal bacteria or the target infectious agents, leading to development of phage-resistant mutants which can impact the balance surrounding beneficial bacteria in the niche, as well as the effects of antagonistic phage co-evolution (the reciprocal evolution of host resistance and parasite infectivity) on the gut microbiota. To obtain regulatory approvals, therefore, outstanding questions remain to be answered: does external phage therapy affect the microbiome? Specifically, does it affect non-target commensals? Do phage-resistant mutants emerge, and what is their impact on the microbiome? The literature in this field is scarce and contradictory, and despite impressive recent attempts, there remains a desperate need for more indepth investigations from the scientific community.

6. Conclusions

Phage therapy has the potential to cure drug-resistance pathogenic bacteria as well as patients that are intolerant to antibiotics. However, it is critical to further investigate and understand the detailed mechanistic interaction of bacteriophage as a therapeutic agent with the human microbiota, and specifically, any possible perturbations towards the niche microbiota as a result of the strong selective pressure exerted by lytic phage on their host communities. More mechanistic investigations that shed light on the nature of host-phage dynamics in the context of the niche microbiota will effectively illuminate and push forward the process to obtain regulatory approval for phage therapy in Western medicine.

Author Contributions: Conceptualization—Z.H.; writing-original draft preparation—S.D.G.; writing-review and major editing—Z.H.; visualization—S.D.G.; supervision—Z.H.; funding acquisition—Z.H.

Funding: This research was funded by Natural Sciences and Engineering Research Council of Canada (NSERC) and Farncombe Family Digestive Health Research Institute.

Conflicts of Interest: The authors declare no conflict of interest.

References

1. D'Herelle, F.; Malone, R.H.; Lahiri, M.N. Studies on Asiatic cholera. *Indian Med. Res. Mem.* **1927**, *14*, 195–219.
2. Aminov, R. History of antimicrobial drug discovery: Major classes and health impact. *Biochem. Pharm.* **2017**, *133*, 4–19. [CrossRef] [PubMed]
3. Fleming, A. On the Antibacterial Action of Cultures of a Penicillium, with Special Reference to their Use in the Isolation of *B. influenzæ*. *Br. J. Exp. Pathol.* **1929**, *10*, 226–236. [CrossRef]
4. Summers, W.C. The strange history of phage therapy. *Bacteriophage* **2012**, *2*, 130–133. [CrossRef] [PubMed]
5. Michail, S.; Durbin, M.; Turner, D.; Griffiths, A.M.; Mack, D.R.; Hyams, J.; Leleiko, N.; Kenche, H.; Stolfi, A.; Wine, E. Alterations in the gut microbiome of children with severe ulcerative colitis. *Inflamm. Bowel Dis.* **2011**, *18*, 1799–1808. [CrossRef] [PubMed]
6. Scher, J.U.; Abramson, S.B. The microbiome and rheumatoid arthritis. *Nat. Rev. Rheumatol.* **2011**, *7*, 569. [CrossRef]
7. Khan, A.A.; Shrivastava, A.; Khurshid, M. Normal to cancer microbiome transformation and its implication in cancer diagnosis. *Biochim. Biophys. Acta Rev. Cancer* **2012**, *1826*, 331–337. [CrossRef] [PubMed]
8. Alam, R.; Abdolmaleky, H.M.; Zhou, J.R. Microbiome, inflammation, epigenetic alterations, and mental diseases. *Am. J. Med. Genet. Part B Neuropsychiatr. Genet.* **2017**, *174*, 651–660. [CrossRef]
9. Ubeda, C.; Pamer, E.G. Antibiotics, microbiota, and immune defense. *Trends Immunol.* **2012**, *33*, 459–466. [CrossRef]
10. Heintz-Buschart, A.; Pandey, U.; Wicke, T.; Sixel-Döring, F.; Janzen, A.; Sittig-Wiegand, E.; Trenkwalder, C.; Oertel, W.H.; Mollenhauer, B.; Wilmes, P. The nasal and gut microbiome in Parkinson's disease and idiopathic rapid eye movement sleep behavior disorder. *Mov. Disord.* **2018**, *33*, 88–98. [CrossRef]
11. Shin, H.; Price, K.; Albert, L.; Dodick, J.; Park, L.; Dominguez-Bello, M.G. Changes in the eye microbiota associated with contact lens wearing. *MBio* **2016**, *7*, e00198-16. [CrossRef]
12. Dudek, N.K.; Sun, C.L.; Burstein, D.; Kantor, R.S.; Goltsman DS, A.; Bik, E.M.; Thomas, B.C.; Banfield, J.F.; Relman, D.A. Novel Microbial Diversity and Functional Potential in the Marine Mammal Oral Microbiome. *Curr. Biol.* **2017**, *27*, 3752–3762. [CrossRef] [PubMed]
13. Ma, B.; Forney, L.J.; Ravel, J. Vaginal Microbiome: Rethinking Health and Disease. *Annu. Rev. Microbiol.* **2012**, *66*, 371–389. [CrossRef]
14. Hosseinidoust, Z.; Tufenkji, N.; Van De Ven, T.G. Predation in homogeneous and heterogeneous phage environments affects virulence determinants of *Pseudomonas aeruginosa*. *Appl. Environ. Microbiol.* **2013**, *79*, 2862–2871. [CrossRef]
15. Hosseinidoust, Z.; Tufenkji, N.; van de Ven, T.G. Formation of biofilms under phage predation: Considerations concerning a biofilm increase. *Biofouling* **2013**, *29*, 457–468. [CrossRef]
16. Hosseinidoust, Z.; Van De Ven, T.G.; Tufenkji, N. Evolution of *Pseudomonas aeruginosa* virulence as a result of phage predation. *Appl. Environ. Microbiol.* **2013**, *79*, 6110–6116. [CrossRef]
17. Bobay, L.M.; Ochman, H. Biological species in the viral world. *Proc. Natl. Acad. Sci. USA* **2018**, *115*, 6040–6045. [CrossRef]
18. Armon, R. Soil Bacteria and Bacteriophages. In *Biocommunication in Soil Microorganisms*; Witzany, G., Ed.; Springer: Berlin/Heidelberg, Germany, 2011; pp. 67–112.
19. Perez Sepulveda, B.; Redgwell, T.; Rihtman, B.; Pitt, F.; Scanlan, D.J.; Millard, A. Marine phage genomics: The tip of the iceberg. *FEMS Microbiol. Lett.* **2016**, *363*, fnw158. [CrossRef]
20. Navarro, F.; Muniesa, M. Phages in the Human Body. *Front. Microbiol.* **2017**, *8*, 566. [CrossRef]
21. Calendar, R.L. *The Bacteriophages*, 2nd ed.; Oxford University Press: New York, NY, USA, 2006.
22. Hatfull, G.F. Bacteriophage genomics. *Curr. Opin. Microbiol.* **2008**, *11*, 447–453. [CrossRef]
23. Ackermann, H.W. Chapter 1—Bacteriophage Electron Microscopy. In *Advances in Virus Research*; Łobocka, M., Szybalski, W.T., Eds.; Academic Press: Cambridge, MA, USA, 2012; Volume 82, pp. 1–32.
24. Grose, J.H.; Casjens, S.R. Understanding the enormous diversity of bacteriophages: The tailed phages that infect the bacterial family *Enterobacteriaceae*. *Virology* **2014**, *468*, 421–443. [CrossRef]

25. Hyman, P.; Abedon, S.T. Bacteriophage host range and bacterial resistance. *Adv. Appl. Microbiol.* **2010**, *70*, 217–248. [CrossRef]
26. Dy, R.L.; Richter, C.; Salmond, G.P.; Fineran, P.C. Remarkable Mechanisms in Microbes to Resist Phage Infections. *Annu. Rev. Virol.* **2014**, *1*, 307–331. [CrossRef]
27. Adriaenssens, E.; Brister, J.R. How to Name and Classify Your Phage: An Informal Guide. *Viruses* **2017**, *9*, 70. [CrossRef]
28. Simmonds, P.; Aiewsakun, P. Virus classification—Where do you draw the line? *Arch. Virol.* **2018**, *163*, 2037–2046. [CrossRef]
29. Hobbs, Z.; Abedon, S.T. Diversity of phage infection types and associated terminology: The problem with 'Lytic or lysogenic'. *FEMS Microbiol. Lett.* **2016**, *363*, fnw047. [CrossRef]
30. Campbell, A. The future of bacteriophage biology. *Nat. Rev. Genet.* **2003**, *4*, 471. [CrossRef]
31. Hu, B.; Margolin, W.; Molineux, I.J.; Liu, J. The Bacteriophage T7 Virion Undergoes Extensive Structural Remodeling During Infection. *Science* **2013**, *339*, 576–579. [CrossRef]
32. Dou, C.; Xiong, J.; Gu, Y.; Yin, K.; Wang, J.; Hu, Y.; Zhou, D.; Fu, X.; Qi, S.; Zhu, X.; et al. Structural and functional insights into the regulation of the lysis–lysogeny decision in viral communities. *Nat. Microbiol.* **2018**, *3*, 1285–1294. [CrossRef]
33. Goerke, C.; Köller, J.; Wolz, C. Ciprofloxacin and Trimethoprim Cause Phage Induction and Virulence Modulation in *Staphylococcus aureus*. *Antimicrob. Agents Chemother.* **2006**, *50*, 171–177. [CrossRef]
34. Parkinson, J.S. Classic Spotlight: The Discovery of Bacterial Transduction. *J. Bacteriol.* **2016**, *198*, 2899–2900. [CrossRef]
35. Ptashne, M. *A Genetic Switch: Phage Lambda Revisited*; Cold Spring Harbor Laboratory Press: Cold Spring Harbor, NY, USA, 2004.
36. Chatterjee, S.; Rothenberg, E. Interaction of bacteriophage l with its *E. coli* receptor, LamB. *Viruses* **2012**, *4*, 3162–3178. [CrossRef]
37. Straus, S.K.; Bo, H.E. Filamentous Bacteriophage Proteins and Assembly. *Subcell. Biochem.* **2018**, *88*, 261–279. [CrossRef]
38. Abedon, S.T. Phage Therapy: Various Perspectives on How to Improve the Art. In *Host-Pathogen Interactions: Methods and Protocols*; Medina, C., López-Baena, F.J., Eds.; Springer: New York, NY, USA, 2018; pp. 113–127.
39. Monteiro, R.; Pires, D.P.; Costa, A.R.; Azeredo, J. Phage Therapy: Going Temperate? *Trends Microbiol.* **2018**. [CrossRef]
40. Furfaro, L.L.; Payne, M.S.; Chang, B.J. Bacteriophage Therapy: Clinical Trials and Regulatory Hurdles. *Front. Cell. Infect. Microbiol.* **2018**, *8*. [CrossRef]
41. Calendar, R. *The Bacteriophages*; Oxford University Press: Oxford, UK, 2006; p. 746.
42. Yap, M.L.; Rossmann, M.G. Structure and function of bacteriophage T4. *Future Microbiol.* **2014**, *9*, 1319–1327. [CrossRef]
43. Biozentrum, University of Basel/Sceince Photo Library, TEM Of T4 Bacteriophage. 2013. Available online: https://www.sciencephoto.com/media/249780/view (accessed on 18 June 2019).
44. Yong, E. Viruses in the gut protect from infection. *Nature* **2013**. [CrossRef]
45. Sabehi, G.; Shaulov, L.; Silver, D.H.; Yanai, I.; Harel, A.; Lindell, D. A novel lineage of myoviruses infecting cyanobacteria is widespread in the oceans. *Proc. Natl. Acad. Sci. USA* **2012**, *109*, 2037–2042. [CrossRef]
46. Twort, F.W. An investigation on the nature of ultra-microscopic viruses. *Lancet* **1915**, *2*, 1241–1243. [CrossRef]
47. d'Herelle, F. An invisible antagonist microbe of dysentery bacillus. *C. R. Hebd. Seances Acad. Sci.* **1917**, *165*, 373–375.
48. Duckworth, D.H. Who discovered bacteriophage? *Bacteriol. Rev.* **1976**, *40*, 793. [PubMed]
49. Summers, W.C. How Bacteriophage Came to Be Used by the Phage Group. *J. Hist. Biol.* **1993**, *26*, 255–267. [CrossRef] [PubMed]
50. Barrangou, R.; Fremaux, C.; Deveau, H.; Richards, M.; Boyaval, P.; Moineau, S.; Romero, D.A.; Horvath, P. CRISPR Provides Acquired Resistance Against Viruses in Prokaryotes. *Science* **2007**, *315*, 1709–1712. [CrossRef] [PubMed]
51. d'Herelle, F.; Smith, G.H. *The Bacteriophage and Its Behavior*; Williams & Wilkins: Baltimore, MA, USA, 1926.
52. Eaton, M.D.; Bayne-Jones, S. Bacteriophage therapy: Review of the principles and results of the use of bacteriophage in the treatment of infections. *J. Am. Med. Assoc.* **1934**, *103*, 1769–1776. [CrossRef]

53. Wittebole, X.; De Roock, S.; Opal, S.M. A historical overview of bacteriophage therapy as an alternative to antibiotics for the treatment of bacterial pathogens. *Virulence* **2014**, *5*, 226–235. [CrossRef]
54. Kuchment, A. They're Not a Panacea: Phage Therapy in the Soviet Union and Georgia. In *The Forgotten Cure: The Past and Future of Phage Therapy*; Kuchment, A., Ed.; Springer: New York, NY, USA, 2012; pp. 53–62.
55. Kutateladze, M.; Adamia, R. Phage therapy experience at the Eliava Institute. *Med. Mal. Infect.* **2008**, *38*, 426–430. [CrossRef]
56. Stone, R. Bacteriophage therapy: Stalin's forgotten cure. *Science* **2002**, *298*, 728–731. [CrossRef]
57. Clardy, J.; Fischbach, M.A.; Currie, C.R. The natural history of antibiotics. *Curr. Biol.* **2009**, *19*, R437–R441. [CrossRef]
58. Freedberg, D.E.; Salmasian, H.; Cohen, B.; Abrams, J.A.; Larson, E.L. Receipt of Antibiotics in Hospitalized Patients and Risk for *Clostridium difficile* Infection in Subsequent Patients Who Occupy the Same Bed. *JAMA Intern. Med.* **2016**, *176*, 1801–1808. [CrossRef]
59. Weledji, E.P.; Assob, J.C.; Nsagha, D.S.; Weledji, E.K. Pros, cons and future of antibiotics. *New Horiz. Transl. Med.* **2017**, *4*, 9–14. [CrossRef]
60. Chan, B.K.; Abedon, S.T.; Loc-Carrillo, C. Phage cocktails and the future of phage therapy. *Future Microbiol.* **2013**, *8*, 769–783. [CrossRef]
61. Davies, J.; Davies, D. Origins and evolution of antibiotic resistance. *Microbiol. Mol. Biol. Rev.* **2010**, *74*, 417–433. [CrossRef]
62. El Haddad, L.; Harb, C.P.; A Gebara, M.; A Stibich, M.; Chemaly, R.F. A Systematic and Critical Review of Bacteriophage Therapy Against Multidrug-resistant ESKAPE Organisms in Humans. *Clin. Infect. Dis.* **2018**, *69*, 167–178. [CrossRef]
63. Loc-Carrillo, C.; Abedon, S.T. Pros and cons of phage therapy. *Bacteriophage* **2011**, *1*, 111–114. [CrossRef]
64. Kristian, S.A.; Timmer, A.M.; Liu, G.Y.; Lauth, X.; Sal-Man, N.; Rosenfeld, Y.; Shai, Y.; Gallo, R.L.; Nizet, V. Impairment of innate immune killing mechanisms by bacteriostatic antibiotics. *FASEB J.* **2007**, *21*, 1107–1116. [CrossRef]
65. Moldovan, R.; Chapman-McQuiston, E.; Wu, X.L. On kinetics of phage adsorption. *Biophys. J.* **2007**, *93*, 303–315. [CrossRef]
66. Payne, R.; Jansen, V.A. Phage therapy: The peculiar kinetics of self-replicating pharmaceuticals. *Clin. Pharm. Ther.* **2000**, *68*, 225–230. [CrossRef]
67. Pirnay, J.-P.; Blasdel, B.G.; Bretaudeau, L.; Buckling, A.; Chanishvili, N.; Clark, J.R.; Corte-Real, S.; Debarbieux, L.; Dublanchet, A.; De Vos, D.; et al. Quality and Safety Requirements for Sustainable Phage Therapy Products. *Pharm. Res.* **2015**, *32*, 2173–2179. [CrossRef]
68. Gill, J.J.; Hyman, P. Phage Choice, Isolation, and Preparation for Phage Therapy. *Curr. Pharm. Biotechnol.* **2010**, *11*, 2–14. [CrossRef]
69. Tanji, Y.; Shimada, T.; Fukudomi, H.; Miyanaga, K.; Nakai, Y.; Unno, H. Therapeutic use of phage cocktail for controlling *Escherichia coli* O157:H7 in gastrointestinal tract of mice. *J. Biosci. Bioeng.* **2005**, *100*, 280–287. [CrossRef]
70. Cheng, M.; Zhang, L.; Zhang, H.; Li, X.; Wang, Y.; Xia, F.; Wang, B.; Cai, R.; Guo, Z.; Zhang, Y.; et al. An Ointment Consisting of the Phage Lysin LysGH15 and Apigenin for Decolonization of Methicillin-Resistant *Staphylococcus aureus* from Skin Wound. *Viruses* **2018**, *10*, 244. [CrossRef]
71. Leung, S.S.Y.; Parumasivam, T.; Gao, F.G.; Carrigy, N.B.; Vehring, R.; Finlay, W.H.; Morales, S.; Britton, W.J.; Kutter, E.; Chan, H.K. Production of Inhalation Phage Powders Using Spray Freeze Drying and Spray Drying Techniques for Treatment of Respiratory Infections. *Pharm. Res.* **2016**, *33*, 1486–1496. [CrossRef]
72. Malik, D.J.; Sokolov, I.J.; Vinner, G.K.; Mancuso, F.; Cinquerrui, S.; Vladisavljevic, G.T.; Clokie, M.R.; Garton, N.J.; Stapley, A.G.; Kirpichnikova, A. Formulation, stabilisation and encapsulation of bacteriophage for phage therapy. *Adv. Colloid Interface Sci.* **2017**, *249*, 100–133. [CrossRef]
73. Labrie, S.J.; Samson, J.E.; Moineau, S. Bacteriophage resistance mechanisms. *Nat. Rev. Microbiol.* **2010**, *8*, 317. [CrossRef]
74. Comeau, A.M.; Tétart, F.; Trojet, S.N.; Prère, M.F.; Krisch, H.M. Phage-Antibiotic Synergy (PAS): β-Lactam and Quinolone Antibiotics Stimulate Virulent Phage Growth. *PLoS ONE* **2007**, *2*, e799. [CrossRef]
75. Abedon, S.T. Ecology of anti-biofilm agents I: Antibiotics versus bacteriophages. *Pharmaceuticals* **2015**, *8*, 525–558. [CrossRef]

76. Harper, D.R.; Parracho, H.M.R.T.; Walker, J.; Sharp, R.; Hughes, G.; Werthén, M.; Lehman, S.; Morales, S. Bacteriophages and Biofilms. *Antibiotics* **2014**, *3*, 270–284. [CrossRef]
77. Abedon, S.T. Ecology of anti-biofilm agents II: Bacteriophage exploitation and biocontrol of biofilm bacteria. *Pharmaceuticals* **2015**, *8*, 559–589. [CrossRef]
78. Międzybrodzki, R.; Borysowski, J.; Weber-Dąbrowska, B.; Fortuna, W.; Letkiewicz, S.; Szufnarowski, K.; Zdzisław, P.; Paweł, R.; Marlena, K.; Elżbieta, W.; et al. Chapter 3—Clinical Aspects of Phage Therapy. In *Advances in Virus Research*; Łobocka, M., Szybalski, W., Eds.; Academic Press: Cambridge, MA, USA, 2012; Volume 83, pp. 73–121.
79. Miedzybrodzki, R.; Fortuna, W.; Weber-Dabrowska, B.; Górski, A. Phage therapy of *staphylococcal* infections (including MRSA) may be less expensive than antibiotic treatment. *Postepy Hig. Med. Dosw.* **2007**, *61*, 461–465.
80. Ross, A.; Ward, S.; Hyman, P. More is better: Selecting for broad host range bacteriophages. *Front. Microbiol.* **2016**, *7*, 1–6. [CrossRef]
81. Alam Sarker, S.; Sultana, S.; Reuteler, G.; Moine, D.; Descombes, P.; Charton, F.; Bourdin, G.; McCallin, S.; Ngom-Bru, C.; Neville, T.; et al. Oral Phage Therapy of Acute Bacterial Diarrhea With Two Coliphage Preparations: A Randomized Trial in Children From Bangladesh. *EBioMedicine* **2016**, *4*, 124–137. [CrossRef]
82. Rose, T.; Verbeken, G.; De Vos, D.; Merabishvili, M.; Vaneechoutte, M.; Lavigne, R.; Jennes, S.; Zizi, M.; Pirnay, J.P. Experimental phage therapy of burn wound infection: Difficult first steps. *Int. J. Burn. Trauma* **2014**, *4*, 66–73.
83. Smithyman, A.; Speck, P. Safety and efficacy of phage therapy via the intravenous route. *FEMS Microbiol. Lett.* **2015**, *363*, 1–5. [CrossRef]
84. Sarhan, W.A.; Azzazy, H.M. Phage approved in food, why not as a therapeutic? *Expert Rev. Anti Infect. Ther.* **2015**, *13*, 91–101. [CrossRef]
85. Aleshkin, A.V.; Rubalskii, E.O.; Volozhantsev, N.V.; Verevkin, V.V.; Svetoch, E.A.; Kiseleva, I.A.; Bochkareva, S.S.; Borisova, O.Y.; Popova, A.V.; Bogun, A.G.; et al. A small-scale experiment of using phage-based probiotic dietary supplement for prevention of *E. coli* traveler's diarrhea. *Bacteriophage* **2015**, *5*, e1074329. [CrossRef]
86. Fish, R.; Kutter, E.; Wheat, G.; Blasdel, B.; Kutateladze, M.; Kuhl, S. Compassionate Use of Bacteriophage Therapy for Foot Ulcer Treatment as an Effective Step for Moving Toward Clinical Trials. *Methods Mol. Biol.* **2018**, *1693*, 159–170. [CrossRef]
87. Schooley, R.T.; Biswas, B.; Gill, J.J.; Hernandez-Morales, A.; Lancaster, J.; Lessor, L.; Barr, J.J.; Reed, S.L.; Rohwer, F.; Benler, S.; et al. Development and Use of Personalized Bacteriophage-Based Therapeutic Cocktails To Treat a Patient with a Disseminated Resistant Acinetobacter baumannii Infection. *Antimicrob. Agents Chemother.* **2017**, 61. [CrossRef]
88. Koskella, B.; Brockhurst, M.A. Bacteria–phage coevolution as a driver of ecological and evolutionary processes in microbial communities. *FEMS Microbiol. Rev.* **2014**, *38*, 916–931. [CrossRef]
89. Bryan, D.; El-Shibiny, A.; Hobbs, Z.; Porter, J.; Kutter, E.M. Bacteriophage T4 Infection of Stationary Phase, *E. coli*: Life after Log from a Phage Perspective. *Front. Microbiol.* **2016**, *7*, 1391. [CrossRef]
90. Sunderland, K.S.; Yang, M.; Mao, C. Phage-Enabled Nanomedicine: From Probes to Therapeutics in Precision Medicine. *Angew. Chem. Int. Ed.* **2017**, 1964–1992. [CrossRef]
91. Jault, P.; Leclerc, T.; Jennes, S.; Pirnay, J.P.; Que, Y.A.; Resch, G.; Rousseau, A.F.; Ravat, F.; Carsin, H.; Le Floch, R.; et al. Efficacy and tolerability of a cocktail of bacteriophages to treat burn wounds infected by *Pseudomonas aeruginosa* (PhagoBurn): A randomised, controlled, double-blind phase 1/2 trial. *Lancet Infect. Dis.* **2018**. [CrossRef]
92. McCallin, S.; Alam Sarker, S.; Barretto, C.; Sultana, S.; Berger, B.; Huq, S.; Krause, L.; Bibiloni, R.; Schmitt, B.; Reuteler, G.; et al. Safety analysis of a Russian phage cocktail: From MetaGenomic analysis to oral application in healthy human subjects. *Virology* **2013**, *443*, 187–196. [CrossRef]
93. Nahum, G.G.; Uhl, K.; Kennedy, D.L. Antibiotic Use in Pregnancy and Lactation: What Is and Is Not Known About Teratogenic and Toxic Risks. *Obs. Gynecol.* **2006**, *107*, 1120–1138. [CrossRef]
94. Bacteriophage. Available online: https://www.microgen.ru/en/products/bakteriofagi/ (accessed on 26 November 2019).
95. Chan, B.K.; Sistrom, M.; Wertz, J.E.; Kortright, K.E.; Narayan, D.; Turner, P.E. Phage selection restores antibiotic sensitivity in MDR Pseudomonas aeruginosa. *Sci. Rep.* **2016**, *6*, 26717. [CrossRef]

96. Chan, B.K.; E Turner, P.; Kim, S.; Mojibian, H.R.; A Elefteriades, J.; Narayan, D. Phage treatment of an aortic graft infected with Pseudomonas aeruginosa. *Evol. Med. Public Health* **2018**, *2018*, 60–66. [CrossRef]
97. Dedrick, R.M.; Guerrero-Bustamante, C.A.; Garlena, R.A.; Russell, D.A.; Ford, K.; Harris, K.; Gilmour, K.C.; Soothill, J.; Jacobs-Sera, D.; Schooley, R.T.; et al. Engineered bacteriophages for treatment of a patient with a disseminated drug-resistant Mycobacterium abscessus. *Nat. Med.* **2019**, *25*, 730–733. [CrossRef]
98. Criscuolo, E.; Spadini, S.; Lamanna, J.; Ferro, M.; Burioni, R. Bacteriophages and Their Immunological Applications against Infectious Threats. *J. Immunol. Res.* **2017**, *2017*, 13. [CrossRef]
99. Roach, D.R.; Donovan, D.M. Antimicrobial bacteriophage-derived proteins and therapeutic applications. *Bacteriophage* **2015**, *5*, e1062590. [CrossRef]
100. Standard Treatment Associated with Phage Therapy Versus Placebo for Diabetic Foot Ulcers Infected by *S. aureus*. Available online: https://clinicaltrials.gov/ct2/show/NCT02664740 (accessed on 17 June 2019).
101. AmpliPhi Biosciences Corporation. Individual Patient Expanded Access for AB-SA01, An Investigational Anti-*Staphylococcus aureus* Bacteriophage Therapeutic. Available online: https://clinicaltrials.gov/ct2/show/NCT03395769 (accessed on 26 June 2019).
102. Corporation AmpliPhi Biosciences. Individual Patient Expanded Access for AB-PA01, an Investigational Anti-Pseudomonas Aeruginosa Bacteriophage Therapeutic. Available online: https://clinicaltrials.gov/ct2/show/NCT03395743 (accessed on 18 June 2019).
103. Safety and Efficacy of EcoActive on Intestinal Adherent Invasive E coli in Patients with Inactive Crohn's Disease. Available online: https://clinicaltrials.gov/ct2/show/NCT03808103 (accessed on 18 June 2019).
104. Experimental Phage Therapy of Bacterial Infections. Available online: https://clinicaltrials.gov/ct2/show/NCT00945087 (accessed on 19 June 2019).
105. Langlet, J.; Gaboriaud, F.; Gantzer, C.; Duval, J.F.L. Impact of chemical and structural anisotropy on the electrophoretic mobility of spherical soft multilayer particles: The case of bacteriophage MS2. *Biophys. J.* **2008**, *94*, 3293–3312. [CrossRef]
106. Wright, A.; Hawkins, C.; Änggård, E.; Harper, D. A controlled clinical trial of a therapeutic bacteriophage preparation in chronic otitis due to antibiotic-resistant Pseudomonas aeruginosa; a preliminary report of efficacy. *Clin. Otolaryngol.* **2009**, *34*, 349–357. [CrossRef]
107. Leitner, L.; Sybesma, W.; Chanishvili, N.; Goderdzishvili, M.; Chkhotua, A.; Ujmajuridze, A.; Schneider, M.P.; Sartori, A.; Mehnert, U.; Bachmann, L.M.; et al. Bacteriophages for treating urinary tract infections in patients undergoing transurethral resection of the prostate: A randomized, placebo-controlled, double-blind clinical trial. *BMC Urol.* **2017**, *17*, 90. [CrossRef]
108. A Prospective, Randomized, Double-Blind Controlled Study of WPP-201 for the Safety and Efficacy of Treatment of Venous Leg Ulcers (WPP-201). Available online: https://clinicaltrials.gov/ct2/show/NCT00663091 (accessed on 18 June 2019).
109. Fernández, L.; Rodriguez, A.; García, P. Phage or foe: An insight into the impact of viral predation on microbial communities. *ISME J.* **2018**, *12*, 1171–1179. [CrossRef]
110. Scanlan, P.D. Bacteria-Bacteriophage Coevolution in the Human Gut: Implications for Microbial Diversity and Functionality. *Trends Microbiol.* **2017**, *25*, 614–623. [CrossRef]
111. Janzen, D.H. When is it Co-evolution? *Evolution* **1980**, *34*, 611–612. [CrossRef]
112. Mizoguchi, K.; Morita, M.; Fischer, C.R.; Yoichi, M.; Tanji, Y.; Unno, H. Coevolution of bacteriophage PP01 and *Escherichia coli* O157:H7 in continuous culture. *Appl. Environ. Microbiol.* **2003**, *69*, 170–176. [CrossRef]
113. Hynes, A.P.; Rousseau, G.M.; Agudelo, D.; Goulet, A.; Amigues, B.; Loehr, J.; Romero, D.A.; Fremaux, C.; Horvath, P.; Doyon, Y.; et al. Widespread anti-CRISPR proteins in virulent bacteriophages inhibit a range of Cas9 proteins. *Nat. Commun.* **2018**, *9*, 2919. [CrossRef]
114. Levin, B.R.; Bull, J.J. Population and evolutionary dynamics of phage therapy. *Nat. Rev. Microbiol.* **2004**, *2*, 166–173. [CrossRef]
115. Azam, A.H.; Tanji, Y. Bacteriophage-host arm race: An update on the mechanism of phage resistance in bacteria and revenge of the phage with the perspective for phage therapy. *Appl. Microbiol. Biotechnol.* **2019**, *103*, 2121–2131. [CrossRef]
116. Lopez Pascua, L.; Gandon, S.; Buckling, A. Abiotic heterogeneity drives parasite local adaptation in coevolving bacteria and phages. *J. Evol. Biol.* **2012**, *25*, 187–195. [CrossRef]
117. Mavrich, T.N.; Hatfull, G.F. Bacteriophage evolution differs by host, lifestyle and genome. *Nat. Microbiol.* **2017**, *2*, 17112. [CrossRef]

118. Chan, B.K.; Abedon, S.T. Phage therapy pharmacology: Phage cocktails. In *Advances in Applied Microbiology*; Elsevier: Amsterdam, The Netherlands, 2012; Volume 78, pp. 1–23.
119. Avrani, S.; Lindell, D. Convergent evolution toward an improved growth rate and a reduced resistance range in Prochlorococcus strains resistant to phage. *Proc. Natl. Acad. Sci. USA* **2015**, *112*, E2191–E2200. [CrossRef]
120. Yang, H.; Schmitt-Wagner, D.; Stingl, U.; Brune, A. Niche heterogeneity determines bacterial community structure in the termite gut (*Reticulitermes santonensis*). *Environ. Microbiol.* **2005**, *7*, 916–932. [CrossRef]
121. Schrag, S.J.; Mittler, J.E. Host-parasite coexistence: The role of spatial refuges in stabilizing bacteria-phage interactions. *Am. Nat.* **1996**, *148*, 348–377. [CrossRef]
122. Heilmann, S.; Sneppen, K.; Krishna, S. Coexistence of phage and bacteria on the boundary of self-organized refuges. *Proc. Natl. Acad. Sci. USA* **2012**, *109*, 12828–12833. [CrossRef]
123. Brockhurst, M.A.; Buckling, A.; Rainey, P.B.; Brockhurst, M. Spatial heterogeneity and the stability of host-parasite coexistence. *J. Evol. Biol.* **2006**, *19*, 374–379. [CrossRef]
124. De Sordi, L.; Khanna, V.; Debarbieux, L. The Gut Microbiota Facilitates Drifts in the Genetic Diversity and Infectivity of Bacterial Viruses. *Cell Host Microbe* **2017**, *22*, 801–808. [CrossRef]
125. Duerkop, B.A. Bacteriophages shift the focus of the mammalian microbiota. *PLoS Path.* **2018**, *14*, e1007310. [CrossRef]
126. Sender, R.; Fuchs, S.; Milo, R. Revised Estimates for the Number of Human and Bacteria Cells in the Body. *PLoS Biol.* **2016**, *14*, e1002533. [CrossRef]
127. Lewis, J.D.; Chen, E.Z.; Baldassano, R.N.; Otley, A.R.; Griffiths, A.M.; Lee, D.; Bittinger, K.; Bailey, A.; Friedman, E.S.; Hoffmann, C.; et al. Inflammation, Antibiotics, and Diet as Environmental Stressors of the Gut Microbiome in Pediatric Crohn's Disease. *Cell Host Microbe* **2015**, *18*, 489–500. [CrossRef]
128. Morgan, X.C.; Tickle, T.L.; Sokol, H.; Gevers, D.; Devaney, K.L.; Ward, D.V.; A Reyes, J.; A Shah, S.; Leleiko, N.; Snapper, S.B.; et al. Dysfunction of the intestinal microbiome in inflammatory bowel disease and treatment. *Genome Biol.* **2012**, *13*, R79. [CrossRef]
129. Qin, J.; Li, Y.; Cai, Z.; Li, S.; Zhu, J.; Zhang, F.; Liang, S.; Zhang, W.; Guan, Y.; Shen, D.; et al. A metagenome-wide association study of gut microbiota in type 2 diabetes. *Nature* **2012**, *490*, 55–60. [CrossRef]
130. Turnbaugh, P.J.; Ley, R.E.; Mahowald, M.A.; Magrini, V.; Mardis, E.R.; Gordon, J.I. An obesity-associated gut microbiome with increased capacity for energy harvest. *Nature* **2006**, *444*, 1027. [CrossRef]
131. Hill, J.M.; Clement, C.; Pogue, A.I.; Bhattacharjee, S.; Zhao, Y.; Lukiw, W.J. Pathogenic microbes, the microbiome, and Alzheimer's disease (AD). *Front. Aging Neurosci.* **2014**, *6*, 127.
132. Love, B.L.; Mann, J.R.; Hardin, J.W.; Lu, Z.K.; Cox, C.; Amrol, D.J. Antibiotic prescription and food allergy in young children. *Allergy Asthma Clin. Immunol. Off. J. Can. Soc. Allergy Clin. Immunol.* **2016**, *12*, 41. [CrossRef]
133. Maurice, C.F.; Haiser, H.J.; Turnbaugh, P.J. Xenobiotics shape the physiology and gene expression of the active human gut microbiome. *Cell* **2012**, *152*, 39–50. [CrossRef]
134. Langdon, A.; Crook, N.; Dantas, G. The effects of antibiotics on the microbiome throughout development and alternative approaches for therapeutic modulation. *Genome Med.* **2016**, *8*, 1283. [CrossRef]
135. Maier, L.; Pruteanu, M.; Kuhn, M.; Zeller, G.; Telzerow, A.; Anderson, E.E.; Brochado, A.R.; Fernandez, K.C.; Dose, H.; Mori, H.; et al. Extensive impact of non-antibiotic drugs on human gut bacteria. *Nature* **2018**, *555*, 623–628. [CrossRef]
136. Lugli, G.A.; Milani, C.; Turroni, F.; Tremblay, D.; Ferrario, C.; Mancabelli, L.; Duranti, S.; Ward, D.V.; Ossiprandi, M.C.; Moineau, S.; et al. Prophages of the genus *Bifidobacterium* as modulating agents of the infant gut microbiota. *Environ. Microbiol.* **2016**, *18*, 2196–2213. [CrossRef]
137. Allen, H.K.; Looft, T.; Bayles, D.O.; Humphrey, S.; Levine, U.Y.; Alt, D.; Stanton, T.B. Antibiotics in feed induce prophages in swine fecal microbiomes. *MBio* **2011**, *2*, e00260-11. [CrossRef]
138. Duerkop, B.A.; Clements, C.V.; Rollins, D.; Rodrigues, J.L.; Hooper, L.V. A composite bacteriophage alters colonization by an intestinal commensal bacterium. *Proc. Natl. Acad. Sci. USA* **2012**, *109*, 17621–17626. [CrossRef]
139. Rajpal, D.K.; Brown, J.R. Modulating the human gut microbiome as an emerging therapeutic paradigm. *Sci. Prog.* **2013**, *96*, 224–236. [CrossRef]
140. Paule, A.; Frezza, D.; Edeas, M. Microbiota and Phage Therapy: Future Challenges in Medicine. *Med. Sci.* **2018**, *6*, 86. [CrossRef]
141. Viertel, T.M.; Ritter, K.; Horz, H.P. Viruses versus bacteria–novel approaches to phage therapy as a tool against multidrug-resistant pathogens. *J. Antimicrob. Chemother.* **2014**, *69*, 2326–2336. [CrossRef]

142. Górski, A.; Dąbrowska, K.; Międzybrodzki, R.; Weber-Dąbrowska, B.; Łusiak-Szelachowska, M.; Jończyk-Matysiak, E.; Borysowski, J. Phages and immunomodulation. *Future Microbiol.* **2017**, *12*, 905–914. [CrossRef]
143. Kuramitsu, H.K.; He, X.; Lux, R.; Anderson, M.H.; Shi, W. Interspecies interactions within oral microbial communities. *Microbiol. Mol. Biol. Rev.* **2007**, *71*, 653–670. [CrossRef]
144. Hibbing, M.E.; Fuqua, C.; Parsek, M.R.; Peterson, S.B. Bacterial competition: Surviving and thriving in the microbial jungle. *Nat. Rev. Microbiol.* **2010**, *8*, 15. [CrossRef]
145. Hsu, B.B.; Gibson, T.E.; Yeliseyev, V.; Liu, Q.; Lyon, L.; Bry, L.; Silver, P.A.; Gerber, G.K. Dynamic Modulation of the Gut Microbiota and Metabolome by Bacteriophages in a Mouse Model. *Cell Host Microbe* **2019**, *25*, 803–814. [CrossRef]
146. Galtier, M.; De Sordi, L.; Maura, D.; Arachchi, H.; Volant, S.; Dillies, M.A.; Debarbieux, L. Bacteriophages to reduce gut carriage of antibiotic resistant uropathogens with low impact on microbiota composition. *Environ. Microbiol.* **2016**, *18*, 2237–2245. [CrossRef]
147. Reyes, A.; Wu, M.; McNulty, N.P.; Rohwer, F.L.; Gordon, J.I. Gnotobiotic mouse model of phage–bacterial host dynamics in the human gut. *Proc. Natl. Acad. Sci. USA* **2013**, *110*, 20236–20241. [CrossRef]
148. Weiss, M.; Denou, E.; Bruttin, A.; Serra-Moreno, R.; Dillmann, M.L.; Brüssow, H. In vivo replication of T4 and T7 bacteriophages in germ-free mice colonized with Escherichia coli. *Virology* **2009**, *393*, 16–23. [CrossRef]
149. Jończyk, E.; Kłak, M.; Międzybrodzki, R.; Górski, A. The influence of external factors on bacteriophages. *Folia Microbiol.* **2011**, *56*, 191–200. [CrossRef]
150. Langlet, J.; Gaboriaud, F.; Gantzer, C. Effects of pH on plaque forming unit counts and aggregation of MS2 bacteriophage. *J. Appl. Microbiol.* **2007**, *103*, 1632–1638. [CrossRef]
151. Ly-Chatain, M.H. The factors affecting effectiveness of treatment in phages therapy. *Front. Microbiol.* **2014**, *5*, 51. [CrossRef]
152. Evans, D.F.; Pye, G.; Bramley, R.; Clark, A.G.; Dyson, T.J.; Hardcastle, J.D. Measurement of gastrointestinal pH profiles in normal ambulant human subjects. *Gut* **1988**, *29*, 1035–1041. [CrossRef]
153. Dressman, J.B.; Berardi, R.R.; Dermentzoglou, L.C.; Russell, T.L.; Schmaltz, S.P.; Barnett, J.L.; Jarvenpaa, K.M. Upper gastrointestinal (GI) pH in young, healthy men and women. *Pharm. Res.* **1990**, *7*, 756–761. [CrossRef]
154. Dini, C.; Islan, G.A.; De Urraza, P.J.; Castro, G.R. Novel Biopolymer Matrices for Microencapsulation of Phages: Enhanced Protection Against Acidity and Protease Activity. *Macromol. Biosci.* **2012**, *12*, 1200–1208. [CrossRef]
155. Leung, V.; Szewczyk, A.; Chau, J.; Hosseinidoust, Z.; Groves, L.; Hawsawi, H.; Anany, H.; Griffiths, M.W.; Ali, M.M.; Filipe, C.D.M. Long-Term Preservation of Bacteriophage Antimicrobials Using Sugar Glasses. *ACS Biomater. Sci. Eng.* **2017**, *4*, 3802–3808. [CrossRef]
156. Łusiak-Szelachowska, M.; Weber-Dąbrowska, B.; Jończyk-Matysiak, E.; Wojciechowska, R.; Górski, A. Bacteriophages in the gastrointestinal tract and their implications. *Gut Pathog.* **2017**, *9*, 44. [CrossRef]
157. Barr, J.J. A bacteriophages journey through the human body. *Immunol. Rev.* **2017**, *279*, 106–122. [CrossRef]

antibiotics

MDPI

Review

Bacteriophages in the Dairy Environment: From Enemies to Allies

Lucía Fernández *, Susana Escobedo, Diana Gutiérrez, Silvia Portilla, Beatriz Martínez , Pilar García and Ana Rodríguez

Instituto de Productos Lácteos de Asturias (IPLA-CSIC), Paseo Río Linares s/n, Villaviciosa, 33300 Asturias, Spain; s.escobedo@ipla.csic.es (S.E.); dianagufer@ipla.csic.es (D.G.); silvia.portilla@ipla.csic.es (S.P.); bmf1@ipla.csic.es (B.M.); pgarcia@ipla.csic.es (P.G.); anarguez@ipla.csic.es (A.R.)

* Correspondence: lucia.fernandez@ipla.csic.es; Tel.: +34-985-892-131

Academic Editor: Christopher C. Butler

Received: 5 October 2017; Accepted: 6 November 2017; Published: 8 November 2017

Abstract: The history of dairy farming goes back thousands of years, evolving from a traditional small-scale production to the industrialized manufacturing of fermented dairy products. Commercialization of milk and its derived products has been very important not only as a source of nourishment but also as an economic resource. However, the dairy industry has encountered several problems that have to be overcome to ensure the quality and safety of the final products, as well as to avoid economic losses. Within this context, it is interesting to highlight the role played by bacteriophages, or phages, viruses that infect bacteria. Indeed, bacteriophages were originally regarded as a nuisance, being responsible for fermentation failure and economic losses when infecting lactic acid bacteria, but are now considered promising antimicrobials to fight milk-borne pathogens without contributing to the increase in antibiotic resistance.

Keywords: bacteriophages; dairy industry; pathogens; lactic acid bacteria; fermentation failure; biofilms; antimicrobial resistance

1. Introduction

1.1. Origins and Industrialization of Dairy Production

Archaeological evidence indicates that already in ancient times, the people of Mesopotamia learned to domesticate milk-producing animals, using and preserving milk for nourishment [1]. Thousands of years later, milk is still the most consumed dairy product worldwide, playing a fundamental role in the diet of all populations [2,3]. It is precisely from the exercise of milk extraction by man that the dairy industry was developed [1,4]. Indeed, cheese and yogurt, the first dairy derivatives, were accidentally discovered as a result of the difficulties encountered to transport and preserve milk. From that time to the present, there has been a continuous development of new and improved dairy products. One of the most striking features of the traditional dairy industry is the manner in which chemical, microbiological, physical, and engineering principles were integrated to allow the manufacture of high quality and safe products. This multidisciplinary strategy has led to the wide variety of products available today. Nowadays, aspects like the availability and presentation of products are very important for the consumer. An example of this is the diversification of dairy products by the inclusion of fruits and cereals [3,5,6]. Moreover, the creation of new and sophisticated products that contribute to improving the health of final users, the so-called functional foods, is on high demand [2,5]. Some examples include products with added vitamins and minerals or those supplemented with living beneficial microorganisms (probiotics). Besides dairy products, the technological development of the dairy industry has made it possible to separate solids from milk, and subsequently transform these components into raw material for other food industry sectors [4,7]. It is also worth noting that the diversity of dairy products varies considerably from region to region depending on dietary habits, available milk-making technologies, market demand, and sociocultural circumstances [8].

1.2. Economic Importance of the Dairy Industry in Different Countries

The dairy sector is a dynamic global industry that plays an important economic role in the agricultural sector of most industrialized and developing countries [8,9]. Currently, in the face of rising global demand and imminent industrial globalization, there has been an increase in both the scope and the intensity of world trade of dairy products [8]. Based on data estimates by the Food and Agricultural Organization (FAO), world milk production for 2016 was 817 million tons. In addition, the expected increase in global demand and production of dairy products until 2025 is estimated to be around 6–20 percent [9]. The most important milk producers are Europe, Asia and the Americas. More specifically, the European Union (EU) is the largest producing economic region worldwide, while India is the largest producer as a country [10]. According to the International Dairy Federation [9], milk production has increased by 50 percent in the last three decades, with a total of 150 million smallholders around the world participating in this activity. On the other hand, developed countries account for one-third of the world milk production, while the remaining two-thirds correspond to developing countries. In developing countries, however, growth in the dairy sector is limited by refrigeration, marketing and transportation problems as well as nutritional and zootechnical issues [8]. Thus, smallholders often lack the necessary skills to manage their farms as companies because they have limited access to animal health services, genetic improvement and training of personnel, which results in low yields and poor quality milk. In addition, the economic importance of dairy production both nationally and internationally is directly related to the sustainability of pasture production areas and the size of herds [11]. Other important factors that influence the success of the dairy sector are the degree of government intervention through subsidies and the demand in the export markets. Furthermore, the success of dairy development programs in different countries also depends to a large extent on traditional habits of consumption of dairy products [7,10]. Nonetheless, food safety remains a key global challenge in the dairy industry of any country to prevent economic losses and health concerns. Within this context, bacteriophages (or phages) have consistently played a significant role in the success of the dairy industry. Indeed, bacterial fermentation processes are threatened by contamination of raw milk with phages that infect lactic acid bacteria. This makes necessary the development of techniques to ensure control of the phage load in starting materials and equipment. In contrast, more recently, phages have been proposed as biocontrol agents to eliminate pathogenic or spoilage bacteria in dairy products. This review aims to summarize and discuss both the negative and positive impact of phages in dairy settings, depending on their specific bacterial hosts.

2. Bacteriophages as Unwanted Guests

Phage infection of dairy starter cultures remains the main cause of fermentation failures in the dairy industry. Phage outbreaks can lead to substantial economic losses due to manufacturing delays, waste of ingredients, lower quality product, growth of spoilage and pathogenic microorganisms or even total production loss [12]. Close monitoring of entry routes, quick and effective phage detection methods and control measurements are currently applied to reduce the risk of phage propagation within dairy settings (Figure 1).

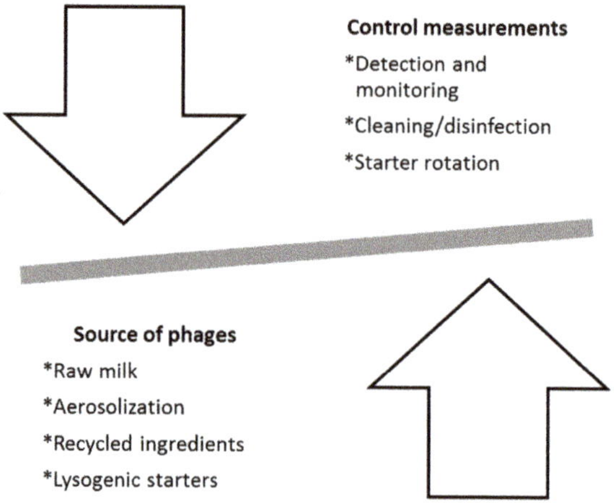

Figure 1. Factors that contribute to the presence of phages in dairy settings.

2.1. Sources of Contamination

The sources of phage entry into dairy plant facilities and dissemination routes must be identified in order to implement corrective actions to limit their propagation. Due to the wide diversity of phages present in raw milk, either as free virions or as prophages in wild lactic acid bacteria (LAB) strains, milk is considered to be the primary entry route for phages into the dairy environment [13]. As much as 10% of milk samples obtained from different dairies in Spain yielded viable *Lactococcus lactis* phages, while lactococcal and streptococcal phages were detected in 37% of raw milk samples used for yogurt production [14,15].

Personnel and equipment movement, raw materials handling, air displacements around contaminated surfaces and liquid splashes can aerosolize viruses and cause dissemination of phage particles in the air to the entire factory environment [16]. Concentrations ranging from 10^2 PFU/m^3 to 10^8 PFU/m^3 in air have been detected in different areas of a cheese manufacturing plant during the fermentation process [17]. A variety of samplers are now available for viral detection in the air; however, there is no standard sampling procedure [18]. In many cases, these devices may have damaging effects on the virus structure that can lead to false-negative results; that is why analytical methods that are independent of viral infectivity, such as quantitative PCR (qPCR), are more suitable for the analysis of air samples [19]. Other reservoirs of phages include materials and equipment used in the manufacturing process as well as surfaces in dairy facilities. Phages can be found in places where conditions for development of their host are favorable and where cleaning and disinfection are difficult.

A common practice in the manufacturing of yogurt and other fermented products consists in the utilization of reconstituted milk from powder and whey proteins obtained from cheese production to increase the product yield and improve the texture and nutritional value of the final products [20]. However, whey proteins may protect phages during heat; there is a correlation between thermal stability of molecular structures and their ability to protect lactococcal virulent phage P1532 from thermal treatments [21]. In addition, whey protein concentrate often contains high temperature-resistant phages, which are able to survive pasteurization and contaminate starters during the manufacturing process [14]. Furthermore, separation and concentration steps of the whey products, consisting in ultrafiltration and microfiltration, may also increase significantly phage titers in these ingredients [22].

LAB strains used as starter cultures can also be a source of phages since they may contain temperate phages integrated into the bacterial chromosome. Lysogeny is widely distributed among dairy lactococci, lactobacilli and with lower incidence in *Streptococcus thermophilus* strains [23,24]. Prophages may be induced and enter into the lytic life cycle under stress conditions such as heat, salts, bacteriocins, starvation, ultraviolet light or may also occur naturally with a frequency of even up to 9% [25–27].

2.2. Detection and Elimination

Great research efforts have focused on early detection of infective phages in dairy manufacturing. Phage monitoring methods include microbiological and molecular assays designed for rapid, low cost and high sensitive evaluation [28].

One of the most common methods for the detection of phages from industrial dairy plants is the activity test based on the acidification rate of milk that provides a reliable indication of their presence when acid production slows down. Acidification can be evaluated by pH measurements, color change of an indicator compound or variations in the electrical conductance of milk [29]. Another method is the double layer plaque assay, which allows a quantitative analysis of infective phage levels, but requires availability of a sensitive strain [30,31]. Flow cytometry can also be used for detection lysed bacterial cells that are found late in the lytic cycle, allowing an accurate and rapid monitoring of phage contamination [32].

Because microbiological tests are time consuming and mostly rely on the availability of single indicator strains, a number of alternative molecular methods focused on detecting the presence of phage particles or their components (DNA, proteins) have been developed. Immunological assays are based on the use of specific antibodies against principal structural proteins of the virion, while viral DNA can be detected with specific DNA-hybridization probes or by polymerase chain reaction [28]. PCR methods have been successfully adapted to detect and identify phages in different stages of dairy product manufacture. In a single reaction, multiplex PCR test allows the detection of several of the most common phages infecting LAB, such as *L. lactis* phage species P335, 936, and c2 and phages infecting *S. thermophilus* and *Lactobacillus delbrueckii* [15,33]. More sensitive than conventional PCR, real-time qPCR can be used to estimate the copy number of a target gene, allowing quantitative viral contamination diagnosis. By using different fluorogenic reporters in the same reaction it is possible to develop multiplex qPCR to detect different targets [34]. qPCR suppliers constantly offer new solutions to get automated systems adapted to industrial needs. Recently, phage metagenomics studies have been conducted to assess the biodiversity and dynamics of phage populations in dairy settings, providing a rational basis for suitable control strategies [35].

2.3. Control Methods

Significant progress in the control of phage populations within the dairy sector has been made in order to keep these bacterial viruses at bay. Although cleaning of equipment and facilities can remove a large proportion of microorganisms, the presence of residual LAB may increase the risk of phage contamination. The role of disinfection is to kill microorganisms that survived the cleaning procedures, reducing the spread of phages within the facility. Disinfectants active against bacteria are not always efficient to inactivate phages [36]. Several biocides used in the dairy industry as well as cleaning procedures have been tested for viral effectiveness on different phages infecting LAB strains. Peracetic acid and sodium hypochlorite containing products are shown to be the most efficient biocides for inactivation of phage particles, while ethanol and isopropanol were usually not effective [37]. The majority of disinfectants consist of several biocides and they must ensure the lack of negative impact on the final product and be able to degrade into harmless final compounds. Combining biocides and heat or using them at extreme pH conditions have shown to give the best results [38]. Photocatalysis intended to destroy fungi, bacteria and spores in the air has been recently explored for inactivating viruses infecting *Lactobacillus casei*, *Lb. delbrueckii* and *Lactobacillus plantarum* [39]. Photocatalytic reaction has shown to completely eliminate two 936-type phages, CHD and QF9 within 120 and 60 min of exposure; respectively [39]. Of note, UV-A radiation assayed by the authors has the advantage of safe use, thus allowing their application for long periods even in the presence of personnel.

The viral load of the ingredients used in dairy production should be reduced as much as possible. Although heating can reduce the activity of phage particles, many LAB phages are not inactivated by classical pasteurization procedures (63 °C for 30 min or 72 °C for 15 s). Therefore, emerging non-thermal technologies such as pulsed electric field, high hydrostatic pressure and high pressure homogenization as well as the combination with heat are currently being explored for inactivating phages [40]. It is important to take into consideration that phages also react differently to heat depending on the medium. Moreover, protective effects due to the presence of proteins, salt or fat have been reported [21,22].

Phage inhibitory media have been developed for starter propagation in dairy plants. The addition of components that inhibit or delay phage propagation such as chelating agents, sodium tripolyphosphate or purified phage peptides can help protect from further phage infection [41–43].

Rotation of defined phage-free cultures is an efficient phage control method to avoid recontamination by the same phage and the build-up of specific phages. A follow up is necessary in order to detect the emergence of new virulent phages to adjust the strain rotation protocol. Recently, a multiplex PCR method based on the genetic locus of the cell wall polysaccharide that acts as phage receptor for many lactococcal phages has been developed to predict phage susceptibility and aid to design suitable starter rotation schemes [44].

The availability of alternative phage resistant starters is of paramount importance and many efforts are being made to search for potential new starter bacteria with different phage sensitivity profiles or to engineer phage-resistant starters. Bacteria have developed natural defense mechanisms against phage infection based on adsorption inhibition, blockage of phage DNA injection, restriction-modification, abortive infection and CRISPR-Cas systems [45]. Many of these systems are plasmid encoded and can be moved from one strain to another for genetically improving dairy starters. Isolation of spontaneous bacteriophage insensitive mutants (BIMs) is a feasible alternative for bacteria without conjugative plasmids, and involves no genetic manipulation. On the other hand, construction of genetically engineered strains has been intensively studied. Several genetic tools, based in the LAB native phage defense mechanisms as well as phage elements have been designed. Examples of these engineered antiphage approaches include cloning of replication origin, antisense RNA technology, phage triggered suicide systems, overproduction of phage proteins, DARPins and neutralizing antibody fragments [12]. Nevertheless, legislation and consumers' concerns regarding genetically modified organisms (GMOs) makes its application to dairy industry difficult.

3. Problems Associated with Bacterial Contamination

3.1. Foodborne Infectious Diseases in Dairy Products

Ensuring access to safe food products remains one of the major global health challenges. Indeed, foodborne diseases constitute a sanitary and economic burden in countries all over the world. To be effective, food safety measures require the participation of all the different actors along the food supply chain, "from farm to fork", including farmers, manufacturers, vendors and consumers. This has become particularly difficult in our global market economy, as these different actors are often far away from each other, frequently across national borders. In this context, adequate regulatory frameworks need to be in place to ensure that the required safety standards are met throughout the process. Nonetheless, foodborne infections are still a major health care concern, with a total of 600 million people falling ill and 420,000 dying every year from eating contaminated food [46].

Dairy products can get contaminated at different points along the production chain (Figure 2). For instance, raw milk can carry microorganisms from the udder or teat canal, the milking equipment, storage containers, the animal's or handler's skin, etc. [47]. Since some of these microbes can be human pathogens, milk can be a potential source of infections if consumed unpasteurized. These pathogens may even persist in aged products made from raw milk, like some traditionally-manufactured cheeses [48]. Pasteurization, on the other hand, can kill most potentially dangerous microorganisms present in milk [47]. However, outbreaks may still occur due to improper pasteurization or post-pasteurization contamination of the milk. Indeed, proper cleaning and hygiene procedures are essential to prevent milk-borne infections.

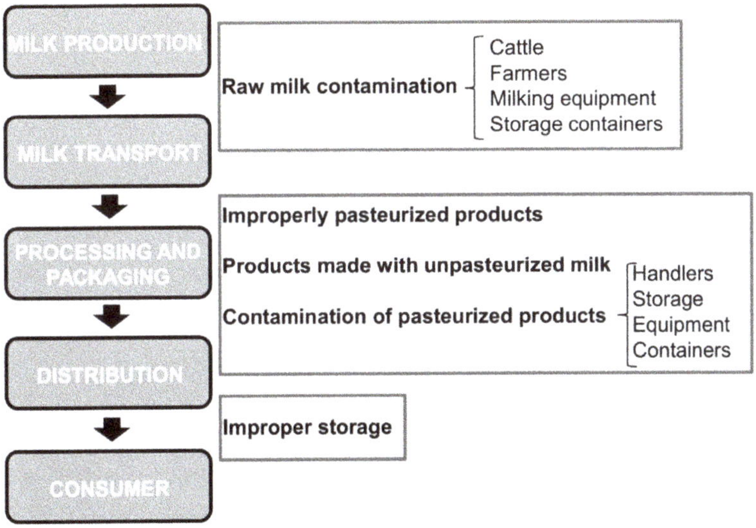

Figure 2. Schematic representation of different points of the dairy supply chain susceptible to microbial contamination.

The pathogens commonly found in the dairy environment include viruses, parasites, fungi and bacteria [49]. Some of the most notorious bacterial pathogens are *Brucella* spp., *Campylobacter jejuni*, *Bacillus cereus*, Shiga toxin-producing *Escherichia coli* (*E. coli* O157:H7), *Staphylococcus aureus*, *Listeria monocytogenes*, *Coxiella burnietti*, *Mycobacterium tuberculosis*, *Mycobacterium bovis*, *Salmonella* spp. and *Yersinia enterocolitica*. Consumption of unpasteurized milk and its derived products is the main source of contamination for most of these pathogens [50–62]. Although unpasteurized milk is not easily available to consumers, it is still consumed by dairy farmers and raw-milk health advocates [51,63]. The human pathogenic bacterium *S. aureus* is one of the microorganisms responsible for mastitis in dairy cows and can also be a source of raw milk contamination [64]. However, this microbe can frequently contaminate food after pasteurization as a result of improper handling during production. *S. aureus* is also problematic due to the production by some strains of heat stable enterotoxins that cannot be easily destroyed by cooking the product [65]. As a result, contaminated products will remain dangerous even after the bacterium has been killed, potentially leading to intoxications.

Taking all of this into account, it is evident that proper hygiene and disinfection measures are essential along the dairy production chain, from the handling of dairy cows to the final product before it reaches the consumer. On top of that, consumers need to be aware that following the instructions for preservation of dairy products and obeying expiry dates are important to ensure their safety.

3.2. Antimicrobial Resistance in the Dairy Environment

Antimicrobials have been overused and misused in human and veterinary medicine ever since their introduction in the clinic. One of the main consequences of this has been the spread of antibiotic resistance determinants amongst microorganisms, including human pathogens, even in environments where antimicrobials themselves were not present [66]. This increase in antibiotic resistance has ultimately led to a decrease in the efficacy of routine disinfection regimes. Indeed, strains belonging to some species have acquired resistance to almost all antibiotics available in the market. The so-called "superbugs" have raised the alarms within the medical and scientific community at large as an indicator that the antibiotic era might be coming to an end. From a less dramatic perspective, perhaps superbugs remind us of the need to understand resistance mechanisms and develop new antimicrobials.

The use of antibiotics in the context of the dairy industry is subject to strict regulations, which are in place to avoid the presence of antibiotic residues in milk aimed for human consumption. For instance, in the US, safety standards for milk are specified in the Grade "A" Pasteurized Milk Ordinance and the Regulation EC 853/2004 defines food safety standards for foodstuffs in the EU [67,68]. In the dairy environment, antimicrobials are used for the treatment of infections in cattle, as growth promoters and as prophylactic agents. The most prevalent infectious illness affecting dairy cattle is mastitis, followed by respiratory infections, lameness, infections of the reproductive system and diarrhea/gastrointestinal tract infections [69]. In many cases, cows require antibiotic treatment with cephalosporins and tetracycline being the most frequently used for mastitis and lameness, respectively [69]. Also, farmers often administer antibiotics to prevent infections, usually penicillin G or dihydrostreptomycin, following the end of the lactation period, the so-called dry cow therapy [69]. The most common routes of antibiotic administration in cows are intramammary and intramuscular [70].

Generalized used of antimicrobials in agriculture and animal farming is considered a potential risk factor for the increased prevalence of antimicrobial resistance in bacteria from food-producing animals [71,72]. Thus, antibiotic pressure would favor the selection and spread of resistance markers by horizontal transfer [73–75]. It must be pointed out, however, that there is no definitive scientific evidence of a direct link between the two. Nevertheless, there have been numerous studies that tried to determine whether antibiotic resistance increased in microorganisms from dairy environments as a result of antimicrobial exposure. However, the results obtained have shown contradictory information. Thus, some studies point that there is an increase in antibiotic resistance over time under antibiotic pressure, while others show no change whatsoever, with differences observed for certain species or antimicrobials [76,77]. Also, some studies have assessed whether there are differences in the amount of antimicrobial resistant organisms in conventional versus organic (antibiotic-free) dairies. For instance, Pol and Ruegg [78] observed that some microorganism-antibiotic combinations were indeed dependent on the farm type while others showed no difference.

Due to the concern regarding antibiotic resistance in pathogenic bacteria, there has been a boom in research regarding the development of novel antimicrobials and new disinfection regimes. Amongst these therapeutic alternatives, phages have been gaining particular attention, as we will discuss below.

4. Bacteriophages as Unexpected Allies

4.1. *Phages as Disinfectants and Preservatives in the Dairy Industry*

As we mentioned previously, foodborne diseases continue to be a hurdle for human health and those associated to dairy industries are not an exception. Thus, many pathogenic bacteria can spread along the food chain from "farm to fork". In this regard, phages can be used as antimicrobials and biocontrol agents in food industries to prevent and control step by step the pathogenic bacterial contamination during food production (Figure 3). The use of phages has some advantages over conventional disinfectants such as their narrow host range, targeting specifically bacteria from one species or genus, being also effective against bacteria resistant to antibiotics. Moreover, phages have been described as safe for humans, animals, plants and the environment [79]. Besides, they do not cause equipment or surface damage or alter the organoleptic properties of food.

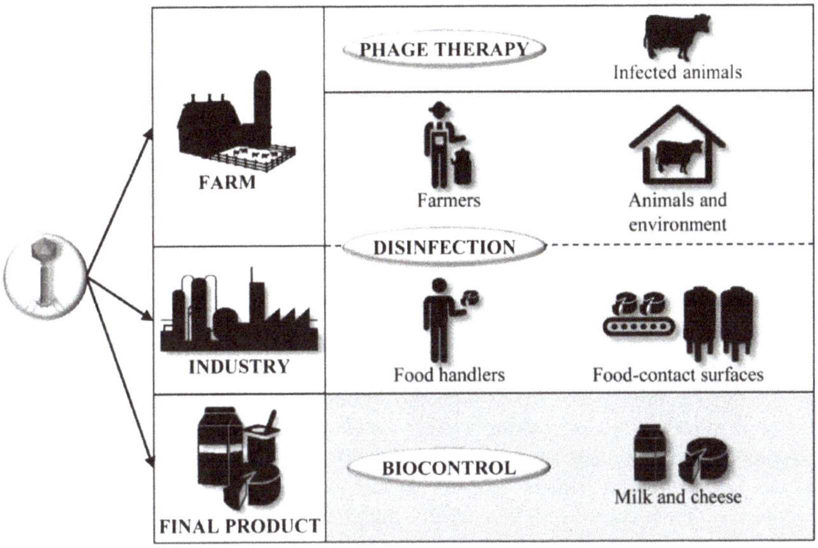

Figure 3. Principal points of disinfection and biocontrol along the dairy chain (from "farm to fork"), where phages can be applied to ensure dairy safety.

The efficacy of phages as an intervention strategy in primary production to reduce bacterial infections in food-producing animals has been widely demonstrated [80]. Nevertheless, data regarding the use of phages in the dairy industry are still scarce. The treatment of subclinical *S. aureus* mastitis in lactating dairy cattle with phage K resulted in a cure rate of 16.7%, although the difference between the treated and non-treated groups was not statistically significant. This can be the consequence of phage inactivation in the udder due to milk proteins and fats [81]. However, utilization of phages as biocontrol agents in milk seems to be a better approach, since the combination of two temperate phages ΦH5 and ΦA72 inhibited the growth of *S. aureus* at 37 °C in ultra-high-temperature (UHT) and traditionally pasteurized whole-fat milk [82]. Moreover, lytic derivatives of these phages, Φ88 and Φ35 were successfully used to completely remove *S. aureus* during curd manufacturing and also during the maturation of fresh and hard-type cheeses [83,84]. Similarly, the application of listeriaphages in combination with a bacteriocin (coagulin C23) to extended shelf life (ESL) milk contaminated with *L. monocytogenes* prevented bacterial growth at 4 °C after 10 days [85].

In the dairy industry, recurrent contamination comes from inadequate cleaning of the equipment and the growth of pathogenic bacteria forming biofilms. Biofilms are structures where bacterial cells are protected by a surrounding matrix, thus becoming difficult to clean and remove. Several studies using biofilms preformed in laboratory conditions (onto polystyrene) have confirmed the potential of phages for staphylococcal biofilm removal. Phage K and a mixture of derivative phages removed biofilms in a time-dependent manner, with the highest reduction occurring after 72 h at 37 °C [86]. The combination of phage K with another staphylococcal phage (DRA88), completely removed biofilms after 48 h at 37 °C [87]. In a similar way, phages phiIPLA-RODI, phiIPLA-C1C, and a mixture of both phages, achieved a reduction of 2 log units after 8 h of treatment at 37 °C [88]. On the other hand, *E. coli* biofilms formed onto materials typically used in food processing surfaces (stainless steel, ceramic tile and high density polyethylene) were removed below the detection level after treatment with a phage mixture named BEC8 [89]. Biofilms formed by *L. monocytogenes* onto stainless steel were reduced up to 5.4 log-units/cm^2 by phage P100 [90]. In this regard, a commercial phage-based product ListShield^{TM}, developed by Intralytix Inc. (Baltimore, MD, USA), has been proposed as a disinfectant for food facilities and also on cheese surfaces [91].

The potential of phages in the food industry is so extensive that several companies have developed phage-based products against important foodborne pathogens that could be used as disinfectants and as food-processing aids. But only Intralytix Inc. (Baltimore, MD, USA) and Micreos BV (Wageningen, The Netherlands) commercialize phage-based products (ListShield^{TM} and PhageGuard Listex, respectively) that can be applied in dairy settings. PhageGuard Listex can be applied as a surface intervention against *Listeria* contamination on cheese by spraying or by immersion, without affecting the color, texture or taste of the product [92]. These phage-based products provide a basis for the future approval of phages as disinfectants and preservatives, overcoming the specific regulatory shortcomings of each country.

4.2. Regulatory Framework for the Application of Phage-Derived Products in the Food Industry

One of the major difficulties for the use of phages as antimicrobial agents is the lack of a proper regulatory framework for their authorization. Moreover, the European Food Safety Authority (EFSA) expressed concerns regarding the efficacy of phages and the danger of recontamination of the food products [93]. In the case of the dairy industry, and food industry at large, phages have great potential for the control of foodborne pathogens. As mentioned previously, phages can be used as food preservatives or for the disinfection of food-contact surfaces, especially against biofilms. However, depending on their intended use and label claims, the procedure for their approval may vary and, in some cases, be time-consuming and costly. Moreover, legislation can differ considerably from country to country.

Probably the easiest route for placing a phage-based product on the market is for application as a food-processing aid. Indeed, several products have been granted clean label processing in the USA, Canada, Israel, Australia, New Zealand, Switzerland, Norway and the EU (The Netherlands). The first product to be approved by the Food and Drug Administration (FDA) and the US Department of Agriculture (USDA) was LISTEX^{TM} P100, now named PhageGuard Listex, in 2007 (EBI Food Safety, Wageningen, The Netherlands). More recently, three phage-based products manufactured by Intralytix Inc. (Baltimore, MD, USA), have also been approved by the FDA for application in food-processing facilities against *L. monocytogenes*, *E. coli* O157:H7, and several *Salmonella* species.

Another potential application of phage-based products is as food additives. So far, only Intralytics has achieved FDA approval for commercializing the phage product ListShield^{TM} as a food additive.

The approval of phage-based products as surface disinfectants for the food industry is proving to be more complicated than the previously discussed applications. Indeed, only one product, ListShield^{TM}, produced by Intralytix Inc., has been granted approval by the FDA and the Environmental Protection Agency (EPA) in the United States to be used for disinfection of non-food contact surfaces and equipment in food-processing facilities and food establishments. In the EU, use of these products as disinfectants in food environments requires authorization under the current Biocidal Products Regulation 528/12 [94]. Preparation of a dossier for this purpose can be somewhat complicated and, most especially, very expensive as it requires a number of studies demonstrating the safety for humans and the environment as well as the efficacy of the active substance, in this case phages, and the product itself. Analysis of potential resistance development is also quite frequently requested by the authorities.

Overall, despite the obvious difficulties encountered for marketing phage-based products, the need for alternatives to conventional antimicrobials and disinfectants seems to be encouraging progress in this field. Hopefully, this will only be the first step towards the development of a proper legal framework that allows an easier path to authorization and commercialization of phage-based products.

5. Concluding Remarks

One century after phages were first described, there is no doubt regarding their importance in diverse fields including ecology, biotechnology, medicine and industrial activities. The dairy industry provides a perfect example of the diverse ways in which bacterial viruses can affect human activities.

This review intends to compile these different aspects, both positive and negative, and gives an overview of how phages have in some ways shaped the development of a whole industrial sector. Thus, achieving a good understanding of phages that infect lactic acid bacteria has enabled the development and implementation of strategies to limit the economic losses associated to fermentation failures. On the other hand, phages appear as a viable alternative to conventional disinfectants for application in food industrial surfaces and dairy products themselves. In the midst of a crisis of rising resistance rates to antimicrobials, phages are giving new hope in the fight against bacterial pathogens. Nevertheless, it is still necessary to conduct further research and develop the appropriate regulatory framework in order to ensure that phage disinfection procedures are effective, safe and easily available.

Acknowledgments: Our research was funded by grants AGL2015-65673-R (Ministry of Science and Innovation, Spain), BIO2013-46266-R (MINECO, Spain), EU ANIWHA ERA-NET BLAAT, GRUPIN14-139 (Program of Science, Technology and Innovation 2013–2017 and FEDER EU funds, Principado de Asturias, Spain). L.F. and S.P. were respectively awarded a "Marie Curie Clarin-Cofund" grant and a CONCACYT (Mexico) postdoctoral fellowship. P.G., B.M. and A.R. are members of the bacteriophage network FAGOMA II and the FWO Vlaanderen funded "Phagebiotics" research community (WO.016.14).

Author Contributions: Lucía Fernández, Susana Escobedo, Diana Gutiérrez, Silvia Portilla, Beatriz Martínez, Ana Rodríguez and Pilar García wrote the paper.

Conflicts of Interest: The authors declare no conflict of interest.

References

1. Rodríguez González, A.; Roces Rodríguez, C.; Martínez Fernández, B. Chapter 9: Cultivos Iniciadores en Quesería: Tradición y Modernidad. In *Biocontrol en la Industria Láctea*; Roa, I., Pacheco, M., Tabla, R., Rebollo, J.E., Eds.; Bubok Publishing, S.L.: Madrid, Spain, 2014; pp. 110–125, ISBN 978-84-686-5316-7. (In Spanish)
2. Dugdill, B.; Bennett, A.; Phelan, J.; Scholten, B.A. Chapter 8: Dairy-industry development programmes: Their role in food and nutrition security and poverty reduction. In *Milk and Dairy Products in Human Nutrition*; Muehlhoff, E., Bennett, A., McMahon, D., Eds.; FAO: Rome, Italy, 2013; pp. 313–348, ISBN 978-92-5-107863-1.
3. Kongerslev Thorning, T.; Raben, A.; Tholstrup, T.; Soedamah-Muthu, S.S.; Givens, I.; Astrup, A. Milk and dairy products: Good or bad for human health? An assessment of the totality of scientific evidence. *Food Nutr. Res.* **2016**, *60*. [CrossRef]
4. Chandan, R.C. Dairy Industry: Chapter 2: Production and Consumption Trends. In *Dairy Processing & Quality Assurance*; Chandan, R.C., Kilara, A., Schah, N.P., Eds.; Wiley-Blackwell: Ames, IA, USA, 2008; pp. 41–58, ISBN 978-0813827568.
5. Moncada Jiménez, A.; Pelayo Consuegra, B.H. Chapter 4: El Proceso Industrial de los Productos Lácteos. In *El libro blanco de la Leche y los Productos Lácteos*; Cámara Nacional de Industriales de la Leche: Mexico City, Mexico, 2011; pp. 52–65.
6. Visioli, F.; Strata, A. Milk, dairy products, and their functional effects in humans: A narrative review of recent evidence. *Adv. Nutr.* **2014**, *5*, 131–143. [CrossRef] [PubMed]
7. Chandan, R.C. Chapter 1: Dairy Processing and Quality Assurance: An Overview. In *Dairy Processing & Quality Assurance*; Chandan, R.C., Kilara, A., Schah, N.P., Eds.; Wiley-Blackwell: Ames, IA, USA, 2008; pp. 1–38, ISBN 978-0813827568.
8. OECD/FAO. *OECD-FAO Agricultural Outlook 2016–2025*; OECD Publishing: Paris, France, 2016; ISBN 9789264253223.
9. International Dairy Federation. *The IDF Guide on Biodiversity for the Dairy Sector*; Bulletin of the International Dairy Federation; International Dairy Federation: Schaerbeek, Belgium, 2017; Volume 488.
10. FAO. *Food Outlook Biannual Report on Global Food Markets*; FAO: Rome, Italy, 2016; ISSN 0251-1959.
11. Hemme, T.; Uddin, M.M.; Oghaiki Asaah Ndambi, O.A. Benchmarking cost of milk production in 46 countries. *J. Rev. Glob. Econ.* **2014**, *3*, 254–270.
12. Samson, J.E.; Moineau, S. Bacteriophages in food fermentations: New frontiers in a continuous arms race. *Annu. Rev. Food Sci. Technol.* **2013**, *4*, 347–368. [CrossRef] [PubMed]
13. Kleppen, H.P.; Bang, T.; Nes, I.F.; Holo, H. Bacteriophages in milk fermentations: Diversity fluctuations of normal and failed fermentations. *Int. Dairy J.* **2011**, *21*, 592–600. [CrossRef]

14. Madera, C.; Monjardin, C.; Suarez, J.E. Milk contamination and resistance to processing conditions determine the fate of *Lactococcus lactis* bacteriophages in dairies. *Appl. Environ. Microbiol.* **2004**, *70*, 7365–7371. [CrossRef] [PubMed]
15. Del Rio, B.; Binetti, A.G.; Martin, M.C.; Fernandez, M.; Magadan, A.H.; Alvarez, M.A. Multiplex PCR for the detection and identification of dairy bacteriophages in milk. *Food Microbiol.* **2007**, *24*, 75–81. [CrossRef] [PubMed]
16. Verreault, D.; Moineau, S.; Duchaine, C. Methods for sampling of airborne viruses. *Microbiol. Mol. Biol. Rev.* **2008**, *72*, 413–444. [CrossRef] [PubMed]
17. Neve, H.; Berger, A.; Heller, K.J. A method for detecting and enumerating airborne virulent bacteriophage of dairy starter cultures. *Kieler Milchwirtschaftliche Forschungsberichte* **1995**, *47*, 193–207.
18. Verreault, D.; Gendron, L.; Rousseau, G.M.; Veillette, M.; Masse, D.; Lindsley, W.G.; Moineau, S.; Duchaine, C. Detection of airborne lactococcal bacteriophages in cheese manufacturing plants. *Appl. Environ. Microbiol.* **2011**, *77*, 491–497. [CrossRef] [PubMed]
19. Verreault, D.; Rousseau, G.M.; Gendron, L.; Massé, D.; Moineau, S.; Duchaine, C. Comparison of polycarbonate and polytetrafluoroethylene filters for sampling of airborne bacteriophages. *Aerosol Sci. Technol.* **2010**, *44*, 197–201. [CrossRef]
20. Ipsen, R. Microparticulated whey proteins for improving dairy product texture. *Int. Dairy J.* **2017**, *67*, 73–79. [CrossRef]
21. Geagea, H.; Gomaa, A.I.; Remondetto, G.; Moineau, S.; Subirade, M. Investigation of the protective effect of whey proteins on lactococcal phages during heat treatment at various pH. *Int. J. Food Microbiol.* **2015**, *210*, 33–41. [CrossRef] [PubMed]
22. Atamer, Z.; Samtlebe, M.; Neve, H.; Heller, K.; Hinrichs, J. Review: Elimination of bacteriophages in whey and whey products. *Front. Microbiol.* **2013**, *4*. [CrossRef] [PubMed]
23. Sun, X.; Van Sinderen, D.; Moineau, S.; Heller, K.J. Impact of lysogeny on bacteria with a focus on Lactic Acid Bacteria. In *Contemporary Trends in Bacteriophage Research*; Adams, H.T., Ed.; Nova Science Publishers, Inc.: New York, NY, USA, 2009; pp. 309–336, ISBN 978-1-60692-181-4.
24. Brüssow, H.; Frémont, M.; Bruttin, A.; Sidoti, J.; Constable, A.; Fryder, V. Detection and classification of *Streptococcus thermophilus* bacteriophages isolated from industrial milk fermentation. *Appl. Environ. Microbiol.* **1994**, *60*, 4537–4543. [PubMed]
25. Lunde, M.; Aastveit, A.H.; Blatny, J.M.; Nes, I.F. Effects of diverse environmental conditions on ϕLC3 prophage stability in *Lactococcus lactis*. *Appl. Environ. Microbiol.* **2005**, *71*, 721–727. [CrossRef] [PubMed]
26. Madera, C.; Garcia, P.; Rodriguez, A.; Suarez, J.E.; Martinez, B. Prophage induction in *Lactococcus lactis* by the bacteriocin Lactococcin 972. *Int. J. Food Microbiol.* **2009**, *129*, 99–102. [CrossRef] [PubMed]
27. Lunde, M.; Blatny, J.M.; Lillehaug, D.; Aastveit, A.H.; Nes, I.F. Use of real-time quantitative PCR for the analysis of ϕLC3 prophage stability in lactococci. *Appl. Environ. Microbiol.* **2003**, *69*, 41–48. [CrossRef] [PubMed]
28. Magadán, A.H.; Ladero, V.; Martínez, N.; del Río, B.; Martín, M.C.; Alvarez, M.A. Detection of bacteriophages in milk. In *Handbook of Dairy Foods Analysis*; Nollet, L.M.L., Toldrá, F., Eds.; CRC Press, Taylor & Francis Group: Boca Raton, FL, USA, 2009; pp. 469–482, ISBN 978-1-4200-4631-1.
29. Marcó, M.B.; Moineau, S.; Quiberoni, A. Bacteriophages and dairy fermentations. *Bacteriophage* **2012**, *2*, 149–158. [CrossRef] [PubMed]
30. Lillehaug, D. An improved plaque assay for poor plaque-producing temperate lactococcal bacteriophages. *J. Appl. Microbiol.* **1997**, *83*, 85–90. [CrossRef] [PubMed]
31. Cormier, J.; Janes, M. A double layer plaque assay using spread plate technique for enumeration of bacteriophage MS2. *J. Virol. Methods* **2014**, *196*, 86–92. [CrossRef] [PubMed]
32. Michelsen, O.; Cuesta-Dominguez, A.; Albrechtsen, B.; Jensen, P.R. Detection of bacteriophage-infected cells of *Lactococcus lactis* by using flow cytometry. *Appl. Environ. Microbiol.* **2007**, *73*, 7575–7581. [CrossRef] [PubMed]
33. Labrie, S.; Moineau, S. Multiplex PCR for detection and identification of lactococcal bacteriophages. *Appl. Environ. Microbiol.* **2000**, *66*, 987–994. [CrossRef] [PubMed]
34. Del Río, B.; Martín, M.C.; Martínez, N.; Magadán, A.H.; Alvarez, M.A. Multiplex fast real-time polymerase chain reaction for quantitative detection and identification of cos and pac *Streptococcus thermophiles* bacteriophages. *Appl. Environ. Microbiol.* **2008**, *74*, 4779–4781. [CrossRef] [PubMed]

35. Muhammed, M.K.; Kot, W.; Neve, H.; Mahony, J.; Castro-Mejía, J.L.; Krych, L.; Hansen, L.H.; Nielsen, D.S.; Sørensen, S.J.; Heller, K.J.; et al. Metagenomic analysis of dairy bacteriophages: Extraction method and pilot study on whey samples derived from using undefined and defined mesophilic starter cultures. *Appl. Environ. Microbiol.* **2017**, *83*. [CrossRef] [PubMed]
36. Campagna, C.; Villion, M.; Labrie, S.J.; Duchaine, C.; Moineau, S. Inactivation of dairy bacteriophages by commercial sanitizers and disinfectants. *Int. J. Food Microbiol.* **2014**, *171*, 41–47. [CrossRef] [PubMed]
37. Guglielmotti, D.M.; Mercanti, D.J.; Reinheimer, J.A.; Quiberoni, A.L. Review: Efficiency of physical and chemical treatments on the inactivation of dairy bacteriophages. *Front. Microbiol.* **2011**, *2*, 282–297. [CrossRef] [PubMed]
38. Murphy, J.; Mahony, J.; Bonestroo, M.; Nauta, A.; van Sinderen, D. Impact of thermal and biocidal treatments on lactococcal 936-type phages. *Int. Dairy J.* **2014**, *34*, 56–61. [CrossRef]
39. Marcó, M.B.; Quiberoni, A.; Negro, A.C.; Reinheimer, J.A.; Alfano, O.M. Evaluation of the photocatalytic inactivation efficiency of dairy bacteriophages. *Chem. Eng. J.* **2011**, *172*, 987–993. [CrossRef]
40. Capra, M.J.; Patrignani, F.; Guerzoni, M.E.; Laciotti, R. Non-thermal technologies: Pulsed electric field, high hydrostatic pressure and high pressure homogenization. Application on virus inactivation. In *Bacteriophages in Dairy Processing*; Nova Science Publishers, Inc.: New York, NY, USA, 2012; pp. 215–238, ISBN 978-1-61324-517-0.
41. Mahony, J.; Tremblay, D.M.; Labrie, S.J.; Moineau, S.; van Sinderen, D. Investigating the requirement for calcium during lactococcal phage infection. *Int. J. Food Microbiol.* **2015**, *201*, 47–51. [CrossRef] [PubMed]
42. Carminati, D.; Giraffa, G.; Quiberoni, A.; Binetti, A.; Suárez, V.; Reinheimer, J. Advances and trends in starter cultures for dairy fermentations. In *Biotechnology of Lactic Acid Bacteria: Novel Applications*; Mozzi, F., Raya, R., Vignolo, G., Eds.; Wiley-Blackwell: Ames, IA, USA, 2010; pp. 177–192, ISBN 9781118868409.
43. Hicks, C.L.; Clark-Safko, P.A.; Surjawan, I.; O'Leary, J. Use of bacteriophage derived peptides to delay phage infections. *Food Res. Int.* **2004**, *37*, 115–122. [CrossRef]
44. Mahony, J.; Kot, W.; Murphy, J.; Ainsworth, S.; Neve, H.; Hansen, L.H.; Heller, K.J.; Sørensen, S.J.; Hammer, K.; Cambillau, C.; et al. Investigation of the relationship between lactococcal host cell wall polysaccharide genotype and 936 phage receptor binding protein phylogeny. *Appl. Environ. Microbiol.* **2013**, *79*, 4385–4392. [CrossRef] [PubMed]
45. Labrie, S.J.; Samson, J.E.; Moineau, S. Bacteriophage resistance mechanisms. *Nat. Rev. Microbiol.* **2010**, *8*, 317–327. [CrossRef] [PubMed]
46. World Health Organization. *WHO Estimates of the Global Burden of Foodborne Diseases*; World Health Organization: Geneva, Switzerland, 2015; ISBN 978 92 4 156516 5.
47. Rampling, A. The microbiology of milk and milk products. In *Topley and Wilson's Principles of Bacteriology, Virology, and Immunity*, 8th ed.; Parker, M.T., Collier, L.H., Eds.; B.C. Decker: Philadelphia, PA, USA, 1990; pp. 265–287.
48. Altekruse, S.F.; Timbo, B.B.; Mowbray, J.C.; Bean, N.H.; Potter, M.E. Cheese associated outbreaks of human illness in the United States, 1973 to 1992: Sanitary manufacturing processes protect consumers. *J. Food Prot.* **1998**, *61*, 1405–1407. [CrossRef] [PubMed]
49. Dhanashekar, R.; Akkinepalli, S.; Nellutla, A. Milk-borne infections. An analysis of their potential effect on the milk industry. *Germs* **2012**, *2*, 101–109. [CrossRef] [PubMed]
50. Costard, S.; Espejo, L.; Groenendaal, H.; Zagmutt, F.J. Outbreak-related disease burden associated with consumption of unpasteurized cow's milk and cheese, United States, 2009–2014. *Emerg. Infect. Dis.* **2017**, *23*, 957–964. [CrossRef] [PubMed]
51. Claeys, W.L.; Cardoen, S.; Daube, G.; De Block, J.; Dewettinck, K.; Dierick, K.; De Zutter, L.; Huyghebaert, A.; Imberechts, H.; Thiange, P.; et al. Raw or heated cow milk consumption: Review of risks and benefits. *Food Control* **2013**, *31*, 251–262. [CrossRef]
52. Christidis, T.; Pintar, K.D.; Butler, A.J.; Nesbitt, A.; Thomas, M.K.; Marshall, B.; Pollari, F. *Campylobacter* spp. prevalence and levels in raw milk: A systematic review and meta-analysis. *J. Food Prot.* **2016**, *79*, 1775–1783. [CrossRef] [PubMed]
53. Jamali, H.; Paydar, M.; Radmehr, B.; Ismail, S. Prevalence, characterization, and antimicrobial resistance of *Yersinia* species and *Yersinia enterocolitica* isolated from raw milk in farm bulk tanks. *J. Dairy Sci.* **2015**, *98*, 798–803. [CrossRef] [PubMed]

54. Bernardino-Varo, L.; Quiñones-Ramírez, E.I.; Fernández, F.J.; Vázquez-Salinas, C. Prevalence of *Yersinia enterocolitica* in raw cow's milk collected from stables of Mexico City. *J. Food Prot.* **2013**, *76*, 694–698. [CrossRef] [PubMed]
55. Chmielewski, T.; Tylewska-Wierzbanowska, S. Q fever at the turn of the century. *Pol. J. Microbiol.* **2012**, *61*, 81–93. [PubMed]
56. Mailles, A.; Rautureau, S.; Le Horgne, J.M.; Poignet-Leroux, B.; d'Arnoux, C.; Dennetière, G.; Faure, M.; Lavigne, J.P.; Bru, J.P.; Garin-Bastuji, B. Re-emergence of brucellosis in cattle in France and risk for human health. *Euro Surveill.* **2012**, *17*. [CrossRef]
57. Ning, P.; Guo, M.; Guo, K.; Xu, L.; Ren, M.; Cheng, Y.; Zhang, Y. Identification and effect decomposition of risk factors for *Brucella* contamination of raw whole milk in China. *PLoS ONE* **2013**, *8*. [CrossRef] [PubMed]
58. Pearson, L.J.; Marth, E.H. *Listeria monocytogenes*—Threat to a safe food supply: A review. *J. Dairy Sci.* **1990**, *73*, 912–928. [CrossRef]
59. Swaminathan, B.; Gerner-Smidt, P. The epidemiology of human listeriosis. *Microbes Infect.* **2007**, *9*, 1236–1243. [CrossRef] [PubMed]
60. Bolaños, C.A.D.; Paula, C.L.; Guerra, S.T.; Franco, M.M.J.; Ribeiro, M.G. Diagnosis of mycobacteria in bovine milk: An overview. *Rev. Inst. Med. Trop. Sao Paulo* **2017**, *59*. [CrossRef] [PubMed]
61. Doyle, M.P. *Escherichia coli* O157:H7 and its significance in foods. *Int. J. Food Microbiol.* **1991**, *12*, 289–301. [CrossRef]
62. Honish, L.; Predy, G.; Hislop, N.; Chui, L.; Kowalewska-Grochowska, K.; Trottier, L.; Kreplin, C.; Zazulak, I. An outbreak of *E. coli* O157:H7 hemorrhagic colitis associated with unpasteurized gouda cheese. *Can. J. Public Health* **2005**, *96*, 182–184. [PubMed]
63. Buzby, J.C.; Gould, L.H.; Kendall, M.E.; Jones, T.F.; Robinson, T.; Blayney, D.P. Characteristics of consumers of unpasteurized milk in the United States. *J. Consum. Aff.* **2013**, *47*, 153–166. [CrossRef]
64. Zecconi, A. *Staphylococcus aureus* mastitis: What we need to control them. *Israel J. Vet. Med.* **2010**, *65*, 93–99.
65. Schelin, J.; Wallin-Carlquist, N.; Cohn, M.T.; Lindqvist, R.; Barker, G.C.; Radstrom, P. The formation of *Staphylococcus aureus* enterotoxin in food environments and advances in risk assessment. *Virulence* **2011**, *2*, 580–592. [CrossRef] [PubMed]
66. Martínez, J.L. Natural antibiotic resistance and contamination by antibiotic resistance determinants: The two ages in the evolution of resistance to antimicrobials. *Front. Microbiol.* **2012**, *3*. [CrossRef] [PubMed]
67. United States Food and Drug Administration, Department of Health and Human Services. Grade "A" Pasteurized Milk Ordinance. 2015 Revision. Available online: https://www.fda.gov/downloads/food/guidanceregulation/guidancedocumentsregulatoryinformation/milk/ucm513508.pdf (accessed on 1 October 2017).
68. European Parliament and Council. Regulation EU No 853/2004 laying down specific hygiene rules for on the hygiene of foodstuffs. *Off. J. Eur. Union* **2004**, *139*, 55–205.
69. United States Department of Agriculture, Animal Plant Health Inspection Service National Animal Health Monitoring System. Antibiotic Use on U.S. Dairy Operations, 2002 and 2007. 2008. Available online: https://www.aphis.usda.gov/animal_health/nahms/dairy/downloads/dairy07/Dairy07_is_AntibioticUse.pdf (accessed on 1 October 2017).
70. United States Department of Agriculture, Animal Plant Health Inspection Service National Animal Health Monitoring System. Injection Practices on U.S. Dairy Operations, 2007. 2009. Available online: https://www.aphis.usda.gov/animal_health/nahms/dairy/downloads/dairy07/Dairy07_is_InjectionPrac.pdf (accessed on 1 October 2017).
71. World Health Organization. *WHO Global Principles for the Containment of Antimicrobial Resistance in Animals Intended for Food*; Report of a WHO Consultation, 5–9 June 2000; World Health Organization: Geneva, Switzerland, 2000.
72. World Health Organization. *Monitoring Antimicrobial Usage in Food Animals for the Protection of Human Health*; Report of a WHO Consultation, Oslo, Norway, 10–13 September 2001; World Health Organization: Geneva, Switzerland, 2002.
73. Witte, W. Medical consequences of antimicrobial use in agriculture. *Science* **1998**, *279*, 996–997. [CrossRef] [PubMed]

74. O'Brien, T.F. Emergence, spread, and environmental effect of antimicrobial resistance: How use of an antimicrobial anywhere can increase resistance to any antimicrobial anywhere else. *Clin. Infect. Dis.* **2002**, *34*, S78–S84. [CrossRef] [PubMed]
75. Molbak, K. Spread of resistant bacteria and resistance genes from animals to humans—The public health consequences. *J. Vet. Med. B Infect. Dis. Vet. Public Health* **2004**, *51*, 364–369. [CrossRef] [PubMed]
76. Erskine, R.J.; Walker, R.D.; Bolin, C.A.; Bartlett, P.C.; White, D.G. Trends in antibacterial susceptibility of mastitis pathogens during a seven-year period. *J. Dairy Sci.* **2002**, *85*, 1111–1118. [CrossRef]
77. Rajala-Schultz, P.J.; Smith, K.L.; Hogan, J.S.; Love, B.C. Antimicrobial susceptibility of mastitis pathogens from first lactation and older cows. *Vet. Microbiol.* **2004**, *102*, 33–42. [CrossRef] [PubMed]
78. Pol, M.; Ruegg, P.L. Treatment practices and quantification of antimicrobial drug usage in conventional and organic dairy farms in Wisconsin. *J. Dairy Sci.* **2007**, *90*, 249–261. [CrossRef]
79. Bruttin, A.; Brussow, H. Human volunteers receiving *Escherichia coli* phage T4 orally: A safety test of phage therapy. *Antimicrob. Agents Chemother.* **2005**, *49*, 2874–2878. [CrossRef] [PubMed]
80. Carvalho, C.; Costa, A.R.; Silva, F.; Oliveira, A. Bacteriophages and their derivatives for the treatment and control of food-producing animal infections. *Crit. Rev. Microbiol.* **2017**, *43*, 583–601. [CrossRef] [PubMed]
81. Gill, J.J.; Pacan, J.C.; Carson, M.E.; Leslie, K.E.; Griffiths, M.W.; Sabour, P.M. Efficacy and pharmacokinetics of bacteriophage therapy in treatment of subclinical *Staphylococcus aureus* mastitis in lactating dairy cattle. *Antimicrob. Agents Chemother.* **2006**, *50*, 2912–2918. [CrossRef] [PubMed]
82. García, P.; Madera, C.; Martínez, B.; Rodríguez, A.; Suárez, J.E. Prevalence of bacteriophages infecting *Staphylococcus aureus* in dairy samples and their potential as biocontrol agents. *J. Dairy Sci.* **2009**, *92*, 3019–3026. [CrossRef] [PubMed]
83. García, P.; Madera, C.; Martínez, B.; Rodríguez, A. Biocontrol of *Staphylococcus aureus* in curd manufacturing processes using bacteriophages. *Int. Dairy J.* **2007**, *17*. [CrossRef]
84. Bueno, E.; García, P.; Martínez, B.; Rodríguez, A. Phage inactivation of *Staphylococcus aureus* in fresh and hard-type cheeses. *Int. J. Food Microbiol.* **2012**, *158*, 23–27. [CrossRef] [PubMed]
85. Rodríguez-Rubio, L.; García, P.; Rodríguez, A.; Billington, C.; Hudson, J.A.; Martínez, B. Listeriaphages and coagulin C23 act synergistically to kill *Listeria monocytogenes* in milk under refrigeration conditions. *Int. J. Food Microbiol.* **2015**, *205*, 68–72. [CrossRef] [PubMed]
86. Kelly, D.; McAuliffe, O.; Ross, R.P.; Coffey, A. Prevention of *Staphylococcus aureus* biofilm formation and reduction in established biofilm density using a combination of phage K and modified derivatives. *Lett. Appl. Microbiol.* **2012**, *54*, 286–291. [CrossRef] [PubMed]
87. Alves, D.R.; Gaudion, A.; Bean, J.E.; Perez Esteban, P.; Arnot, T.C.; Harper, D.R.; Kot, W.; Hansen, L.H.; Enright, M.C.; Jenkins, A.T. Combined use of bacteriophage K and a novel bacteriophage to reduce *Staphylococcus aureus* biofilm formation. *Appl. Environ. Microbiol.* **2014**, *80*, 6694–6703. [CrossRef] [PubMed]
88. Gutiérrez, D.; Vandenheuvel, D.; Martínez, B.; Rodríguez, A.; Lavigne, R.; García, P. Two phages, philPLA-RODI and philPLA-C1C, lyse mono- and dual-species staphylococcal biofilms. *Appl. Environ. Microbiol.* **2015**, *81*, 3336–3348. [CrossRef] [PubMed]
89. Viazis, S.; Akhtar, M.; Feirtag, J.; Diez-Gonzalez, F. Reduction of *Escherichia coli* O157:H7 viability on hard surfaces by treatment with a bacteriophage mixture. *Int. J. Food Microbiol.* **2011**, *145*, 37–42. [CrossRef] [PubMed]
90. Soni, K.A.; Nannapaneni, R. Removal of *Listeria monocytogenes* biofilms with bacteriophage P100. *J. Food Prot.* **2010**, *73*, 1519–1524. [CrossRef] [PubMed]
91. Intralytix Inc. Available online: http://www.intralytix.com (accessed on 1 October 2017).
92. PhageGuard. Available online: https://www.phageguard.com (accessed on 1 October 2017).
93. EFSA. Scientific opinion on the evaluation of the safety and efficacy of Listex™ P100 for the removal of surface contamination of raw fish. *EFSA J.* **2012**, *10*. [CrossRef]
94. European Parliament and Council. Regulation EU No 528/2012 concerning the making available on the market and use of biocidal products. *Off. J. Eur. Union* **2012**, *167*, 1–123.

antibiotics

MDPI

Review

Bacteriophage Interactions with Marine Pathogenic Vibrios: Implications for Phage Therapy

Panos G. Kalatzis [1,2], Daniel Castillo [1], Pantelis Katharios [2] and Mathias Middelboe [1,*]

1 Marine Biological Section, University of Copenhagen, DK-3000 Helsingør, Denmark; panos.kalatzis@bio.ku.dk (P.G.K.); daniel.castillo@bio.ku.dk (D.C.)
2 Institute of Marine Biology, Biotechnology and Aquaculture, Hellenic Centre for Marine Research, 71500 Crete, Greece; katharios@hcmr.gr
* Correspondence: mmiddelboe@bio.ku.dk; Tel.: +45-3532-1991

Received: 1 February 2018; Accepted: 21 February 2018; Published: 24 February 2018

Abstract: A global distribution in marine, brackish, and freshwater ecosystems, in combination with high abundances and biomass, make vibrios key players in aquatic environments, as well as important pathogens for humans and marine animals. Incidents of *Vibrio*-associated diseases (vibriosis) in marine aquaculture are being increasingly reported on a global scale, due to the fast growth of the industry over the past few decades years. The administration of antibiotics has been the most commonly applied therapy used to control vibriosis outbreaks, giving rise to concerns about development and spreading of antibiotic-resistant bacteria in the environment. Hence, the idea of using lytic bacteriophages as therapeutic agents against bacterial diseases has been revived during the last years. Bacteriophage therapy constitutes a promising alternative not only for treatment, but also for prevention of vibriosis in aquaculture. However, several scientific and technological challenges still need further investigation before reliable, reproducible treatments with commercial potential are available for the aquaculture industry. The potential and the challenges of phage-based alternatives to antibiotic treatment of vibriosis are addressed in this review.

Keywords: marine vibrios; bacteriophages; phage therapy; biological control; aquaculture; interactions; vibriosis

---

1. Vibrios in Marine Ecosystems

The Vibrionaceae family, and more specifically, the genus *Vibrio*, encompasses genetically and metabolically diverse, heterotrophic bacteria that can thrive in a great range of habitats. The particularly versatile features of vibrios have made them ubiquitous components of world's marine and even brackish or freshwater ecosystems [1]. The relatively high abundance (often 10^3 to 10^4 cells per mL) and biomass of vibrios in the oceans [2], makes them important players in marine biogeochemical cycling. Key traits supporting this are (a) their ability to survive for a long time under nutrient-limited conditions [3,4], (b) their ability to maintain high ribosome content, which helps them achieve a fast recovery from starvation as soon as carbon sources become available [5–7], and c) their chemotactic response in finding nutrient sources [8–11].

The vast majority of vibrios occupy ecological niches associated with attachment to living organisms, which provide them protection and nutrients [12–14]. However, vibrios also occur as free-living cells in the water column [15,16]. Among several environmental variables that have been examined, salinity and temperature have been consistently linked to the observed variation in the total *Vibrio* abundance in the water column [17,18]. For example, *V. vulnificus* could tolerate a broad range in salinity from 5 to 38 ppt [19], while *V. cholerae* can grow in salinities of up to 45 ppt, if the nutrient concentration is high [20]. High temperature significantly boosts the growth of vibrios and the increased sea surface temperature has been suggested to promote a long-term increase in *Vibrio* abundance [21,22].

Their opportunistic lifestyle features, as well as their easy cultivation under laboratory conditions, have made them ideal models for investigations of bacterial population biology and genomics, disease dynamics, bacteria-phage interactions, and quorum sensing (QS) [23–26].

Several *Vibrio* species are pathogenic, and constitute a serious threat for human health. Over 80 species have been described, and at least 12 of them are known human pathogens [27,28]. *V. cholerae*, the causative agent of epidemic cholera, was introduced in Europe via sea trade routes from Asia, and was a devastating disease during 1817–1923 [29]. *V. cholerae*, as well as the seafood poisoning agents *V. parahaemolyticus* and *V. vulnificus* [30], have aroused significant attention among the scientific community, especially today, when increases in *Vibrio*-associated disease outbreaks in response to elevated ocean temperatures [31–33] emphasize the increasing importance of vibrio pathogens in a future warmer climate [34,35].

2. Vibrios in Aquaculture

According to Food and Agriculture Organization (FAO) [36], aquaculture is one of the most rapidly growing sectors for animal food production, supporting approximately 50% of the global human fish consumption. Vibrios have been characterized as the "scourge" of marine fish and shellfish, since several members of the genus can be the causative agents of a fatal disease, commonly known as vibriosis [37]. Sudden vibriosis outbreaks have been causing severe losses in biomass, with significant economic consequences for the aquaculture industry [38]. Furthermore, lower growth rate of sick fish and shellfish, excessive waste of fish feeds, and finally, the increased skepticism of consumers about aquaculture's quality and credibility, are also important consequences of vibriosis.

Sustainability in aquaculture demands a thorough and sophisticated disease management plan in which the issue of pathogenic vibrios should be an integral part. The last report of the World Bank about prospects for fisheries and aquaculture [39] is a case in point, since it was reported that *Vibrio*-caused disease designated as early mortality syndrome (EMS), or else, acute hepatopancreatic necrosis disease (AHNPD), is a rapidly emerging disease, and a serious setback to the shrimp rearing industries of Asia and America [40,41]. FAO has drawn special attention to vibriosis [36], because the distribution of vibrios is being shifted according to the changing warming patterns, hence, outbreaks tend to be observed even in temperate or cold regions [42].

V. anguillarum, initially reported as *Bacillus anguillarum* [43], used to be the first isolated *Vibrio* to which "Red Pest of eels" was attributed, during early 1900s [44]. Although it still remains a serious threat for aquaculture [45,46], a plethora of other *Vibrio* species have been recorded in the literature as causative agents of vibriosis in aquaculture. *V. harveyi*, *V. parahaemolyticus*, *V. alginolyticus*, *V. vulnificus*, and *V. splendidus* [28,47–52] are the most important, while the list is expanding with the discovery of new pathogenic species, such as *V. owensii* [53].

Chemical stressors, such as poor water quality and diet composition, biological stressors, such as population density and presence of other micro- or macro-organisms, and physical stressors, such as temperature above 15 °C, are the most important factors triggering vibriosis outbreaks [46,54]. Although the regulatory mechanisms of virulence in vibrios still need to be elucidated, virulence-related factors and genes that have been found in several pathogenic marine *Vibrio*. Iron uptake systems of *V. ordalii*, *V. vulnificus*, *V. alginolyticus*, and *V. anguillarum* have been recorded to contribute to their virulence, by binding the iron attached to the siderophore proteins of their hosts [46,50,55]. Extracellularly secreted proteins can have proteolytic, hydrolytic, hemolytic, and cytotoxic activity in several pathogenic *Vibrio*, such as *V. anguillarum*, *V. alginolyticus*, *V. harveyi*, *V. splendidus*, and *V. pelagius* [56–60]. However, the presence of virulence genes alone is not always a sufficient condition for a virulent phenotype. For instance, both virulent and avirulent *V. harveyi* and *V. campbellii* do carry virulence genes. It has been found that virulence can be coordinated via cell to cell communication, regulated by the presence of specific signal molecules [61]. The three-channel QS system of *V. campbellii*, previously described as *V. harveyi* [62], is a well-described case of QS-regulated virulence system using three different signals [63]. It was also recently shown that the virulence of *V. anguillarum* against European seabass (*Dicentrarchus labrax*) larvae is regulated by the indole signaling molecule pathway [64]. Clinical signs of vibriosis (Figure 1) include lethargic behavior, loss of appetite, unusual swimming behavior close to the water surface, increased mucus secretion, as well as petechiae and hemorrhages on their skin. Additional symptoms of the disease commonly observed are intestinal necrosis, anemia, ascetic fluid, petechial hemorrhages in the muscle wall, and liquid in the swim bladder [65,66].

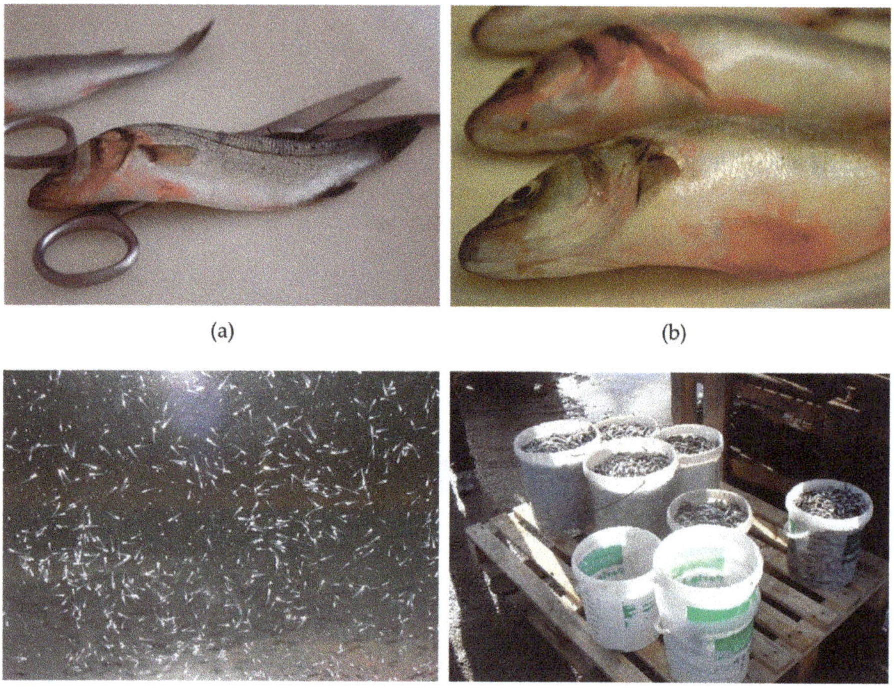

(a) (b)

(c) (d)

Figure 1. Massive mortalities caused by vibriosis in different developmental stages. (**a**,**b**) cultured European seabass, *Dicentrarchus labrax*, (**c**) cultured European seabass, *Dicentrarchus labrax* fry and (**d**) cultured gilthead sea bream, *Sparus aurata* larvae in the hatchery.

A vibriosis outbreak may have catastrophic consequences for both the cultured animals and the producer, hence, implementation of preventive strategies is the safest way to cope with such potential events. Development of vaccines against vibriosis has so far been quite successful, since it has managed to significantly prevent the outbreaks of the disease [67], yet there are still important issues to be addressed. While fish are still in the hatchery, their immune system is not completely developed yet, thus vaccination is inefficient at this stage. Additionally, vaccination of juvenile fish by injection is difficult, so they are vaccinated orally and/or by immersion, resulting in low efficacy and short protection [68]. However, this is not the case for all vibrios, since immersion vaccination against *V. anguillarum* has been shown to provide a high level of protection. The biggest problem, however, is the lack of commercially available vaccines for the majority of the pathogenic *Vibrio* species other than *V. anguillarum*. Administration of antibiotics is therefore the most commonly applied strategy to tackle vibriosis outbreaks. However, if applied in marine hatcheries, antibiotics disturb the natural microbial balance in the water, as well as the developing microbiota of the larvae [69]. Furthermore, the excessive amount of antibiotics that have been used, not only for treatment, but even for prophylaxis during the last decades, has become a constantly growing problem for human and animal health, as well as for the environment [70]. There is a fundamental difficulty in controlling the amount and types of antibiotics that are applied, since the regulations for their usage can vary broadly among different countries. Development of multi-drug resistant strains, disturbance of natural microbiota, environmental residues, and public health issues, are only some of the most important problems caused by the excessive use of chemotherapy [71], and new alternatives are necessary.

3. Lytic Bacteriophages against Marine Vibrios

The use of bacteriophages against pathogenic bacteria in aquaculture was first introduced experimentally in Japan against *Lactococcus garvieae* in 1999 [72], and it has since been a topic of great interest for the scientific community [73–76]. Vibrios have been one of the main targets for bacteriophage isolation because of their high pathogenicity, broad presence, and ability to infect cultured fish and shellfish at various culture stages. Several potent phages have been tested against vibriosis causative agents, such as *V. harveyi*, *V. parahaemolyticus*, *V. alginolyticus*, *V. splendidus*, *V. anguillarum*, and *V. coralliilyticus* (Table 1), leading, in all cases, to increased survival rates of the cultured animals.

Table 1. Phage therapy trials against causative agents of vibriosis in experimental aquaculture setups.

Cultured Animal	Causative Agent	Reference
Penaeus monodon	*V. harveyi*	[77–81]
Haliotis laevigata		[82]
Panulirus ornatus		[83]
Ostrea plicaltula	*V. parahaemolyticus*	[84]
Litopenaeus vannamei		[85]
Apostichopus japonicus	*V. alginolyticus*	[86]
Apostichopus japonicus	*V. splendidus*	[87]
Apostichopus japonicas	*V. cyclitrophicus*	[88]
Salmo salar	*V. anguillarum*	[89]
Danio rerio		[90]
Acropora millepora	*V. coralliilyticus*	[91]

Biological treatment of *V. harveyi*-caused vibriosis has been quite successful in *Penaeus monodon* shrimp hatcheries. Vinod and colleagues [77] performed both short-term and long-term phage treatment trials using a broad host range, lytic siphovirus. During the short-term trials (48 h), the lytic vibriophage was administered as phage suspension at low multiplicity of infection (MOI = 1) to post-larval shrimps (18 days) that were previously infected by *V. harveyi*. Both single-dose (0 h) and double-dose (0 and 24 h) phage administration, led to 70% shrimp survival along with a 2-log reduction of *V. harveyi*, and 80% shrimp survival along with a 3-log reduction of *V. harveyi*, respectively. By contrast, controls without phage treatment showed only 25% survival and a 1-log increase of *V. harveyi*. During long-term trials (17 days), 35,000 naturally *V. harveyi*-infected nauplii were treated with the lytic vibriophage on a daily basis, and their average survival was 86%, compared to only 17% in the non-treated nauplii. Compared to antibiotics, which only led to a 40% survival, phage therapy provided better protection for infected shrimp. Similarly, Karunasagar and colleagues performed large-scale phage trials in a commercial shrimp hatchery using two lytic *V. harveyi*-specific broad host range bacteriophages. Phage application yielded 88% and 86% shrimp survival for each of the phages, while in antibiotic-treated (oxytetracycline and kanamycin) tanks, shrimp survival was 68% and 65%, respectively [79].

Phage therapy applications have shown promising results in other commercial species, such as sea cucumber, *Apostichopus japonicus* [87]. Three lytic bacteriophages (PVS-1, PVS-2 and PVS-3) were in all cases effective when tested in vitro against four pathogenic *V. splendidus* strains. Focusing on the preventive aspect of phage therapy, the authors prepared six different diets: a non-supplemented diet serving as control, an antibiotic-supplemented, three diets supplemented with single phages, and a diet that was supplemented with a cocktail of the three phages. Juvenile sea cucumbers were then fed on a daily basis. After 60 days, the animals were challenged by immersion in seawater containing *V. splendidus* for 2 days, and their survival rates were monitored for 10 days. The survival was 18% for the control diet, 82% for the antibiotics-supplemented diet, 65%, 58%, and 50% for the individual phage-supplemented diets and 82% for the phage cocktail-supplemented diet. In the same study, *V. splendidus* strain VS-ABTNL was injected in two groups of healthy sea cucumbers, while the control group was injected with sterile seawater. A phage cocktail was subsequently injected in one of the infected groups, and the survival rates were monitored for the following 10 days. All animals in the control group survived (100%), only 20% in the non-treated group and 80% survived in the phage treated group. Thus, it was concluded that phage cocktails could successfully protect *A. japonicus* against *V. splendidus* infection, and that both injection and immersion worked as delivery routes of phages. *V. splendidus* is a rapidly emerging pathogen, and the attempts for isolation of lytic phages have attracted a keen interest lately [92]. Similar phage trials in *A. japonicus* cultures have been performed against *V. alginolyticus* and *V. cyclitrophicus* using a mixture of two vibriophages and one vibriophage, respectively. In the former case, phage treatment at MOI = 10 led to 73% survival rates of the sea cucumbers, compared to only 3% survival that was observed in the non-treated group [86]. In the latter, the survival rate of juvenile *A. japonicus* was enhanced from 18% to 81% when fed with phage-containing feed, to 63% when injected with purified phage virions and to 58% when immersed in the phage-containing bath [88].

The bacteriophage CHOED has been tested for conferring protection against vibriosis in Atlantic salmon (*Salmo salar*) [89]. The presence of CHOED at MOI of 1 and 20 provided 100% protection of the fish against *V. anguillarum*, whereas untreated fish suffered over 90% mortality. When *S. salar* was challenged with *V. anguillarum* in aquaculture conditions, the administration of CHOED at MOI of 100 resulted in 100% fish survival 20 days after exposure to the pathogen, compared to only 60% survival in the non-treated fish.

The in vitro use of a phage cocktail with VP-1, VP-2, and VP-3 against *V. parahaemolyticus*, has been significantly more effective than using individual phages, albeit VP-3 was mainly responsible for the cocktail's lytic activity [93]. Although the efficacy of the phages contained in the cocktail can vary, multivalent phage cocktails can be effective against several pathogenic strains of the host and they can greatly delay the development of resistance due to the different phage components. Moreover, the idea of a phage cocktail allows the use of lytic phages with narrow host ranges, since several of them can be combined to produce a much broader lytic spectrum [94].

Phage delivery methods are of vital importance for a successful therapy, and depend on the presence of the phages at the area of infection in a titer above the therapeutic threshold. Ryan and colleagues have reviewed the phage delivery routes in human phage therapy trials, and they concluded that parenteral injection is the most successful route of phage administration, because the phages can immediately reach the systemic circulation [95]. In several aquaculture phage therapy trials, administration of bacteriophages via injection has also been the most successful route of delivery, since bacteriophages could be detected in the fish tissues for several days after administration [76,96]. However, parenteral injection, apart from the fact that it is rather stressful for the animals, has significant limitations in its practical application when (1) fish or shellfish are too small or too numerous or (2) continuous treatment is required. In the majority of the in vivo trials, phages are added to the water simultaneously, or right after the bacteria. This method reduces the number of pathogens used for the challenge, which in turn results in lower infection rate. The oral route of delivery, the immersion in phage bath, and the addition of phages to the surrounding water are very common methods that often lead to high protection against bacterial pathogens [74,75,97] and greatly increase the applicability of phage therapy. Especially, administration of phages via phage-coated feed has been shown to be an efficient delivery method, resulting in constant, high abundance of phages in the fish organs for several weeks [97]. A variety, though, of delivery routes has been suggested in aquaculture phage trials, because bacterial infections can occur during all the developmental stages of cultured organisms; from the eggs to the broodstock [98].

Reducing the number of vibrios in aquaculture environment is another strategy that has been examined. Pathogenic vibrios are present in live feeds offered to fish or invertebrate larvae, and the feed is thus a major source of pathogens entering the marine hatcheries [93,99]. Preventive administration of bacteriophages can be applied either directly to the environment of the cultured animal, preferably during early growth stages, or to the live prey, to control the source of pathogens to the hatchery facilities. *V. anguillarum*, *V. alginolyticus*, and *V. splendidus* are some common examples of pathogenic vibrios which are entering the aquaculture environment through live feeds, such as *Artemia salina* and *Brachionus plicatilis* (Figure 2) [52,100–102].

(a) (b)

Figure 2. Facilities for live feed production from a commercial fish farm unit. (**a**) *Artemia salina* in culture tanks with vigorous aeration, where the native presumptive *Vibrio* load is regularly estimated between 10^7 and 10^8 cells per mL; (**b**) *Brachionus plicatilis* culture tanks, where the native presumptive *Vibrio* load is regularly between 10^2 and 10^8 cells per mL [103].

The *V. alginolyticus*-specific broad host range lytic phages *φ*St2 and *φ*Grn1 have been successfully used as a "smart" disinfectant that selectively reduces vibrios in live feeds. Since *V. alginolyticus* is a prevalent component in live feeds, such as *Artemia* and rotifers, a scheme of precautionary phage administration in *Artemia salina* live feed cultures was evaluated. A combination of these two phages was administered in *A. salina* live prey at MOI: 100, leading to a significant reduction of the native *Vibrio* load by 1.3 log units, suggesting a decrease in the risk of a vibriosis outbreak in the marine hatchery [104]. Further research on *φ*St2 and *φ*Grn1 has revealed that during infection, these phages are able to hijack and reprogram the host's metabolic machinery, in order to meet their augmented demands for energy and nucleotide biosynthesis, making their therapeutic potential highly efficient [105].

The Profile of a Good Candidate: KVP40 Case

Several issues are important when selecting safe and efficient candidates for phage therapy and next generation sequencing technology plays an important role in revealing the genomic composition and the lytic nature of viruses, which are main criteria in the selection process. The absence of genes related either to lysogeny or to any known toxins [106] confirmed the lytic nature of the vibriophage KVP40, making it a proper candidate for phage therapy trials against vibriosis. Phage KVP40 [106,107], is a myovirus classified in a group which has been designated as "schizoT4like" or "KVP40-like" [108]. It was isolated against a clinical *V. parahaemolyticus* strain, however, it has a broad host range, able to infect several other strains of eight different *Vibrio* species: *V. alginolyticus*, *V. cholerae*, *V. parahaemolyticus*,

V. anguillarum, *V. splendidus*, *V. mimicus*, *V. natriegens*, and *V. fluvialis* [25,107]. The broad lytic spectrum and efficiency against several causative agents of vibriosis emphasizes the potential of KVP40 to control vibriosis in aquaculture settings.

The bacterial receptor that KVP40 recognizes in order to infect its hosts is the universal outer membrane protein K (OmpK), which is very common among *Vibrio* species [109]. Targeting a broadly distributed receptor is a key point when looking for a broad lytic spectrum bacteriophage; however, as it will be further discussed below, this can also result in the development of several defense strategies from the bacterial host in order to reduce the cost of resistance. In addition, KVP40 is a phage with a large genome (244,835 bp) able to take advantage of its host's metabolic machinery in order to maximize the efficiency of the infection, and thus, the overall impact of phage therapy. Lytic bacteriophages can manipulate and reprogram the host's metabolic machinery in order to support and facilitate their own DNA replication and protein synthesis, which are necessary for the packaging and release of the new virions [110,111]. They can mediate a transition from a host-oriented to a phage-oriented metabolism [112,113] during infection, since the interactions of their early phage genes with DNA metabolism-involved host proteins, cease the host replication [110]. KVP40 was found to encode a functional NAD^+ salvage pathway, which can boost its own replication during infection. This pathway is also conserved in other large genome phages that carry similar genes involved in nucleotide metabolism [114]. Last but not least, many phages, including KVP40, carry a high number of tRNAs, which may provide the phage with a small degree of autonomy when it comes to the translation of its own genes [115].

4. Issues Raised in Phage Therapy

As evident from above, phage therapy is definitely an attractive alternative to combat pathogenic bacteria, which may be used not only as a treatment, but also to prevent infections. However, there are several important constraints, such as the phage efficacy under aquaculture conditions, administration methods and persistence of phages in the system, the possibility of unwanted phage-encoded properties and, perhaps most importantly, the development of phage resistance, that need to be evaluated before a phage therapy application scheme can be considered successful.

4.1. Phage Therapy from the Lab to the Field

During the stages of a therapeutic phage suspension development in the lab, host specificity, life cycle parameters and lytic nature of the phage, are the main prerequisites that need to be covered. The selection of appropriate phages that are going to be used alone or forming a phage cocktail is also crucial for the outcome of the phage therapy. However, despite the promising results that some lytic bacteriophages have shown under laboratory conditions, application of phage-based treatment in aquaculture settings is associated with a number of additional challenges that need to be addressed.

Previously, reporting of phages with low in vivo activity has been one reason for questioning globally the actual efficacy of phage therapy against bacterial infections in animals and humans [116,117]. The optimal phage delivery method (injection, oral, immersion) may vary between different aquaculture settings, and should be carefully determined in each case. For instance, although injection has been mentioned as the most effective delivery route [95], immersion of the cultured animals in phage-containing water has been also quite effective, since bacteria begin their infection cycle from adhering to the mucosa of the fish, which constitutes the first physical and chemical barrier of fish against pathogens [118]. Marine fish species drink water to maintain their internal ionic balance, and therefore, phages of the water will have the opportunity to encounter pathogenic bacteria for which the infection route is through the fish intestinal mucosa. Even when bacteria attack the intestinal mucosa, fish drink a lot of water, so phages still encounter intestinal bacteria [97]. In vitro results based on immersion are very often similar to those obtained in vivo, since this approach, in both cases, is based on phage-bacteria interactions that take place in a phage-containing suspension [119]. Quantification of the viruses, in the animal tissues or in the aquatic environment where therapy was applied, will define the efficacy of the delivery route. However, repeated phage administration using either delivery route has been the most effective way to maintain a high bacteriophage titer in the system [77,87,97]. Oral administration via phage-coated fish pellets is a quite feasible and effective way to keep a constant phage input to the system with minimum effort, and easily incorporated into the daily routine of the fish farm [97]. Furthermore, as the pathogens may be present in different stages of the production process, it is important to consider where in the production the addition of phages is expected to most efficiently reduce the pathogen (i.e., disinfection of live feed, disinfection of fish eggs, treatment of infected fish, etc.).

4.2. Concerns about Phage-Treated Organisms

A bacterial lysate might contain endotoxins which, if not removed, may be fatal for the cultured organism [120,121]. The phage stocks that are administered to the cultured organisms should therefore be meticulously prepared to remove bacterial debris, secondary metabolites, enzymes, etc., that might potentially be toxic for the fish or shellfish [122]. Endotoxin-free phage suspensions are regularly produced today [123,124], eliminating potential side effects that may create unnecessary consideration to legislation and public opinion about phage therapy. Another concern about phage therapy in organisms such as fish, which have an adaptive immune system, is the potential immunological response of the phage-treated organism, that might trigger the production of phage-neutralizing antibodies, decreasing in vivo phage efficacy [125,126]. This possibility in aquaculture has been examined after phage-coated feed administered in yellowtail, *Seriola lalandi* [72] and intramuscular phage injection in ayu, *Plecoglossus altivelis* [127], however, phage-neutralizing antibodies were not detected in the studies. Production of such antibodies after phage administration in aquaculture is not yet documented in the literature [74].

4.3. Development of Resistance

Development of resistance is probably the most significant limitation in the whole concept of phage therapy. In the ocean, phages and their bacterial hosts are in a perpetual arms race, under strong evolutionary pressure [128,129]. Although the use of phage cocktails can reduce or delay the emergence of resistant strains [93,130], bacteria have developed several strategies (Figure 3) to cope with their viral predators [131–133].

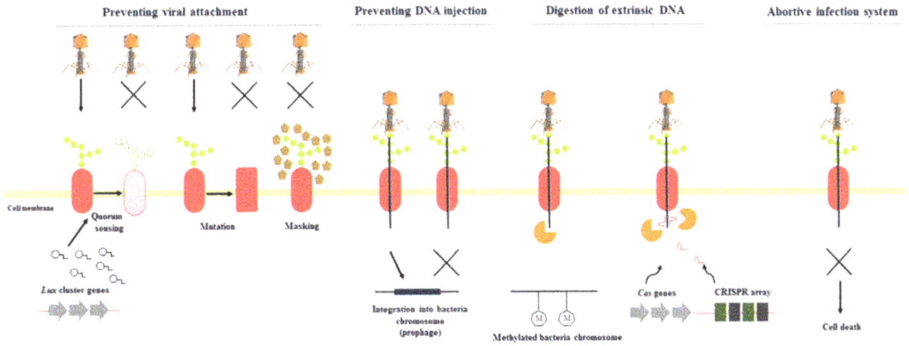

Figure 3. Overview of the main phage defense mechanisms in bacteria. Prevention of viral attachment on the bacterial surface can be achieved by mutating or masking the receptors, as well as downregulation of receptor expression, orchestrated by quorum sensing (QS). DNA injection may be successfully averted by superinfection exclusion (Sie) mechanisms. If phage DNA enters the bacterial host, its digestion can be catalyzed by R-M mechanism and CRISPR-Cas arrays systems. Deliberate death of the infected cell (abortive infection) constitutes another strategy against viral predators, where prevention of phage proliferation reduces spreading of the infection to the rest of the population.

4.3.1. Preventing Viral Attachment

The most crucial step for the successful infection of a bacterial host by a phage is its adsorption by the host through a specific reaction between the phage receptor-binding protein and the bacterial cell surface receptor. There is a great variety of components on the bacterial surfaces that are targets for phage attachment, such as proteins, polysaccharides, and lipopolysaccharides [134]. However, bacteria have developed strategies to effectively prevent phage adsorption events by (a) modifying the phage receptors by mutational changes, (b) masking receptors by producing an extracellular matrix, (c) producing competitive inhibitors, or (d) losing or downregulating the expression of the receptor [131,133]. In several cases the regulation of surface modifications is orchestrated by QS [135]. For instance, in the case of KVP40 phage infecting *V. anguillarum*, QS is used by bacteria in order to select between two different protection mechanisms according to their population density. At low host densities, the OmpK receptor for KVP40 is fully expressed, and the bacteria are protected from infection by forming aggregates and biofilms. At high host densities, the OmpK receptor is downregulated through a QS regulation pathway, making *V. anguillarum* less susceptible to the phage [24]. In *V. cholerae*, a different mechanism has been described, where surface modification of the phage receptor prevents attachment of the lytic phage ICP2 [136]. However, bacteriophages can regain their ability to attach on their targets by modifying their receptor-binding proteins and getting access to the modified bacterial cell surface receptors [137].

4.3.2. Preventing DNA injection

Even if the adsorption of the phage has been successful, bacteria have developed strategies to prevent the injection of the incoming viral DNA. Superinfection exclusion (Sie) systems are based on proteins related to cell surface modifications or inhibition of replication, and are often encoded by prophages or plasmids. Sie systems can provide immunity to the prophage-carrying bacterium against a second potential infection by similar bacteriophages [133,138,139]. A subcategory of Sie is the repressor-mediated immunity, where the repressor protein retains the prophage in the lysogenic cycle while providing immunity against an infective phage carrying the same type of repressor [140]. In a recent study, several temperate *V. anguillarum*-specific bacteriophages, designated as H20-like phages, were shown to contain a lambda-like cI repressor gene in their genomes. It was suggested that this mechanism possibly confers repressor-mediated immunity to other H20-like phages in their *V. anguillarum* host [141]. A Sie system-encoding prophage has recently been reported to confer phage resistance in the *V. cholerae* strain 919TP [142]. Compared to receptor alteration strategies, which prevent phage adsorption, and thus only protect the individual resistant cell, the Sie systems immobilize and inactivate the phage, thereby reducing the infection load on the remaining population [143].

4.3.3. Digesting Extrinsic DNA

In cases where the phage DNA enters the bacterial host, bacteria have developed several mechanisms for its inactivation: (a) Restriction-modification (R-M) systems and (b) CRISPR-Cas system [131]. R-M systems (type I, II, III, and IV) are composed by a methyltransferase and a restriction endonuclease, which catalyze the methylation of the bacterial DNA and the cleavage of the viral, unmethylated DNA, respectively [144–146]. Almost all bacterial genera carry R-M systems [133], and it is hypothesized that high levels of horizontal gene transfer (HGT) are responsible for their spreading and evolution among prokaryotes [145]. Vibriophage KVP40 has been reported to be restricted and modified by the R-M system of at least five *Vibrio* species [147]. However, phages have simultaneously evolved to evade the omnipresent R-M systems. Approximately, 20% of the available phage genomes carry methyltransferase-encoding genes, suggesting the ability to protect their own genome through methylation [141,148]. Methyltransferases can also affect bacterial virulence [149] or they may function as transcriptional regulators by either activating or repressing bacterial genes [150,151]. N6-adenine methyltransferase was previously found in the temperate vibriophage VHML, where it was linked to the virulence of *V. harveyi* host strain upon integration [152]. H20-like vibriophages were also found to carry a N6-adenine methyltransferase gene; however, its exact function still needs to be explored [141]. Phages may methylate parts of their genome, preventing its degradation by the host's restriction enzymes, whereas methylation could modify the properties of the host. It has been reported that viral infection of specific bacterial hosts subsequently affected the host range of the newly produced virions [153,154], suggesting that specific differences in the methylation of viral DNA during phage production affects the infectivity of the produced phages. Integrating conjugative elements (ICEs), ICE*Vsp*Por3 and ICE*Val*Spa1, that were recently identified in *V. splendidus* and *V. alginolyticus*, respectively, harbor genes encoding distinct R-M systems, which are able to confer protection against viral infection when expressed in *Escherichia coli* [155].

CRISPR-Cas has been described as an adaptive immune system of bacteria. This system is composed of (1) *cas* genes, which are responsible for the expression of the protein machinery that performs the immune response, and (2) the CRISPR loci, composed of 21–48 bp direct repeats interspaced by non-repetitive spacers of 26–72 bp, which provide genetic memory of previous viral infections. Although there are two classes, six types, and 16 subtypes of CRISPR-Cas systems [156], all of them are based on three common functional phases: (1) adaptation-spacer(s) from invasive DNA are acquired and then integrated into the CRISPR loci of the bacterial genome, (2) expression—transcription of the CRISPR loci that encode a CRISPR RNA (crRNA) molecule which will be combined with Cas protein forming crRNA–Cas complexes, and (3) interference—the crRNA-Cas complexes attach and digest complementary nucleic acids, providing immunity to the bacterial host [131]. Among the sequenced bacterial genomes, CRISPR loci are found in approximately 40% of them [157]. Although CRISPR-Cas systems are highly sophisticated, their conferred immunity can be bypassed by the bacteriophages by mutations or deletions in the targeted proto-spacer in the phage genome. Single nucleotide mutations in the protospacer genomic region of *S. thermophilus* phages were able to circumvent their bacterial host's CRISPR-Cas defense system [158,159]. Screening of the 1935 publicly available *Vibrio* genomes using CRISPRfinder [160] showed that CRISPR(s) were present in 278 (14.4%) genomes. Most of the CRISPR-containing *Vibrio* strains had one CRISPR array, but some carried up to 11 [161]. In total, 388 CRISPR arrays were identified in all the *Vibrio* genomes The CRISPR prevalence in the *Vibrio* genus is thus substantially lower than the reported 50% in bacteria in general [156]. Within the 28 genome-sequenced *V. anguillarum* strains, only one strain, *V. anguillarum* PF7, contained CRISPR. Out of the two CRISPR-Cas arrays of *V. anguillarum* PF7, eighteen spacers were >95% similar to genomic parts of the H20-like vibriophages, with eight of them being 100% identical [141]. The knowledge of the contribution of CRISPR-Cas systems in vibrios' defense is sporadic, and needs to be more thoroughly evaluated. It is also worth mentioning that the *V. cholerae* phage ICP1 is the first bacteriophage recorded to encode its own CRISPR-Cas arrays as a counter-defense against *V. cholerae* phage inhibitory chromosomal island [162].

4.3.4. Abortive Infection System

Bacterial infection by phages can sometimes be non-productive, even though it leads to the death of bacterial host. This kind of abortive infection may be the result of "lysis from without" (LO), where extensive simultaneous infections (i.e., very high MOI) may destroy the cell membrane. In another type of abortive infection leading to non-productive phage infections, the infected bacteria can enter a programmed cell death, and thereby prevent the spreading of the infection to the neighboring cells. This altruistic behavior is orchestrated by the abortive infection (Abi) system [131,163]. Although the best characterized Abi system is the Rex system, which is found in phage λ-lysogenized *E. coli* preventing infection by other coliphages [164,165], and most Abis have been identified in plasmids of Gram-positive strains [163], *V. cholerae* has also been documented as a carrier of such a system [166]. However, it has been proven that some Abis, such as TA system ToxIN described in *Pectobacterium atrosepticum*, may act also through toxin-antitoxin (TA) mechanisms, aborting phage infection [167]. TAs were initially found in plasmids, but are now known to be abundant in bacterial genomes. The components of these systems, a toxin and an antitoxin, neutralize each other, keeping the balance that maintains bacterial life. Upon phage infection, toxin and antitoxin production stops, yet antitoxin degrades faster, allowing the toxin to kill the bacterial cell [168]. For instance, MosAT TA system in *V. cholerae* resembles AbiE, a broadly distributed TA-Abi system in bacterial genomes inducing bacteriostasis and conferring phage resistance [168]. A high number of both confirmed and putative TA systems have been reported in *V. cholerae* and *V. parahaemolyticus* [169,170], however, their potential functionality as Abi systems still needs to be elucidated.

Even though the result of abortive infections is lysis of the bacterial cell, the phenomenon limits self-adjusting properties of the phages because there is no virion production. The implication of this defense mechanism in vibrios needs to be evaluated in vitro, during the assessment of the viral lytic spectrum.

4.3.5. Resistance Comes at a Cost

Thinking of the variety of phage defense mechanisms identified in bacteria, one might wonder what determines their distribution among bacteria, and why not all bacteria carry all of them. The explanation is that development of resistance comes at a fitness cost for the bacteria [171], so they need to be parsimonious when they invest on resistance strategies. Phage-binding bacterial receptors, which serve as recognition points for the phages, are often also associated with important cell functions, such as nutrient uptake, motility, and ability to attach on surfaces [172–174]. Therefore, surface modifications may significantly reduce bacterial fitness, including a reduced colonization ability [175,176] and increased susceptibility to other phages [177]. Loss of virulence and gliding motility has been observed in phage resistant *Flavobacterium columnare* [178], and similar results were recorded in phage-resistant *Flavobacterium psychrophilum* strains, which have shown decreased hemolytic activity, gelatinase activity, and total protease activity, as well as mutations in significant virulence genes [179]. Accordingly, in vibrios, phage-resistant strains have been shown to be less harmful against their eukaryotic host. In experimentally challenged pipefish (*Syngnathus typhle*), three *Vibrio* spp. Isolates, representing different phage susceptibility, showed positive correlation between phage susceptibility and virulence [180]. Further, loss of virulence was observed in four KVP40-resistant *V. anguillarum* strains as demonstrated by reduced mortality of cod (*Gadus morhua*) larvae challenged with phage-resistant clones, compared to a control group challenged with the wild type *V. anguillarum* strain [181].

Bacteria and phages are in a perpetual arms race, so phages also evolve counter-defensive strategies to circumvent bacterial defense mechanisms. The fitness cost, including loss of virulence, which is often associated with resistance, constitutes a barrier to the prevalence of these defense mechanisms [131], and selection pressure on different anti-phage strategies depends on the trade-off between mortality imposed by phages and fitness cost of the defense strategy, under the given environmental conditions. Hence, constitutive defense strategies, such as mutating the bacterial receptors and inducible defense strategies, such as CRISPR-Cas systems, may prevail under different conditions and for different phage-host interactions. In the former case, modification of the receptors is permanently associated with a substantial fixed cost that directly affects bacterial fitness, while in the latter case, bacterial CRISPR-Cas systems may be elicited only upon viral infection [182]. The force of infection and the nutrient availability are usually the most important factors that will determine the bacterial decision between constitutive and inducible defense mechanisms [182–184]. The CRISPR-Cas system is favored when the frequency of infection is low and the nutrients are limited, because it is associated with a lower cost for the bacterial cell. Viruses can, though, quickly mutate and escape, suggesting that these systems would be most effective when exposed to low phage diversity, due to relatively limited capacity for spacer acquisition. Accordingly, genomic mapping of CRISPR spacers and viral genomes has shown that only recently acquired resistance was functional for phage defense [185]. On the other hand, surface modifications are favored when frequency of infection is high and nutrients are abundant, because bacteria need to be in an "always on" defensive position, even if the costs are high [182].

Acquiring resistance against more than one bacteriophage might increase the fitness cost for the host, since it will need to modify different phage-binding surface receptors [175,186]. Therefore, coping with phage cocktails implies a higher fitness cost for the bacteria. Furthermore, the accumulating indications that phage-resistance leads to dominance of less virulent phenotypes, suggest that the problems with resistance associated with phage treatment in aquaculture may be limited. Indeed, the fitness cost of phage immunity needs to be comprehensively explored, with special emphasis on the virulence properties of phage resistant pathogens, potentially allowing the prediction of the implications of the development of presumably less virulent phage resistant bacteria for the survival of phage treated animals.

5. Temperate Vibriophages and Lysogenic Conversion

Another concern when selecting phages for therapeutic purposes is the risk of using phages encoding unwanted genes that may spread in the pathogen population. Consequently, it must be explicit that only lytic bacteriophages should be considered as potential candidates.

However, temperate bacteriophages are also a vital part of the natural virome, and key players in driving bacterial evolution by disseminating genetic information through horizontal gene transfer (HGT). HGT may take place in both lytic and temperate phage with the form of generalized transduction, whereas the events of specialized transduction and lysogenic conversion are restricted only to temperate phages [187,188]. The high rate of generalized transduction events, which also applies to lytic phages, has lately raised significant concerns against phage therapy, since virulence or resistance genetic element might spread among the pathogenic bacteria [189]. Complementary researches on *E. coli* and *Streptococcus pyogenes*, though, have shown that antibiotic resistance genes may be disseminated only through temperate transducing bacteriophages [190–192]. However, it was recently found that two lytic bacteriophages against *E. coli*, designated as "superspreaders", could promote extensive plasmid transformation, and therefore efficiently disperse antibiotic resistance genes [193]. Hence, it is crucial that such phages are avoided in phage therapy or other medical applications.

A serious constraint in phage therapy is, therefore, the unsafe use of temperate phages as therapeutic agents. The process through which prophages integrate in their host's genome and transfer genes whose expression may render increased bacterial fitness, either directly (e.g., phage-encoded toxins) or indirectly (e.g., increased fitness during infection) is designated as lysogenic conversion. This process possesses a dominant role in conferring enhanced fitness to the prophage-carrying bacteria. The importance of this process is highlighted in vibrios, since there is a plethora of vibriophages where lysogenic conversion has positively affected, both directly and indirectly, the fitness of their lysogenized hosts. A classic example of lysogenic conversion in vibrios is the phage-mediated production of cholera toxin, by the filamentous phage CTXΦ [194]. However, there are several examples of lysogenic conversion in other *Vibrio* pathogens. *V. harveyi* strains carrying the prophage VHML, were able to metabolize fewer nutrient sources than their uninfected counterparts [195]. Switching off unnecessary bacterial functions would make lysogenized *V. harveyi* strains less energy-consuming, hence more competitive under nutrient-limited environments [196]. The same temperate phage VHML was reported as being responsible for conferring virulence to the *V. harveyi* strain 642, since avirulent *V. harveyi* strains were converted to virulent, when infected by phage VHML [78,197]. A similar observation was made in the case of the prophage VOB, which was integrated in the genome of its *V. owensii* host. After the induction of VOB in the lab, it was co-cultured with naïve *V. harveyi* and *V. campbellii*. The avirulent vibrios were lysogenized by the induced VOB and they became virulent, causing increased mortality to *Penaeus monodon*. It was concluded that VOB was responsible for some of the virulence of *V. owensii*, as well as for the acquired virulence of *V. harveyi* and *V. campbellii* lysogens [198]. In a very recent study, prophage-like elements that were identified in the genomes of *V. anguillarum* strains T265 and Ba35, contained genes related to zonula occludens toxin (Zot), implying the contribution of the prophage to bacterial virulence [199]. Prophages K139 and VIPΦ have also been reported to increase the pathogenesis of their *V. cholerae* hosts by increasing both virulence and colonization ability [200,201]. Lysogenic conversion may be implicated also in bacterial host's defense against viral infection. Phage-induced lysis of some cells could release prophage-encoded toxins, such as colicins, that might eliminate competitors, helping the rest of the bacterial population to take advantage of the environmental niche and resources [202,203]. In a recently published study, the tail length tape measure gene that was identified in the genome of the H20-like temperate vibriophages resembled the structure of channel forming toxin colicin Ia, hence, an additional role of the gene in bacterial competition was suggested by the authors [141]. Furthermore, Sie and Abi defense systems that were previously mentioned, are also among the beneficial effects that prophages impart to their hosts [204].

Although temperate bacteriophages are not suitable for therapeutic purposes, lysogenic lifestyle is the symbiotic aspect of the virus-bacteria interactions which has been the outcome of their refined co-evolutionary relationships [205]. Already in 1961, Campbell suggested a beneficiary contribution of prophages to their host by stating "One therefore must look for possible means by which the phage might impart a selective advantage to its host" [206].

6. Perspectives on Phage Therapy Today

Wherever bacteria thrive, so do predatory phages. During 2017, we celebrated 100 years from the discovery of the bacteriophages and the idea of using specific bacteriophages as a weapon to biologically control pathogenic bacteria. Phage therapy approaches against bacterial infections have been revived, primarily due to major problems with antibiotic-resistant bacteria we are facing as a result of excessive usage of antibiotics. In addition, the increasing temperature in the oceans, the fatal effects of vibriosis on the global aquaculture industry, as well as a plethora of different vibrios that may trigger the disease has further emphasized the need for exploring the potential of phages to control vibrio pathogens.

While the initial idea behind phage therapy was treatment of diseases, as in the case of antibiotics, future work should include prophylactic use of lytic bacteriophages to reduce the pathogen load and reduce the risk of infection. In marine aquaculture, addition of bacteriophages to live feeds, such as *Artemia* and rotifers, may be an efficient way to selectively disinfect the life feeds immediately prior to entering fish or invertebrate production cycle. Since both *Artemia* and rotifers are produced in batch cultures with a short retention time, the risk of resistance development is minimal, as opposed to providing phages prophylactically in the feeds of fish, or directly in rearing tanks, which would allow for phage-bacteria co-evolution. However, application of lytic phages as fish feed additives may be an efficient way to prevent the pathogens from establishing in the fish organs. Recirculating aquaculture systems (RAS) could be ideal environments for the application of phage therapy, since the water exchange is quite limited. Additionally, the combinatory usage of bacteriophages, together with another ecologically friendly alternative such as probiotic bacteria, constitutes a strategy that would be expected to be highly effective against bacterial diseases. Combining biological approaches with different targets and modes of action may minimize the risk of future resistance development, as has been seen in human medicine, where combined drugs are successful in antibacterial and antiviral treatment.

More research is still required to optimize the phage application under field conditions (phage composition, timing of application, delivery etc.) and to eliminate the potential risk factors associated with phage application (dispersal of unwanted genes, effects on fish microbiota). Further investigation of the naturally occurring phages in the cultured animals' microbiota is going to unravel their role in the organism's protection against bacterial diseases, and evaluate the possibility of them being used in a more targeted phage therapy scheme. This requires extensive sequencing of viral genomes and analyses for presence of genetic elements that might potentially interfere with the bacterial fitness or affect the organism's health. Such knowledge would also provide us a window for visualizing a future of molecularly engineered lytic virions. Consequently, more in vitro and in vivo test trials are required before the final release, and any side effects need to be meticulously recorded. However, facing a future with increasing problems with antibiotic resistant pathogens, exploring phage-based alternatives is now more necessary than ever.

Acknowledgments: The study was supported by the Danish Council for Strategic Research (ProAqua project 12-132390), The Danish Council for Independent Research (DFF—7014-00080), and by the Greek National Strategic Reference Framework 2007–2013 (co-funded by European Social Fund and Greek National Funds) FISHPHAGE project 131, www.gsrt.gr. We would like to thank Panos Varvarigos, DVM and Nancy Dourala, DVM for their contribution in Figure 1c,d and Figure 2a,b.

Author Contributions: Panos G. Kalatzis collected, analyzed, interpreted the data and drafted the manuscript. Daniel Castillo designed Figure 3. Daniel Castillo, Pantelis Katharios and Mathias Middelboe interpreted the data and revised the manuscript. All authors read and approved the manuscript.

Conflicts of Interest: The authors declare no conflict of interest.

References

1. Thompson, F.L.; Iida, T.; Swings, J. Biodiversity of vibrios. *Microbiol. Mol. Biol. Rev.* **2004**, 403–431. [CrossRef] [PubMed]
2. Yooseph, S.; Nealson, K.H.; Rusch, D.B.; McCrow, J.P.; Dupont, C.L.; Kim, M.; Johnson, J.; Montgomery, R.; Ferriera, S.; Beeson, K.; et al. Genomic and functional adaptation in surface ocean planktonic prokaryotes. *Nature* **2010**, *468*, 60–66. [CrossRef] [PubMed]
3. Ramaiah, N.; Ravel, J.; Straube, W.L.; Hill, R.T.; Colwell, R.R. Entry of *Vibrio harveyi* and *Vibrio fischeri* into the viable but nonculturable state. *J. Appl. Microbiol.* **2002**, *93*, 108–116. [CrossRef] [PubMed]
4. Armada, S.P.; Farto, R.; Pérez, M.J.; Nieto, T.P. Effect of temperature, salinity and nutrient content on the survival responses of *Vibrio splendidus* biotype I. *Microbiology* **2003**, *149*, 369–375. [CrossRef] [PubMed]
5. Flardh, K.; Cohen, P.S.; Kjelleberg, S. Ribosomes exist in large excess over the apparent demand for protein synthesis during carbon starvation in marine *Vibrio* sp. strain CCUG 15956. *J. Bacteriol.* **1992**, *174*, 6780–6788. [CrossRef] [PubMed]
6. Kramer, J.G.; Singleton, F.L. Variations in rRNA content of marine *Vibrio* spp. during starvation- survival and recovery. *Appl. Environ. Microbiol.* **1992**, *58*, 201–207. [PubMed]
7. Eilers, H.; Pernthaler, J.; Amann, R. Succession of pelagic marine bacteria during enrichment: A close look at cultivation-induced shifts. *Appl. Environ. Microbiol.* **2000**, *66*, 4634–4640. [CrossRef] [PubMed]
8. Yu, C.; Bassler, B.L.; Roseman, S. Chemotaxis of the marine bacterium *Vibrio furnissii* to sugars. A potential mechanism for initiating the chitin catabolic cascade. *J. Biol. Chem.* **1993**, *268*, 9405–9409. [PubMed]
9. Gosink, K.K.; Kobayashi, R.; Kawagishi, I.; Häse, C.C. Analyses of the roles of the three *cheA* homologs in chemotaxis of *Vibrio cholerae*. *J. Bacteriol.* **2002**, *184*, 1767–1771. [CrossRef] [PubMed]
10. Larsen, M.H.; Blackburn, N.; Larsen, J.L.; Olsen, J.E. Influences of temperature, salinity and starvation on the motility and chemotactic response of *Vibrio anguillarum*. *Microbiology* **2004**, *150*, 1283–1290. [CrossRef] [PubMed]
11. Grimes, D.J.; Johnson, C.N.; Dillon, K.S.; Flowers, A.R.; Noriea, N.F.; Berutti, T. What genomic sequence information has revealed about *Vibrio* ecology in the ocean-a review. *Microb. Ecol.* **2009**, *58*, 447–460. [CrossRef] [PubMed]
12. Andrews, J.H.; Harris, R.F. The ecology and biogeography of microorganisms on plant surfaces. *Annu. Rev. Phytopathol.* **2000**, 145–180. [CrossRef] [PubMed]
13. Simidu, U.; Ashino, K.; Kaneko, E. Bacterial flora of phyto- and zoo- plankton in the inshore water of Japan. *Can. J. Microbiol.* **1971**, *19*, 1157–1160. [CrossRef]
14. Hollants, J.; Leliaert, F.; De Clerck, O.; Willems, A. What we can learn from sushi: A review on seaweed-bacterial associations. *FEMS Microbiol. Ecol.* **2013**, *83*, 1–16. [CrossRef] [PubMed]
15. Froelich, B.; Ayrapetyan, M.; Oliver, J.D. Integration of *Vibrio vulnificus* into marine aggregates and its subsequent uptake by *Crassostrea virginica* oysters. *Appl. Environ. Microbiol.* **2013**, *79*, 1454–1458. [CrossRef] [PubMed]
16. Lyons, M.M.; Lau, Y.T.; Carden, W.E.; Ward, J.E.; Roberts, S.B.; Smolowitz, R.; Vallino, J.; Allam, B. Characteristics of marine aggregates in shallow-water ecosystems: Implications for disease ecology. *Ecohealth* **2007**, *4*, 406–420. [CrossRef]

17. Takemura, A.F.; Chien, D.M.; Polz, M.F. Associations and dynamics of Vibrionaceae in the environment, from the genus to the population level. *Front. Microbiol.* **2014**, *5*, 1–26. [CrossRef] [PubMed]
18. Thompson, J.R.; Randa, M.A.; Marcelino, L.A.; Tomita-Mitchell, A.; Lim, E.; Polz, M.F. Diversity and dynamics of a north atlantic coastal *Vibrio* community. *Appl. Environ. Microbiol* **2004**, *70*, 4103–4110. [CrossRef] [PubMed]
19. Kaspar, C.W.; Tamplin, M.L. Effects of temperature and salinity on the survival of *Vibrio vulnificus* in seawater and shellfish. *Appl. Environ. Microbiol.* **1993**, *59*, 2425–2429. [PubMed]
20. Singleton, F.L.; Attwell, R.; Jangi, S.; Colwell, R.R. Effects of temperature and salinity on *Vibrio cholerae* growth. *Appl. Environ. Microbiol.* **1982**, *44*, 1047–1058. [PubMed]
21. Vezzulli, L.; Grande, C.; Reid, P.C.; Hélaouët, P.; Edwards, M.; Höfle, M.G.; Brettar, I.; Colwell, R.R.; Pruzzo, C. Climate influence on *Vibrio* and associated human diseases during the past half-century in the coastal North Atlantic. *Proc. Natl. Acad. Sci. USA* **2016**, *113*, E5062–E5071. [CrossRef] [PubMed]
22. Vezzulli, L.; Höfle, M.; Pruzzo, C.; Pezzati, E.; Brettar, I. Effects of global warming on *Vibrio* ecology. *Microbiol. Spectr.* **2015**, *3*. [CrossRef] [PubMed]
23. Letchumanan, V.; Pusparajah, P.; Tan, L.T.H.; Yin, W.F.; Lee, L.H.; Chan, K.G. Occurrence and antibiotic resistance of *Vibrio parahaemolyticus* from shellfish in Selangor, Malaysia. *Front. Microbiol.* **2015**, *6*, 1–11. [CrossRef] [PubMed]
24. Tan, D.; Svenningsen, S.L.; Middelboe, M. Quorum sensing determines the choice of antiphage defense strategy in *Vibrio anguillarum*. *mBio* **2015**, *6*, 1–10. [CrossRef] [PubMed]
25. Tan, D.; Gram, L.; Middelboe, M. Vibriophages and their interactions with the fish pathogen *Vibrio anguillarum*. *Appl. Environ. Microbiol.* **2014**, *80*, 3128–3140. [CrossRef] [PubMed]
26. Le Roux, F.; Wegner, K.M.; Polz, M.F. Oysters and vibrios as a model for disease dynamics in wild animals. *Trends Microbiol.* **2016**, *24*, 568–580. [CrossRef] [PubMed]
27. Oliver, J.; Pruzzo, C.; Vezzulli, L.; Kaper, J. Vibrio species. In *Food Microbiology: Fundamentals and Frontiers*; Doyele, M., Buchanan, R., Eds.; ASM Press: Washington, WA, USA, 2013; pp. 401–440.
28. Plaza, N.; Castillo, D.; Pérez-Reytor, D.; Higuera, G.; García, K.; Bastías, R. Bacteriophages in the control of pathogenic vibrios. *Electron. J. Biotechnol.* **2018**, *31*, 24–33. [CrossRef]
29. Epstein, P.R. Algal blooms in the spread and persistence of cholera. *BioSystems* **1993**, *31*, 209–221. [CrossRef]
30. Thompson, J.R.; Marcelino, L.; Polz, M.F. Diversity, sources and detection of human bacterial pathogens in the marine environment. In *Oceans and Health: Pathogens in the Marine Environment*; Springer: New York, NY, USA, 2005; pp. 29–69. ISBN 0-387-23709-7.
31. Kimes, N.E.; Grim, C.J.; Johnson, W.R.; Hasan, N.A.; Tall, B.D.; Kothary, M.H.; Kiss, H.; Munk, A.C.; Tapia, R.; Green, L.; et al. Temperature regulation of virulence factors in the pathogen *Vibrio coralliilyticus*. *ISME J.* **2012**, *6*, 835–846. [CrossRef] [PubMed]
32. Vezzulli, L.; Brettar, I.; Pezzati, E.; Reid, P.C.; Colwell, R.R.; Höfle, M.G.; Pruzzo, C. Long-term effects of ocean warming on the prokaryotic community: Evidence from the vibrios. *ISME J.* **2012**, *6*, 21–30. [CrossRef] [PubMed]
33. Baker-Austin, C.; Trinanes, J.A.; Taylor, N.G.H.; Hartnell, R.; Siitonen, A.; Martinez-Urtaza, J. Emerging *Vibrio* risk at high latitudes in response to ocean warming. *Nat. Clim. Chang.* **2013**, *3*, 73–77. [CrossRef]
34. Lipp, E.K.; Huq, A.; Colwell, R.R. Effects of global climate on infectious disease: The cholera model. *Clin. Microbiol. Rev.* **2002**, *15*, 757–770. [CrossRef] [PubMed]
35. Froelich, B.A.; Noble, R.T. *Vibrio* bacteria in raw oysters: Managing risks to human health. *Philos. Trans. R. Soc. Lond. B Biol. Sci.* **2016**, *371*. [CrossRef] [PubMed]
36. Food and Agriculture Organization of the United Nations. *The State of World Fisheries and Aquaculture—Contributing to Food Security and Nutrition for All*; FAO Report; FAO: Rome, Italy, 2016; ISBN 9789251091852.
37. Austiin, B.; Austin, D.A. Vibrionaceae representatives. In *Bacterial Fish Pathogens, Disease of Frmed and Wild Fish*; Springer: Dordrecht, The Netherlands, 2012; pp. 369–389, ISBN 978-94-007-4884-2.
38. Toranzo, A.E.; Magariños, B.; Romalde, J.L. A review of the main bacterial fish diseases in mariculture systems. *Aquaculture* **2005**, *246*, 37–61. [CrossRef]
39. The World Bank. Fish to 2030: Prospects for fisheries and aquaculture. *Agric. Environ. Serv. Discuss. Pap.* **2013**, *3*, 102.
40. De Schryver, P.; Defoirdt, T.; Sorgeloos, P. Early mortality syndrome outbreaks: A microbial management issue in shrimp farming? *PLoS Pathog.* **2014**, *10*, 10–11. [CrossRef] [PubMed]

41. Zorriehzahra, M.J.; Banaederakhshan, R. Early Mortality Syndrome (EMS) as new emerging threat in shrimp industry. *Adv. Anim. Vet. Sci.* **2015**, 2309–2331. [CrossRef]
42. Rowley, A.F.; Cross, M.E.; Culloty, S.C.; Lynch, S.A.; Mackenzie, C.L.; Morgan, E.; O'Riordan, R.M.; Robins, P.E.; Smith, A.L.; Thrupp, T.J.; et al. The potential impact of climate change on the infectious diseases of commercially important shellfish populations in the Irish Sea—A review. *ICES J. Mar. Sci.* **2014**, *71*, 741–759. [CrossRef]
43. Canestrini, G. La malatti dominate delle anguille. *Atti Inst. Veneto Serv.* **1893**, *7*, 809–814.
44. Bergman, A.M. Die rote Beulenkrankheit des Aals. *Bericht aus der Königlichen Bayer. Versuchsstation* **1909**, *2*, 10–54.
45. Rønneseth, A.; Castillo, D.; D'Alvise, P.; Tønnesen, Ø.; Haugland, G.; Grotkjaer, T.; Engell-Sørensen, K.; Nørremark, L.; Bergh, Ø.; Wergeland, H.I.; et al. Comparative assessment of *Vibrio* virulence in marine fish larvae. *J. Fish Dis.* **2017**. [CrossRef] [PubMed]
46. Frans, I.; Michiels, C.W.; Bossier, P.; Willems, K.A.; Lievens, B.; Rediers, H. *Vibrio anguillarum* as a fish pathogen: Virulence factors, diagnosis and prevention. *J. Fish Dis.* **2011**, *34*, 643–661. [CrossRef] [PubMed]
47. Austin, B.; Zhang, X.H. *Vibrio harveyi*: A significant pathogen of marine vertebrates and invertebrates. *Lett. Appl. Microbiol.* **2006**, *43*, 119–124. [CrossRef] [PubMed]
48. Wang, R.; Zhong, Y.; Gu, X.; Yuan, J.; Saeed, A.F.; Wang, S. The pathogenesis, detection, and prevention of *Vibrio parahaemolyticus*. *Front. Microbiol.* **2015**, *6*, 1–13. [CrossRef] [PubMed]
49. Zorilla, I.; Chabrillon, M.; Arijo, S.; Dıaz-Rozales, P.; Martinez-Manzanares, E.; Balebona, M.C.; Morinigo, M.A. Bacteria recovered from diseased cultured gilthead sea bream (*Sparus aurata* L.) in southwestern Spain. *Aquaculture* **2003**, *218*, 11–20. [CrossRef]
50. Balebona, M.C.; Andreu, M.J.; Bordas, M.A.; Zorrilla, I.; Moriñigo, M.A.; Borrego, J.J. Pathogenicity of *Vibrio alginolyticus* for cultured gilt-head sea bream (*Sparus aurata* L.). *Appl. Environ. Microbiol.* **1998**, *64*, 4269–4275. [PubMed]
51. Fouz, B.; Amaro, C. Isolation of a new serovar of *Vibrio vulnificus* pathogenic for eels cultured in freshwater farms. *Aquaculture* **2003**, *217*, 677–682. [CrossRef]
52. Thomson, R.; Macpherson, H.L.; Riaza, A.; Birkbeck, T.H. *Vibrio splendidus* biotype 1 as a cause of mortalities in hatchery-reared larval turbot, *Scophthalmus maximus* (L.). *J. Appl. Microbiol.* **2005**, *99*, 243–250. [CrossRef] [PubMed]
53. Cano-Gómez, A.; Goulden, E.F.; Owens, L.; Høj, L. *Vibrio owensii* sp. nov., isolated from cultured crustaceans in Australia. *FEMS Microbiol. Lett.* **2010**, *302*, 175–181. [CrossRef] [PubMed]
54. Austin, B.; Austin, D.A. *Bacterial Fish Pathogens: Diseases of Farmed and Wild Fish*, 4th ed.; Springer-Praxis Publishing: New York, NY, USA, 2007.
55. Biosca, E.G.; Amaro, C. Toxic and enzymatic activities of *Vibrio vulnificus* biotype 2 with respect to host specificity. *Appl. Environ. Microbiol.* **1996**, *62*, 2331–2337. [PubMed]
56. Zhang, X.-H.; Austin, B. Pathogenicity of *Vibrio harveyi* to salmonids. *J. Fish Dis.* **2000**, *23*, 93–102. [CrossRef]
57. Binesse, J.; Delsert, C.; Saulnier, D.; Champomier-Vergès, M.C.; Zagorec, M.; Munier-Lehmann, H.; Mazel, D.; Le Roux, F. Metalloprotease Vsm is the major determinant of toxicity for extracellular products of *Vibrio splendidus*. *Appl. Environ. Microbiol.* **2008**, *74*, 7108–7117. [CrossRef] [PubMed]
58. Gómez-León, J.; Villamil, L.; Lemos, M.L.; Novoa, B.; Figueras, A. Isolation of *Vibrio alginolyticus* and *Vibrio splendidus* from aquacultured carpet shell clam (*Ruditapes decussatus*) larvae associated with mass mortalities. *Appl. Environ. Microbiol.* **2005**, *71*, 98–104. [CrossRef] [PubMed]
59. Kanemori, Y.; Nakai, T.; Muroga, K. The Role of Extracellular Protease Produced by *Vibrio anguillarum*. *Fish Pathol.* **1987**, *22*, 153–158. [CrossRef]
60. Farto, R.; Pérez, M.J.; Fernández-Briera, A.; Nieto, T.P. Purification and partial characterisation of a fish lethal extracellular protease from *Vibrio pelagius*. *Vet. Microbiol.* **2002**, *89*, 181–194. [CrossRef]
61. Defoirdt, T. Virulence mechanisms of bacterial aquaculture pathogens and antivirulence therapy for aquaculture. *Rev. Aquac.* **2014**, *6*, 100–114. [CrossRef]
62. Lin, B.; Wang, Z.; Malanoski, A.P.; O'Grady, E.A.; Wimpee, C.F.; Vuddhakul, V.; Alves, N.; Thompson, F.L.; Gomez-Gil, B.; Vora, G.J. Comparative genomic analyses identify the *Vibrio harveyi* genome sequenced strains BAA-1116 and HY01 as *Vibrio campbellii*. *Environ. Microbiol. Rep.* **2010**, *2*, 81–89. [CrossRef] [PubMed]

63. Darshanee Ruwandeepika, H.A.; Sanjeewa Prasad Jayaweera, T.; Paban Bhowmick, P.; Karunasagar, I.; Bossier, P.; Defoirdt, T. Pathogenesis, virulence factors and virulence regulation of vibrios belonging to the harveyi clade. *Rev. Aquac.* **2012**, *4*, 59–74. [CrossRef]
64. Li, X.; Yang, Q.; Dierckens, K.; Milton, D.L.; Defoirdt, T. RpoS and indole signaling control the virulence of *Vibrio anguillarum* towards gnotobiotic sea bass (*Dicentrarchus labrax*) larvae. *PLoS ONE* **2014**, *9*, 1–7. [CrossRef] [PubMed]
65. Soumya Haldar, S.C. *Vibrio* related diseases in aquaculture and development of rapid and accurate identification methods. *J. Mar. Sci. Res. Dev.* **2012**, *s1*. [CrossRef]
66. Diggles, B.K.; Carson, J.; Hine, P.M.; Hickman, R.W.; Tait, M.J. *Vibrio* species associated with mortalities in hatchery-reared turbot (*Colistium nudipinnis*) and brill (*C. guntheri*) in New Zealand. *Aquaculture* **2000**, *183*, 1–12. [CrossRef]
67. Colquhoun, D.J.; Lillehaug, A. Vaccination against vibriosis. In *Fish Vaccination*; John Wiley & Sons, Ltd.: Chichester, UK, 2014; pp. 172–184, ISBN 9781118806913.
68. Embregts, C.W.E.; Forlenza, M. Oral vaccination of fish: Lessons from humans and veterinary species. *Dev. Comp. Immunol.* **2016**, *64*, 118–137. [CrossRef] [PubMed]
69. Olafsen, J.A. Interactions between fish larvae and bacteria in marine aquaculture. *Aquaculture* **2001**, *200*, 223–247. [CrossRef]
70. Cabello, F.C. Heavy use of prophylactic antibiotics in aquaculture: A growing problem for human and animal health and for the environment. *Environ. Microbiol.* **2006**, *8*, 1137–1144. [CrossRef] [PubMed]
71. Perreten, V. Resistance in the food chain and in bacteria from animals: Relevance to human infections. In *Frontiers in Antimicrobial Resistance*; White, D., Alekshun, M., McDermott, P., Eds.; American Society for Microbiology: Washington, DC, USA, 2005; pp. 446–464.
72. Nakai, T.; Sugimoto, R.; Park, K.H.; Matsuoka, S.; Mori, K.; Nishioka, T.; Maruyama, K. Protective effects of bacteriophage on experimental *Lactococcus garvieae* infection in yellowtail. *Dis. Aquat. Organ.* **1999**, *37*, 33–41. [CrossRef] [PubMed]
73. Defoirdt, T.; Sorgeloos, P.; Bossier, P. Alternatives to antibiotics for the control of bacterial disease in aquaculture. *Curr. Opin. Microbiol.* **2011**, *14*, 251–258. [CrossRef] [PubMed]
74. Oliveira, J.; Castilho, F.; Cunha, A.; Pereira, M.J. Bacteriophage therapy as a bacterial control strategy in aquaculture. *Aquac. Int.* **2012**, *20*, 879–910. [CrossRef]
75. Richards, G.P. Bacteriophage remediation of bacterial pathogens in aquaculture: A review of the technology. *Bacteriophage* **2014**, *4*, e975540. [CrossRef] [PubMed]
76. Nakai, T.; Park, S.C. Bacteriophage therapy of infectious diseases in aquaculture. *Res. Microbiol.* **2002**, *153*, 13–18. [CrossRef]
77. Vinod, M.G.; Shivu, M.M.; Umesha, K.R.; Rajeeva, B.C.; Krohne, G.; Karunasagar, I.; Karunasagar, I. Isolation of *Vibrio harveyi* bacteriophage with a potential for biocontrol of luminous vibriosis in hatchery environments. *Aquaculture* **2006**, *255*, 117–124. [CrossRef]
78. Oakey, H.J.; Owens, L. A new bacteriophage, VHML, isolated from a toxin-producing strain of *Vibrio harveyi* in tropical Australia. *J. Appl. Microbiol.* **2000**, *89*, 702–709. [CrossRef] [PubMed]
79. Karunasagar, I.; Shivu, M.M.; Girisha, S.K.; Krohne, G.; Karunasagar, I. Biocontrol of pathogens in shrimp hatcheries using bacteriophages. *Aquaculture* **2007**, *268*, 288–292. [CrossRef]
80. Phumkhachorn, P.; Rattanachaikunsopon, P. Isolation and partial characterization of a bacteriophage infecting the shrimp pathogen *Vibrio harveyi*. *Afr. J. Microbiol.* **2010**, *4*, 1794–1800.
81. Stalin, N.; Srinivasan, P. Efficacy of potential phage cocktails against *Vibrio harveyi* and closely related *Vibrio* species isolated from shrimp aquaculture environment in the south east coast of India. *Vet. Microbiol.* **2017**, *207*, 83–96. [CrossRef] [PubMed]
82. Wang, Y.; Barton, M.; Elliott, L.; Li, X.; Abraham, S.; Dea, M.O.; Munro, J. Bacteriophage therapy for the control of *Vibrio harveyi* in greenlip abalone (*Haliotis laevigata*). *Aquaculture* **2017**, *473*, 251–258. [CrossRef]
83. Crothers-Stomps, C.; Høj, L.; Bourne, D.G.; Hall, M.R.; Owens, L. Isolation of lytic bacteriophage against *Vibrio harveyi*. *J. Appl. Microbiol.* **2010**, *108*, 1744–1750. [CrossRef] [PubMed]
84. Rong, R.; Lin, H.; Wang, J.; Khan, M.N.; Li, M. Reductions of *Vibrio parahaemolyticus* in oysters after bacteriophage application during depuration. *Aquaculture* **2014**, *418–419*, 171–176. [CrossRef]
85. Lomelí-Ortega, C.O.; Martínez-Díaz, S.F. Phage therapy against *Vibrio parahaemolyticus* infection in the whiteleg shrimp (*Litopenaeus vannamei*) larvae. *Aquaculture* **2014**, *434*, 208–211. [CrossRef]

86. Zhang, J.; Cao, Z.; Li, Z.; Wang, L.; Li, H.; Wu, F.; Jin, L.; Li, X.; Li, S.; Xu, Y. Effect of bacteriophages on *Vibrio alginolyticus* infection in the sea cucumber, *Apostichopus japonicus* (Selenka). *J. World Aquac. Soc.* **2015**, *46*, 149–158. [CrossRef]
87. Li, Z.; Li, X.; Zhang, J.; Wang, X.; Wang, L.; Cao, Z.; Xu, Y. Use of phages to control *Vibrio splendidus* infection in the juvenile sea cucumber *Apostichopus japonicus*. *Fish Shellfish Immunol.* **2016**, *54*, 302–311. [CrossRef] [PubMed]
88. Li, Z.; Zhang, J.; Li, X.; Wang, X.; Cao, Z.; Wang, L.; Xu, Y. Efficiency of a bacteriophage in controlling *Vibrio* infection in the juvenile sea cucumber *Apostichopus japonicus*. *Aquaculture* **2016**, *451*, 345–352. [CrossRef]
89. Higuera, G.; Bastías, R.; Tsertsvadze, G.; Romero, J.; Espejo, R.T. Recently discovered *Vibrio anguillarum* phages can protect against experimentally induced vibriosis in Atlantic salmon, *Salmo salar*. *Aquaculture* **2013**, *392–395*, 128–133. [CrossRef]
90. Silva, Y.J.; Costa, L.; Pereira, C.; Mateus, C.; Cunha, A.; Calado, R.; Gomes, N.C.M.; Pardo, M.A.; Hernandez, I.; Almeida, A. Phage therapy as an approach to prevent *Vibrio anguillarum* infections in fish larvae production. *PLoS ONE* **2014**, *9*, e114197. [CrossRef] [PubMed]
91. Cohen, Y.; Joseph Pollock, F.; Rosenberg, E.; Bourne, D.G. Phage therapy treatment of the coral pathogen *Vibrio coralliilyticus*. *Microbiologyopen* **2013**, *2*, 64–74. [CrossRef] [PubMed]
92. Katharios, P.; Kalatzis, P.G.; Kokkari, C.; Sarropoulou, E.; Middelboe, M. Isolation and characterization of a N4-like lytic bacteriophage infecting *Vibrio splendidus*, a pathogen of fish and bivalves. *PLoS ONE* **2017**, *12*, e0190083. [CrossRef] [PubMed]
93. Mateus, L.; Costa, L.; Silva, Y.J.; Pereira, C.; Cunha, A.; Almeida, A. Efficiency of phage cocktails in the inactivation of *Vibrio* in aquaculture. *Aquaculture* **2014**, *424–425*, 167–173. [CrossRef]
94. Doss, J.; Culbertson, K.; Hahn, D.; Camacho, J.; Barekzi, N. A review of phage therapy against bacterial pathogens of aquatic and terrestrial organisms. *Viruses* **2017**. [CrossRef] [PubMed]
95. Ryan, E.M.; Gorman, S.P.; Donnelly, R.F.; Gilmore, B.F. Recent advances in bacteriophage therapy: How delivery routes, formulation, concentration and timing influence the success of phage therapy. *J. Pharm. Pharmacol.* **2011**, *63*, 1253–1264. [CrossRef] [PubMed]
96. Madsen, L.; Bertelsen, S.K.; Dalsgaard, I.; Middelboe, M. Dispersal and survival of *Flavobacterium psychrophilum* phages *in vivo* in rainbow trout and *in vitro* under laboratory conditions: Implications for their use in phage therapy. *Appl. Environ. Microbiol.* **2013**, *79*, 4853–4861. [CrossRef] [PubMed]
97. Christiansen, R.H.; Dalsgaard, I.; Middelboe, M.; Lauritsen, A.H.; Madsen, L. Detection and quantification of *Flavobacterium psychrophilum*-specific bacteriophages *in vivo* in rainbow trout upon oral administration: Implications for disease control in aquaculture. *Appl. Environ. Microbiol.* **2014**, *80*, 7683–7693. [CrossRef] [PubMed]
98. Nakai, T. Application of bacteriophages for control of infectious diseases in aquaculture. In *Bacteriophages in the Control of Food- and Waterborne Pathogens*; Sabour, P.M., Griffiths, M.W., Eds.; American Society for Microbiology Press: Washington, DC, USA, 2010; pp. 257–272.
99. Sharma, S.; Chatterjee, S.; Datta, S.; Prasad, R.; Dubey, D.; Prasad, R.K.; Vairale, M.G. Bacteriophages and its applications: An overview. *Folia Microbiol. (Praha)* **2017**, *62*, 17–55. [CrossRef] [PubMed]
100. Prol-García, M.J.; Planas, M.; Pintado, J. Different colonization and residence time of *Listonella anguillarum* and *Vibrio splendidus* in the rotifer *Brachionus plicatilis* determined by real-time PCR and DGGE. *Aquaculture* **2010**, *302*, 26–35. [CrossRef]
101. Snoussi, M.; Chaieb, K.; Mahmoud, R.; Bakhrouf, A. Quantitative study, identification and antibiotics sensitivity of some Vibrionaceae associated to a marine fish hatchery. *Ann. Microbiol.* **2006**, *56*, 289–293. [CrossRef]
102. Høj, L.; Bourne, D.G.; Hall, M.R. Localization, abundance and community structure of bacteria associated with *Artemia*: Effects of nauplii enrichment and antimicrobial treatment. *Aquaculture* **2009**, *293*, 278–285. [CrossRef]
103. Dourala, N.; (Fish Health Manager, Selonda S.A., Greece). Personal communication, 2018.
104. Kalatzis, P.G.; Bastías, R.; Kokkari, C.; Katharios, P. Isolation and characterization of two lytic bacteriophages, φSt2 and φGrn1; phage therapy application for biological control of *Vibrio alginolyticus* in aquaculture live feeds. *PLoS ONE* **2016**, e0151101. [CrossRef] [PubMed]

105. Skliros, D.; Kalatzis, P.G.; Katharios, P.; Flemetakis, E. Comparative functional genomic analysis of two *Vibrio* phages reveals complex metabolic interactions with the host cell. *Front. Microbiol* **2016**, *7*, 1–13. [CrossRef] [PubMed]
106. Miller, E.S.; Heidelberg, J.F.; Eisen, J.A.; Nelson, W.C.; Durkin, A.S.; Ciecko, A.; Feldblyum, T.V.; White, O.; Paulsen, I.T.; Nierman, W.C.; et al. Complete genome sequence of the broad host-range vibriophage KVP40: Comparative genomics of a T4-related bacteriophage. *J. Bacteriol.* **2003**, *185*, 5220–5233. [CrossRef] [PubMed]
107. Matsuzaki, S.; Tanaka, S.; Koga, T.; Kawata, T.A. Broad host-range vibriophage, KVP40, isolated from sea water. *Microbiol. Immunol.* **1992**, *36*, 93–97. [CrossRef] [PubMed]
108. Lavigne, R.; Darius, P.; Summer, E.J.; Seto, D.; Mahadevan, P.; Nilsson, A.S.; Ackermann, H.W.; Kropinski, A.M. Classification of Myoviridae bacteriophages using protein sequence similarity. *BMC Microbiol.* **2009**, *9*, 224. [CrossRef] [PubMed]
109. Inoue, T.; Matsuzaki, S.; Tanaka, S. A 26-kDa outer membrane protein, OmpK, common to *Vibrio* species is the receptor for a broad-host-range vibriophage, KVP40. *FEMS Microbiol. Lett.* **1995**, *125*, 101–105. [CrossRef] [PubMed]
110. Drulis-Kawa, Z.; Majkowska-Skrobek, G.; Maciejewska, B.; Delattre, A.-S.; Lavigne, R. Learning from bacteriophages-advantages and limitations of phage and phage-encoded protein applications. *Curr. Protein Pept. Sci.* **2012**, *13*, 699–722. [CrossRef] [PubMed]
111. Chevallereau, A.; Blasdel, B.G.; De Smet, J.; Monot, M.; Zimmermann, M.; Kogadeeva, M.; Sauer, U.; Jorth, P.; Whiteley, M.; Debarbieux, L.; Lavigne, R. Next-Generation "-omics" approaches reveal a massive alteration of host RNA metabolism during bacteriophage infection of *Pseudomonas aeruginosa*. *PLoS Genet.* **2016**, *12*, 1–20. [CrossRef] [PubMed]
112. Roucourt, B.; Lavigne, R. The role of interactions between phage and bacterial proteins within the infected cell: A diverse and puzzling interactome. *Environ. Microbiol.* **2009**, *11*, 2789–2805. [CrossRef] [PubMed]
113. Miller, E.S.; Kutter, E.; Mosig, G.; Kunisawa, T.; Rüger, W.; Arisaka, F.; Ru, W. Bacteriophage T4 genome. *Microbiol. Mol. Biol. Rev.* **2003**, *67*, 86–156. [CrossRef] [PubMed]
114. Lee, J.Y.; Li, Z.; Miller, E.S. *Vibrio* phage KVP40 encodes a functional NAD^+ salvage pathway. *J. Bacteriol.* **2017**, *199*, 1–18. [CrossRef] [PubMed]
115. Bailly-Bechet, M.; Vergassola, M.; Rocha, E. Causes for the intriguing presence of tRNAs in phages. *Genome Res.* **2007**, *17*, 1486–1495. [CrossRef] [PubMed]
116. Smith, H.W.; Huggins, M.B. Successful treatment of experimental *Escherichia coli* infections in mice using phage; its general superiority over antibiotics. *J. Gen. Microbiol.* **1982**, *128*, 307–318. [CrossRef] [PubMed]
117. Barrow, P.A.; Soothill, J.S. Bacteriophage therapy and prophylaxis: Rediscovery and renewed assessment of potential. *Trends Microbiol.* **1997**, *5*, 268–271. [CrossRef]
118. Shephard, K.L. Functions for fish mucus. *Rev. Fish Biol. Fish.* **1994**, *4*, 401–429. [CrossRef]
119. Summers, W.C. Bacteriophage therapy. *Annu. Rev. Microbiol.* **2001**, 437–451. [CrossRef] [PubMed]
120. Gorbet, M.B.; Sefton, M.V. Endotoxin: The uninvited guest. *Biomaterials* **2005**, *26*, 6811–6817. [CrossRef] [PubMed]
121. Opal, S.M. Endotoxins and other sepsis triggers. *Endotoxemia Endotoxin Shock Dis. Diagnosis Ther.* **2010**, *167*, 14–24. [CrossRef]
122. Boratyński, J.; Syper, D.; Weber-Dąbrowska, B.; Łusiak-Szelachowska, M.; Poźniak, G.; Górski, A. Preparation of endotoxin-free bacteriophages. *Cell. Mol. Biol. Lett.* **2004**, *9*, 253–259. [PubMed]
123. Cooper, C.J.; Denyer, S.P.; Maillard, J.Y. Stability and purity of a bacteriophage cocktail preparation for nebulizer delivery. *Lett. Appl. Microbiol.* **2014**, *58*, 118–122. [CrossRef] [PubMed]
124. Szermer-Olearnik, B.; Boratyński, J. Removal of endotoxins from bacteriophage preparations by extraction with organic solvents. *PLoS ONE* **2015**, *10*, e0122672. [CrossRef] [PubMed]
125. Pirisi, A. Phage therapy-advantages over antibiotics? *Lancet* **2000**, *356*, 1418. [CrossRef]
126. Sulakvelidze, A.; Alavidze, Z.; Morris, J.G. Bacteriophage therapy. *Antimicrob. Agents Chemother.* **2001**, *45*, 649–659. [CrossRef] [PubMed]
127. Park, S.C.; Nakai, T. Bacteriophage control of *Pseudomonas plecoglossicida* infection in ayu *Plecoglossus altivelis*. *Dis. Aquat. Organ.* **2003**, *53*, 33–39. [CrossRef] [PubMed]
128. Suttle, C.A. Viruses in the sea. *Nature* **2005**, *437*, 356–361. [CrossRef] [PubMed]
129. Samson, J.E.; Magadán, A.H.; Sabri, M.; Moineau, S. Revenge of the phages: Defeating bacterial defences. *Nat. Rev. Microbiol.* **2013**, *11*, 675–687. [CrossRef] [PubMed]

130. Chan, B.K.; Abedon, S.T.; Loc-carrillo, C. Phage cocktails and the future of phage therapy. *Future Microbiol.* **2013**, 769–783. [CrossRef] [PubMed]
131. Houte, S. van; Buckling, A.; Westra, E.R. Evolutionary ecology of prokaryotic immune mechanisms. *Microbiol. Mol. Biol. Rev.* **2016**, *80*, 745–763. [CrossRef] [PubMed]
132. Westra, E.R.; Swarts, D.C.; Staals, R.H.J.; Jore, M.M.; Brouns, S.J.J.; van der Oost, J. The CRISPRs, they are A-Changin': How prokaryotes generate adaptive immunity. *Annu. Rev. Genet.* **2012**, *46*, 311–339. [CrossRef] [PubMed]
133. Labrie, S.J.; Samson, J.E.; Moineau, S. Bacteriophage resistance mechanisms. *Nat. Rev. Microbiol.* **2010**, *8*, 317–327. [CrossRef] [PubMed]
134. Rakhuba, D.V.; Kolomiets, E.I.; Szwajcer Dey, E.; Novik, G.I. Bacteriophage receptors, mechanisms of phage adsorption and penetration into host cell. *Polish J. Microbiol.* **2010**, *59*, 145–155.
135. Høyland-Kroghsbo, N.M.; Mærkedahl, R.B.; Svenningsen, S.L. A quorum-sensing-induced bacteriophage defense mechanism. *mBio* **2013**, *4*. [CrossRef] [PubMed]
136. Seed, K.D.; Yen, M.; Shapiro, B.J.; Hilaire, I.J.; Charles, R.C.; Teng, J.E.; Ivers, L.C.; Boncy, J.; Harris, J.B.; Camilli, A. Evolutionary consequences of intra-patient phage predation on microbial populations. *eLife* **2014**, *3*, e03497. [CrossRef] [PubMed]
137. Chatterjee, S.; Rothenberg, E. Interaction of bacteriophage λ with Its *E. coli* receptor, LamB. *Viruses* **2012**, *4*, 3162–3178. [CrossRef] [PubMed]
138. Garvey, P.; Hill, C.; Fitzgerald, G.F. The lactococcal plasmid pNP40 encodes a third bacteriophage resistance mechanism, one which affects phage DNA penetration. *Appl. Environ. Microbiol.* **1996**, *62*, 676–679. [PubMed]
139. McGrath, S.; Fitzgerald, G.F.; Van Sinderen, D. Identification and characterization of phage-resistance genes in temperate lactococcal bacteriophages. *Mol. Microbiol.* **2002**, *43*, 509–520. [CrossRef] [PubMed]
140. Pope, W.H.; Jacobs-Sera, D.; Russel, D.A.; Peebles, C.L.; Al-Atrache, Z.; Alcoser, T.A.; Alexander, L.M.; Alfano, M.B.; Alford, S.T.; Amy, N.E.; et al. Expanding the diversity of mycobacteriophages: Insights into genome architecture and evolution. *PLoS ONE* **2011**, *6*. [CrossRef] [PubMed]
141. Kalatzis, P.G.; Rørbo, N.; Castillo, D.; Mauritzen, J.J.; Jørgensen, J.; Kokkari, C.; Zhang, F.; Katharios, P.; Middelboe, M. Stumbling across the same phage: Comparative genomics of widespread temperate phages infecting the fish pathogen *Vibrio anguillarum*. *Viruses* **2017**, *9*. [CrossRef] [PubMed]
142. Shen, X.; Zhang, J.; Xu, J.; Du, P.; Pang, B.; Li, J.; Kan, B. The resistance of *Vibrio cholerae* O1 El Tor strains to the typing phage 919TP, a member of K139 phage family. *Front. Microbiol.* **2016**, *7*, 1–9. [CrossRef] [PubMed]
143. Seed, K.D. Battling Phages: How bacteria defend against viral attack. *PLoS Pathog.* **2015**, *11*, 1–5. [CrossRef] [PubMed]
144. Roberts, R.J.; Belfort, M.; Bestor, T.; Bhagwat, A.S.; Bickle, T.A.; Bitinaite, J.; Blumenthal, R.M.; Degtyarev, S.K.; Dryden, D.T.F.; Dybvig, K.; et al. A nomenclature for restriction enzymes, DNA methyltransferases, homing endonucleases and their genes. *Nucleic Acids Res.* **2003**, *31*, 1805–1812. [CrossRef] [PubMed]
145. Oliveira, P.H.; Touchon, M.; Rocha, E.P.C. The interplay of restriction-modification systems with mobile genetic elements and their prokaryotic hosts. *Nucleic Acids Res.* **2014**, *42*, 10618–10631. [CrossRef] [PubMed]
146. Vasu, K.; Nagaraja, V. Diverse functions of restriction-modification systems in addition to cellular defense. *Microbiol. Mol. Biol. Rev.* **2013**, *77*, 53–72. [CrossRef] [PubMed]
147. Matsuzaki, S.; Inoue, T.; Tanaka, S. Evidence for the existence of a restriction-modification system common to several species of the family Vibrionaceae. *FEMS Microbiol. Lett.* **1992**, *94*, 191–194. [CrossRef]
148. Murphy, J.; Mahony, J.; Ainsworth, S.; Nauta, A.; van Sinderen, D. Bacteriophage orphan DNA methyltransferases: Insights from their bacterial origin, function, and occurrence. *Appl. Environ. Microbiol.* **2013**, *79*, 7547–7555. [CrossRef] [PubMed]
149. Wion, D.; Casadesús, J. N6-methyl-adenine: An epigenetic signal for DNA–protein interactions. *Nat. Rev. Microbiol.* **2006**, *4*, 183–192. [CrossRef] [PubMed]
150. Low, D.A.; Weyand, N.J.; Mahan, M.J. Roles of DNA adenine methylation in regulating bacterial gene expression and virulence. *Infect. Immun.* **2001**, *69*, 7197–7204. [CrossRef] [PubMed]
151. Portillo, F.G.-D.; Pucciarelli, M.G.; Casadesus, J. DNA adenine methylase mutants of *Salmonella typhimurium* show defects in protein secretion, cell invasion, and M cell cytotoxicity. *Proc. Natl. Acad. Sci. USA* **1999**, *96*, 11578–11583. [CrossRef]
152. Oakey, H.J.; Cullen, B.R.; Owens, L. The complete nucleotide sequence of the *Vibrio harveyi* bacteriophage VHML. *J. Appl. Microbiol.* **2002**, *93*, 1089–1098. [CrossRef] [PubMed]

153. Loenen, W.A.M.; Dryden, D.T.F.; Raleigh, E.A.; Wilson, G.G.; Murrayy, N.E. Highlights of the DNA cutters: A short history of the restriction enzymes. *Nucleic Acids Res.* **2014**, *42*, 3–19. [CrossRef] [PubMed]
154. Luria, S.E. Host-induced modifications of viruses. *Cold Spring Harb. Symp. Quant. Biol.* **1953**, *18*, 237–244. [CrossRef] [PubMed]
155. Balado, M.; Lemos, M.L.; Osorio, C.R. Integrating conjugative elements of the SXT/R391 family from fish-isolated vibrios encode restriction-modification systems that confer resistance to bacteriophages. *FEMS Microbiol. Ecol.* **2013**, *83*, 457–467. [CrossRef] [PubMed]
156. Makarova, K.S.; Wolf, Y.I.; Alkhnbashi, O.S.; Costa, F.; Shah, S.A.; Saunders, S.J.; Barrangou, R.; Brouns, S.J.J.; Charpentier, E.; Haft, D.H.; et al. An updated evolutionary classification of CRISPR–Cas systems. *Nat. Rev. Microbiol.* **2015**, *13*, 722–736. [CrossRef] [PubMed]
157. Van der Oost, J.; Jore, M.M.; Westra, E.R.; Lundgren, M.; Brouns, S.J.J. CRISPR-based adaptive and heritable immunity in prokaryotes. *Trends Biochem. Sci.* **2009**, *34*, 401–407. [CrossRef] [PubMed]
158. Deveau, H.; Barrangou, R.; Garneau, J.E.; Labonté, J.; Fremaux, C.; Boyaval, P.; Romero, D.A.; Horvath, P.; Moineau, S. Phage response to CRISPR-encoded resistance in *Streptococcus thermophilus*. *J. Bacteriol.* **2008**, *190*, 1390–1400. [CrossRef] [PubMed]
159. Andersson, A.F.; Banfield, J.F. Virus population dynamics and acquired virus resistance in natural microbial communities. *Science* **2008**, *320*, 1047–1050. [CrossRef] [PubMed]
160. Grissa, I.; Vergnaud, G.; Pourcel, C. CRISPRFinder: A web tool to identify clustered regularly interspace short palindromic repeats. *Nucleic Acids Res.* **2007**, *35*, 52–57. [CrossRef] [PubMed]
161. Jørgensen, J. CRISPR-Cas in the Fish Pathogen *Vibrio anguillarum*. Master's Thesis, University of Copenhagen, Nørregade, Denmark, 2017.
162. Seed, K.D.; Lazinski, D.W.; Calderwood, S.B.; Camilli, A. A bacteriophage encodes its own CRISPR/Cas adaptive response to evade host innate immunity. *Nature* **2013**, *494*, 489–491. [CrossRef] [PubMed]
163. Chopin, M.C.; Chopin, A.; Bidnenko, E. Phage abortive infection in lactococci: Variations on a theme. *Curr. Opin. Microbiol.* **2005**, *8*, 473–479. [CrossRef] [PubMed]
164. Molineux, I.J. Host-parasite interactions: Recent developments in the genetics of abortive phage infections. *New Biol.* **1991**, *3*, 230–236. [PubMed]
165. Snyder, L. Phage-exclusion enzymes: A bonanza of biochemical and cell biology reagents? *Mol. Microbiol.* **1995**, *15*, 415–420. [CrossRef] [PubMed]
166. Chowdhury, R.; Biswas, S.K.; Das, J. Abortive replication of choleraphage phi 149 in *Vibrio cholerae* biotype el tor. *J. Virol.* **1989**, *63*, 392–397. [PubMed]
167. Fineran, P.C.; Blower, T.R.; Foulds, I.J.; Humphreys, D.P.; Lilley, K.S.; Salmond, G.P.C. The phage abortive infection system, ToxIN, functions as a protein-RNA toxin-antitoxin pair. *Proc. Natl. Acad. Sci. USA* **2009**, *106*, 894–899. [CrossRef] [PubMed]
168. Dy, R.L.; Przybilski, R.; Semeijn, K.; Salmond, G.P.C.; Fineran, P.C. A widespread bacteriophage abortive infection system functions through a Type IV toxin-antitoxin mechanism. *Nucleic Acids Res.* **2014**, *42*, 4590–4605. [CrossRef] [PubMed]
169. Iqbal, N.; Guérout, A.M.; Krin, E.; Le Roux, F.; Mazel, D. Comprehensive functional analysis of the 18 *Vibrio cholerae* N16961 toxin-antitoxin systems substantiates their role in stabilizing the superintegron. *J. Bacteriol.* **2015**, *197*, 2150–2159. [CrossRef] [PubMed]
170. Hino, M.; Zhang, J.; Takagi, H.; Miyoshi, T.; Uchiumi, T.; Nakashima, T.; Kakuta, Y.; Kimura, M. Characterization of putative toxin/antitoxin systems in *Vibrio parahaemolyticus*. *J. Appl. Microbiol.* **2014**, *117*, 185–195. [CrossRef] [PubMed]
171. Bohannan, B.J.M.; Kerr, B.; Jessup, C.M.; Hughes, J.B.; Sandvik, G. Trade-offs and coexistence in microbial microcosms. *Antonie van Leeuwenhoek, Int. J. Gen. Mol. Microbiol.* **2002**, *81*, 107–115. [CrossRef]
172. Lenski, R.E.; Levin, B.R. Constraints on the coevolution of bacteria and virulent phage: A model, some experiments, and predictions for natural communities. *Am. Nat.* **1985**, *125*, 585–602. [CrossRef]
173. Middelboe, M.; Holmfeldt, K.; Riemann, L.; Nybroe, O.; Haaber, J. Bacteriophages drive strain diversification in a marine *Flavobacterium*: Implications for phage resistance and physiological properties. *Environ. Microbiol.* **2009**, *11*, 1971–1982. [CrossRef] [PubMed]
174. Middelboe, M. Bacterial growth rate and marine virus–host dynamics. *Microb. Ecol.* **2000**, *40*, 114–124. [CrossRef] [PubMed]

175. Koskella, B.; Lin, D.M.; Buckling, A.; Thompson, J.N. The costs of evolving resistance in heterogeneous parasite environments. *Proc. R. Soc. B Biol. Sci.* **2012**, *279*, 1896–1903. [CrossRef] [PubMed]
176. Castillo, D.; Christiansen, R.H.; Espejo, R.; Middelboe, M. Diversity and geographical distribution of *Flavobacterium psychrophilum* isolates and their phages: Patterns of susceptibility to phage infection and phage host range. *Microb. Ecol.* **2014**, *67*, 748–757. [CrossRef] [PubMed]
177. Marston, M.F.; Pierciey, F.J.; Shepard, A.; Gearin, G.; Qi, J.; Yandava, C.; Schuster, S.C.; Henn, M.R.; Martiny, J.B.H. Rapid diversification of coevolving marine *Synechococcus* and a virus. *Proc. Natl. Acad. Sci. USA* **2012**, *109*, 4544–4549. [CrossRef] [PubMed]
178. Laanto, E.; Bamford, J.K.H.; Laakso, J.; Sundberg, L.R. Phage-driven loss of virulence in a fish pathogenic bacterium. *PLoS ONE* **2012**, *7*. [CrossRef] [PubMed]
179. Castillo, D.; Christiansen, R.H.; Dalsgaard, I.; Madsen, L.; Middelboe, M. Bacteriophage resistance mechanisms in the fish pathogen *Flavobacterium psychrophilum*: Linking genomic mutations to changes in bacterial virulence factors. *Appl. Environ. Microbiol.* **2015**, *81*, 1157–1167. [CrossRef] [PubMed]
180. Wendling, C.C.; Piecyk, A.; Refardt, D.; Chibani, C.; Hertel, R.; Liesegang, H.; Bunk, B.; Overmann, J.; Roth, O. Tripartite species interaction: Eukaryotic hosts suffer more from phage susceptible than from phage resistant bacteria. *BMC Evol. Biol.* **2017**, *17*, 98. [CrossRef] [PubMed]
181. Rørbo, N.; Rønneseth, A.; Kalatzis, P.G.; Barker Rasmussen, B.; Engell-Sørensen, K.; Kleppen, H.P.; Wergeland, H.I.; Gram, L.; Middelboe, M. Potential of phage therapy in preventing *Vibrio anguillarum* infections in cod and turbot larvae. *Antibiotics* **2018**. under review.
182. Westra, E.R.; Van houte, S.; Oyesiku-Blakemore, S.; Makin, B.; Broniewski, J.M.; Best, A.; Bondy-Denomy, J.; Davidson, A.; Boots, M.; Buckling, A. Parasite exposure drives selective evolution of constitutive versus inducible defense. *Curr. Biol.* **2015**, *25*, 1043–1049. [CrossRef] [PubMed]
183. Tollrian, R.; Harvell, D. *The Ecology and Evolution of Inducible Defenses*; Princeton University Press: Princeton, NJ, USA, 1999.
184. Iranzo, J.; Lobkovsky, A.E.; Wolf, Y.I.; Koonin, E.V. Evolutionary dynamics of the prokaryotic adaptive immunity system CRISPR-Cas in an explicit ecological context. *J. Bacteriol.* **2013**, *195*, 3834–3844. [CrossRef] [PubMed]
185. Lundgren, M. Exploring the ecological function of CRISPR-Cas virus defense. *Commun. Integr. Biol.* **2016**, *9*, e1216740. [CrossRef] [PubMed]
186. Stoddard, L.I.; Martiny, J.B.H.; Marston, M.F. Selection and characterization of cyanophage resistance in marine *Synechococcus* strains. *Appl. Environ. Microbiol.* **2007**, *73*, 5516–5522. [CrossRef] [PubMed]
187. Touchon, M.; Moura de Sousa, J.A.; Rocha, E.P. Embracing the enemy: The diversification of microbial gene repertoires by phage-mediated horizontal gene transfer. *Curr. Opin. Microbiol.* **2017**, *38*, 66–73. [CrossRef] [PubMed]
188. Davies, E.V.; Winstanley, C.; Fothergill, J.L.; James, C.E. The role of temperate bacteriophages in bacterial infection. *FEMS Microbiol. Lett.* **2016**, *363*, 1–10. [CrossRef] [PubMed]
189. Matilla, M.A.; Fang, X.; Salmond, G.P. Viunalikeviruses are environmentally common agents of horizontal gene transfer in pathogens and biocontrol bacteria. *ISME J.* **2014**, *8*, 2143–2147. [CrossRef] [PubMed]
190. Billard-Pomares, T.; Fouteau, S.; Jacquet, M.E.; Roche, D.; Barbe, V.; Castellanos, M.; Bouet, J.Y.; Cruveiller, S.; Médigue, C.; Blanco, J.; et al. Characterization of a P1-like bacteriophage carrying an SHV-2 extended-spectrum β-lactamase from an *Escherichia coli* strain. *Antimicrob. Agents Chemother.* **2014**, *58*, 6550–6557. [CrossRef] [PubMed]
191. Goh, S.; Hussain, H.; Chang, B.J.; Emmett, W.; Riley, T.V.; Mullany, P. Phage φC2 mediates transduction of Tn 6215, encoding erythromycin resistance, between *Clostridium difficile* strains. *mBio* **2013**, *4*, 1–7. [CrossRef] [PubMed]
192. Iannelli, F.; Santagati, M.; Santoro, F.; Oggioni, M.R.; Stefani, S.; Pozzi, G. Nucleotide sequence of conjugative prophage Φ1207.3 (formerly Tn1207.3) carrying the mef(A)/msr(D) genes for efflux resistance to macrolides in *Streptococcus pyogenes*. *Front. Microbiol.* **2014**, *5*, 1–7. [CrossRef] [PubMed]
193. Keen, E.C.; Bliskovsky, V.V.; Malagon, F.; Baker, J.D.; Prince, J.S.; Klaus, J.S.; Adhya, S.L. Novel "Superspreader" bacteriophages promote horizontal gene transfer by transformation. *mBio* **2017**, *8*, 1–12. [CrossRef] [PubMed]
194. Waldor, M.K.; Mekalanos, J.J. Lysogenic conversion by a filamentous phage encoding cholera toxin. *Science* **1996**, *272*, 1910–1914. [CrossRef] [PubMed]

195. Vidgen, M.; Carson, J.; Higgins, M.; Owens, L. Changes to the phenotypic profile of *Vibrio harveyi* when infected with the *Vibrio harveyi* myovirus-like (VHML) bacteriophage. *J. Appl. Microbiol.* **2006**, *100*, 481–487. [CrossRef] [PubMed]
196. Paul, J.H. Prophages in marine bacteria: Dangerous molecular time bombs or the key to survival in the seas? *ISME J.* **2008**, *2*, 579–589. [CrossRef] [PubMed]
197. Munro, J.; Oakey, J.; Bromage, E.; Owens, L. Experimental bacteriophage-mediated virulence in strains of *Vibrio harveyi*. *Dis. Aquat. Organ.* **2003**, *54*, 187–194. [CrossRef] [PubMed]
198. Busico-Salcedo, N.; Owens, L. Virulence changes to harveyi clade bacteria infected with bacteriophage from *Vibrio owensii*. *Indian J. Virol.* **2013**, *24*, 180–187. [CrossRef] [PubMed]
199. Castillo, D.; Alvise, P.D.; Xu, R.; Zhang, F.; Middelboe, M.; Gram, L. Comparative genome analyses of *Vibrio anguillarum* strains reveal a link with pathogenicity traits. *mSystems* **2017**, *2*, e00001-17. [CrossRef] [PubMed]
200. Reidl, J.; Mekalanos, J.J. Characterization of *Vibrio cholerae* bacteriophage K139 and use of a novel mini-transposon to identify a phage-encoded virulence factor. *Mol. Microbiol.* **1995**, *18*, 685–701. [CrossRef] [PubMed]
201. Karaolis, D.K.R.; Somara, S.; Maneval, D.R.; Johnson, J.A.; Kaper, J.B. A bacteriophage encoding a pathogenicity island, a type-IV pilus and a phage receptor in cholera bacteria. *Nature* **1999**, *399*, 375–379. [CrossRef] [PubMed]
202. Nedialkova, L.P.; Sidstedt, M.; Koeppel, M.B.; Spriewald, S.; Ring, D.; Gerlach, R.G.; Bossi, L.; Stecher, B. Temperate phages promote colicin-dependent fitness of *Salmonella enterica* serovar Typhimurium. *Environ. Microbiol.* **2016**, *18*, 1591–1603. [CrossRef] [PubMed]
203. Van Raay, K.; Kerr, B. Toxins go viral: Phage-encoded lysis releases group B colicins. *Environ. Microbiol.* **2016**, *18*, 1308–1311. [CrossRef] [PubMed]
204. Bondy-Denomy, J.; Davidson, A.R. When a virus is not a parasite: The beneficial effects of prophages on bacterial fitness. *J. Microbiol.* **2014**, *52*, 235–242. [CrossRef] [PubMed]
205. Chen, Y.; Golding, I.; Sawai, S.; Guo, L.; Cox, E.C. Population fitness and the regulation of *Escherichia coli* genes by bacterial viruses. *PLoS Biol.* **2005**, *3*, 1276–1282. [CrossRef] [PubMed]
206. Campbell, A. Conditions for the existence of bacteriophage. *Evolution* **1961**, *15*, 153–165. [CrossRef]

antibiotics

MDPI

Review

Engineering of Phage-Derived Lytic Enzymes: Improving Their Potential as Antimicrobials

Carlos São-José

Research Institute for Medicines (iMed.ULisboa), Faculty of Pharmacy, Universidade de Lisboa, Av. Prof. Gama Pinto, 1649-003 Lisboa, Portugal; csaojose@ff.ul.pt; Tel.: +351-217-946-420

Received: 2 February 2018; Accepted: 20 March 2018; Published: 22 March 2018

Abstract: Lytic enzymes encoded by bacteriophages have been intensively explored as alternative agents for combating bacterial pathogens in different contexts. The antibacterial character of these enzymes (enzybiotics) results from their degrading activity towards peptidoglycan, an essential component of the bacterial cell wall. In fact, phage lytic products have the capacity to kill target bacteria when added exogenously in the form of recombinant proteins. However, there is also growing recognition that the natural bactericidal activity of these agents can, and sometimes needs to be, substantially improved through manipulation of their functional domains or by equipping them with new functions. In addition, often, native lytic proteins exhibit features that restrict their applicability as effective antibacterials, such as poor solubility or reduced stability. Here, I present an overview of the engineering approaches that can be followed not only to overcome these and other restrictions, but also to generate completely new antibacterial agents with significantly enhanced characteristics. As conventional antibiotics are running short, the remarkable progress in this field opens up the possibility of tailoring efficient enzybiotics to tackle the most menacing bacterial infections.

Keywords: endolysin; lysin; lytic enzyme; peptidoglycan hydrolase; antimicrobial; antibacterial; antibiotic resistance; antimicrobial resistance; bacteriophage

1. Introduction

Alexander Fleming himself was the first to recognize that bacteria could easily develop resistance to penicillin after prolonged exposure to the antibiotic. Since their introduction in clinical practice in the 1940s, antibiotics have been overused and misused both in humans and animals. Over the years this led to the uncontrolled emergence and spread of antibiotic-resistant determinants in almost all bacterial pathogens, some of which are becoming highly refractory to all current antibiotics [1]. As a consequence, we risk entering a post-antibiotic era where we will no longer be able to efficiently treat common bacterial infections [2]. Several international studies anticipate truly catastrophic scenarios on a global scale if effective solutions to tackle antimicrobial resistance are not rapidly found, with tens of million deaths per year and costs ascending to trillions of USD by 2050 [3,4]. This threat, associated with a very limited pipeline of truly new therapies from the pharmaceutical industry [5], has been driving research on alternative antimicrobials, preferentially on those with new modes of action to minimize resistance development. Among the most promising alternatives or complements to conventional antibiotics are phage-derived lytic enzymes [6]. Other phage-encoded enzymes with potential applications as antibacterial weapons will not be discussed in this review. These include, for example, the polysaccharide depolymerases that degrade bacterial capsules, biofilms, and the outer membrane lipopolysaccharide of Gram-negative bacteria [7–10].

Phage lytic enzymes (PLEs) harbor at least one domain responsible for the enzymatic cleavage of peptidoglycan, also known as murein, which is the major structural component of the bacterial cell wall (CW). The peptidoglycan macromolecule forms a sacculus that surrounds the bacterial cytoplasmic membrane (CM) and confers the necessary mechanical resistance to avoid cell lysis as a result of turgor pressure [11]. Therefore, uncontrolled breakdown of the murein structure typically results in osmotic cell lysis. The vast majority of known phages (tailed phages) need to breach the rigid CW barrier in two essential steps of the infection cycle; first, to deliver the virus genetic material into host cells, and then to allow escape of the virion progeny from infected cells. The PLEs that participate in the entry step attack the CW from the outside of the cell (from without). They are transported in the virus particle and can therefore be called virion-associated lysins (VALs) [10,12]. Those responsible for virion release after virus multiplication are synthesized in the host cell during the course of infection. At the appropriate time for lysis to occur, they attack the CW from within and have thus been called endolysins [13]. Although in the literature the term endolysin is very frequently shortened to lysin, this simplification will not be used in this review to avoid confusion. Likewise, VALs are also known as virion-associated peptidoglycan hydrolases (VAPGH), tail-associated muralytic enzymes (TAME), tail-associated lysins (TAL), exolysins, or structural lysins. As argued by Lakta et al. [10] though, virion-associated lysin is probably the best suited designation and will be the one adopted here.

In the natural context of phage infection, the CW peptidoglycan is cleaved by PLEs in a controlled manner, first to avoid compromising host cell viability during phage DNA delivery and then to make sure that lysis does not occur before completion of the replicative cycle (see Section 2 for details). However, when added exogenously to bacteria in the form of purified proteins, the murein-degrading activity of PLEs can lead to rapid osmotic lysis and consequently to cell death. This property has been the basis for the exploration of PLEs as antibacterial agents, as part of a broad group of lytic enzymes defined as "enzybiotics" [14]. The therapeutic potential of endolysins as antibacterial agents has been intensively studied for about twenty years now, both in vitro and in animal models of infection/colonization, with very promising results (for reviews see [15–18]). Since the outer membrane of Gram-negative bacteria most frequently blocks the access of lytic enzymes added from without to the murein layer, the initial stages of the enzybiotics field relied mainly on endolysins targeting Gram-positive pathogens. Recent developments, however, have opened venues for great expansion of the field to Gram-negative pathogens (Section 5.5). Interest in VALs as enzybiotics emerged more recently [10,19], but, as discussed below, they exhibit properties that may confer them some advantages when compared to endolysins. Besides clinical use, PLEs may also find pathogen control applications in other fields such as diagnostics, food industry, and agriculture [20–22].

When employed as antibacterial agents, both VALs and endolysins will have to exert their lytic activity in conditions substantially different from those found in their native context of action. These "unnatural" conditions may negatively impact the performance of these lytic agents. In fact, the activity of VALs at the initial stages of phage infection is most often assisted by other virion components and by dynamic rearrangements of the virus structure. On the other hand, endolysins are naturally designed to attack the CW from within and after the action of a second phage-encoded lysis product: the holin (see below). In a therapeutic perspective, recombinant PLEs will also have to display their killing activity in complex environments such as animal tissues, mucosal membranes, blood and body fluids. In addition to the optimization of killing efficiency, it may be necessary to modify PLEs to ameliorate other properties such as their spectrum of activity, solubility, stability, and half-life in infected hosts. Moreover, the proteinaceous nature of PLEs and their capacity to induce lysis of target bacteria is expected to generate host immune responses, with possible adverse effects, along with the production of antibodies that might neutralize the therapeutic agents. Interestingly, the available studies addressing these potential problems, including several safety studies in humans, reported no serious adverse effects and low impact of anti-PLE antibodies in antimicrobial activity (for a review of immunity and safety issues, and of endolysins in clinical stages, see [23]). Nevertheless, host immune responses must always be considered, and some PLEs may require intervention to improve their therapeutic potential in this regard.

Based on several examples, the following sections will present how PLE features can be improved through protein modification and engineering strategies. Since the rational for several of the PLE tailoring approaches resulted from key knowledge on the mode of action and biochemical characteristics of the lytic enzymes, these aspects we will be covered first.

2. Mode of Action of VALs and Endolysins during Phage Infection

Understanding the biological context and mode of action of PLEs may provide valuable information for their exploration as enzybiotics. VALs are typically associated with the phage DNA injection machinery. Most frequently, they correspond to individual components or to domains of phage tail substructures, like the tape measure protein, central fibers, tail tip knobs, and tail tip puncturing devices, but they can also be capsid inner proteins [10,12,19]. Specific binding of phage tail receptor binding proteins (RBPs) to host cell surface receptors triggers major virion conformational changes that place VALs in close contact with the bacterial CW [24]. This frequently involves the protrusion and insertion in the bacterial cell envelope of tail structures that carry VALs. A landmark example is provided by myoviruses like phage T4 and their contractile tails. Upon tail sheath contraction, the tail tube with a puncturing device at its tip penetrates the cell envelope [25]. One of the proteins composing the piercing apparatus in T4 is gp5, which has muralytic activity [26,27]. Some siphoviruses (long, non-contractile tails) were shown to eject and insert in the cell envelope the internal tail tube formed by the tape measure protein, which may also be endowed with peptidoglycan-cleaving activity [28,29]. After irreversible binding to cell receptors, podoviruses (short tails) like phage T7 eject the proteins composing the capsid inner core to form an extended tail tube that spans the cell envelope. One of the T7 core proteins is gp16 that has CW-degrading activity ([30] and references therein). Other podoviruses like φ29 simply seem to drill through the CW, helped by the peptidoglycan hydrolysis activity of one of the components of the tail tip knob [31]. In conclusion, independently of the mechanism of action, VALs are thought to promote a local digestion of the peptidoglycan structure to allow penetration or extension of the tail tube across the CW and its subsequent fusion to the CM. Tail tube insertion in the CM is then followed by the translocation of the viral DNA into the bacterial cytoplasm [32] (Figure 1a).

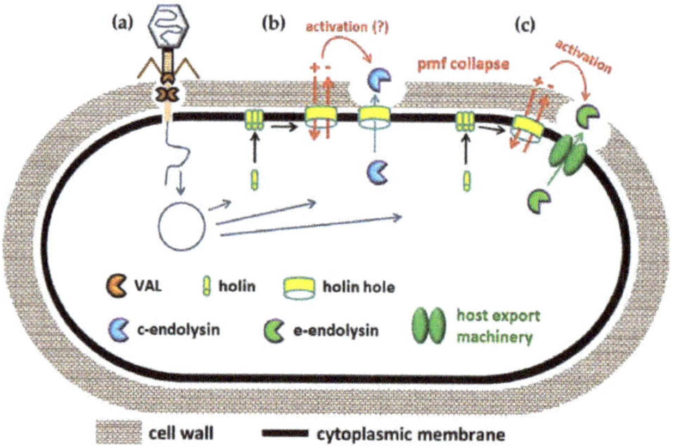

Figure 1. Natural context of action of phage lytic enzymes (PLEs). (**a**) Virion-associated lysins (VALs) promote a local digestion of the cell wall (CW) peptidoglycan to assist penetration of the phage tail tube and passage of the viral DNA to the host cell cytoplasm. After phage genome expression, infected cells must lyse to release the newly-formed virus particles. This is achieved thanks to the peptidoglycan-degrading activity of endolysins; (**b**) Most known endolysins gain access to the CW compartment through the holin channels (c-endolysins); (**c**) Some, however, are exported (e-endolysins) via host cell machineries (e.g., the bacterial Sec system). Holin-mediated dissipation of the cytoplasmic membrane proton-motive force (pmf) is an essential requirement for activation of e-endolysins, while it may also potentiate the lytic activity of c-endolysins (see text).

Endolysins are produced in the cytoplasm of infected cells and need, therefore, to overcome the CM barrier in order to reach the CW and exert their lytic action. These enzymes can be generally classified in canonical endolysins (c-endolysins) or exported endolysins (e-endolysins), depending on the pathway they follow to reach the CW. In the first case, endolysins require the action of a second phage-encoded function, the holin, to be able to escape the cytoplasm. Holins build and oligomerize in the CM, and at a genetically programed time they are triggered to form holes, which cause the lethal collapse of the CM proton-motive force (pmf) [33]. Additionally, these holes are large enough to allow escape of the cytoplasm-accumulated c-endolysin to the CW, which is an essential requirement for occurring lysis. This mechanism defines the so-called canonical lysis model and is prototyped by *Escherichia coli* phage λ [33] (Figure 1b). The endolysins that have been explored as enzybiotics are c-endolysins.

The currently known non-canonical lysis systems employ e-endolysins [13,34,35]. Here, the lytic enzymes are exported to the CW compartment by engaging host cell transport machineries, most frequently the general secretion pathway (the Sec system). E-endolysins start to accumulate in the CW during the viral reproductive cycle, but their lytic activity is inhibited by mechanisms that depend on an energized CM. In fact, although holins do not participate in endolysin channelling to the CW, they still maintain the key role of defining the lysis timing thanks to their pmf-dissipating action, which abolishes the mechanisms that restrain the lytic enzymes (Figure 1c). Interestingly, it was recently shown that energized bacterial cells can also counteract the lytic action of c-endolysins when artificially exported to the CW (from within), or when added from without as recombinant proteins. These results suggest that in canonical lysis, holins may also "activate" endolysins as result of their pmf-collapsing activity [36].

A notion that easily emerges from the comparison of the mode of action of the two types of PLEs is that VALs are "naturally designed" to act on the CW of viable and dividing bacteria, whereas in their natural context both c- and e-endolysins cut the peptidoglycan of cells that are first killed by the holin.

3. Enzymatic Activity of PLEs

The antibacterial potential of PLEs resides in their capacity to cleave different bonds of the CW peptidoglycan network. The murein polymer has as repeating unit a disaccharide made of *N*-acetylglucosamine (NAG) and *N*-acetylmuramic acid (NAM), linked by glycosidic bonds β (1 $\rightarrow$ 4). Neighbouring glycan strands are cross-linked by penta/tetrapeptide side stems that are attached to NAM via amide bonds. The most frequent type of cross-linking involves the amino acid residues at positions 3 (often L-Lys or meso-diaminopimelic acid (m-DAP)) and 4 (D-Ala) of adjacent peptide chains. The peptide stems are interconnected by a direct interpeptide bond in most Gram-negative bacteria and few Gram-positive species, whereas peptide chain cross-linking occurs via an interpeptide bridge in the majority of Gram-positive bacteria. The different peptidoglycan types among bacteria derive mostly from variations within the peptide moiety, notably in the amino acidic composition of the interpeptide bridges [11,37] (Figure 2).

The catalytic domains (CDs) of PLEs, also referred to as enzymatically active domains (EADs), can be classified into three major categories according to their murein cleavage specificity: glycosidases, amidases, and endopeptidases [10,15]. Glycosidases cleave one of the two glycosidic bonds in the glycan chain and can be subdivided into *N*-acetyl-β-D-glucosaminidases (glucosaminidases), *N*-acetyl-β-D-muramidases (muramidases or lysozymes), and lytic transglycosylases (Figure 2). Amidases (*N*-acetylmuramoyl-L-alanine amidases) hydrolyse the amide bond connecting NAM to the first amino acid residue of the peptide stem (generally L-Ala), while endopeptidases cleave within or between the peptide strands (Figure 2). All CDs seem to break the peptidoglycan though a hydrolysis mechanism, except lytic transglycosylases [15].

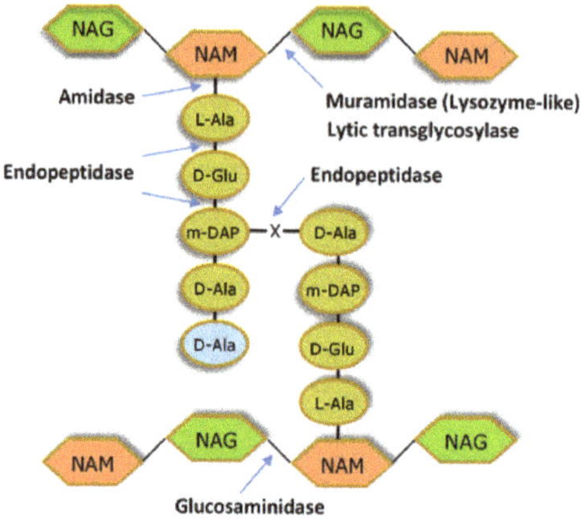

Figure 2. Basic structure of the bacterial cell wall peptidoglycan. The possible enzymatic activities of PLEs and the bonds they cleave are indicated. Typically, PLEs carry one or two catalytic domains displaying one of the indicated enzymatic activities. m-DAP is found in the peptide chains of the peptidoglycan of most Gram-negative bacteria, *Bacillus* spp. and *Listeria* spp., which present also direct m-DAP-D-Ala bonding between adjacent stem peptides. In most Gram-positive bacteria, m-DAP is replaced by L-Lys. Cross-linking between this residue and D-Ala of a neighbor peptide chain usually occurs by an interpeptide bridge of variable amino acidic composition (X). Despite some variation observed among isolates of the same bacterial species, examples of X bridges are $(Gly)_5$ found in *Staphylococcus aureus*, L-Ala-L-Ala in *Enterococcus faecalis* and *Streptococcus pyogenes*, D-Asp in *E. faecium*, and L-Ser-L-Ala in *S. pneumoniae*. The D-Ala residue in light blue may be lost after peptidoglycan maturation.

The most common CDs in known endolysins specify muramidase or amidase activity, whereas VALs seem to carry preferentially glycosidase and endopeptidase activities (the latter almost exclusively present in phages infecting Gram-positive bacteria). The CD may impact the range of activity of a given PLE, depending on whether it targets widely conserved peptidoglycan bonds (such as the glycosidic bonds) or linkages that are specific to particular CW types (like the pentaglycine peptide bridge of staphylococcal CW).

The increasing number of PLE sequences deposited in protein databases has enabled the organization of CDs and CW binding domains (see next) into different families and superfamilies [15,38]. This classification coupled to powerful bioinformatics tools [39] frequently allow the inclusion of PLE CDs into known superfamilies/families and infer about the type of muralytic activity. This analysis, however, should be confirmed experimentally to avoid erroneous assumptions. For example, CDs of the CHAP family most often display endopeptidase activity [40,41], but they can also work as amidases [42] or even exhibit both types of activity [43]. Another interesting observation is that CDs presenting the same murein cleavage specificity can be grouped into distinct families according to their sequence relatedness.

4. Domain Architecture of PLEs

In their simplest form, PLEs are monomeric, globular proteins that essentially correspond to the CD responsible for cleaving a specific peptidoglycan bond. In fact, this simple structure is predominant among endolysins of phages infecting Gram-negative bacteria. In the next step of complexity, endolysins may possess a cell wall binding domain (CWBD or CBD) connected to the CD by a linker segment. The CWBD has high affinity to a particular cell envelope component, and, in general, it contributes positively to the lytic action of endolysins against natural target bacteria. According to the level of conservation of the cell envelope ligand, the CWBD will variably influence the endolysin spectrum of activity. It should be noted, however, that depending on the endolysin and/or target bacteria, the presence of CWBD may or may not be essential for enzybiotic applications (see Section 5.2). Well-known examples of CW binding motifs targeting the murein or other polymeric components of the CW include LysM, SH3 (with different subtypes), the cell binding domains (CBDs) of listerial endolysins, and the choline-binding repeats (ChBRs) of streptococcal endolysins [44–47]. Modular endolysins with an N-terminal CD region coupled to a C-terminal CWBD are typical of phages infecting Gram-positive hosts and mycobacteria [15,48]. The modular character of endolysin CDs and CWBDs was early recognized in pioneer studies of the lytic enzymes of *S. pneumoniae* and its phages, which showed that natural chimeric enzymes are generated through evolution by the interchange of cleavage and binding modules [47,49].

In addition to the two basic domain configurations referred to above, other endolysin architectures exist that essentially vary regarding the presence, number, and relative position of CDs and CWBDs (for reviews on the diversity of endolysin domain architectures see [15,38,50]). For example, endolysins equipped with two CDs that target distinct peptidoglycan bonds or that carry tandem repetitions of CW binding motifs are quite common in Gram-positive systems. Moreover, a considerable number of endolysins with peptidoglycan binding modules at their N-terminus have already been described for phages infecting Gram-negative hosts ([51,52] and references therein). It is also worth noting that some endolysins seem to work as hetero-oligomeres, in which CD-containing subunits associate with several independently produced copies of the CWBD. In the case of the streptococcal multimeric endolysin PlyC, the dual-CD subunit (PlyCA) and the 8-mer CWBD (PlyCB) are expressed from separate genes [42,53]. In the case of the enterococcal Lys170 and of the clostridial CD27L and CPT1L lytic enzymes, a single gene with an internal translation start site generates both the full-length endolysin and independent CWBD subunits, which then associate [54,55]. Figure 3a illustrates the CD and CWBD arrangements found among endolysins that have been studied as antibacterial agents. Besides CD and CWBD modules, most known endolysins involved in non-canonical lysis modes (e-endolysins, see Section 2) are endowed with N-terminal secretion signals such as Sec-type signal peptides or "signal-anchor-release" (SAR) domains [13,35].

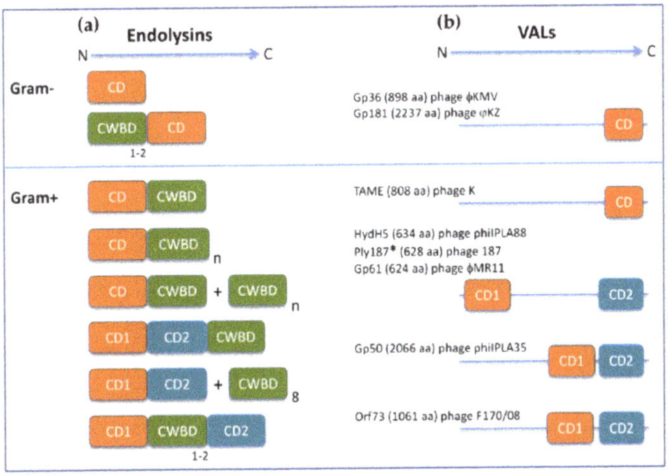

Figure 3. Domain architecture of endolysins (**a**) and VALs (**b**) that have been explored as enzybiotics, from Gram-positive and Gram-negative systems. CD, catalytic domain; CWBD, cell wall binding domain. The cardinals indicate the copy number of CW binding motifs composing the CWBD. The "n" letter indicates that a variable number of CW binding motifs may compose the CWBD (2 to 7 copies). These may be present either as tandem repetitions (in monomeric enzymes) or as oligomers when the CWBD subunit is independently produced by in-frame, alternative start sites (see text). The subunits of hetero-oligomeric endolysins are separated by the "+" sign. The presented VALs are from phages infecting *Pseudomonas aeruginosa* [56,57], *S. aureus* [58–61], and *E. faecalis* [62]. * Ply187 was firstly described as an endolysin. Schemes of phage lytic enzymes are not drawn to scale.

VALs are much less studied than endolysins regarding their biochemical properties, murein degrading activity, and antimicrobial potential. In addition to the function associated with murein cleavage, they may play a role in the morphogenesis and stability of phage virions. VALs tend to be multidomain and much larger than endolysins, and some are found as oligomers in the virus particle. Still, the VAL CDs responsible for cutting peptidoglycan are related to their counterparts in endolysins. In contrast to VALs of phages infecting Gram-negative bacteria, those of Gram-positive systems frequently carry two CDs, with distinct cleavage specificities, and with variable modular organization. The vast majority of known VALs do not exhibit any obvious CWBD under in silico analysis. The presence of this domain is probably dispensable, because VAL/CW contact is guaranteed by the tight interaction between phage tails and host cell surface receptors. The absence of CWBD in VALs constitutes a major difference relative to endolysins (for reviews on VALs features, see [10,19]). Figure 3b presents the domain architecture of some VALs that have been explored (or their CDs) as enzybiotics.

5. Improving the Potential of PLEs as Antibacterials through Protein Engineering

As referred earlier, over the past twenty years numerous studies have supported that native PLEs, particularly endolysins targeting Gram-positive bacteria, are capable of antibacterial activity when added from without. More recently, however, a growing number of reports have described enzybiotics with improved features as result of PLEs modification and engineering of new derivatives. Among the upgraded properties, we can find the enhancement of killing activity against bacteria growing in different conditions and environments, the expansion of the natural bactericidal spectrum, and the extension of PLEs as efficient antibacterial agents towards Gram-negative pathogens. At the production level, different strategies have been followed to increase enzybiotic solubility when heterologously expressed (almost always in *E. coli*) and subsequent stability. Remarkably, as we will see next, several of these improvements may be achieved by a single engineering approach.

The seminal studies demonstrating the modular character of the functional domains of peptidoglycan hydrolases from Gram-positive systems were the basis for the most widely used engineering strategy applied to PLEs, that is, the shuffling and fusion of CDs and CWBDs of different origin to generate chimeric lytic enzymes (also known as chimeolysins). The design of chimeolysins has been much helped by the increasing availability of PLEs tridimensional structures and bioinformatics tools. These normally allow identification and delimitation of enzyme functional domains at the sequence level, and fairly good prediction of their biochemical properties. The native features of PLEs functional domains are usually preserved after being combined to generate certain chimeolysins. This has been explored to tailor enzybiotics with specific properties.

Other tailoring approaches applied to PLEs include fusion of full-length lytic enzymes, domain deletion, addition or duplication, random or site-directed mutagenesis, fusion to peptides, and the combination of few of these [17,23,63]. Examples of PLE engineering or modification strategies and how they enhance specific enzybiotic features are provided next.

5.1. Generation of Chimeolysins with Increased Lytic Spectrum and Activity

The combination of PLEs functional domains of heterologous origin has been extensively used to produce chimeric enzymes with altered properties. Table 1 lists key examples of chimeolysins that clearly exhibited improvements when compared to their parental PLEs or to other lytic enzymes. When a CW binding module recognizes a ligand that is specific to a certain CW type, it might be expected that its fusion to a heterologous CD will retarget the activity of the latter, as long as the peptidoglycan bond matching the CD specificity is available in the new CW [64]. This was well demonstrated in a work with two *Listeria monocytogenes* phage endolysins, Ply118 and PlyPSA, in which individual CD and CWBD modules were swapped or combined to generate fusions with improved capacity to label and lyse *Listeria* cells [65]. Interestingly, one of the fusions (EAD118_III_CBDPSA) showed 3-fold higher activity against the *Listeria* serovar that was naturally targeted by the parental PlyPSA. However, in some cases the chimeolysin may also keep the activity of the parental enzyme that provided the CD, thus resulting in an expansion of the lytic spectrum. This was observed when the endopeptidase CD of the endolysin of the streptococcal prophage λSA2 was fused to the SH3b-type CWBD of either LysK or lysostaphin. LysK is the endolysin of the staphylococcal phage K [66], and this enzyme is known for its strong staphylolytic activity. Lysostaphin is a potent anti-staphylococcal bacteriolysin (exolysin) produced by *Staphylococcus simulans*, which targets the pentaglycine interpeptide bridge of the *S. aureus* CW ([45] and references therein). The endopeptidase-SH3b fusions (λSA2-E-Lyso-SH3b and λSA2-E-LysK-SH3b) exhibited a ~5-fold increase in staphylolytic activity when compared to the parental λSA2 endolysin, while retaining significant streptolytic activity [67]. Although not as effective as lysostaphin, the fusions were bactericidal against *S. aureus* mastitis isolates in processed cow milk and in a mouse model of mastitis [68].

Several chimeric lytic enzymes with improved enzybiotic properties have been obtained by the aleatory exchange of PLE functional modules, normally coupled to high throughput screening methods. Yang et al. [69] developed a strategy to rapidly screen a chimeolysin library containing combinations of different CDs (7 donors) and CWBDs (3 donors). One of the chimeolysins, ClyR, which was particularly active against *S. dysgalactiae* and very stable under different storage conditions, was composed of PlyCAC, the amidase CD of the endolysin PlyC, fused to PlySb, the CWBD of the endolysin PlySs2 (from an *S. suis* prophage). Probably because of the wide binding capacity of PlySb, ClyR showed broader lytic spectrum than PlyC, being active against several streptococcal species (including *S. pneumoniae*), *E. faecalis*, and *S. aureus*. ClyR exhibited also higher lytic activity than other known streptococcal endolysins and was capable of killing mastitis-causing streptococci in pasteurized milk. Moreover, the chimeolysin protected mice from lethal *S. agalactiae* systemic infection and exhibited antibiofilm activity towards *S. mutans* in a murine model of dental colonization [69,70].

Table 1. Examples of PLE engineering through domain shuffling and resulting improvements.

Chimeolysin	CD Source	CWBD Source	Susceptible Bacteria	In Vivo Assay(s)	Outcome	Reference
EAD118_III_CBDPSA	Endopeptidase CD of Ply118 (endolysin *L. monocytogenes* phage A118)	PlyPSA (endolysin *L. monocytogenes* phage PSA)	*L. monocytogenes*	Not reported	3-fold higher activity compared with parental PlyPSA	[65]
λSA2-E-Lyso-SH3b and λSA2-E-LysK-SH3b	Endopeptidase CD of λSA2 (endolysin *S. agalactiae* prophage λSA2)	SH3b-like CWBD of Lysostaphin or of LysK (endolysin of *S. aureus* phage K)	Staphylococci, streptococci	Mouse model of mastitis	Efficient activity extended to *S. aureus* while retaining significant streptolytic activity	[67,68]
ClyR	Amidase CD (CHAP) of PlyC (endolysin streptococcal phage C_1)	PlySs2 (endolysin *S. suis* prophage)	Several streptococcal species (including *S. pneumoniae*), *E. faecalis*, *S. aureus*	Murine models of *S. agalactiae* systemic infection and of *S. mutans* dental colonization	Higher activity and broader lytic spectrum than the parental and other streptococcal endolysins. Stable under storage	[69,70]
Cpl-711	Muramidase CD of Cpl-7S (improved variant of pneumococcal endolysin Cpl-7, see below)	Cpl-1 (endolysin pneumococcal phage Cpl-1)	*S. pneumoniae*, *S. mitis*	Murine bacteraemia model	Greater killing and antibiofilm activity than parental endolysins in vitro. Superior protection compared with Cpl-1 in a mouse model of bacteraemia	[71]
Csl2	Muramidase CD of Cpl-7 (endolysin pneumococcal phage Cp-7)	LySMP (endolysin *S. suis* phage SMP)	*S. suis*, *S. pseudopneumonia*, *S. mitis*, *S. oralis*	Adult zebrafish model of infection	Superior bactericidal and antibiofilm activity than parental LysSMP	[72]
PL3	Amidase CD of Pal (endolysin pneumococcal phage Dp-1)	First two choline-binding repeats of Pal and the last four of LytA (major pneumococcal autolysin)	*S. pneumonia* and other choline-containing streptococci	Zebrafish embryo infection model	Superior bactericidal activity than parental enzymes and high stability	[73]
CHAPSH3b	Endopeptidase CD (CHAP) of HydH5 (VAL *S. aureus* phage philPLA88)	SH3b-like CWBD of lysostaphin (bacteriolysin *S. simulans*)	*S. aureus*, *S. epidermidis*	Not reported	Thermostability. Much higher activity than the parental HydH5	[59,74–76]
P128	Putative endopeptidase (CHAP) of Orf56 (VAL *S. aureus* phage K)	SH3b-like CWBD of lysostaphin (bacteriolysin *S. simulans*)	*S. aureus*, *S. epidermidis*, *S. carnosus*, *S. simulans*	Rat nasal colonization model (*S. aureus* USA 300)	P128 has much higher killing activity than the isolated CHAP CD of Orf56. Effective antibiofilm activity. Better thermostability than lysostaphin	[58,77,78]

Table 1. *Cont.*

Chimeolysin	CD Source	CWBD Source	Susceptible Bacteria	In Vivo Assay(s)	Outcome	Reference
Ply187AN-KSH3b	Putative endopeptidase CD of Ply187 (PLE from *S. aureus* phage 187)	SH3b-like CWBD of LysK (endolysin of *S. aureus* phage K)	*S. aureus* and other staphylococcal species	Mouse model of *S. aureus* Endophthalmitis	More active than native Ply187 and Ply187AN truncated Enzyme. Effective antibiofilm activity	[79,80]
EC300	Putative endopeptidase CD (M23) of Orf73 (putative VAL *E. faecalis* phage F170/08)	Oligomerization-prone CWBD of Lys170 (endolysin *E. faecalis* phage F170/08)	Multidrug-resistant *E. faecalis*, including VRE	Not reported	In contrast to the parental endolysin, EC300 lysis *E. faecalis* actively growing in rich medium	[62]
ClyS	Endopeptidase CD of PlyTW (endolysin *S. aureus* phage Twort)	Endolysin *S. aureus* phage phiNM3 (highly soluble CWBD not related to the very common SH3b)	*S. aureus*, *S. sciuri*, *S. simulans*, *S. epidermidis*	Different murine colonization/infection models (nasal, skin and systemic)	Broad-spectrum activity and high solubility when compared to most staphylococcal endolysins	[81,82]
Lys168-87	Putative endopeptidase CD of Lys168 (endolysin *E. faecalis* phage F168/08)	Putative CWBD of Lys87 (endolysin *S. aureus* phage F87s/06)	Staphylococci, *E. faecalis*, *E. faecium*, *S. pyogenes*	Not reported	High solubility compared to most native PLEs targeting *S. aureus*. Expanded spectrum of activity	[83]
PlyGVE2CpCWB	Amidase CD of PlyGVE2 (endolysin *Geobacillus* phage ΦGVE2)	PlyCP26F (endolysin *C. perfringens* phage ΦCP26F)	*C. perfringens*	Not reported	Better thermostability than parental PlyCP26F	[84]

In contrast to the previous example, the design of some chimeolysins has been rationalized based on previously known features of PLE functional domains. Cpl-711 was constructed by combining the greater affinity of the CW binding module of the pneumococcal endolysin Cpl-1 with the highly active muramidase CD of another pneumococcal endolysin, Cpl-7. The CD source for Cpl-711 construction was in fact Cpl-7S, an artificial variant of Cpl-7 with modifications introduced in its CWBD (see Section 5.2) but carrying an unchanged CD. Compared to the parental endolysins, Cpl-711 exhibited enhanced killing and antibiofilm activity in vitro and superior protection in a mouse model of pneumococcal bacteraemia (compared to Cpl-1) [71]. When this study was published, Cpl-711 was considered the most lethal anti-pneumococcal lytic enzyme with muramidase activity. Using an analogous rational, the very active CD of Cpl-7 was retargeted to the zoonotic pathogen *S. suis* through its fusion with the CWBD of LysMP, the endolysin of *S. suis* phage SMP. The generated chimeolysin, Csl2, displayed superior bactericidal and antibiofilm activity when compared to LysSMP. At the highest dose, Csl2 fully protected adult zebrafish from lethal *S. suis* infection [72].

In a very elegant study employing several endolysins, Low et al. [85] concluded that the enzymes capable of efficient lysis in absence of CWBD (see Section 5.2) carried CDs with intrinsic affinity to the CW. The binding capacity and lytic activity of isolated CDs seemed to be impaired as the negative net charge of their primary structure increased. In accordance to this, Blázquez et al. [73] coupled the less negatively charged CD of the pneumococcal endolysin Pal with the high affinity CWBD of LytA, the major pneumococcal autolysin. The resulting chimera, PL3, killed pneumococci in vitro more efficiently than the parental enzymes, while being also bactericidal against other choline-containing streptococci. In addition, PL3 showed remarkable stability at 37 °C, it resisted lyophilisation and it fully protected zebrafish embryos from *S. pneumoniae* deadly infection. PL3 conferred higher protection than other pneumococcal endolysins in this infection model.

VALs have also been used to generate chimeric lytic enzymes by functioning as an alternative source of CDs. Compared to those derived from endolysins, CDs from VALs may present increased thermostability [56,57,59]. Moreover, by following their native mode of action, VALs may be better prepared to promote lysis from without than endolysins, since the latter are naturally designed to attack the CW from within and after holin-mediated cell death [36,62]. A limitation, however, that may negatively impact VALs lytic performance is the fact that they usually lack CWBD. This has been solved by equipping VALs, or their isolated CDs, with CWBDs of heterologous origin. In fact, Rodríguez-Rubio et al. [74] demonstrated that just by adding the SH3b-type CWBD of lysostaphin to the C-terminus of HydH5, the very thermostable VAL of *S. aureus* phage philPLA88, was sufficient to provoke a marked increase of the staphylolytic activity of the virion lysin (chimeolysin HydH5SH3b). However, an even better chimeolysin was obtained when the same CWBD was fused to the isolated endopeptidase domain (CHAP family) of HydH5 (CD1 in Figure 3b). This chimeric lytic enzyme (CHAPSH3b) paralleled the very potent staphylolytic activity of lysostaphin in several assays. CHAPSH3b showed also the capacity to disrupt and inhibit the formation of *S. aureus* biofilms [76]. Of note, this chimeolysin proved very efficacious in eradicating *S. aureus* present in raw and pasteurized milk [75]. Interestingly, another CHAPSH3b-like chimera recently emerged from a screening of a library of 170 recombinant lytic enzymes that aimed at finding constructs with high killing activity against *S. aureus* in milk. In this case, the CHAP CD was from the endolysin LysK [86].

Another well-studied chimeolysin derived from a VAL is P128, which was built by fusing the putative endopeptidase CD of the tail-associated muralytic enzyme (TAME or Orf56) of phage K to the CWBD of lysostaphin [58]. The presence of this CWBD resulted in more than 100-fold higher bactericidal activity when compared to the isolated CD. P128 could rapidly lyse several staphylococcal species, including a representative panel of typed methicillin-resistant *S. aureus* (MRSA). A P128 concentration of 10 μg/mL was sufficient to reduce the cell counts of *S. aureus* clinical strains between 2 to 4 orders of magnitude (initially at 10^8 CFU/mL). Within the same concentration range, P128 was also efficient at eradicating *S. aureus* biofilms and exhibited much greater thermostability than lysostaphin [77,78].

The putative CHAP-type endopeptidase CD of Ply187 (CD1 in Figure 3b) has been used to generate several chimeolysins through its combination with different CWBDs. It should be noted that Ply187 was initially described as the endolysin of the *S. aureus* phage 187 [87]. Although still referred to as endolysin in some literature, Ply187 is most probably a VAL. In fact, it shares CD organization and significant amino acidic identity with the VALs HydH5 (see above) and gp49 (from *S. aureus* phage phi11) [59,88]. The most probable endolysin of phage 187 is Orf16, a putative 251 aa peptidase lying downstream the putative holin Orf63 [89]. Addition of the SH3b-type CWBD of LysK to the CHAP CD of Ply187 resulted in a 10-fold increment of specific activity, making the chimeolysin, Ply187AN-KSH3b, more potent than LysK in some in vitro assays. As described for some of the chimeric enzymes abovementioned, Ply187AN-KSH3b was also active in milk and exhibited antibiofilm activity [79,80]. Analogous chimeolysins sharing the Ply187 CHAP domain but harboring distinct CWDBs include ClyH, Ply187AN-V12C, and ClyF. ClyH carries the CWBD of the staphylococcal phage phiNM3 endolysin. It displayed higher lytic activity than lysostaphin and than the parental lytic enzymes in vitro, it efficiently degraded MRSA biofilms, and conferred great protection to mice against systemic MRSA infection [90,91]. The V12C CWBD, from the enterococcal endolysin PlyV12, allowed expansion of the Ply187AN lytic activity to different streptococcal and enterococcal species [92]. ClyF emerged from the chimeolysin library that originated ClyR (see above), but in this case after screening for constructs showing high activity towards *S. aureus*. As for ClyR, ClyF carried the CWBD PlySb of the streptococcal endolysin PlySs2, but its lytic spectrum was limited to staphylococcal species (planktonic and biofilm growth). ClyF demonstrated superior staphylolytic activity in different in vitro environments and in murine models of bacteremia and burn wound infection [93]. Moreover, PlySb conferred enhanced thermostability and pH tolerance to the chimeolysin when compared to the isolated Ply187 CHAP domain [93].

Building on the idea that VAL CDs may be better adapted to induce lysis of actively growing bacteria, Proença et al. [62] noticed that several VALs and a few bacteriolysins (like lysostaphin and enterolysin A) shared a CD that is very unusual in endolysins (specifically, the endopeptidase domain of the M23 family). One putative VAL carrying such domain is Orf73 from the *E. faecalis* phage F170/08 (CD1 in Figure 3b). Fusion of the multimerization-prone, high-affinity CWBD of the cognate endolysin Lys170 [54] to the M23 CD of Orf73 generated the chimeolysin EC300 (holoenzyme with ~70 kDa) [62]. Both Lys170 and EC300 could efficiently lyse logarithmic phase *E. faecalis* cells resuspended in a nutrient-depleted buffered solution. However, when the lytic agents were added directly to cells growing in a rich culture medium, growth inhibition and culture lysis was only observed with EC300. The chimeolysin efficacy under growth-promoting conditions was verified for a panel of multidrug-resistant strains, including vancomycin-resistant *E. faecalis* (VRE). In these conditions, the endolysin could only elicit cell lysis if cultures were concomitantly treated with a membrane pmf-dissipating drug. Such inverse correlation between highly energized cells and their susceptibility to endolysins was later observed for other endolysins [36].

Finally, although not very frequent, augmented lytic activity and expanded antibacterial spectrum can be achieved by fusing PLEs (or their CDs) to other full-length lytic enzymes. This was the basis for the generation of the chimeric enzymes B30-443-Lyso and B30-182-Lyso, which combined the *S. agalactiae* phage B30 endolysin, or its endopeptidase domain, respectively, with mature lysostaphin. While the native endolysin lytic range was confined to *Streptococcus* species, the fusions were also active against *S. aureus*. Although with some impact in their lytic performance, the fusions could still lyse *S. agalactiae* and *S. aureus* pathogens in whey [94]. Note that B30-443-Lyso is a triple-CD fusion: two from B30 (CHAP peptidase and muramidase) and one from lysostaphin (M23 peptidase). As we will see in Section 5.4, the emergence of bacterial resistance against enzybiotics with multiple CDs seems to occur at very low levels.

5.2. Other Enginnering Approaches to Expand the Lytic Spectrum and Activity of PLEs

Several alternative strategies to the swapping of functional domains have been followed to obtain PLEs derivatives with improved features (Table 2). After the generation of chimeric lytic enzymes, perhaps the next most frequently used engineering approach applied to PLEs involves domain deletions. It could be expected that any deletion affecting PLE domains directly involved in peptidoglycan binding or cleavage would produce a negative impact on lytic activity. Although this was proven true when the CWBD was removed from certain endolysins targeting *L. monocytogenes*, *B. anthracis*, *S. aureus*, and *S. suis* [95–98], in other cases elimination of one CD (in dual-CD PLEs) and/or CWBDs proved to be either innocuous or to benefit PLE properties. Deletion of the binding domain was reported to produce no major effect on the in vitro lytic activity of the clostridial endolysin CS74L [99], of the streptococcal endolysin B30 [94], and of the staphylococcal endolysin LysK [100]. In fact, the two latter examples correspond to dual-CD endolysins in which the first lytic domain is an endopeptidase and the second a muramidase (B30) or an amidase (LysK). Strikingly, these two endolysins seemed also to support elimination of the second CD along with the CWBD [94,100,101]. The truncated LysK composed exclusively of the N-terminal endopeptidase domain (C165 or $CHAP_k$) was reported to have enhanced activity [100], including against *S. aureus* biofilms [102], and the capacity to eliminate this bacterium from the nares of mice [103]. However, as highlighted by Becker et al. [97], this type of deletion analysis should be taken with care, as discrepant results may arise depending on the methods used to measure lytic activity. This is well illustrated by the LysK dependence on its CWBD for lytic activity, which seems to vary according to the assay conditions [104] More consistent results were obtained concerning the poor contribution of the amidase CD for the in vitro activity of LysK-like endolysins [105–107], and in some case its deletion resulted even in heightened lytic activity [97].

As mentioned above, some endolysin CDs may dispense the presence of a CWBD for cutting efficiently the CW murein, especially if they display a positive net charge [85]. Removal of the CWBD may actually expand the lytic spectrum and activity of endolysins, as shown for the truncated derivatives $PlyL^{CAT}$ and $PlyBa04^{CAT}$ that are lytic against different *Bacillus* species [85,108]. Full-length PlyL only lysed efficiently *B. cereus* and *B. megaterium*, whereas full-length PlyBa04 could only provoke lysis of *B. anthracis* and *B. cereus*. The reasons for the increment of lytic active following CWBD deletion are not fully understood, but probably size reduction may facilitate the truncated lytic enzymes in cutting through the peptidoglycan mesh [85].

Table 2. Other examples of PLE engineering strategies and major outcomes.

Engineering Approach	Example(s) [1]	Engineering Details [1]	Susceptible Bacteria	In Vivo Assay(s)	Outcome	Reference
Fusion to lytic enzymes	B30-443-Lyso B30-182-Lyso	Fusion of *S. agalactiae* phage B30 endolysin (or of its endopeptidase CD) to *S. simulans* lysostaphin	Several streptococcal species, including pathogens and dairy bacteria. *S. aureus*	Not reported	Lytic spectrum extended to *S. aureus* and increased activity (B30-182-Lyso)	[94]
Domain deletion	$CHAP_K$	$CHAP_K$ corresponds to the endopeptidase (CHAP) CD of LysK (first 165 aa of de endolysin of *S. aureus* phage K)	*S. aureus*	*S aureus* elimination in the nares of mice	Higher lytic activity than LysK	[100,102,103]
	$PlyL^{CAT}$ (amidase) $PlyBa04^{CAT}$ (muramidase)	Deletion of the C-ter CWBDs of PlyL and PlyBa04, the endolysins of *B. anthracis* λBa02 Prophage and *B. anthracis* phage Ba04, respectively	*B. cereus*, *B. megaterium*, *B. anthracis*, *B. subtilis*	Not reported	Extended lytic spectrum. Enhanced lytic activity (especially against *B. subtilis* in the case of $PlyL^{CAT}$)	[85,108]
	$CD27L_{1\text{-}179}$ (N-ter amidase CD)	Deletion of the C-ter CWBD of the clostridial endolysin CD27L	*Clostridium* spp. (including *C. difficile*), *Bacillus* spp., *Listeria* spp.	Not reported	Increased lytic activity and spectrum extended to two additional *Listeria* sp.	[109]
	PlyGBS94	PlyGBS94 corresponds to the first 146 aa of native PlyGBS (endolysin *S. agalactiae* phage NCTC 11261), carrying only a endopeptidase CD	Group B streptococci (*Streptococcus agalactiae*)	Not reported	~25-fold increase of specific activity	[110]
	λSa2-ECC	Deletion of C-ter glycosidase CD of λSA2 endolysin (*S. agalactiae* prophage λSA2)	Several streptococcal species and few *S. aureus* strains	Not reported	Increased activity towards certain streptococcal strains and few *S. aureus* strains	[111]
Domain addition	HydH5SH3b	Addition of lysostaphin CWBD (SH3b) to VAL HydH5 of *S. aureus* phage philPLA88	*S. aureus*, *S. epidermidis*	Not reported	Higher activity than the parental HydH5	[74]
Domain duplication	EAD_CBD500-500	Extra copy of CWBD added to Ply500 (endolysin *L. monocytogenes* phage A500)	Essentially *Listeria* spp.	Not reported	Much higher affinity improves endolysin activity at high salt concentrations	[65]
Random mutagenesis	PlyGBS90-1	Frameshift mutation truncates PlyGBS at aa 141 and adds 13 aa	Group B streptococci (*Streptococcus agalactiae*)	Decolonization in a mouse vaginal model	~28-fold increase of specific activity, although less stable than native PlyGBS in certain conditions. Improved killing activity in vivo	[110]
	29C3 mutant of PlyC	Mutation-prone PCR of PlyCA subunit of PlyC (endolysin streptococcal phage C_1)	*S. pyogenes*	Not reported	The 29C3 mutant exhibits higher thermostability than PlyC, which should translate into extended shelf life	[112]

Table 2. *Cont.*

Engineering Approach	Example(s) [1]	Engineering Details [1]	Susceptible Bacteria	In Vivo Assay(s)	Outcome	Reference
Site-directed mutagenesis	Cpl-7S	15 aa substitutions added positive charges to the CWBD of pneumococcal endolysin Cpl-7 (from −14.93 to +3.0 at neutral pH)	*S. pneumoniae*, *E. faecalis*, *S. mitis*, *S. pyogenes*, and, to a lesser extent, *S. dysgalactiae* and *S. iniae*. *E. coli* and *P. putida* in presence of carvacrol	Zebrafish embryo infection model (*S. pneumoniae* and *S. pyogenes*)	Improved killing activity compared to the native Cpl-7 endolysin	[113]
	(PlyC)T406R	T406R substitution in PlyCA subunit of PlyC (endolysin streptococcal phage C_1)	*S. pyogenes*	Not reported	Thermostabilization of PlyC (16-fold increase of half-life at 45 °C), although with moderate loss of lytic activity in vitro	[114]
Multimerization	Cpl-1 dimer	Cpl-$1^{C45S,D324C}$. Introduction of Cys residues at aa position 324 allowed intermolecular disulphide bonding. The C45S substitution avoided unwanted interactions with this Cys residue	*S. pneumoniae*	Not reported	2-fold increase of antipneumococcal activity and ~10-fold decrease in plasma clearance (mice) compared to native Cpl-1	[115]
Mixed approaches	L98WCD27$L_{1\text{-}179}$	Deletion of CD27L C-ter CWBD and L98W mutation in CD27L CD	*Clostridium* spp. (including *C. difficile*), *Bacillus* spp., *Listeria* spp.	Not reported	The L98W mutation further increased lytic activity of CD27$L_{1\text{-}179}$ against *L. monocytogenes*	[109]
	K-L K-L-PTD L-K L-K-PTD (triple-CD PLEs, i.e., 3 distinct CDs)	LysK/Lysostaphin chimeras added or not of protein transduction domains (PTD). K-L: CHAP-Amidase CDs of LysK fused to lysostaphin. L-K: LysK CDs inserted between the CD (M23) and CWBD (SH3b) of lysostaphin	*S. aureus* and coagulase negative staphylococci	Decolonization in rat nasal model. Murine model of mastitis	The presence of 3 distinct CDs in the chimeras reduces emergence of resistant strains. Superior killing activity of L-K in rat nasal model	[116]
	CHAP-Amidase	Codon-optimized CHAP and amidase CDs of LysK (endolysin *S. aureus* phage K) connected by the linker GSH_6GS. No CWBD	*S. aureus*, *S. epidermidis*, *E. faecium*, and *E. faecalis*	Not reported	Enhanced production, stability, and solubility by improving codon-usage and the properties of primary, secondary, and tertiary structures	[117]

[1] N-ter: N-terminal; C-ter: C-terminal.

On the other hand, expansion of the lytic spectrum to other bacterial species might be explained if, in the native enzymes, the CD is inhibited by intramolecular interactions with the CWBD. In this case, relief of this inhibition would require CWBD binding to the cognate CW [108,118].

Elimination of the endolysin CWBD may result in increased activity without changing significantly the lytic spectrum, indicating that specificity to certain bacteria may be an intrinsic feature of some CDs. This was observed for the clostridial endolysin CD27L and its truncation product CD27L$_{1\text{-}179}$, which essentially corresponded to the amidase CD of the enzyme. Although exhibiting enhanced lytic activity, CD27L$_{1\text{-}179}$ basically lysed the same range of bacteria as CD27L, except for two *Listeria* species that could only be lysed by the truncated mutant. Interestingly, the substitution of the conserved residue Leu 98 of the enzyme's CD by its equivalent present in the listerial endolysin PlyPSA (a tryptophan residue) resulted in L98WCD27L$_{1\text{-}179}$, which presented a cumulative augmentation in lysis efficiency towards certain *L. monocytogenes* serovars [109]. Similar results were reported for the truncation product PlyCD$_{1\text{-}174}$ of the clostridial endolysin PlyCD [119].

Cheng and Fischetti [110] subjected PlyGBS, a streptococcal endolysin analogous to B30 (see above), to a random mutagenesis protocol for the isolation of mutants with increased lytic activity against group B streptococci (GBS). A frameshift mutation produced a truncated mutant, PlyGBS90-1, which essentially corresponded to the N-terminal endopeptidase CD (CHAP family). Besides retaining the original lytic spectrum, PlyGBS90-1 exhibited a ~28-fold increase in specific activity when compared to native PlyGBS and also improved decolonization activity in a mouse vaginal model. A similar mutant, PlyGBS94, which resulted from directed elimination of the central muramidase CD and of the putative C-terminal CWBD, gave similar results in vitro [110]. Removal of the low-activity, C-terminal glycosidase CD of the endolysin λSA2 resulted also in increased streptolytic activity, as long as the 2 central CWB motifs were preserved attached to the N-terminal endopeptidase CD (λSa2-ECC truncated product) [111].

Increasing the affinity of PLEs to target cells, by modifying CDs and/or CWBDs, in principle should result in improved lytic efficacy. Schmelcher et al. [65] showed that the listerial endolysin Ply500 equipped with an extra copy of its natural CWBD enhanced affinity by ~50-fold, which translated into greater capacity of the endolysin to act at high salt concentrations.

Despite its very active CD, the native endolysin Cpl-7 presents much lower bacteriolytic activity when compared to other pneumococcal enzybiotics, such as the endolysin Cpl-1 [113]. Díez-Martínez et al. [113] noticed that a major difference between these two lytic enzymes related to their net charge, which was much more negative in Cpl-7 at neutral pH. Negatively charged residues were particularly concentrated in the Cpl-7 CWBD, which is composed of three tandem repeats of the CW_7 binding motif. Inspired again by the work of Low et al. [85], the authors hypothesized that inverting the charge of the binding module could result in increased lytic activity. Five carefully selected amino acid substitutions were introduced per repeat, changing the net charge of the whole CWBD from −14.93 to +3.0. Overall, the mutagenized endolysin, Cpl-7S, demonstrated improved killing capacity against pneumococcal and non-pneumococcal species when compared to the wild-type Cpl-7, paralleling the Cpl-1 bactericidal activity on some pneumococcal stains (note that Cpl-1 is specific to pneumococci). A single dose of Cpl-7S was very effective in protecting from death zebrafish embryos infected with *S. pneumoniae* or *S. pyogenes* [113].

Enhanced lysis/killing of bacteria has also been described as the result of a synergistic effect when combining lytic agents displaying different enzymatic specificities, or when combining CW degrading enzymes (native or engineered) with conventional antibiotics or antimicrobial peptides. The reasons for this synergy are still poorly understood, particularly those resulting from the simultaneous action of lytic agents and antibiotics. It is conceivable that the CW peptidoglycan is more efficiently destroyed when different bonds of its structure are attacked at the same time, an effect that may also be potentiated by the facilitated access of the lytic enzymes to their substrate in these conditions. Most antimicrobial peptides act by damaging the bacterial CM, often accompanied by pmf collapse [120], and, as explained above, this condition has been shown to greatly enhance the lytic action of endolysins. Discussing the strategies relying on the simultaneous use of different antibacterial agents is out of the scope of this review. Nevertheless, they show great promise as they may translate into resensitization of bacteria to current antibiotics and into much lower doses of the individual agents being required in treatments (synergy studies summarized in [23,121]).

5.3. Improving the Production, Solubility, and Stability of PLEs

The engineering strategies used to enhance activity and widen the lytic spectrum of PLEs have also proved useful for ameliorating the heterologous production, solubility, or stability of some potential enzybiotics. A problem that was commonly reported for natural *S. aureus* phage endolysins was their poor solubility when heterologously overproduced. Recent optimizations of protein production and purification conditions allowed overcoming the insolubility problem at least partially. In some cases, however, chimeolysin engineering was followed to obtain highly soluble staphylolytic enzymes. One of the first reported examples was ClyS, a chimeric enzybiotic composed of the endopeptidase CD of PlyTW, the endolysin of *S. aureus* phage Twort, which was fused to the CWBD of the *S. aureus* phage phiNM3 endolysin [81]. In addition to high solubility, probably conferred by the very soluble CWBD, ClyS had a broad lytic spectrum among *Staphylococcus* species and demonstrated efficacy in murine models of MRSA colonization or infection [81,82]. By following an inverse approach, Fernandes et al. [83] successfully produced two chimeolysins by fusing the highly soluble CDs of two different *E. faecalis* endolysins (Lys168 and Lys170) to the CWBD of Lys87, a broadly active staphylococcal endolysin with solubility issues. Remarkably, the resulting chimeras Lys168-87 and Lys170-87 showed not only to be broadly active against a large cohort of *S. aureus* isolates, which included representatives of the most relevant MRSA pandemic clones, but also against other staphylococcal, streptococcal, and enterococcal species.

Engineered chimeolysins may also yield enzybiotics that are more thermostable and therefore that may better preserve activity during storage. Besides the cases of high thermostability (or good stability under different storage conditions) mentioned in the previous sections (see ClyR, PL3, CHAPSH3b, P128, and ClyF), other cases of heat-stable enzybiotics have been described as result of chimeolysin engineering. For example, with the goal of producing a stable enzybiotic to combat *C. perfringens* in poultry, Swift et al. [84] assembled the amidase CD of the thermostable endolysin PlyGVE2 (from *Geobacillus* phage ΦGVE2) with the CWBD of PlyCP26F (endolysin of *C. perfringens* phage ΦCP26F). The produced chimera, PlyGVE2CpCWB, preserved more than 57% of its activity after 30 min at 55 °C, whereas the parental endolysin PlyCP26F was completely inactivated by this treatment.

The presently available bioinformatics tools may also be of great help for optimization of protein heterologous expression. A detailed in silico analysis of the primary, secondary, and tertiary structures of the staphylococcal endolysin LysK allowed precise sequence delimitation of the enzyme's CHAP peptidase and amidase CDs, and elimination of putative protein segments contributing to instability/insolubility [117]. The *E. coli* codon-optimized coding sequences of the two CDs were then connected by an artificial linker (GSH_6GS), which contained the hexahistidine tag for protein purification by affinity chromatography. Moreover, the CHAP-Amidase fusion was N-terminally fused to the signal peptide sequence of PelB, which allowed targeting of the enzybiotic to the *E. coli* periplasm during overproduction. This strategy resulted in high production of a soluble (~12 mg soluble protein per liter of culture), stable, and highly active enzybiotic, with a lytic spectrum covering *S. aureus*, *S. epidermidis*, *E. faecalis*, and *E. faecium* [117].

As described above, the multimeric endolysin PlyC is composed of two subunits, the dual-CD PlyCA and PlyCB, with the latter forming the octameric, ring-shaped CWBD. PlyC loses activity rapidly at 45 °C, which may hint for a reduced shelf life. Interestingly, the poor thermal stability of PlyC is conferred by the CD subunit, since PlyCB resists up to ~90 °C [112]. PlyCA was subjected to a random mutagenesis method based on error-prone PCR, followed by a screening for mutants with enhanced thermostability. One of the selected variants, 29C3, showed more than 2-fold increase in kinetic stability at 45 °C. This translated into better preservation of activity than the native endolysin when tested at different temperatures and incubation periods [112]. In a subsequent study, the same authors performed a computational modelling study to find PlyCA residues that, upon change, would likely result in a ΔG decrease (stabilizing mutations). One of the mutants, PlyC (PlyCA) T406R, was confirmed experimentally to have a denaturation temperature increased by ~2.2 °C and a kinetic stability augmented 16-fold over the wild type PlyC [114].

Many enzybiotics are expected to have a relatively short half-life after reaching the blood stream, since proteins below 45–50 kDa tend to be rapidly cleared from plasma by renal filtration [122]. Aiming at extending the half-life of the endolysin Cpl-1, Resch et al. [115] engineered a dimeric version (Cpl-$1^{C45S,D324C}$) of the lytic enzyme by eliminating and introducing appropriate Cys residues in the primary sequence. Dimerization occurred through intermolecular disulphide bonding involving the Cys residue inserted at position 324, yielding a 74 kDa enzybiotic. The Cpl-1 dimer showed a ~10-fold decrease in plasma clearance (in mice) compared to native Cpl-1, while doubling the specific activity of Cpl-1. Conjugation with non-immunogenic polymers like polyethylene glycol (PEGylation) or polycationic polymers such as poly-L-lysines has proved successful in extending the half-life of many biological molecules, including the lytic enzyme lysostaphin, and in decreasing immunogenicity, proteolysis, and instability [123–125]. PEGylation of Cpl-1, however, resulted in abolishment of the endolysin lytic activity in vitro [126], suggesting some limitations in the application of this strategy with this kind of enzymes. Currently known multimeric PLEs (native or engineered, see above) have a molecular weight above the renal filtration cut-off, meaning that they may be available in circulation for longer time.

5.4. *Minimizing Development of Resistance to PLEs*

The presently available studies support that emergence of resistance to endolysins should occur at much reduced levels when compared to antibiotics and other lytic enzymes like exolysins (studies summarized in [15,127]). It has been hypothesized that endolysins probably evolved to target essential CW components that cannot be easily modified without seriously compromising bacterial fitness, thus ensuring virion release and phage survival at the end of infection [127]. Nevertheless, in a recent study Becker et al. [116] hypothesized that the chances of resistance development could be further reduced if multiple, different CDs were assembled in a single enzybiotic. This was tested using fusions incorporating the two CDs of LysK and the CD of lysostaphin, in two possible configurations, K-L or L-K. In K-L, a LysK segment encompassing its two CDs was fused to the N-terminus of lysostaphin. In the L-K configuration, the same dual-CD segment was inserted between the CD and the SH3b-type CWBD of lysostaphin. The chimeolysins carried thus the endopeptidase (CHAP) and amidase CDs of LysK and the endopeptidase CD (M23) of lysostaphin, while sharing the same C-terminal CWBD (SH3b from the exolysin). Depending on the assay, the triple-CD fusions showed slightly decreased or equivalent in vitro activity when compared to the parental lytic enzymes. However, when the *S. aureus* strain Newman was exposed to sub-lethal doses of the four agents during 10 growth rounds, either in liquid cultures or in solid medium, the emergence of resistance (assessed as the MIC fold-increase) was much lower with the chimeolysins, especially compared to lysostaphin. The fusions were effective as the parental enzymes at eradicating biofilms, whereas the L-K fusion exhibited superior killing activity in a rat nasal model.

In a previous study, Rodríguez-Rubio et al. [128] evaluated the emergence of *S. aureus* resistance to four lytic enzymes: lysostaphin, the LysK-like endolysin LysH5 (from phage philPLA88), and the fusions HydH5Lyso and HydH5SH3b. The latter chimeolysins were generated by adding lysostaphin, or its SH3b CWBD, respectively, to the dual-CD VAL (HydH5) of phage philPLA88 (Figure 3b). While *S. aureus* cells resistant to lysostaphin were rapidly isolated after 1 or 2 growth cycles, none could be detected after exposure to the endolysin or to the chimeolysins in 10 rounds of subculturing, either in liquid or solid medium.

5.5. Enhancing PLEs as Antibacterials towards Gram-Negative Bacteria

As stated above, the outer membrane (OM) of Gram-negative bacteria (and also that of mycobacteria) was long seen as a major obstacle impeding access of exogenously-added PLEs to the murein layer of the CW. Interestingly, in recent years a growing number of endolysins have been described that have some intrinsic capacity to cross the OM (reviewed in [23,52]). Endolysin C-termini that are highly positively charged and/or that form amphipathic helices appear to be common requirements for traversing the Gram-negative OM. In some cases, when produced as synthetic peptides, these C-terminal segments were even shown to be bactericidal [129,130]. Nevertheless, when compared to endolysins of phages infecting Gram-positive bacteria, often much higher concentrations of these OM-crossing lytic enzymes are required for significant cell killing. Therefore, efforts have been made to find agents that could act either attached or in conjunction with PLEs to facilitate their access to the peptidoglycan (reviewed in [23,52]). Regarding protein engineering strategies, two principal types of fusions have been successfully employed to improve endolysin OM penetration (Table 3). In the first case, the lytic enzymes are fused to domains that target them to specific receptor/transport systems of the OM. The second approach relies on the fusion of endolysins to OM-destabilizing peptides, generating the so-called Artilysins® [52].

In the first example of transporter-mediated crossing of the OM, Lukacik et al. [131] fused the N-terminal binding domain of pesticin, a lytic bacteriocin produced by *Yersinia pestis*, to the N-terminus of *E. coli* phage T4 lysozyme. The bacteriocin binding domain specifically targets the OM transporter FyuA, which is responsible for the uptake of the toxin. FyuA is a major virulence factor present in *Y. pestis* and some pathogenic *E. coli* strains. *Y. pestis* produces an immunity protein (Pim) that binds the muramidase CD of pesticin, thus conferring protection against its own pesticin. The authors showed that bacterial cells expressing FyuA were killed by the pesticin-T4 lysozyme hybrid, although not as effectively as with the native bacteriocin. In contrast, the same cells were unaffected by the addition of purified T4 lysozyme. Since lethality was the result of peptidoglycan degradation in the periplasm, the binding domain was considered responsible for FyuA-mediated transport of the fusion across the OM. Most importantly, the fusion could kill Pim-producing cells, because the immunity protein was unable to bind the T4 lysozyme moiety, thereby including *Y. pestis* in its spectrum of activity. An analogous approach was used to promote OM translocation of Lysep3, the endolysin of *E. coli* phage vB_EcoM-ep3. In this case, a protein segment encompassing the translocation and receptor binding domains of colicin A was fused to the N-terminus of Lysep3. In contrast to the endolysin, the colicin-Lysep3 fusion could kill *E. coli* cells, most likely because it was translocated to the periplasm by the TolB machinery after binding to the OM receptor BtuB [132].

Table 3. Fusion of PLEs with domains or peptides that promote crossing of the OM barrier.

Engineering Approach	Example(s)	Engineering Details [1]	Susceptible Bacteria	In Vivo Assay(s)	Outcome	Reference
Fusion to domains targeting OM receptor/transport systems	Pesticin-T4 lysozyme hybrid	Pesticin (bacteriocin) domain targeting FyuA (OMP) fused to the N-ter of *E. coli* phage T4 endolysin	FyuA-expressing pathogenic bacteria (*Y. pestis*, *Y. pseudotuberculosis*, uropathogenic *E. coli*)	-	The hybrid protein crosses the OM through FyuA-mediated transport	[131]
Fusion to domains or peptides that destabilize the OM	LoGT-001 LoGT-008	LoGT-001: PCNP (polycationic nonapeptide) connected to the N-ter of OBPgp279 (endolysin *P. fluorescens* phage OBP) LoGT-008: PCNP connected to the N-ter of PVP-SE1gp146 (endolysin *S. enterica* phage PVP-SE1)	*P. aeruginosa*. Other Artilysins of the LoGT series also killed effectively *A. baumannii* and *E. coli* (≥1 Log reduction). Killing of *S. Typhimurium* required EDTA	*C. elegans* infection assay (LoGT-008)	The PCNP tag increased the intrinsic antibacterial of two modular endolysins (OBPgp279 and PVPSE1gp146) by facilitation OM crossing	[133]
	Art-175	Antimicrobial peptide SMAP-29 fused to the N-ter of mutated KZ144 (endolysin *P. aeruginosa* phage φKZ)	*P. aeruginosa* (and few other *Pseudomonas* spp.), *K. pneumoniae*, *A. baumannii*, colistin-resistant *E. coli*	-	In contrast to KZ144, Art-175 crosses the outer membrane and efficiently kills target cells. Capacity to eliminate *P. aeruginosa* and *A. baumannii* persister cells. Art-175 outcompetes conventional antibiotics in bactericidal activity against *A. baumannii*	[134–136]
	Lysep3-D8	Lysep3 (endolysin *E. coli* phage vB_EcoM-ep3) fused to region D8 of the endolysin of *B. amyloliquefaciens* phage Morita2001	*E. coli*, *P. aeruginosa* (3 strains), *A. baumannii* (1 strain), *Streptococcus* sp. (1 strain)	-	In contrast to isolated Lyse3 and D8, Lysep3-D8 has bactericidal activity	[137]

[1] N-ter: N-terminus.

Despite the elegance of the previous approaches, some potential limitations are foreseen. In order to act, these enzybiotics require the presence of the corresponding receptors in the OM of target cells. Although this selectivity may avoid the collateral damage to the natural microbiota (particularly in the case of pesticin-T4 lysozyme), it may also restrict the application to certain bacterial strains. Moreover, as highlighted by Briers and Lavigne [52], point mutations may easily arise in the receptors/transporters that might impair uptake of the hybrid lytic enzymes. These limitations are less likely to occur with Artilysins.

In the basic Artilysin technology, one peptide with the capacity to destabilize the negatively-charged outer leaflet of the OM, which is essentially composed of lipopolysaccharides (LPS) and some phospholipids, is fused to a given endolysin. Regarding their biochemical properties, the peptides can be polycationic, hydrophobic, or amphipathic, with some actually deriving or mimicking natural antimicrobial peptides (AMPs). Although destabilizing the OM by different mechanisms, all these peptides promote OM crossing of the endolysins without requiring the presence of dedicated cell envelope receptors/transporters [52].

In one of the first studies to produce effective Artilysins, seven different putative OM-destabilizing peptides were fused to the N-terminus of two modular endolysins, OBPgp279 and PVP-SE1gp146 (from *Pseudomonas fluorescens* and *Salmonella enterica* phages, respectively) [133]. Of the 14 possible combinations, those that mostly increased the bactericidal activity of OBPgp279 and PVP-SE1gp146, evaluated against *P. aeruginosa*, were LoGT-001 and LoGT-008, respectively. These Artilysins carried a polycationic nonapeptide (PNCP, aa sequence KRKKRKKRK) fused to the N-terminus of the endolysins. The killing effect of both Artilysins was highly potentiated in presence of 0.5 mM EDTA, a well-known OM permeabilizer. Increasing the distance between PCNP and OBPgp27 with a 16-aa segment of alternating Ala and Gly residues (LoGT-023) further improved the antibacterial effect against *P. aeruginosa* and expanded efficient killing activity to *A. baumannii*. The Artilysin with higher bactericidal activity towards *E. coli* in absence of EDTA was LoGT-037, which carried PCNP and HPP (hydrophobic pentapeptide FFVAP) tandemly fused to the N-terminus of OBPgp279. These results show the power and versatility of the Artilysin technology. One of the Artilysins, LoGT-008, proved its efficacy in rescuing human keratinocytes and *Caenorhabditis elegans* from lethal infection with the highly virulent *P. aeruginosa* strain PA14. The endolysin Lysep3 (see above) could also kill *E. coli* from without after the addition of a 15-aa polycationic peptide to its C-terminus [138].

Art-175 is an example of an Artilysin that was generated by fusing a natural AMP to an endolysin. Specifically, the broad-spectrum sheep myeloid antimicrobial peptide 29 (SMAP-29), a 29-aa α-helical cationic peptide produced by sheep leukocytes, was fused to the endolysin KZ144 of *P. aeruginosa* phage φKZ [134]. Actually, Art-175 carries a mutated version of KZ144, obtained after three Cys to Ser substitutions. This modification impaired oligomerization and conferred structural stability to the Artilysin. In contrast to KZ144 that poorly killed *P. aeruginosa* (~0.5 Log reduction), Art-175 had nearly the same lethal effect as SMAP-29, with almost complete elimination of bacterial cells (>4 Log reduction). Art-175 had a very wide spectrum of activity among environmental and clinical *P. aeruginosa* isolates, including multidrug-resistant strains. Other susceptible pathogens included *Klebsiella pneumoniae* and *Salmonella enterica* serovar Enteritidis. Importantly, the lethal character of Art-175 depended on the lytic action of the KZ144 moiety and did not rely on a potential bactericidal effect of the attached SMAP-29. In fact, the Artilysin had no significant antibacterial activity against *S. aureus*, whereas the isolated SMAP-29 efficiently killed this bacterial species. As expected from the different mode of action of enzybiotics versus antibiotics, Art-175 efficiently eliminated *P. aeruginosa* persister cells, in contrast to ciprofloxacin (both tested at 10 and 30X the MIC) [134].

Remarkably, a subsequent study revealed that Art-175 also displayed potent bactericidal activity against multidrug-resistant *Acinetobacter baumannii*, outcompeting ciprofloxacin and tobramycin in time-kill assays with stationary-phase cultures [135]. As observed for *P. aeruginosa*, persister cells of *A. baumannii* were efficiently eliminated by Art-175. Curiously, the endolysin KZ144 exhibited also antipersister activity, although with about half the efficacy of the Artilysin. Of note, the *P. aeruginosa*

OM lipoprotein OprI was shown to be responsible for the susceptibility of this bacterium to SMAP-29 (as well as to other AMPs), instead of the surface LPS [139]. Mutations in OprI or in the functional homologues in other bacteria could therefore impair Art-175 uptake. However, repeated exposure (20 cycles) of either *P. aeruginosa* or *A. baumannii* to subinhibitory concentrations of Art-175 did not lead to resistance development other than a 2-fold increase of the MIC of some strains. In contrast, similar experiments with ciprofloxacin led to the rapid isolation of highly resistant bacteria [134,135]. Quite recently, Art-175 was also shown to be bactericidal against colistin-resistant, *mcr-1*-positive *E. coli* isolates. This shows that the modifications of the lipid A moiety of LPS responsible for colistin resistance have no impact on the antibacterial activity of Art-175 [136].

One of the endolysins (OBPgp279) that was used to generate the LoGT Artilysin series (see above) was also modified with an AMP-derived peptide, in this case the first eight amino acid residues of cecropin A, which is produced by the cecropia moth [140]. The modified endolysin (PlyA) demonstrated efficient and wide bactericidal activity in vitro against several clinical isolates of *A. baumannii* and *P. aeruginosa*, with the former species being in general more susceptible. However, in contrast to Art-175, PlyA was poorly active against cells in stationary growth phase and required the OM permeabilizer EDTA for effective killing.

As explained above, several endolysins from Gram-negative systems with intrinsic capacity to traverse the OM seem to depend on positively charged and/or amphipathic C-termini for crossing this cell barrier. Interestingly, at least in one case OM-crossing properties were also attributed to the C-terminus of an endolysin naturally designed to act on a Gram-positive bacterium, the endolysin of *Bacillus amyloliquefaciens* phage Morita2001 ([141] and references therein). The Morita2001 endolysin, which is 98% identical to that of the *B. subtilis* phage φ29, is a typical Gram-positive endolysin with an N-terminal lysozyme CD linked to a C-terminal CWBD made of two tandem LysM motifs [38]. In this lytic enzyme, the LysM motifs are rich in positive (mostly lysine) and hydrophobic residues. A deletion analysis study showed that the capacity to penetrate the *P. aeruginosa* OM resided in a C-terminal region denominated D8, which basically encompassed the two LysM motifs [141]. Wang et al. [137] envisaged that D8 could be used to promote the OM translocation of other endolysins. This was tested by fusing D8 to the C-terminus of the endolysin Lysep3 (see above). Differently from the isolated Lysep3 and D8, the fusion Lysep3-D8 was bacteriostatic against several *E. coli* clinical isolates, a reduced number of *P. aeruginosa* and *A. baumannii* strains, and even one *Streptococcus* sp. The bactericidal effect of Lysep3-D8 was confirmed, being particularly obvious for 5 out of 14 *E. coli* isolates (enzybiotic at 100 μg/mL).

In quite an innovative approach, Rodríguez-Rubio et al. [142] decided to test whether the OM-destabilizing peptides used in "Artilysation" of endolysins from Gram-negative systems could also somehow enhance the activity of lytic enzymes directed to Gram-positive bacteria. In this work, the polycationic nonapeptide (PCNP) used to construct the LoGT Artilysin series described above was added to the C-terminus of the streptococcal endolysin λSa2lys (from *S. agalactiae* prophage λSA2). The resulting Artilysin, Art-240, retained the broad anti-streptococcal spectrum of the parental endolysin, but its bactericidal activity was enhanced (killing rate ~2-fold higher). For example, a 2 Log reduction in cellular counts required a 12-fold higher dose of λSa2lys. The superior killing activity of Art-240 was observed over a wide range of pH values and salt concentrations. The authors speculated that the improved killing performance of Art-240 probably resulted from an increased affinity to the streptococcal CW, conferred by the positive amino acid residues of PCNP, since the cell surface is negatively charged due to the presence of anionic teichoic acids.

5.6. Targeting Intracellular Bacteria with PLEs

The use of enzybiotics has been mainly foreseen to combat bacterial pathogens that mostly present an extracellular life style when infecting or colonizing animal hosts, as happens with the species mentioned throughout this review. However, many of these pathogens (e.g., *S. aureus*, *S. pneumoniae*, *S. pyogenes*, *E. coli*) can also have a phase of intracellular inhabitance that can be important for the establishment of infection, to evade antimicrobials and the immune system, or to persist within the host [143]. Somewhat unexpectedly, recent studies have uncovered the capacity of enzybiotics to get inside mammalian cells and kill resident bacteria. The first work used LysK (K), lysostaphin (L), and their fusion derivatives K-L and L-K. These fusions are triple-CD chimeolysins that were described to significantly reduce the chances of emergence of resistant bacteria (see Section 5.4) [116]. The authors assumed that these lytic agents could not enter animal cells. Therefore, following a previous proposal [144], eleven protein transduction domains (PTDs) were individually fused to the C-terminus of the four enzybiotics. PTDs are typically short, highly cationic peptides, some naturally occurring, which promote protein transduction across eukaryotic cell membranes [145]. Curiously, the PTDs only produced the expected effect when added to lysostaphin, with some L-PTDs capable of killing *S. aureus* internalized in three different cell lines. Strikingly, the K-L fusion had the intrinsic capacity to reduce *S. aureus* intracellular counts in two different cells lines, and addition of a PTD did not improve its killing effect. The K-L-PTD1 fusion was nevertheless superior at clearing *S. aureus* in a murine model of mastitis and in biofilm eradication when compared to K-L [116].

The results of the previous work further support that appending cationic peptides to enzybiotics designed to act on Gram-positive bacteria may enhance their antibacterial activity in certain conditions (see Art-240 above). In addition, the study brought to light the possibility of enzybiotic uptake by animal cells. In fact, quite recently Shen et al. [146] discovered that the multimeric streptococcal endolysin PlyC could significantly reduce the counts of *S. pyogenes* cells internalized in different human epithelial cell lines, including primary tonsillar epithelial cells. Two other streptococcal endolysins, B30 and Ply700, failed to significantly reduce intracellular *S. pyogenes* in the same assays. PlyC movement across the membrane of epithelial cells was shown to depend on the PlyCB subunit and on its capacity to bind membrane phospholipids, particularly phosphatidylserine. PlyC uptake depended also on caveolae-mediated endocytosis, and endolysin internalization was not harmful to epithelial cells. These results opened new venues for the enzybiotics field, as it is now attractive to explore the potential of these antibacterial agents as killers of intracellular pathogens. In addition, the identification of enzyme's elements like PlyCB that facilitate transport across animal cell membranes provides new engineering opportunities, as these elements may be used to deliver heterologous CDs or other cargos.

6. Conclusions

Enzybiotics derived from phage lytic products are among the most promising alternatives to fight antibiotic-resistant bacteria. In recent years, there has been great investment in modification and engineering approaches to develop enzybiotics with high therapeutic potential to enter clinical trials. Efforts have been made to obtain products with maximized bactericidal activity in vitro and in vivo, with good coverage of the most relevant clinical strains and with the necessary features for large scale production, formulation, and storage. Among the multiple strategies that have been followed, of particular note is the engineering of chimeolysins, the modification of PLEs net charge, and the Artilysin technology. Chimeolysins are among the most potent enzybiotics produced thus far and have been improved by the recent incorporation of VAL CDs in their design. Several studies point to an increase in the affinity and activity of PLEs when their net charge is progressively shifted from negative to positive, an effect that is explained by the fact that the bacterial cell envelope is usually negatively charged. Artilysation of PLEs was a crucial advancement to make enzybiotics a credible alternative to fight Gram-negative pathogens. Interestingly, there is a good chance that this technology may be adapted to improve also enzybiotics targeting Gram-positive bacteria. Finally, the recent discovery that PLEs can cross the membrane of animal cells and kill residing bacteria opens the possibility of using these agents to target obligatory or facultative intracellular pathogens.

Conflicts of Interest: The author declares no conflict of interest.

References

1. Ventola, C.L. The antibiotic resistance crisis: Part 1: Causes and threats. *Pharm. Ther.* **2015**, *40*, 277–283.
2. WHO. The World Is Running Out of Antibiotics, WHO Report Confirms. 2017. Available online: http://www.who.int/mediacentre/news/releases/2017/running-out-antibiotics/en/ (accessed on 3 January 2018).
3. O'Neill, J. Tackling Drug-Restistant Infections Globally: Final Report and Recommendations. 2016. Available online: https://amr-review.org/sites/default/files/160525_Final%20paper_with%20cover.pdf (accessed on 1 March 2018).
4. Adeyi, O.O.; Baris, E.; Jonas, O.B.; Irwin, A.; Berthe, F.C.J.; Le Gall, F.G.; Marquez, P.V.; Nikolic, I.A.; Plante, C.A.; Schneidman, M.; et al. *Drug-Resistant Infections: A Threat to Our Economic Future*; Final Report; World Bank Group: Washington, DC, USA, 2017.
5. Kmietowicz, Z. Few novel antibiotics in the pipeline, WHO warns. *BMJ* **2017**, *358*, j4339. [CrossRef] [PubMed]
6. Czaplewski, L.; Bax, R.; Clokie, M.; Dawson, M.; Fairhead, H.; Fischetti, V.A.; Foster, S.; Gilmore, B.F.; Hancock, R.E.; Harper, D.; et al. Alternatives to antibiotics-a pipeline portfolio review. *Lancet Infect. Dis.* **2016**, *16*, 239–251. [CrossRef]
7. Yan, J.; Mao, J.; Xie, J. Bacteriophage polysaccharide depolymerases and biomedical applications. *BioDrugs* **2014**, *28*, 265–274. [CrossRef] [PubMed]
8. Drulis-Kawa, Z.; Majkowska-Skrobek, G.; Maciejewska, B. Bacteriophages and phage-derived proteins—Application approaches. *Curr. Med. Chem.* **2015**, *22*, 1757–1773. [CrossRef] [PubMed]
9. Pires, D.P.; Oliveira, H.; Melo, L.D.; Sillankorva, S.; Azeredo, J. Bacteriophage-encoded depolymerases: Their diversity and biotechnological applications. *Appl. Microbiol. Biotechnol.* **2016**, *100*, 2141–2151. [CrossRef] [PubMed]
10. Latka, A.; Maciejewska, B.; Majkowska-Skrobek, G.; Briers, Y.; Drulis-Kawa, Z. Bacteriophage-encoded virion-associated enzymes to overcome the carbohydrate barriers during the infection process. *Appl. Microbiol. Biotechnol.* **2017**, *101*, 3103–3119. [CrossRef] [PubMed]
11. Vollmer, W.; Blanot, D.; de Pedro, M.A. Peptidoglycan structure and architecture. *FEMS Microbiol. Rev.* **2008**, *32*, 149–167. [CrossRef] [PubMed]
12. Moak, M.; Molineux, I.J. Peptidoglycan hydrolytic activities associated with bacteriophage virions. *Mol. Microbiol.* **2004**, *51*, 1169–1183. [CrossRef] [PubMed]
13. Catalão, M.J.; Gil, F.; Moniz-Pereira, J.; São-José, C.; Pimentel, M. Diversity in bacterial lysis systems: Bacteriophages show the way. *FEMS Microbiol. Rev.* **2013**, *37*, 554–571. [CrossRef] [PubMed]
14. Nelson, D.; Loomis, L.; Fischetti, V.A. Prevention and elimination of upper respiratory colonization of mice by group a streptococci by using a bacteriophage lytic enzyme. *Proc. Natl. Acad. Sci. USA* **2001**, *98*, 4107–4112. [CrossRef] [PubMed]
15. Nelson, D.C.; Schmelcher, M.; Rodriguez-Rubio, L.; Klumpp, J.; Pritchard, D.G.; Dong, S.; Donovan, D.M. Endolysins as antimicrobials. *Adv. Virus Res.* **2012**, *83*, 299–365. [CrossRef] [PubMed]
16. Pastagia, M.; Schuch, R.; Fischetti, V.A.; Huang, D.B. Lysins: The arrival of pathogen-directed anti-infectives. *J. Med. Microbiol.* **2013**, *62*, 1506–1516. [CrossRef] [PubMed]
17. Roach, D.R.; Donovan, D.M. Antimicrobial bacteriophage-derived proteins and therapeutic applications. *Bacteriophage* **2015**, *5*, e1062590. [CrossRef] [PubMed]
18. Haddad Kashani, H.; Schmelcher, M.; Sabzalipoor, H.; Seyed Hosseini, E.; Moniri, R. Recombinant endolysins as potential therapeutics against antibiotic-resistant *Staphylococcus aureus*: Current status of research and novel delivery strategies. *Clin. Microbiol. Rev.* **2018**, *31*. [CrossRef] [PubMed]
19. Rodríguez-Rubio, L.; Martínez, B.; Donovan, D.M.; Rodríguez, A.; García, P. Bacteriophage virion-associated peptidoglycan hydrolases: Potential new enzybiotics. *Crit. Rev. Microbiol.* **2013**, *39*, 427–434. [CrossRef] [PubMed]
20. Oliveira, H.; Azeredo, J.; Lavigne, R.; Kluskens, L.D. Bacteriophage endolysins as a response to emerging foodborne pathogens. *Trends Food Sci. Technol.* **2012**, *28*, 103–115. [CrossRef]
21. Schmelcher, M.; Loessner, M.J. Bacteriophage endolysins: Applications for food safety. *Curr. Opin. Biotechnol.* **2016**, *37*, 76–87. [CrossRef] [PubMed]

22. Rodríguez-Rubio, L.; Gutiérrez, D.; Donovan, D.M.; Martínez, B.; Rodríguez, A.; García, P. Phage lytic proteins: Biotechnological applications beyond clinical antimicrobials. *Crit. Rev. Biotechnol.* **2016**, *36*, 542–552. [CrossRef] [PubMed]
23. Gerstmans, H.; Criel, B.; Briers, Y. Synthetic biology of modular endolysins. *Biotechnol. Adv.* **2017**. [CrossRef] [PubMed]
24. Fokine, A.; Rossmann, M.G. Molecular architecture of tailed double-stranded DNA phages. *Bacteriophage* **2014**, *4*, e28281. [CrossRef] [PubMed]
25. Hu, B.; Margolin, W.; Molineux, I.J.; Liu, J. Structural remodeling of bacteriophage t4 and host membranes during infection initiation. *Proc. Natl. Acad. Sci. USA* **2015**, *112*, E4919–E4928. [CrossRef] [PubMed]
26. Leiman, P.G.; Shneider, M.M. Contractile tail machines of bacteriophages. *Adv. Exp. Med. Biol.* **2012**, *726*, 93–114. [CrossRef] [PubMed]
27. Taylor, N.M.I.; van Raaij, M.J.; Leiman, P.G. Contractile injection systems of bacteriophages and related systems. *Mol. Microbiol.* **2018**. [CrossRef] [PubMed]
28. Davidson, A.R.; Cardarelli, L.; Pell, L.G.; Radford, D.R.; Maxwell, K.L. Long noncontractile tail machines of bacteriophages. *Adv. Exp. Med. Biol.* **2012**, *726*, 115–142. [CrossRef] [PubMed]
29. Cumby, N.; Reimer, K.; Mengin-Lecreulx, D.; Davidson, A.R.; Maxwell, K.L. The phage tail tape measure protein, an inner membrane protein and a periplasmic chaperone play connected roles in the genome injection process of *E. coli* phage HK97. *Mol. Microbiol.* **2015**, *96*, 437–447. [CrossRef] [PubMed]
30. Hu, B.; Margolin, W.; Molineux, I.J.; Liu, J. The bacteriophage T7 virion undergoes extensive structural remodeling during infection. *Science* **2013**, *339*, 576–579. [CrossRef] [PubMed]
31. Xiang, Y.; Morais, M.C.; Cohen, D.N.; Bowman, V.D.; Anderson, D.L.; Rossmann, M.G. Crystal and cryoem structural studies of a cell wall degrading enzyme in the bacteriophage phi29 tail. *Proc. Natl. Acad. Sci. USA* **2008**, *105*, 9552–9557. [CrossRef] [PubMed]
32. Xu, J.; Xiang, Y. Membrane penetration by bacterial viruses. *J. Virol.* **2017**, *91*. [CrossRef] [PubMed]
33. Young, R. Phage lysis: Do we have the hole story yet? *Curr. Opin. Microbiol.* **2013**, *16*, 790–797. [CrossRef] [PubMed]
34. Frias, M.J.; Melo-Cristino, J.; Ramirez, M. Export of the pneumococcal phage SV1 lysin requires choline-containing teichoic acids and is holin-independent. *Mol. Microbiol.* **2013**, *87*, 430–445. [CrossRef] [PubMed]
35. Young, R. Phage lysis: Three steps, three choices, one outcome. *J. Microbiol.* **2014**, *52*, 243–258. [CrossRef] [PubMed]
36. Fernandes, S.; São-José, C. More than a hole: The holin lethal function may be required to fully sensitize bacteria to the lytic action of canonical endolysins. *Mol. Microbiol.* **2016**, *102*, 92–106. [CrossRef] [PubMed]
37. Labischinski, H.; Maidhof, H. Bacterial peptidoglycan: Overview and evolving concepts. In *Bacterial Cell Wall*; Ghuysen, J.M., Hakenbeck, R., Eds.; Elsevier: Amsterdam, The Netherlands, 1994; pp. 23–38. [CrossRef]
38. Oliveira, H.; Melo, L.D.; Santos, S.B.; Nóbrega, F.L.; Ferreira, E.C.; Cerca, N.; Azeredo, J.; Kluskens, L.D. Molecular aspects and comparative genomics of bacteriophage endolysins. *J. Virol.* **2013**, *87*, 4558–4570. [CrossRef] [PubMed]
39. Marchler-Bauer, A.; Lu, S.; Anderson, J.B.; Chitsaz, F.; Derbyshire, M.K.; DeWeese-Scott, C.; Fong, J.H.; Geer, L.Y.; Geer, R.C.; Gonzales, N.R.; et al. CDD: A conserved domain database for the functional annotation of proteins. *Nucleic Acids Res.* **2011**, *39*, D225–D229. [CrossRef] [PubMed]
40. Navarre, W.W.; Ton-That, H.; Faull, K.F.; Schneewind, O. Multiple enzymatic activities of the murein hydrolase from staphylococcal phage φ11. Identification of a D-alanyl-glycine endopeptidase activity. *J. Biol. Chem.* **1999**, *274*, 15847–15856. [CrossRef] [PubMed]
41. Pritchard, D.G.; Dong, S.; Baker, J.R.; Engler, J.A. The bifunctional peptidoglycan lysin of *Streptococcus agalactiae* bacteriophage B30. *Microbiology* **2004**, *150*, 2079–2087. [CrossRef] [PubMed]
42. Nelson, D.; Schuch, R.; Chahales, P.; Zhu, S.; Fischetti, V.A. Plyc: A multimeric bacteriophage lysin. *Proc. Natl. Acad. Sci. USA* **2006**, *103*, 10765–10770. [CrossRef] [PubMed]
43. Linden, S.B.; Zhang, H.; Heselpoth, R.D.; Shen, Y.; Schmelcher, M.; Eichenseher, F.; Nelson, D.C. Biochemical and biophysical characterization of PlyGRCS, a bacteriophage endolysin active against methicillin-resistant *Staphylococcus aureus*. *Appl. Microbiol. Biotechnol.* **2015**, *99*, 741–752. [CrossRef] [PubMed]

44. Mesnage, S.; Dellarole, M.; Baxter, N.J.; Rouget, J.B.; Dimitrov, J.D.; Wang, N.; Fujimoto, Y.; Hounslow, A.M.; Lacroix-Desmazes, S.; Fukase, K.; et al. Molecular basis for bacterial peptidoglycan recognition by LysM domains. *Nat. Commun.* **2014**, *5*, 4269. [CrossRef] [PubMed]
45. Gründling, A.; Schneewind, O. Cross-linked peptidoglycan mediates lysostaphin binding to the cell wall envelope of *Staphylococcus aureus*. *J. Bacteriol.* **2006**, *188*, 2463–2472. [CrossRef] [PubMed]
46. Eugster, M.R.; Haug, M.C.; Huwiler, S.G.; Loessner, M.J. The cell wall binding domain of listeria bacteriophage endolysin PlyP35 recognizes terminal GlcNac residues in cell wall teichoic acid. *Mol. Microbiol.* **2011**, *81*, 1419–1432. [CrossRef] [PubMed]
47. López, R.; García, E.; García, P.; García, J.L. The pneumococcal cell wall degrading enzymes: A modular design to create new lysins? *Microb. Drug Resist.* **1997**, *3*, 199–211. [CrossRef] [PubMed]
48. Payne, K.M.; Hatfull, G.F. Mycobacteriophage endolysins: Diverse and modular enzymes with multiple catalytic activities. *PLoS ONE* **2012**, *7*, e34052. [CrossRef] [PubMed]
49. Díaz, E.; López, R.; García, J.L. Chimeric phage-bacterial enzymes: A clue to the modular evolution of genes. *Proc. Natl. Acad. Sci. USA* **1990**, *87*, 8125–8129. [CrossRef] [PubMed]
50. Schmelcher, M.; Donovan, D.M.; Loessner, M.J. Bacteriophage endolysins as novel antimicrobials. *Future Microbiol.* **2012**, *7*, 1147–1171. [CrossRef] [PubMed]
51. Rodríguez-Rubio, L.; Gerstmans, H.; Thorpe, S.; Mesnage, S.; Lavigne, R.; Briers, Y. DUF3380 domain from a salmonella phage endolysin shows potent *N*-Acetylmuramidase activity. *Appl. Environ. Microbiol.* **2016**, *82*, 4975–4981. [CrossRef] [PubMed]
52. Briers, Y.; Lavigne, R. Breaking barriers: Expansion of the use of endolysins as novel antibacterials against Gram-negative bacteria. *Future Microbiol.* **2015**, *10*, 377–390. [CrossRef] [PubMed]
53. McGowan, S.; Buckle, A.M.; Mitchell, M.S.; Hoopes, J.T.; Gallagher, D.T.; Heselpoth, R.D.; Shen, Y.; Reboul, C.F.; Law, R.H.; Fischetti, V.A.; et al. X-ray crystal structure of the streptococcal specific phage lysin PlyC. *Proc. Natl. Acad. Sci. USA* **2012**, *109*, 12752–12757. [CrossRef] [PubMed]
54. Proença, D.; Velours, C.; Leandro, C.; Garcia, M.; Pimentel, M.; São-José, C. A two-component, multimeric endolysin encoded by a single gene. *Mol. Microbiol.* **2015**, *95*, 739–753. [CrossRef] [PubMed]
55. Dunne, M.; Leicht, S.; Krichel, B.; Mertens, H.D.; Thompson, A.; Krijgsveld, J.; Svergun, D.I.; Gómez-Torres, N.; Garde, S.; Uetrecht, C.; et al. Crystal structure of the CTP1L endolysin reveals how its activity is regulated by a secondary translation product. *J. Biol. Chem.* **2016**, *291*, 4882–4893. [CrossRef] [PubMed]
56. Lavigne, R.; Briers, Y.; Hertveldt, K.; Robben, J.; Volckaert, G. Identification and characterization of a highly thermostable bacteriophage lysozyme. *Cell. Mol. Life Sci.* **2004**, *61*, 2753–2759. [CrossRef] [PubMed]
57. Briers, Y.; Miroshnikov, K.; Chertkov, O.; Nekrasov, A.; Mesyanzhinov, V.; Volckaert, G.; Lavigne, R. The structural peptidoglycan hydrolase gp181 of bacteriophage φkz. *Biochem. Biophys. Res. Commun.* **2008**, *374*, 747–751. [CrossRef] [PubMed]
58. Paul, V.D.; Rajagopalan, S.S.; Sundarrajan, S.; George, S.E.; Asrani, J.Y.; Pillai, R.; Chikkamadaiah, R.; Durgaiah, M.; Sriram, B.; Padmanabhan, S. A novel bacteriophage tail-associated muralytic enzyme (TAME) from phage K and its development into a potent antistaphylococcal protein. *BMC Microbiol.* **2011**, *11*, 226. [CrossRef] [PubMed]
59. Rodríguez, L.; Martínez, B.; Zhou, Y.; Rodríguez, A.; Donovan, D.M.; García, P. Lytic activity of the virion-associated peptidoglycan hydrolase HydH5 of *Staphylococcus aureus* bacteriophage vB_SauS-phiIPLA88. *BMC Microbiol.* **2011**, *11*, 138. [CrossRef] [PubMed]
60. Rashel, M.; Uchiyama, J.; Takemura, I.; Hoshiba, H.; Ujihara, T.; Takatsuji, H.; Honke, K.; Matsuzaki, S. Tail-associated structural protein gp61 of *Staphylococcus aureus* phage φMR11 has bifunctional lytic activity. *FEMS Microbiol. Lett.* **2008**, *284*, 9–16. [CrossRef] [PubMed]
61. Rodríguez-Rubio, L.; Gutiérrez, D.; Martínez, B.; Rodríguez, A.; Götz, F.; García, P. The tape measure protein of the *Staphylococcus aureus* bacteriophage vB_SauS-phiIPLA35 has an active muramidase domain. *Appl. Environ. Microbiol.* **2012**, *78*, 6369–6371. [CrossRef] [PubMed]
62. Proença, D.; Leandro, C.; Garcia, M.; Pimentel, M.; São-José, C. EC300: A phage-based, bacteriolysin-like protein with enhanced antibacterial activity against *Enterococcus faecalis*. *Appl. Microbiol. Biotechnol.* **2015**, *99*, 5137–5149. [CrossRef] [PubMed]
63. Yang, H.; Yu, J.; Wei, H. Engineered bacteriophage lysins as novel anti-infectives. *Front. Microbiol.* **2014**, *5*, 542. [CrossRef] [PubMed]

64. Croux, C.; Ronda, C.; López, R.; García, J.L. Interchange of functional domains switches enzyme specificity: Construction of a chimeric pneumococcal-clostridial cell wall lytic enzyme. *Mol. Microbiol.* **1993**, *9*, 1019–1025. [CrossRef] [PubMed]
65. Schmelcher, M.; Tchang, V.S.; Loessner, M.J. Domain shuffling and module engineering of *Listeria* phage endolysins for enhanced lytic activity and binding affinity. *Microb. Biotechnol.* **2011**, *4*, 651–662. [CrossRef] [PubMed]
66. O'Flaherty, S.; Coffey, A.; Meaney, W.; Fitzgerald, G.F.; Ross, R.P. The recombinant phage lysin LysK has a broad spectrum of lytic activity against clinically relevant staphylococci, including methicillin-resistant *Staphylococcus aureus*. *J. Bacteriol.* **2005**, *187*, 7161–7164. [CrossRef] [PubMed]
67. Becker, S.C.; Foster-Frey, J.; Stodola, A.J.; Anacker, D.; Donovan, D.M. Differentially conserved staphylococcal SH3b_5 cell wall binding domains confer increased staphylolytic and streptolytic activity to a streptococcal prophage endolysin domain. *Gene* **2009**, *443*, 32–41. [CrossRef] [PubMed]
68. Schmelcher, M.; Powell, A.M.; Becker, S.C.; Camp, M.J.; Donovan, D.M. Chimeric phage lysins act synergistically with lysostaphin to kill mastitis-causing *Staphylococcus aureus* in murine mammary glands. *Appl. Environ. Microbiol.* **2012**, *78*, 2297–2305. [CrossRef] [PubMed]
69. Yang, H.; Linden, S.B.; Wang, J.; Yu, J.; Nelson, D.C.; Wei, H. A chimeolysin with extended-spectrum streptococcal host range found by an induced lysis-based rapid screening method. *Sci. Rep.* **2015**, *5*, 17257. [CrossRef] [PubMed]
70. Yang, H.; Bi, Y.; Shang, X.; Wang, M.; Linden, S.B.; Li, Y.; Nelson, D.C.; Wei, H. Antibiofilm activities of a novel chimeolysin against streptococcus mutans under physiological and cariogenic conditions. *Antimicrob. Agents Chemother.* **2016**, *60*, 7436–7443. [CrossRef] [PubMed]
71. Díez-Martínez, R.; De Paz, H.D.; García-Fernández, E.; Bustamante, N.; Euler, C.W.; Fischetti, V.A.; Menendez, M.; García, P. A novel chimeric phage lysin with high in vitro and in vivo bactericidal activity against *Streptococcus pneumoniae*. *J. Antimicrob. Chemother.* **2015**, *70*, 1763–1773. [CrossRef] [PubMed]
72. Vázquez, R.; Domenech, M.; Iglesias-Bexiga, M.; Menéndez, M.; García, P. Csl2, a novel chimeric bacteriophage lysin to fight infections caused by *Streptococcus suis*, an emerging zoonotic pathogen. *Sci. Rep.* **2017**, *7*, 16506. [CrossRef] [PubMed]
73. Blázquez, B.; Fresco-Taboada, A.; Iglesias-Bexiga, M.; Menéndez, M.; García, P. PL3 amidase, a tailor-made lysin constructed by domain shuffling with potent killing activity against pneumococci and related species. *Front. Microbiol.* **2016**, *7*, 1156. [CrossRef] [PubMed]
74. Rodríguez-Rubio, L.; Martínez, B.; Rodríguez, A.; Donovan, D.M.; García, P. Enhanced staphylolytic activity of the *Staphylococcus aureus* bacteriophage vB_SauS-phiIPLA88 HydH5 virion-associated peptidoglycan hydrolase: Fusions, deletions, and synergy with LysH5. *Appl. Environ. Microbiol.* **2012**, *78*, 2241–2248. [CrossRef] [PubMed]
75. Rodríguez-Rubio, L.; Martínez, B.; Donovan, D.M.; García, P.; Rodríguez, A. Potential of the virion-associated peptidoglycan hydrolase HydH5 and its derivative fusion proteins in milk biopreservation. *PLoS ONE* **2013**, *8*, e54828. [CrossRef] [PubMed]
76. Fernández, L.; González, S.; Campelo, A.B.; Martínez, B.; Rodríguez, A.; García, P. Downregulation of autolysin-encoding genes by phage-derived lytic proteins inhibits biofilm formation in *Staphylococcus aureus*. *Antimicrob. Agents Chemother.* **2017**, *61*, e02724-16. [CrossRef] [PubMed]
77. Drilling, A.J.; Cooksley, C.; Chan, C.; Wormald, P.J.; Vreugde, S. Fighting sinus-derived *Staphylococcus aureus* biofilms in vitro with a bacteriophage-derived muralytic enzyme. *Int. Forum Allergy Rhinol.* **2016**, *6*, 349–355. [CrossRef] [PubMed]
78. Saravanan, S.R.; Paul, V.D.; George, S.; Sundarrajan, S.; Kumar, N.; Hebbur, M.; Veena, A.; Maheshwari, U.; Appaiah, C.B.; Chidambaran, M.; et al. Properties and mutation studies of a bacteriophage-derived chimeric recombinant staphylolytic protein P128: Comparison to recombinant lysostaphin. *Bacteriophage* **2013**, *3*, e26564. [CrossRef] [PubMed]
79. Mao, J.; Schmelcher, M.; Harty, W.J.; Foster-Frey, J.; Donovan, D.M. Chimeric Ply187 endolysin kills *Staphylococcus aureus* more effectively than the parental enzyme. *FEMS Microbiol. Lett.* **2013**, *342*, 30–36. [CrossRef] [PubMed]
80. Singh, P.K.; Donovan, D.M.; Kumar, A. Intravitreal injection of the chimeric phage endolysin Ply187 protects mice from *Staphylococcus aureus* endophthalmitis. *Antimicrob. Agents Chemother.* **2014**, *58*, 4621–4629. [CrossRef] [PubMed]

81. Daniel, A.; Euler, C.; Collin, M.; Chahales, P.; Gorelick, K.J.; Fischetti, V.A. Synergism between a novel chimeric lysin and oxacillin protects against infection by methicillin-resistant *Staphylococcus aureus*. *Antimicrob. Agents Chemother.* **2010**, *54*, 1603–1612. [CrossRef] [PubMed]
82. Pastagia, M.; Euler, C.; Chahales, P.; Fuentes-Duculan, J.; Krueger, J.G.; Fischetti, V.A. A novel chimeric lysin shows superiority to mupirocin for skin decolonization of methicillin-resistant and -sensitive *Staphylococcus aureus* strains. *Antimicrob. Agents Chemother.* **2011**, *55*, 738–744. [CrossRef] [PubMed]
83. Fernandes, S.; Proença, D.; Cantante, C.; Silva, F.A.; Leandro, C.; Lourenço, S.; Milheiriço, C.; de Lencastre, H.; Cavaco-Silva, P.; Pimentel, M.; et al. Novel chimerical endolysins with broad antimicrobial activity against methicillin-resistant *Staphylococcus aureus*. *Microb. Drug Resist.* **2012**, *18*, 333–343. [CrossRef] [PubMed]
84. Swift, S.M.; Seal, B.S.; Garrish, J.K.; Oakley, B.B.; Hiett, K.; Yeh, H.Y.; Woolsey, R.; Schegg, K.M.; Line, J.E.; Donovan, D.M. A thermophilic phage endolysin fusion to a *Clostridium perfringens*-specific cell wall binding domain creates an anti-*Clostridium* antimicrobial with improved thermostability. *Viruses* **2015**, *7*, 3019–3034. [CrossRef] [PubMed]
85. Low, L.Y.; Yang, C.; Perego, M.; Osterman, A.; Liddington, R. Role of net charge on catalytic domain and influence of cell wall binding domain on bactericidal activity, specificity, and host range of phage lysins. *J. Biol. Chem.* **2011**, *286*, 34391–34403. [CrossRef] [PubMed]
86. Verbree, C.T.; Dätwyler, S.M.; Meile, S.; Eichenseher, F.; Donovan, D.M.; Loessner, M.J.; Schmelcher, M. Corrected and republished from: Identification of peptidoglycan hydrolase constructs with synergistic staphylolytic activity in cow's milk. *Appl. Environ. Microbiol.* **2018**, *84*, e02134-17. [CrossRef] [PubMed]
87. Loessner, M.J.; Gaeng, S.; Scherer, S. Evidence for a holin-like protein gene fully embedded out of frame in the endolysin gene of *Staphylococcus aureus* bacteriophage 187. *J. Bacteriol.* **1999**, *181*, 4452–4460. [PubMed]
88. Rodríguez-Rubio, L.; Quiles-Puchalt, N.; Martínez, B.; Rodríguez, A.; Penadés, J.R.; García, P. The peptidoglycan hydrolase of *Staphylococcus aureus* bacteriophage 11 plays a structural role in the viral particle. *Appl. Environ. Microbiol.* **2013**, *79*, 6187–6190. [CrossRef] [PubMed]
89. Kwan, T.; Liu, J.; DuBow, M.; Gros, P.; Pelletier, J. The complete genomes and proteomes of 27 *Staphylococcus aureus* bacteriophages. *Proc. Natl. Acad. Sci. USA* **2005**, *102*, 5174–5179. [CrossRef] [PubMed]
90. Yang, H.; Zhang, Y.; Yu, J.; Huang, Y.; Zhang, X.E.; Wei, H. Novel chimeric lysin with high-level antimicrobial activity against methicillin-resistant *Staphylococcus aureus* in vitro and in vivo. *Antimicrob. Agents Chemother.* **2014**, *58*, 536–542. [CrossRef] [PubMed]
91. Yang, H.; Zhang, Y.; Huang, Y.; Yu, J.; Wei, H. Degradation of methicillin-resistant *Staphylococcus aureus* biofilms using a chimeric lysin. *Biofouling* **2014**, *30*, 667–674. [CrossRef] [PubMed]
92. Dong, Q.; Wang, J.; Yang, H.; Wei, C.; Yu, J.; Zhang, Y.; Huang, Y.; Zhang, X.E.; Wei, H. Construction of a chimeric lysin Ply187N-V12C with extended lytic activity against staphylococci and streptococci. *Microb. Biotechnol.* **2015**, *8*, 210–220. [CrossRef] [PubMed]
93. Yang, H.; Zhang, H.; Wang, J.; Yu, J.; Wei, H. A novel chimeric lysin with robust antibacterial activity against planktonic and biofilm methicillin-resistant *Staphylococcus aureus*. *Sci. Rep.* **2017**, *7*, 40182. [CrossRef] [PubMed]
94. Donovan, D.M.; Dong, S.; Garrett, W.; Rousseau, G.M.; Moineau, S.; Pritchard, D.G. Peptidoglycan hydrolase fusions maintain their parental specificities. *Appl. Environ. Microbiol.* **2006**, *72*, 2988–2996. [CrossRef] [PubMed]
95. Loessner, M.J.; Kramer, K.; Ebel, F.; Scherer, S. C-terminal domains of *Listeria monocytogenes* bacteriophage murein hydrolases determine specific recognition and high-affinity binding to bacterial cell wall carbohydrates. *Mol. Microbiol.* **2002**, *44*, 335–349. [CrossRef] [PubMed]
96. Porter, C.J.; Schuch, R.; Pelzek, A.J.; Buckle, A.M.; McGowan, S.; Wilce, M.C.; Rossjohn, J.; Russell, R.; Nelson, D.; Fischetti, V.A.; et al. The 1.6 a crystal structure of the catalytic domain of PlyB, a bacteriophage lysin active against *Bacillus anthracis*. *J. Mol. Biol.* **2007**, *366*, 540–550. [CrossRef] [PubMed]
97. Becker, S.C.; Swift, S.; Korobova, O.; Schischkova, N.; Kopylov, P.; Donovan, D.M.; Abaev, I. Lytic activity of the staphylolytic Twort phage endolysin CHAP domain is enhanced by the SH3b cell wall binding domain. *FEMS Microbiol. Lett.* **2015**, *362*, 1–8. [CrossRef] [PubMed]
98. Huang, Y.; Yang, H.; Yu, J.; Wei, H. Molecular dissection of phage lysin PlySs2: Integrity of the catalytic and cell wall binding domains is essential for its broad lytic activity. *Virol. Sin.* **2015**, *30*, 45–51. [CrossRef] [PubMed]

99. Mayer, M.J.; Gasson, M.J.; Narbad, A. Genomic sequence of bacteriophage ATCC 8074-B1 and activity of its endolysin and engineered variants against *Clostridium sporogenes*. *Appl. Environ. Microbiol.* **2012**, *78*, 3685–3692. [CrossRef] [PubMed]
100. Horgan, M.; O'Flynn, G.; Garry, J.; Cooney, J.; Coffey, A.; Fitzgerald, G.F.; Ross, R.P.; McAuliffe, O. Phage lysin LysK can be truncated to its CHAP domain and retain lytic activity against live antibiotic-resistant staphylococci. *Appl. Environ. Microbiol.* **2009**, *75*, 872–874. [CrossRef] [PubMed]
101. Donovan, D.M.; Foster-Frey, J.; Dong, S.; Rousseau, G.M.; Moineau, S.; Pritchard, D.G. The cell lysis activity of the *Streptococcus agalactiae* bacteriophage B30 endolysin relies on the cysteine, histidine-dependent amidohydrolase/peptidase domain. *Appl. Environ. Microbiol.* **2006**, *72*, 5108–5112. [CrossRef] [PubMed]
102. Fenton, M.; Keary, R.; McAuliffe, O.; Ross, R.P.; O'Mahony, J.; Coffey, A. Bacteriophage-derived peptidase $CHAP_K$ eliminates and prevents staphylococcal biofilms. *Int. J. Microbiol.* **2013**, *2013*, 625341. [CrossRef] [PubMed]
103. Fenton, M.; Casey, P.G.; Hill, C.; Gahan, C.G.M.; Ross, R.P.; McAuliffe, O.; O'Mahony, J.; Maher, F.; Coffey, A. The truncated phage lysin $CHAP_K$ eliminates *Staphylococcus aureus* in the nares of mice. *Bioeng. Bugs* **2010**, *1*, 404–407. [CrossRef] [PubMed]
104. Becker, S.C.; Dong, S.; Baker, J.R.; Foster-Frey, J.; Pritchard, D.G.; Donovan, D.M. LysK CHAP endopeptidase domain is required for lysis of live staphylococcal cells. *FEMS Microbiol. Lett.* **2009**, *294*, 52–60. [CrossRef] [PubMed]
105. Sass, P.; Bierbaum, G. Lytic activity of recombinant bacteriophage φ11 and φ12 endolysins on whole cells and biofilms of *Staphylococcus aureus*. *Appl. Environ. Microbiol.* **2007**, *73*, 347–352. [CrossRef] [PubMed]
106. Benešík, M.; Nováček, J.; Janda, L.; Dopitová, R.; Pernisová, M.; Melková, K.; Tišáková, L.; Doškař, J.; Žídek, L.; Hejátko, J.; et al. Role of SH3b binding domain in a natural deletion mutant of Kayvirus endolysin LysF1 with a broad range of lytic activity. *Virus Genes* **2017**. [CrossRef]
107. Zhou, Y.; Zhang, H.; Bao, H.; Wang, X.; Wang, R. The lytic activity of recombinant phage lysin LysKΔamidase against staphylococcal strains associated with bovine and human infections in the Jiangsu province of China. *Res. Vet. Sci.* **2017**, *111*, 113–119. [CrossRef] [PubMed]
108. Low, L.Y.; Yang, C.; Perego, M.; Osterman, A.; Liddington, R.C. Structure and lytic activity of a *Bacillus anthracis* prophage endolysin. *J. Biol. Chem.* **2005**, *280*, 35433–35439. [CrossRef] [PubMed]
109. Mayer, M.J.; Garefalaki, V.; Spoerl, R.; Narbad, A.; Meijers, R. Structure-based modification of a *Clostridium difficile*-targeting endolysin affects activity and host range. *J. Bacteriol.* **2011**, *193*, 5477–5486. [CrossRef] [PubMed]
110. Cheng, Q.; Fischetti, V.A. Mutagenesis of a bacteriophage lytic enzyme PlyGBS significantly increases its antibacterial activity against group B streptococci. *Appl. Microbiol. Biotechnol.* **2007**, *74*, 1284–1291. [CrossRef] [PubMed]
111. Donovan, D.M.; Foster-Frey, J. LambdaSa2 prophage endolysin requires Cpl-7-binding domains and amidase-5 domain for antimicrobial lysis of streptococci. *FEMS Microbiol. Lett.* **2008**, *287*, 22–33. [CrossRef] [PubMed]
112. Heselpoth, R.D.; Nelson, D.C. A new screening method for the directed evolution of thermostable bacteriolytic enzymes. *J. Vis. Exp.* **2012**, 4216. [CrossRef] [PubMed]
113. Díez-Martínez, R.; de Paz, H.D.; de Paz, H.; Bustamante, N.; García, E.; Menéndez, M.; García, P. Improving the lethal effect of Cpl-7, a pneumococcal phage lysozyme with broad bactericidal activity, by inverting the net charge of its cell wall-binding module. *Antimicrob. Agents Chemother.* **2013**, *57*, 5355–5365. [CrossRef] [PubMed]
114. Heselpoth, R.D.; Yin, Y.; Moult, J.; Nelson, D.C. Increasing the stability of the bacteriophage endolysin plyc using rationale-based foldx computational modeling. *Protein Eng. Des. Sel.* **2015**, *28*, 85–92. [CrossRef] [PubMed]
115. Resch, G.; Moreillon, P.; Fischetti, V.A. A stable phage lysin (Cpl-1) dimer with increased antipneumococcal activity and decreased plasma clearance. *Int. J. Antimicrob. Agents* **2011**, *38*, 516–521. [CrossRef] [PubMed]
116. Becker, S.C.; Roach, D.R.; Chauhan, V.S.; Shen, Y.; Foster-Frey, J.; Powell, A.M.; Bauchan, G.; Lease, R.A.; Mohammadi, H.; Harty, W.J.; et al. Triple-acting lytic enzyme treatment of drug-resistant and intracellular *Staphylococcus aureus*. *Sci. Rep.* **2016**, *6*, 25063. [CrossRef] [PubMed]

117. Haddad Kashani, H.; Fahimi, H.; Dasteh Goli, Y.; Moniri, R. A novel chimeric endolysin with antibacterial activity against methicillin-resistant *Staphylococcus aureus*. *Front. Cell. Infect. Microbiol.* **2017**, *7*, 290. [CrossRef] [PubMed]
118. Pohane, A.A.; Joshi, H.; Jain, V. Molecular dissection of phage endolysin: An interdomain interaction confers host specificity in Lysin A of *Mycobacterium* phage D29. *J. Biol. Chem.* **2014**, *289*, 12085–12095. [CrossRef] [PubMed]
119. Wang, Q.; Euler, C.W.; Delaune, A.; Fischetti, V.A. Using a novel lysin to help control *Clostridium difficile* infections. *Antimicrob. Agents Chemother.* **2015**, *59*, 7447–7457. [CrossRef] [PubMed]
120. Ageitos, J.M.; Sánchez-Pérez, A.; Calo-Mata, P.; Villa, T.G. Antimicrobial peptides (AMPS): Ancient compounds that represent novel weapons in the fight against bacteria. *Biochem. Pharmacol.* **2017**, *133*, 117–138. [CrossRef] [PubMed]
121. Wittekind, M.; Schuch, R. Cell wall hydrolases and antibiotics: Exploiting synergy to create efficacious new antimicrobial treatments. *Curr. Opin. Microbiol.* **2016**, *33*, 18–24. [CrossRef] [PubMed]
122. Kontermann, R.E. Strategies for extended serum half-life of protein therapeutics. *Curr. Opin. Biotechnol.* **2011**, *22*, 868–876. [CrossRef] [PubMed]
123. Veronese, F.M.; Mero, A. The impact of PEGylation on biological therapies. *BioDrugs* **2008**, *22*, 315–329. [CrossRef] [PubMed]
124. Walsh, S.; Shah, A.; Mond, J. Improved pharmacokinetics and reduced antibody reactivity of lysostaphin conjugated to polyethylene glycol. *Antimicrob. Agents Chemother.* **2003**, *47*, 554–558. [CrossRef] [PubMed]
125. Filatova, L.Y.; Donovan, D.M.; Becker, S.C.; Lebedev, D.N.; Priyma, A.D.; Koudriachova, H.V.; Kabanov, A.V.; Klyachko, N.L. Physicochemical characterization of the staphylolytic LysK enzyme in complexes with polycationic polymers as a potent antimicrobial. *Biochimie* **2013**, *95*, 1689–1696. [CrossRef] [PubMed]
126. Resch, G.; Moreillon, P.; Fischetti, V.A. PEGylating a bacteriophage endolysin inhibits its bactericidal activity. *AMB Express* **2011**, *1*, 29. [CrossRef] [PubMed]
127. Fischetti, V.A. Bacteriophage lytic enzymes: Novel anti-infectives. *Trends Microbiol.* **2005**, *13*, 491–496. [CrossRef] [PubMed]
128. Rodríguez-Rubio, L.; Martínez, B.; Rodríguez, A.; Donovan, D.M.; Götz, F.; García, P. The phage lytic proteins from the *Staphylococcus aureus* bacteriophage vB_SauS-phiIPLA88 display multiple active catalytic domains and do not trigger staphylococcal resistance. *PLoS ONE* **2013**, *8*, e64671. [CrossRef] [PubMed]
129. Thandar, M.; Lood, R.; Winer, B.Y.; Deutsch, D.R.; Euler, C.W.; Fischetti, V.A. Novel engineered peptides of a phage lysin as effective antimicrobials against multidrug-resistant *Acinetobacter baumannii*. *Antimicrob. Agents Chemother.* **2016**, *60*, 2671–2679. [CrossRef] [PubMed]
130. Peng, S.Y.; You, R.I.; Lai, M.J.; Lin, N.T.; Chen, L.K.; Chang, K.C. Highly potent antimicrobial modified peptides derived from the *Acinetobacter baumannii* phage endolysin LysAB2. *Sci. Rep.* **2017**, *7*, 11477. [CrossRef] [PubMed]
131. Lukacik, P.; Barnard, T.J.; Keller, P.W.; Chaturvedi, K.S.; Seddiki, N.; Fairman, J.W.; Noinaj, N.; Kirby, T.L.; Henderson, J.P.; Steven, A.C.; et al. Structural engineering of a phage lysin that targets Gram-negative pathogens. *Proc. Natl. Acad. Sci. USA* **2012**, *109*, 9857–9862. [CrossRef] [PubMed]
132. Yan, G.; Liu, J.; Ma, Q.; Zhu, R.; Guo, Z.; Gao, C.; Wang, S.; Yu, L.; Gu, J.; Hu, D.; et al. The N-terminal and central domain of colicin A enables phage lysin to lyse *Escherichia coli* extracellularly. *Antonie Van Leeuwenhoek* **2017**, *110*, 1627–1635. [CrossRef] [PubMed]
133. Briers, Y.; Walmagh, M.; Van Puyenbroeck, V.; Cornelissen, A.; Cenens, W.; Aertsen, A.; Oliveira, H.; Azeredo, J.; Verween, G.; Pirnay, J.P.; et al. Engineered endolysin-based "Artilysins" to combat multidrug-resistant gram-negative pathogens. *mBio* **2014**, *5*, e01379-14. [CrossRef] [PubMed]
134. Briers, Y.; Walmagh, M.; Grymonprez, B.; Biebl, M.; Pirnay, J.P.; Defraine, V.; Michiels, J.; Cenens, W.; Aertsen, A.; Miller, S.; et al. Art-175 is a highly efficient antibacterial against multidrug-resistant strains and persisters of *Pseudomonas aeruginosa*. *Antimicrob. Agents Chemother.* **2014**, *58*, 3774–3784. [CrossRef] [PubMed]
135. Defraine, V.; Schuermans, J.; Grymonprez, B.; Govers, S.K.; Aertsen, A.; Fauvart, M.; Michiels, J.; Lavigne, R.; Briers, Y. Efficacy of artilysin art-175 against resistant and persistent *Acinetobacter baumannii*. *Antimicrob. Agents Chemother.* **2016**, *60*, 3480–3488. [CrossRef] [PubMed]
136. Schirmeier, E.; Zimmermann, P.; Hofmann, V.; Biebl, M.; Gerstmans, H.; Maervoet, V.E.; Briers, Y. Inhibitory and bactericidal effect of Artilysin® Art-175 against colistin-resistant mcr-1-positive *Escherichia coli* isolates. *Int. J. Antimicrob. Agents* **2017**. [CrossRef] [PubMed]

137. Wang, S.; Gu, J.; Lv, M.; Guo, Z.; Yan, G.; Yu, L.; Du, C.; Feng, X.; Han, W.; Sun, C.; et al. The antibacterial activity of *E. coli* bacteriophage lysin lysep3 is enhanced by fusing the *Bacillus amyloliquefaciens* bacteriophage endolysin binding domain D8 to the C-terminal region. *J. Microbiol.* **2017**, *55*, 403–408. [CrossRef] [PubMed]
138. Ma, Q.; Guo, Z.; Gao, C.; Zhu, R.; Wang, S.; Yu, L.; Qin, W.; Xia, X.; Gu, J.; Yan, G.; et al. Enhancement of the direct antimicrobial activity of Lysep3 against *Escherichia coli* by inserting cationic peptides into its C-terminus. *Antonie Van Leeuwenhoek* **2017**, *110*, 347–355. [CrossRef] [PubMed]
139. Lin, Y.M.; Wu, S.J.; Chang, T.W.; Wang, C.F.; Suen, C.S.; Hwang, M.J.; Chang, M.D.; Chen, Y.T.; Liao, Y.D. Outer membrane protein I of *Pseudomonas aeruginosa* is a target of cationic antimicrobial peptide/protein. *J. Biol. Chem.* **2010**, *285*, 8985–8994. [CrossRef] [PubMed]
140. Yang, H.; Wang, M.; Yu, J.; Wei, H. Antibacterial activity of a novel peptide-modified lysin against *Acinetobacter baumannii* and *Pseudomonas aeruginosa*. *Front. Microbiol.* **2015**, *6*, 1471. [CrossRef] [PubMed]
141. Orito, Y.; Morita, M.; Hori, K.; Unno, H.; Tanji, Y. *Bacillus amyloliquefaciens* phage endolysin can enhance permeability of *Pseudomonas aeruginosa* outer membrane and induce cell lysis. *Appl. Microbiol. Biotechnol.* **2004**, *65*, 105–109. [CrossRef] [PubMed]
142. Rodríguez-Rubio, L.; Chang, W.L.; Gutiérrez, D.; Lavigne, R.; Martínez, B.; Rodríguez, A.; Govers, S.K.; Aertsen, A.; Hirl, C.; Biebl, M.; et al. 'Artilysation' of endolysin λSa2lys strongly improves its enzymatic and antibacterial activity against streptococci. *Sci. Rep.* **2016**, *6*, 35382. [CrossRef] [PubMed]
143. Silva, M.T. Classical labeling of bacterial pathogens according to their lifestyle in the host: Inconsistencies and alternatives. *Front. Microbiol.* **2012**, *3*, 71. [CrossRef] [PubMed]
144. Borysowski, J.; Górski, A. Fusion to cell-penetrating peptides will enable lytic enzymes to kill intracellular bacteria. *Med. Hypotheses* **2010**, *74*, 164–166. [CrossRef] [PubMed]
145. Dietz, G.P. Cell-penetrating peptide technology to deliver chaperones and associated factors in diseases and basic research. *Curr. Pharm. Biotechnol.* **2010**, *11*, 167–174. [CrossRef] [PubMed]
146. Shen, Y.; Barros, M.; Vennemann, T.; Gallagher, D.T.; Yin, Y.; Linden, S.B.; Heselpoth, R.D.; Spencer, D.J.; Donovan, D.M.; Moult, J.; et al. A bacteriophage endolysin that eliminates intracellular streptococci. *eLife* **2016**, *5*, e13152. [CrossRef] [PubMed]

antibiotics

MDPI

Review

Potential for Bacteriophage Endolysins to Supplement or Replace Antibiotics in Food Production and Clinical Care

Michael J. Love [1], Dinesh Bhandari [1,2], Renwick C. J. Dobson [1,3] and Craig Billington [1,2,*]

1 Biomolecular Interaction Centre and School of Biological Sciences, University of Canterbury, Christchurch 8041, New Zealand; michael.love@pg.canterbury.ac.nz (M.J.L.); dinesh.bhandari@esr.cri.nz (D.B.); renwick.dobson@canterbury.ac.nz (R.C.J.D.)
2 Institute of Environmental Science and Research, Christchurch 8041, New Zealand
3 Department of Biochemistry and Molecular Biology, University of Melbourne, Melbourne 3052, Australia
* Correspondence: craig.billington@esr.cri.nz; Tel.: +64-3-351-0128

Received: 21 December 2017; Accepted: 23 February 2018; Published: 27 February 2018

Abstract: There is growing concern about the emergence of bacterial strains showing resistance to all classes of antibiotics commonly used in human medicine. Despite the broad range of available antibiotics, bacterial resistance has been identified for every antimicrobial drug developed to date. Alarmingly, there is also an increasing prevalence of multidrug-resistant bacterial strains, rendering some patients effectively untreatable. Therefore, there is an urgent need to develop alternatives to conventional antibiotics for use in the treatment of both humans and food-producing animals. Bacteriophage-encoded lytic enzymes (endolysins), which degrade the cell wall of the bacterial host to release progeny virions, are potential alternatives to antibiotics. Preliminary studies show that endolysins can disrupt the cell wall when applied exogenously, though this has so far proven more effective in Gram-positive bacteria compared with Gram-negative bacteria. Their potential for development is furthered by the prospect of bioengineering, and aided by the modular domain structure of many endolysins, which separates the binding and catalytic activities into distinct subunits. These subunits can be rearranged to create novel, chimeric enzymes with optimized functionality. Furthermore, there is evidence that the development of resistance to these enzymes may be more difficult compared with conventional antibiotics due to their targeting of highly conserved bonds.

Keywords: endolysin; antibiotics; antimicrobial resistance; one health; protein engineering

---

1. Introduction

In 2014, the World Health Organization (WHO) calculated the global prevalence of seven antibiotic-resistant bacteria of international concern, and noted very high rates of resistance (up to 84% of all isolates for methicillin, 81% for third-generation cephalosporins, 49% for fluoroquinolones, and 60% for penicillin) in all WHO regions [1]. In response to this unprecedented crisis, in late 2016 the United Nations General Assembly called upon the WHO, the Food and Agriculture Organization of the United Nations, and the World Organisation for Animal Health to develop a global development and stewardship framework [2]. This request recognized that there was a need to co-ordinate action against antimicrobial resistance in humans, agriculture, animals, and the environment by using a One Health [3] approach. One of the key recommendations in the draft framework was to develop new antimicrobial agents for use in these key sectors.

Here, we discuss the potential of cell wall lysis proteins (endolysins) derived from bacteriophages for use as a new class of antimicrobial agents, and evaluate whether they could replace, or supplement, some of the conventional antibiotics used to treat animals and humans, and perhaps even find use in food production and environmental decontamination processes.

Endolysins are enzymes encoded by bacteriophages (phage; obligate viruses of bacteria) which lyse the host bacterial cell. Endolysins degrade the main structural component of the cell wall (peptidoglycan) at the conclusion of the replicative cycle to release newly assembled progeny phage [4] (Figure 1). Recombinantly expressed endolysins display similarly effective lytic abilities to their native counterparts when applied exogenously to susceptible bacteria [5]. This feature underpins the application of endolysins in medicine, food and agriculture.

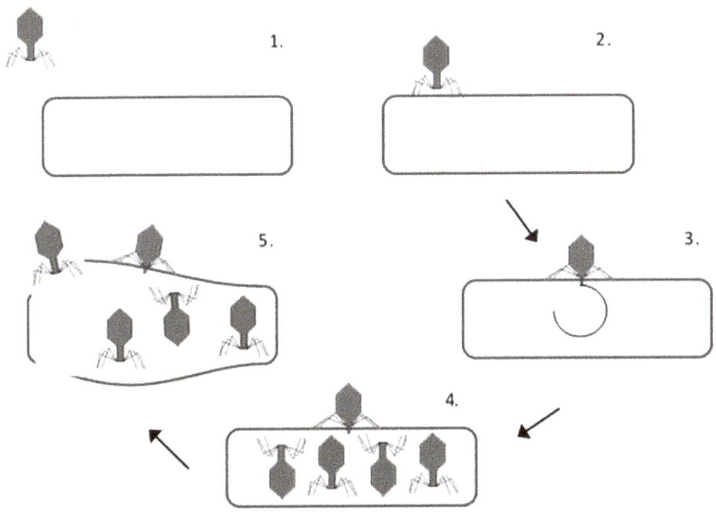

Figure 1. Life cycle of a virulent tailed phage (not to scale). (1) The phage collides with the bacterial cell; (2) the phage binds to cell receptors; (3) the phage is irreversibly bound and injects nucleic acid into the cell via the tail tube, where it is transcribed and translated; (4) many progeny phages are produced within intact cells; (5) endolysins degrade the host bacterial cell wall, which loses its structural integrity and ruptures due to the osmotic pressure, releasing the progeny phages.

Endolysins are predominantly more effective against Gram-positive bacteria than Gram-negative bacteria when applied in this way. The outer membrane of Gram-negative bacteria presents a physical protective barrier against the activity of endolysins [6]. Therefore, endolysin research has mainly focused on Gram-positive bacteria. However, recent work on outer membrane permeabilizers (chemicals and engineered peptides) should increase the prospects of endolysins for treating Gram-negative bacteria.

Numerous types of endolysins have been described, and are typically categorized by the structural bonds in the peptidoglycan that are cleaved by the enzyme [7] (Figure 2). The two alternating glycosidic bonds between the amino sugar moieties, *N*-acetylglucosamine and *N*-acetylmuramic acid (MurNAc), are targeted by different endolysin classes. The *N*-acetylmuramoyl-β-1,4-*N*-acetylglucosamine bond is cleaved by lytic transglycosylases and *N*-acetyl-β-D-muramidases, which are commonly known as lysozymes, while the *N*-acetylglucosaminyl-β-1,4-*N*-acetylmuramine bond is hydrolysed by *N*-acetyl-glucosaminyl-β-D-glucosaminidases. The cleavage of the amide linkage between MurNAc and L-alanine is catalyzed by *N*-acetylmuramoyl-L-alanine amidases. There are different endopeptidases depending on the chemical structure of the peptidoglycan, which is dependent on species and growth conditions. Generally, endopeptidases hydrolyze the peptide bonds between the amino acids that form the cross-linking peptide stems [8–10].

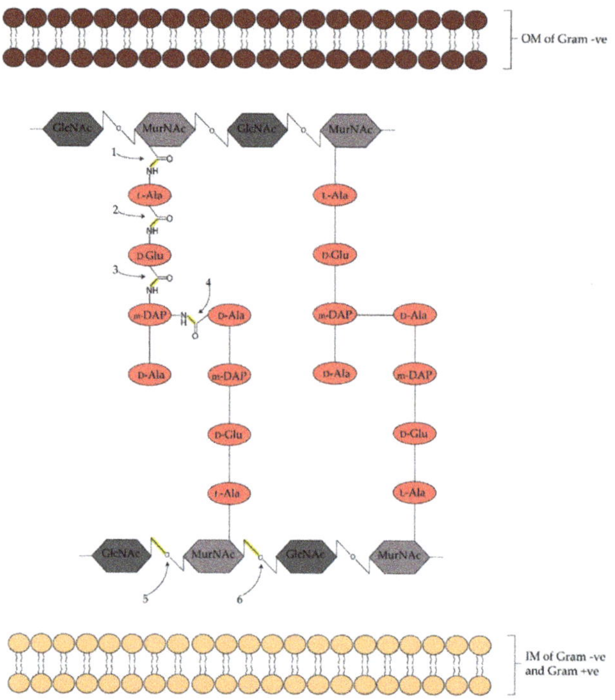

Figure 2. Diagram of the typical cell wall and peptidoglycan structure of bacteria, including the endolysin cleavage sites. The peptidoglycan is composed of repeating sugar units, *N*-acetylglucosamine (GlcNAc) and *N*-acetylmuramic acid (MurNAc), which are cross-linked via an interpeptide bridge between the *meso*-diaminopimelic acid (m-DAP) and D-alanine (D-Ala) residues of adjacent tetrapeptide chains. The chains also contain L-alanine (L-Ala) and D-glutamic acid (D-Glu). Gram-negative bacteria contain an outer membrane (OM) structure not present in Gram-positive bacteria. Both contain an inner membrane (IM) structure. The cleaved bonds and major classifications of endolysin are indicated: (1) *N*-acetylmuramoyl-L-alanine amidase; (2–4) various endopeptidases; (5) N-acetyl-β-D-glucosaminidase; (6) N-acetyl-β-D-muramidase (lysozyme).

A number of promising endolysins have been isolated from phage for application as antimicrobial agents, as described in this review and others [5–9]. However, the bioengineering of modified or novel endolysins also holds promise for the future development of effective tools to kill or detect bacteria. The prospects for engineering are facilitated by the enzymes' structures. The domain structure of endolysins can be modular (Gram-positive bacteria and some Gram-negative species) or globular (most Gram-negative bacteria), with most bioengineering strategies exploiting the modular endolysins. These enzymes usually comprise two distinct domains: an N-terminal enzymatically-active domain and a C-terminal cell wall-binding domain, connected by short, flexible linker regions. N-terminal enzymatically-active domains are responsible for catalyzing the breakdown of specific peptidoglycan bonds, while the C-terminal cell wall-binding domains recognize and bind non-covalently to substrate within the cell wall, resulting in the specificity of the lytic enzymes for the target host [5,8]. In addition, the C-terminal cell wall-binding domain is often required to maintain full lytic activity [11–13]. Interestingly, truncation or deletion of the C-terminal cell wall-binding domain can also result in equal or increased lytic activity [14–16]. In contrast, globular endolysins only contain catalytic domains.

The modular endolysin arrangement can be exploited for bioengineering, as the different domains can be shuffled within the protein, or domains from different endolysins can be combined to generate new enzymes [17,18]. Directed mutagenesis is also an effective strategy, as different amino acids may support improved lytic or binding properties [5,8,17,18]. The seemingly limitless possible permutations of endolysin modular arrangements allow for the development of new enzymes with specific functions or features. Some potential bioengineering strategies and examples of endolysin constructs are shown in Table 1.

Table 1. Possible molecular engineering strategies with potential application(s).

Modification	Property	Example	Reference
Truncation of full-length enzyme	Increased catalytic ability and solubility	CHAPk	Horgan et al. [19]
Fusion of EADs with CBDs of different endolysins	Increased catalytic ability and solubility	ClyS	Daniel et al. [20]
	Increased catalytic ability and broader lytic spectrum	SA2-E-Lyso-SH3b, SA2-E-LysK-SH3b	Schmelcher et al. [21]
	Thermostability	PlyGVE2CpCWB	Swift et al. [22]
Fusion of virion-associated lysin with CBD of endolysin	Increased efficacy	EC300	Proença et al. [23]
Endolysin fusion with OMP	Increased efficacy towards Gram-negative bacteria	OBPgp279, PVP-SE1g-146	Briers et al. [24]
Endolysin fusion with AMP	Increased efficacy towards Gram-negative bacteria	Art-175	Briers et al. [25]
Truncation and site-directed mutagenesis	AMP development	LysAB2	Peng et al. [26]

EAD: enzymatically-active domain; CBD: cell wall-binding domain; OMP: outer-membrane permeabilizer; AMP: antimicrobial peptide.

Both bioengineered and phage-isolated endolysins are promising alternatives to antibiotics. Their specificity allows them to target specific bacterial pathogens without affecting the microflora [11], or alternatively, target a larger spectrum for broader efficacy [27]. At the moment, developed resistance to the activity of the endolysins has not been widely reported, meaning these enzymes could be a long-term solution to antibiotics [28–30]. Endolysins may also have potential as diagnostic tools for bacterial identification [31]. In the following sections, we provide an overview of native and chimeric endolysins with potential therapeutic applications.

2. Endolysins as Human Therapeutics

As the efficacy of antibiotics decreases, once easily-treated bacterial infections will become potentially fatal. This will also have secondary effects in clinical care, such as changing risk-benefit considerations for invasive surgeries. Phage-encoded lytic enzymes have the potential to fulfil the need for novel antibacterial therapeutic agents for use in humans. This new class of antimicrobials has been recognized by the United States of America in the National Action Plan for Combating Antibiotic-resistant Bacteria [32], which identified the use of "phage-derived lysins to kill specific bacteria while preserving the microbiota" as a key strategy to reduce the development of antimicrobial resistance.

Methicillin-resistant *Staphylococcus aureus* (MRSA) is a significant public health concern, causing a range of skin and respiratory infections, as well as food-borne illnesses that are not easily treatable with currently available antibiotics [33]. O'Flaherty et al. [27] treated a human-derived MRSA strain with *Lactococcus lactis* cell lysate containing recombinantly overexpressed endolysin LysK, and observed a 99% reduction in colony-forming units at 1 h post-exposure. However, the researchers had difficulties obtaining soluble protein, which would hinder future applications of LysK, a difficulty that was also encountered in subsequent studies [34,35]. A stability study was conducted on LysK, as medical application requires a stable enzyme [34]. LysK was stabilized in the presence of low molecular weight polyols such as sucrose and glycerol, for example, stability increased 100-fold at 30 °C, and LysK retained 100% activity after storage up to 1 month at room temperature. This stability, under simple condition changes, is useful for developing treatment strategies [34]. LysK contains two catalytic domains: a cysteine- and histidine-dependent amidohydrolase/peptidase (CHAP) domain, and an *N*-acetylmuramoyl-L-alanine amidase domain. In an attempt to overcome the solubility issue, Horgan et al. [19] generated a single-domain truncated LysK mutant, designated CHAPk, containing only the CHAP domain. Soluble CHAPk was easier to obtain than full-length LysK, and displayed at least a two-fold increase in lytic activity against both heat-killed and live staphylococcal cells in vitro. Subsequent studies demonstrated that CHAPk was also effective in vivo, and that the loss of the C-terminal cell wall-binding domain, which directs specificity, resulted in activity against a broader range of targets compared with full-length CHAPk [36,37].

Jun et al. [38] compared LysK with SAL-1, an endolysin that differs by three residues. They also produced six derivatives of SAL-1 containing mutations in each of the three residues to investigate the impact of each mutation. SAL-1 displayed cell-wall hydrolytic activity approximately two-fold greater than that of LysK. The mutation of residue 114 from glutamatic acid in LysK to glutamine in SAL-1 had the largest impact on activity. This residue is located inside the catalytic CHAP domain, and the structurally minor sequence change corresponded to enhanced activity. The combination of such enhanced activity with the identification and mechanistic characterization of key residues of different enzymes is important for rational design and engineering of new endolysins with optimized activity [38]. Compared with LysK, SAL-1 had increased catalytic activity, and high yields of soluble protein were easier to obtain; therefore, SAL-1 may be a more promising antibiotic alternative than LysK [39]. The therapeutic application of SAL-1 is currently being trialed by iNtRON Biotechnology in the form of SAL200, an endolysin-based candidate drug for the treatment of *S. aureus*. A preclinical safety study of SAL200 observed no toxicity in rodent intravenous single- and repeat-dose studies [40]. A repeat-dose experiment was also performed in dogs, with each dog receiving four doses of 0, 0.5, 12.5 and 25 mg/kg in 1-week intervals over four weeks. After ten days, short-lived (i.e., lasting only 30 minutes to 1 hour) and mild clinical signs were observed including, vomiting, subdued behavior and irregular respiration. The transient response of the dogs to SAL200 administration was linked to complement system activation that resulted from antibody production [40]. A follow-up study in monkeys investigated the impact of single-dose escalation (up to 80 mg/kg) or 5-day multiple-dose (up to 40 mg/kg/day) administration of SAL200, with no adverse effects observed [41] SAL200 was further evaluated in a human single dose-escalating (up to 10 mg/kg) study. This study was the first in-human clinical study of a intravenously administered endolysin-based drug [42]. The volunteers had a reasonable tolerability to SAL200, with no significant adverse effects and most of the adverse effects were mild and self-limited. Although an increased concentration of antibodies was observed, the antibody concentration of participants administered 3.0 mg/kg was greater than that of those administered 10 mg/kg, with large variation within the different cohorts. The immunogenicity of SAL200 should, therefore, be a focus of future studies in order to better develop treatment regimes [42]. A phase II clinical study is now being conducted on patients with persistent *S. aureus* bacteremia, with results expected in 2018. Overall, the current evaluations shows a promising future for not just SAL200, but also for the development of other endolysin-based drug treatments.

Biofilm formation in clinical environments and on medical devices can have significant medical implications, as biofilms can harbor pathogenic and multidrug-resistant bacteria. Microorganisms within biofilms are protected by extracellular polymeric substances (EPS), which are a source of environmental contamination when partially dislodged. EPS can contain polysaccharides, proteins, phospholipids, teichoic acids, nucleic acids, and polymers, and protect the biofilm inhabitants by concentrating nutrients, preventing access of biocides, sequestering metals and toxins, and preventing desiccation [43]. Linden et al. [44] found that recombinantly-expressed PlyGRCS (from the phage GRCS) effectively lysed *S. aureus* in a biofilm, as well as in stationary phase. PlyGRCS contains a single enzymatically-active domain that can cleave two different bonds in peptidoglycan. This bifunctional domain could be highly useful in developing endolysins with effective lytic activity.

Rashel et al. [45] found that a dose of the phage ϕMR11-derived lysin MV-L rescued mice from fatal levels of MRSA exposure. In addition, MV-L in combination with vancomycin killed vancomycin-resistant strains. MV-L was specific for *S. aureus* and *Staphylococcus simulans*, with no lytic activity observed against other staphylococcal strains or bacterial species, including *Staphylococcus epidermidis* and *Escherichia coli*. Although excessive exposure to MV-L induced the production of antibodies, no adverse effects on the mice or impact on the efficacy of MV-L was observed.

Daniel et al. [20] demonstrated the potential of bioengineering to generate enzymes with novel and specific lytic activity against MRSA. The endolysin ClyS was constructed from the enzymatically active domain of a *S. aureus* Twort phage lysin fused with the cell wall-binding domain of phiNM3. Mice were exposed to MRSA strains that were resistant to the antibiotic oxacillin. A dose of ClyS increased survival rates to 88%, compared with the 0% survival rate for untreated mice. Treatment of infected mice with a sub-therapeutic concentration of ClyS in combination with oxacillin increased survival rates when compared with each treatment alone. This synergistic relationship with antibiotics may have widespread potential, and reinitiate the use of historical antibiotics that have been discontinued due to resistance concerns.

Schuch et al. [46] further showed this synergistic potential with the lysin CF-301. Mice with staphylococcal-induced bacteremia had a survival time of less than 24 h without treatment. Following individual treatments with CF-301 and daptomycin at 4 h post-inoculation, survival rates after 72 h were measured at 13% and 23%, respectively. Combination therapy yielded a survival rate of 73%. The study further confirmed the efficacy of co-therapy in 16 individual experiments including the antibiotics oxacillin and vancomycin. The immunogenicity of CF-301 was briefly evaluated in vitro; rabbit antisera raised against CF-301 did not inhibit the activity of CF-301 [46]. Despite the in vitro results, the immunogenic nature of CF-301 needs to be studied in a range of model organisms in vivo, because there may be clinically relevant adverse effects. CF-301 also has anti-biofilm activity [47], and clinical phase I trials are now underway to evaluate CF-301 as an alternative to traditional antibiotics, with an expected study completion in late 2018.

Thermal injury patients are usually also immunocompromised, meaning they are more susceptible to bacterial infection, including drug-resistant *S. aureus* strains [48]. Chopra et al. [49] investigated the use of endolysin MR-10 alone and in combination with minocycline to treat burn wound infections in mice. The control mice, inoculated with *S. aureus* and receiving no medical treatment, had a 100% mortality rate within 5 days. Individually, MR-10 (50 μg/ml) and minocycline (50 mg/kg) both resulted in a survival rate of 35% at 5 days post-inoculation, but 100% mortality was observed by day 7. In contrast, 100% survival was observed following treatment with a combination of the therapeutic agents at the same concentrations. These findings further support the future use of endolysins in medicine, especially in co-therapy with existing antibiotics.

Staphefekt is an endolysin bioengineered to selectively target *S. aureus* strains, including MRSA. It is currently available as a component of gels and creams for over-the-counter treatment of infections [50,51]. Its specificity for *S. aureus* is an important feature in the treatment of skin infections, as it prevents the disturbance of commensal bacteria, which can cause further health complications [52]. Evidence for the efficacy of Staphefekt is limited; there are few available publications describing the rationale of engineering or the in vitro properties and structure of the endolysin. However, a recent report by Totté et al. [51] demonstrated the efficacy of Staphefekt in treating three different patients with recurrent *S. aureus*-related dermatoses that had previously been unsuccessfully treated with antibiotics. Despite the limited published evidence, these brief findings suggest a promising future for Staphefekt as a long-term alternative to antibiotics.

Enterococcus faecalis is the third most common cause of life-threatening nosocomial infections [53], and is intrinsically resistant to antibiotics [54]. Although vancomycin is considered a drug of last resort, a growing number of vancomycin-resistant *E. faecalis* strains have been isolated. [55,56].

An endolysin isolated from phage ϕ1, PlyV12, kills a variety of *E. faecalis* strains in vitro, including vancomycin-resistant isolates. PlyV12 also showed a broad spectrum of lethality against a variety of streptococcal and staphylococcal strains, highlighting it as a promising candidate in antibacterial medicine [57]. Son et al. [58] also identified a novel phage, EFAP-1, which was not significantly similar to any previously identified phages. The endolysin of EFAP-1, EFAL-1, exhibited lytic activity against 24 different strains of bacteria, including vancomycin-resistant *E. faecalis* strains and four streptococcal strains, whereas the phage itself only showed activity against *E. faecalis*. There is a lack of in vivo studies of *E. faecalis* endolysins. Such studies are important because *E. faecalis* is also a commensal bacterium found in the human gut, and therefore the potential impact of endolysins on the gut microbiota needs to be understood [52]. Zhang et al. [59] isolated phage IME-EF1 and its endolysin from hospital sewage, and investigated their ability to rescue mice from lethal challenge with *E. faecalis*. Individually, both the phage and the endolysin reduced the bacterial count in the blood of infected animals. However, a 200-μg dose of the endolysin at 30 min post-bacterial inoculation supported a higher survival rate (80%) than that observed following phage treatment alone (60%) [59].

The endolysin LysEF-P10, derived from *E. faecalis* phage EF-P10, has also been studied in a mouse model. A single dose of just 5 μg was enough to protect mice from vancomycin-resistant *E. faecalis* infection, suggesting promising protective efficacy of LysEF-P10 against vancomycin-resistant *E. faecalis* strains. Furthermore, when the mice were subjected to a large dose of 5 mg, no side effects were observed. The administration of EF-P10 stimulated specific antibody production, however, there was no impact on the bactericidal activity of the enzyme. Treatment with EF-P10 did not negatively affect the gut microbiota, owing to the specificity of the lysin. Although *E. faecalis* in the normal gut microbiota may have been targeted, no significant health impact on the mice was observed [60]. Proença et al. [23] constructed a bacteriolysin-like enzyme to target *E. faecalis*. The construct, EC300, was created from the fusion of the peptidase domain from a virion-associated lysin and the cell wall-binding domain of Lys170, both found in phage F170/08. EC300 inhibited the growth of *E. faecalis* in bacterial culture media, whereas the parental endolysin Lys170 showed limited inhibition. The enhanced lytic and killing ability of EC300 highlights the potential for engineered endolysins compared with wild-type enzymes.

Streptococcal infections are associated with a range of clinical manifestations, including strep throat, pneumonia, skin infections, and meningitis [61,62]. Drug resistant streptococcal strains are also increasing in prevalence. Loeffler et al. [28] examined the potential of the endolysin Pal to kill *S. pneumoniae* that had colonized the nasopharynx of mice. Pal was effective against 15 different strains of pneumococci, including some drug-resistant strains, reducing *S. pneumoniae* to undetectable levels within 5 h of treatment. The same research group [63] then examined the antibacterial ability of a previously described lytic enzyme, Cpl-1 [64]. Rabbit antiserum was raised against Cpl-1 and the impact on bacterial lysis measured. Only a small amount of inhibition was measured, showing the antibodies had little effect on the enzymatic activity [63]. In the same study, these findings were corroborated in vivo. Mice that were subjected to several doses of Cpl-1 tested positive for IgG. These immunized mice along with naïve mice were challenged with *S. pneumoniae*. Comparatively, no significant difference was measured with regards to the reduction of bacteria numbers by the enzyme [64]. Mice intravenously infected with *S. pneumoniae* and treated only with buffer had a median survival time of 30.75 h, with no mice surviving at 72 h post-infection. Mice treated with Cpl-1 at 5 and 10 h post-infection had a median survival time of 60 h, although after 96 h, only one mouse survived. Although the potential for complete eradication of the bacteria was shown, the dosage used in the experiments was not high enough, meaning that all animals eventually succumbed to infection. Therefore, for greater efficacy of Cpl-1, a higher dose would be required [63].

Subsequent studies demonstrated the effectiveness of Cpl-1 combination therapy with lytic enzyme Pal, or with antibiotics [65–67]. In vivo mouse model studies demonstrated these Cpl-1 treatments were effective in the treatment of pneumococcal diseases such as sepsis [68], endocarditis [69], meningitis [70], and pneumonia [71,72]. Despite its demonstrable antibacterial activity, a key limitation of Cpl-1 is its half-life in blood of only 20.5 min in mice [63,69]. Resch et al. [72] introduced specific cysteine residues into Cpl-1 to promote disulfide bond formation and subsequent dimerization. Dimerization is required for full activity of LytA, a pneumococcal autolysin [73], with which Cpl-1 shares extensive sequence similarity. Dimerized Cpl-1 displayed a two-fold increase in antimicrobial activity and had a nearly ten-fold decrease in plasma clearance, resulting in an increased half-life. These enhanced properties not only increase the prospects for Cpl-1 application, but also highlight the potential for enhanced activity through structural changes and engineering.

The exogenous treatment of Gram-negative bacteria with endolysins has been limited because of the presence of the outer membrane, which prevents access to the peptidoglycan layer [74]. Overcoming this protective layer is a major obstacle in developing endolysin-based treatments. Most studies have focused on nosocomial pathogens *Acinetobacter baumannii* and *Pseudomonas aeruginosa*, both of which are Gram-negative and capable of forming biofilms. Multidrug-resistant strains of both pathogens are also being increasingly isolated [75–78]. Some endolysins can intrinsically permeate the outer membrane [79–81]. Lai et al. [79] recombinantly expressed LysAB2 from ϕAB2 and applied it to *A. baumannii*. The C-terminus of LysAB2 contains an amphipathic α-helix that interacts with the negatively charged elements of the outer membrane, facilitating the formation of a transmembrane pore [82]. This allows the N-terminus catalytic domain to interact with the peptidoglycan layer and lyse the cell, achieving antibacterial activity. Lood et al. [80] identified and screened 21 different endolysins for sequence diversity and *A. baumannii*-killing activity. The endolysins displayed varying degrees of antibacterial activity, with the lysin PlyF307 exhibiting the greatest activity. The C-terminus of PlyF307 contains a highly positively charged region, which may interact with the outer membrane. PlyF307 successfully killed both planktonic cells and, more importantly, those within biofilms, providing an advantage over antibiotics. PlyF307 also functioned under physiological conditions, rescuing mice treated with lethal doses of *A. baumannii*. There remains a library of lysins for structural and biochemical characterization from this research.

Walmagh et al. [81] showed that OBPgp279, from the phage OBP, can permeate the outer membrane of *P. aeruginosa*. It does not appear to contain an amphipathic helix, and thus the mechanism of permeabilization is unclear. OBPgp279 may therefore contain novel structural elements that could lend themselves to the engineering of new endolysins to target Gram-negative bacteria [81]. The lysin LysPA26 from phage JD010 has antibacterial activity against both planktonic and biofilm-contained *P. aeruginosa* cells, as well as other Gram-negative species such as *E. coli* and *Klebsiella pneumoniae*. However, LysPA26 was ineffective against Gram-positive species, including *S. aureus*. This specificity may allow for selective targeting of Gram-negative species in medical treatments [83].

Strategies employing endolysins in conjunction with antibiotics or outer membrane-permeabilizing agents have been explored. Thummeepak et al. [84] investigated the use of LysABP-01 in combination with colistin for treatment of hospital-isolated strains of *A. baumannii*. Although LysABP-01 alone prevented growth, elevated levels of antibacterial activity were observed in combination with colistin. Additionally, the minimum inhibitory concentrations of LysABP-01 and colistin were reduced 32-fold and by up to eight-fold, respectively, when used in combination, compared with individual application. Endolysin EL188, from phage EL, combined with EDTA reduced *P. aeruginosa* cell counts by up to four log units, whereas EL188 on its own exhibited no antibacterial activity [85]. However, the use of EDTA would be restricted to topical applications because of its ability to inhibit blood coagulation [86].

Artilysin bioengineering has also shown promise in targeting Gram-negative bacteria. Artilysins are created through the fusion of outer membrane-permeabilizing peptides, which interfere with the stabilizing forces within the outer membrane, with endolysins. These fusion proteins allow the uptake of an endolysin across the outer membrane, providing access to the peptidoglycan layer [87]. Initially, two endolysins, OBPgp279 and PVP-SE1g-146, fused with one of seven different peptides were investigated for their antibacterial activity. Although the endolysins exhibited limited antibacterial activity on their own, fusion with polycationic nonapeptide correlated with up to a 2.6 log reduction of *P. aeruginosa* in only 30 min. Moderate antibacterial activity against *A. baumannii*, *E. coli*, and *Salmonella* Typhimurium was also observed. The mode of action was examined using time-lapse microscopy, confirming that the artilysins were passing through the outer membrane and degrading the peptidoglycan [24].

Briers et al. [25] developed Art-175, composed of the antimicrobial peptide (AMP) sheep myeloid 29-amino acid peptide (SMAP-29) fused with endolysin KZ144, to target *P. aeruginosa*. AMPs are involved in the innate immune response, and can move across the outer membrane [88]. Art-175 reduced the *P. aeruginosa* cell count by up to 4 log units compared with untreated controls, and continuous exposure to Art-175 to exert a selection pressure did not elicit the development of resistance. On its own, SMAP-29 is cytotoxic to mammalian cells [89]; however, Art-175 exhibited little toxicity in L-292 mouse connective tissue. As a result of these findings, Peng et al. [26] developed AMPs based on the amino acid sequence of LysAB2. The synthesized AMPs killed *A. baumannii* cells by permeating the outer membrane in vitro. Treatment with the AMPs also increased the survival rate of mice infected with a lethal dose of *A. baumannii* by 60%. This research highlights a novel method of bioengineering endolysins for use as antimicrobials.

3. Endolysins as Veterinary Treatments

In response to widespread concern regarding the overuse of antibiotics in food-producing animals, many major food suppliers are now committed to phasing out prophylactic antibiotic use and the tighter control of therapeutic treatments. This presents obvious challenges for the animal-husbandry industry, which may be overcome by the use of endolysins. Companion and working animals can also be susceptible to recalcitrant microbial infections, including those caused by multidrug-resistant microorganisms, and so may also benefit from endolysin treatment.

Clostridium perfringens is a leading cause of necrotic enteritis and sub-clinical disease in poultry, and can lead to significant economic losses [90]. Swift et al. [22] constructed recombinant endolysin PlyGVE2CpCWB, which shows enhanced thermostability, an important feature for surviving feed heat treatments. The recombinant endolysin contains an amidase domain from an endolysin derived from a thermophilic phage fused with the cell wall-recognition domain from *C. perfringens*-specific phage endolysin PlyCP26F, which is not resistant to high temperatures. PlyGVE2CpCWB inactivated *C. perfringens* in both liquid and solid media at temperatures up to 50 °C, and so may be a promising antimicrobial feed treatment for controlling necrotic enteritis in poultry. This thermostable construct demonstrates the potential for other thermophilic bacteriophage endolysins to be utilized in bioengineering. A different approach to *C. perfringens* control was taken by Gervasi et al. [91,92], whereby an amidase endolysin (CP25L) was cloned and expressed in a *Lactobacillus johnsonii* strain isolated from poultry. In co-cultures, reductions of up to 2.6 $\log_{10}$ CFU·ml^{-1} *C. perfringens* were noted; however, the reduction was inconsistent between experiments, and the effect declined significantly over time. This reduced activity was attributed to a loss in stability of the endolysin in culture, and a reduction in the viability of *L. johnsonii*. Other researchers [93,94] have also performed detailed analyses, including X-ray crystallography, on another *C. perfringens* endolysin, Psm, which may have applications in poultry. Psm is an *N*-acetylmuramidase endolysin with wide activity against *C. perfringens* strains.

Another economically significant disease in animal husbandry is bovine mastitis, which is caused by a variety of bacteria, of which staphylococci and streptococci account for 75% of cases [95]. In studies aimed at treating bovine mastitis caused by *S. aureus*, Schmelcher et al. [21] demonstrated that fusion of an endopeptidase endolysin domain from a *Streptococcus* lambda phage (SA2) with either lysostaphin (SA2-E-Lyso-SH3b) or staphylococcal phage endolysin LysK (SA2-E-LysK-SH3b) could inhibit staphylococci in a murine mammary mastitis model. The extended lytic spectrum targeting multiple genera would be particularly useful for efficiently treating bovine mastitis [96]. Infusion of 25 μg of SA2-E-Lyso-SH3b or SA2-E-LysK-SH3b into the mammary glands reduced *S. aureus* counts by 0.63 and 0.81 $\log_{10}$ CFU·mg^{-1}, respectively. Additional testing of SA2-E-LysK-SH3b and lysostaphin in combination (12.5 μg/gland) revealed a 3.36 $\log_{10}$ CFU·mg^{-1} reduction in the concentration of *S. aureus* compared with the control [21]. Further work by the same group [97], determined the potential of the lambda SA2 and phage B30 (a CHAP endopeptidase) endolysins in combination as a therapeutic treatment of *Streptococcus*-induced mastitis, again in a murine mastitis model. The best results obtained by the study were reductions of 1.5 $\log_{10}$ CFU·mg^{-1} against *Streptococcus uberis* (SA2), 4.6 $\log_{10}$ CFU·mg^{-1} against *Streptococcus agalactiae* (B30), and 2.2 $\log_{10}$ CFU·mg^{-1} against *Streptococcus dysgalactiae* (SA2). More recently, purified endolysin Trx-SA1, isolated from *S. aureus* phage IME-SA1, was used to treat naturally infected cow udders [98]. Udder quarters received an intramammary infusion of 20 mg of Trx-SA1 once per day, and qualitative reductions in somatic cell counts and *S. aureus* numbers were noted over the three-day regime.

Anthrax, a potentially fatal zoonotic disease affecting a wide variety of species, is a threat to wild and farmed animals as well as humans, especially as a biological weapon [99,100]. Schuch et al. [29] reported the usefulness of PlyG lysin, isolated from the γ-phage of *Bacillus anthracis*, in killing vegetative cells and germinating spores of *B. anthracis* and streptomycin-resistant *Bacillus cereus* RSVF1. The researchers screened an expression library of cloned γ-phage DNA sequences and identified a 702-bp open reading frame (ORF) encoding a protein with homology to an amidase-type endolysin. When PlyG was injected intraperitoneally (50 U in 0.5 mL) into mice infected with 6 $\log_{10}$ CFU RSVF1, a notable therapeutic effect was observed, with 68.4% (13/19) of mice showing full recovery. Furthermore, the survival time of the remaining mice was prolonged to 21 h post-infection.

Equine strangles is a highly contagious disease of horses caused by *Streptococcus equi*. The disease progresses as an inflammation of the upper respiratory tract, and leads to abscess formation in the retropharyngeal lymph node [101]. Strangles is a significant economic threat to the horse racing industry, where many high value animals are typically housed in close proximity. Hoopes et al. [102] used PlyC, an unusual multimeric amidase-type endolysin [17], as a disinfectant against *S. equi* and reported it to be 1000 times more active on a per weight basis than the widely used disinfectant Virkon-S. PlyC was effective against >20 clinical isolates of *S. equi*, including both *S. equi* subsp. *equi* and *S. equi* subsp. *zooepidemicus*, demonstrating its sterilizing ability against an eight log CFU·ml^{-1} culture of *S. equi* within 30 min of exposure.

The clinical efficacy of a muramidase as a veterinary treatment for companion animals was demonstrated in a trial by Junjappa et al. [103], where they successfully treated 17 dogs suffering from pyoderma (bacterial skin lesions) caused by MRSA. The skin lesions were treated with a hydrogel containing a chimeric endolysin composed of the cell wall-targeting domain (SH3b) of lysostaphin and the phage K ORF56 muralytic domain [104]. Another important zoonotic pathogen, *Streptococcus suis*, has been linked to arthritis, meningitis, septicemia, and endocarditis in pigs, as well as in humans who have come into contact with infected animals or their byproducts. Wang et al. [105] isolated a phage from *S. suis* (SMP), and then expressed the endolysin LySMP in *E. coli* BL21. The resultant product, following chromatography and treatment with β-mercaptoethanol, killing 15 out of 17 clinical *S. suis* serotype 2 isolates from diseased pigs in China, and had demonstrated activity against *S. suis* serotype 7 and 9 strains, *S. equi* subsp. *zooepidemicus*, and *S. aureus* [105].

There is growing evidence to suggest that food-producing animals are an important global reservoir of vancomycin-resistant enterococci (VRE) [106,107]. This group of potentially invasive microorganisms is resistant to almost all of the available antibiotic regimens recommended for treatment of Gram-positive bacterial infections. In an attempt to combat VRE in food-producing animals, Yoong et al. [57] cloned the PlyV12-encoding gene from enterococcal phage Φ1 into the *E. coli-Bacillus* shuttle vector pDG148, followed by its expression in *Bacillus megaterium* strain WH320. The resultant product, an amidase-type endolysin, had lytic activity against 14 clinical and laboratory *E. faecalis* and *Enterococcus faecium* strains, including two vancomycin-resistant *E. faecalis* and three vancomycin-resistant *E. faecium* strains, in addition to its host, *E. faecalis* V12. Intriguingly, PlyV12 also had a significant killing effect on pathogenic streptococcal strains, including *Streptococcus pyogenes* (group A streptococcus) and group C streptococci.

Diarrheal outbreaks caused by *Clostridium difficile* have frequently been reported in animals, including cattle, horses, and pigs [108,109]. Treatment of *C. difficile* diarrhea with antibiotics is not recommended as it can further exacerbate the disease condition [110]. In a quest for an alternative approach to treat infections caused by *C. difficile*, Mayer et al. [111] sequenced the genome of a temperate phage from *C. difficile*. They identified endolysin gene *cd27l* and cloned it into vectors pET15b and pUK200 to express the gene product in *E. coli* and *L. lactis*, respectively. The purified endolysin was active against 30 diverse strains of *C. difficile*, including those belonging to the major epidemic ribotype, 027 (B1/NAP1). Unlike antibiotics, the endolysin was selective for *C. difficile*, demonstrating no activity against a range of commensal species from within the gastrointestinal tract, including other clostridia, bifidobacteria, and lactobacilli.

Paenibacillus larvae subsp. *larvae* causes American Foulbrood disease in honey bees, which are important insect pollinators of agricultural crops. The disease occurs in honeybee larvae as a result of *P. larvae* spores germinating in the larval midgut and subsequently causing sepsis and death. The use of antibiotics to treat the disease in the USA now requires supervision by a veterinarian, and a withholding period of 4–6 weeks is recommended for honey from treated hives prior to sale (https://www.fda.gov/AnimalVeterinary/ResourcesforYou/AnimalHealthLiteracy/ucm309134.htm). In the European Union, no veterinary medicines containing antibiotics are permitted in beekeeping (http://europroxima.com/european-legislation-regarding-antibiotics-in-honey-an-overview/). For these reasons, endolysins are being investigated as a potential alternative control tool. An amidase endolysin, PlyPl23, has been cloned from a *P. larvae* phage and subsequently expressed [112]. In bee larvae experimentally infected with spores, PlyPl23 effectively decreased the rate of *P. larvae* infection, and no toxic side effects were noted in the larvae. However, the endolysin was not effective until the spores had germinated.

4. Endolysins as Food and Environmental Decontaminants

During post-harvest processing, food is vulnerable to cross-contamination from microbial pathogens, which pose a risk to food safety, as well as from microorganisms that can cause quality or shelf-life defects. Effective interventions for foods and the food processing environment are therefore vital to maintain the integrity of the food supply chain. Endolysins have the potential to be key intervention tools for this purpose.

The use of endolysins to prevent contamination of ready-to-eat foods by the common food and environmental pathogen *Listeria monocytogenes* has been established by groups from around the world [16,113–117]. Zhang et al. [113] cloned an endolysin gene (*lysZ5*) from the genome of *L. monocytogenes* phage FWLLm3 into *E. coli* and tested the sterilization efficacy of the expressed protein (a murine hydrolase) in soya milk contaminated with *L. monocytogenes*. The purified protein had a bactericidal effect on *L. monocytogenes* growing in soya milk, with the pathogen concentration reduced by more than 4 $\log_{10}$ CFU·ml^{-1} after 3 h of incubation at 4 °C. Furthermore, the protein displayed a broad host spectrum, lysing lawn cultures of *L. monocytogenes*, *Listeria innocua*, and *Listeria welshimeri*. In a different approach, van Nassau et al. [114] tested the combined effect of previously characterized endolysins (PlyP40, Ply511, or PlyP825) and high hydrostatic pressure on the survival of *L. monocytogenes*. They reported that the combination of treatments had a synergistic effect, capable of reducing viable cell counts of *L. monocytogenes* by up to 5.5 $\log_{10}$, compared with 0.3 and 0.2 $\log_{10}$ CFU reductions, respectively, when used alone.

Turner et al. [115] and Gaeng et al. [16] demonstrated that the *L. monocytogenes* endolysin gene *ply511* can be cloned and expressed in *Lactobacillus* spp., which have potential as biopreservatives in foods and as a starter culture for fermented milk products. Further to this, Turner et al. [115] combined the cloned A511 phage *ply511* gene with a lysostaphin-encoding *lss* gene from *S. simulans* biovar *staphylolyticus* in-frame with a Sep secretion signal. The resulting construct, Sep-6_His-Ply511, was able to secrete both Ply511 and lysostaphin from *Lactobacillus lactis*, indicating that this recombinant organism could be used for industrial applications as a preservative to prevent contamination of foods with *Staphylococcus* spp. and *L. monocytogenes*.

Staphylococcal food poisoning caused by heat stable enterotoxins produced by *S. aureus* is frequently reported following the consumption of contaminated food and milk products [118]. Chang et al. [119] tested LysSA11, a *S. aureus* phage SA11-derived endolysin, to determine its bactericidal activity in food and on food utensils artificially contaminated with MRSA. Treatment of artificially contaminated ham and pasteurized milk with endolysin for 15 min resulted in 3.1 $\log_{10}$ CFU·cm^{-3} and 1.4 $\log_{10}$ CFU·ml^{-1} reductions in viable MRSA, respectively, at refrigeration temperature (4 °C), and 3.4 $\log_{10}$ CFU·cm^{-3} and 2.0 $\log_{10}$ CFU·ml^{-1} reductions, respectively, at room temperature (25 °C). The same group [120] tested the antibacterial potential of LysSA97 in combination with several active compounds derived from essential oils used by the food industry against *S. aureus*. They reported the superior activity of carvacrol in combination with LysSA97, compared with that of the endolysin alone, in food products including beef and milk contaminated with *S. aureus*. When used alone, LysSA97 and carvacrol reduced *S. aureus* concentrations by 0.8 and 1.0 $\log_{10}$ CFU·ml^{-1}, respectively; however, a synergistic reduction of 4.5 $\log_{10}$ CFU·ml^{-1} was observed when the treatments were combined. Similarly, Obeso et al. [121] and Rodriguez-Rubio et al. [122] demonstrated the potential of phage-derived endolysins LysH5 and HydH5 (a hydrolase), respectively, to protect milk products from *S. aureus* contamination.

Salmonella species are the leading cause of bacterial foodborne illness in the USA and many other countries [123]. *Salmonella* disease outbreaks are associated with a wide variety of food products, including red meats, poultry, and produce [124]. Several recombinant endolysins derived from *Salmonella* phages have been characterized [125–127]. Interestingly, many of these endolysins have activity outside of the host species of the parental phage, particularly when cell membrane-disrupting chemicals are used in conjunction with the enzymes. Lim et al. [125] expressed the endolysins and spanin proteins from *Salmonella* phage SPN1S and observed activity against both *Salmonella* Typhimurium and *E. coli* isolates in buffer containing EDTA to destabilize the cell membranes. Furthermore, some activity was also noted against typhoidal salmonellae, *Shigella*, *Cronobacter*, *Pseudomonas*, and *Vibrio* species. Oliveira et al. [126] characterized a thermostable *Salmonella* endolysin, Lys68, that displayed better activity at neutral pH and a wide temperature tolerance, maintaining 76.7% of its activity after 2 months at 4°C and partial activity following exposure to 100 °C for 30 min. Thermostability is a useful feature for diverse application, such as in heat treatment of food When Lys68 was tested in combination with citric or malic acid against *S.* Typhimurium LT2, up to 5 $\log_{10}$ CFU reductions were achieved, with cells in stationary phase or in biofilms also reduced by up to 1 $\log_{10}$ CFU [126]. More recently, Rodriguez-Rubio et al. [127] reported a *Salmonella* phage endolysin, Gp110, that possessed both a novel enzyme structure and N-acetylmuramidase lysis domain, and had unusually high in vitro activity against *Salmonella* and other Gram-negative pathogens [127].

Around the turn of the century, a rare but frequently fatal disease of neonatal infants was first reported to be associated with contamination of powdered infant milk formula with *Enterobacter* (now *Cronobacter*) *sakazakii* [128]. It is now known that several species of *Cronobacter* can cause a variety of diseases, including sepsis and severe meningitis, in neonates, as well as respiratory and urinary tract infections in elderly and immunocompromised individuals. These opportunistic pathogens are now under much scrutiny because of their ability to survive heat, desiccation, and acid stress, which poses a risk of contamination of various milk powders, herbal teas, and other dried products (https://www.cdc.gov/cronobacter/technical.html). Enderson et al. [129,130] expressed and purified a peptidoglycan hydrolase (LysCs4) from *C. sakazakii* that had the highest sequence similarity to a putative lysozyme from the temperate *Cronobacter* phage ES2. The purified lysozyme could degrade the peptidoglycan from both Gram-negative and Gram-positive bacteria belonging to six different genera, and could lyse outer membrane-permeabilized *C. sakazakii*. Similarly, the previously discussed endolysins SPN1S [125] and Lys68 [126] are active against permeabilized *C. sakazakii* cells.

Several groups have investigated the potential of endolysins active against *B. cereus* as biocides and preservatives for use in the food industry [131–133]. *B. cereus*, a Gram-positive spore-forming bacterium, is known for its ability to cause food poisoning by producing both an emetic toxin and a diarrheal toxin [131]. Loessner et al. [132] isolated and characterized three endolysins (PlyBa, Ply12, and Ply21) from the *B. cereus* phages Bastille, TP21, and 12826, respectively, and tested their efficacy against a range of Gram-positive and Gram-negative bacteria. They reported that all three endolysins were effective against 24 strains of *B. cereus*, along with several strains of *B. thuringiensis*. Ply12 and Ply21 were found to be *N*-acetylmuramoyl-L-alanine amidases, while PlyBa could not be classified at the time, but is also likely to be an amidase (http://www.uniprot.org/uniprot/P89927). Park et al. [133] isolated a putative endolysin gene from the genome of *B. cereus* phage BPS13, and expressed it in *E. coli*. The purified LysBPS13 protein, an amidase, retained its lytic activity against *B. cereus* ATCC 10876 even after incubation at 100 °C for 30 min, demonstrating its potential as a decontaminant in food processing applications. In contrast, Son et al. [131] proposed L-alanoyl-D-glutamate endopeptidase LysB4 as a potential biocontrol agent against *B. cereus* and other pathogenic bacteria. They confirmed the endopeptidase had a broad range of bactericidal activity against Gram-positive bacteria, including *B. cereus*, *B. subtilis*, and *L. monocytogenes*, and also a few Gram-negative bacteria. The endolysin showed optimum lytic activity at pH 8–10 and at 50 °C, making it a suitable candidate for use in the food industry.

In addition to causing diseases in poultry, clostridial species are linked to food spoilage. In the dairy industry, germinated *Clostridium sporogenes* and *Clostridium tyrobutyricum* can contribute to the production of gases and acids that change the structural and sensory qualities of cheeses [134]. Mayer et al. [135] isolated an *N*-acetylmuramoyl-L-alanine amidase, CS74L, from *C. sporogenes* and reported that the purified protein effectively lysed *C. sporogenes* cells when added exogenously. Using the turbidity assay and fresh bacterial cells, the authors also demonstrated that CS74L was active against *C. tyrobutyricum* and *Clostridium acetobutylicum*, making it a candidate biopreservative for use in cheese. The same group also characterized another endolysin isolated from a virulent phage, CPT1l, but this enzyme had a more limited host range [134].

The dairy industry has a long-held interest in utilizing endolysins to control the cheese maturation process. Vasala et al. [136] isolated a muramidase, Mur, from the LL-H phage of *Lactobacillus delbrueckii* subsp. *lactis* that had activity against cell wall preparations of *L. delbrueckii* subsp. *lactis*, *L. delbrueckii* subsp. *bulgaricus*, *Lactobacillus acidophilus*, *Lactobacillus helveticus*, and *Pediococcus damnosus*. Similarly, Deutsch et al. [137] purified endolysin Mur-LH from a phage infecting the Swiss cheese starter *L. helveticus*. This muramidase had activity against other lactobacilli, *Leuconostoc lactis*, *P. acidilactici*, and, surprisingly, against *B. subtilis*. Kashige et al. [138] isolated an *N*-acetylmuramoyl-L-alanine amidase from phage PL-1, which was originally isolated from an abnormal fermentation of a lactic acid beverage produced using a *Lactobacillus casei* strain. There are also many examples of endolysins isolated from lactococci, including from phages P001, C2 US3, and TUC2009 [136]. A survey of 18 *L. lactis* phage endolysins revealed that muramidases and amidases predominate [139].

In addition to the aforementioned enzymes, several other endolysins with potential to be used against a range of foodborne microorganisms in different food types have been identified, and selected examples of these are illustrated in Table 2.

Table 2. Examples of other potential uses of endolysins in foods.

Food	Organism	Endolysin	Reference
Fish	*Shewanella putrefaciens*	ORF62	Han et al. [140]
Vegetable fermentation	*Leuconostoc mesenteroides*	ORF35	Lu et al. [141]
Kimchi	*Lactobacillus plantarum*	SC921 lysin	Yoon et al. [142]
Pears	*Erwinia amylovora*	ΦEa1h lysozyme	Kim et al. [143]
Banana juice	*Salmonella* Typhimurium, *Yersinia enterocolitica*, *Escherichia coli O157:H7*, *Shigella flexneri*	λ lysozyme (with high pressure treatment)	Nakimbugwe et al. [144]

Table 2. *Cont.*

Food	Organism	Endolysin	Reference
Shellfish	*Vibrio parahaemolyticus*	Lysqdvp001	Wang et al. [145]
Lettuce	*Listeria innocua*	Ply500 (with packaging film)	Solanki et al. [146]
Milk	*Listeria monocytogenes*	PlyP825 (with high pressure treatment)	Misiou et al. [147]
Mozzarella cheese	*L. monocytogenes*	PlyP825 (with high pressure treatment)	Misiou et al. [147]

The growth of biofilms in food processing environments leads to an increased risk of microbial contamination of foods [148]. There are some examples of endolysins being used to disrupt biofilms with relevance to the food industry. Gutierrez et al. [149] investigated the activity of three endolysins (LysH5, CHAP-SH3b, and HydH5-SH3b) against biofilms formed by two *S. aureus* isolates from food. Preformed biofilms were treated with 7 μM of the enzymes, with LysH5 and CHAP-SH3b most effective against strains IPLA1 and Sa9, respectively. In another study, an *N*-acetylmuramoyl-L-alanine amidase was used to disrupt *Listeria* biofilms in vitro, but was found to be most effective when used in combination with a protease [150]. The *Salmonella*-phage endolysin Lys68 also reduced the concentration of cells in a *S.* Typhimurium LT2 biofilm by up to 1.2 $\log_{10}$ CFU [128], but this required the presence of either citric or maleic acid to permeabilize the cell membranes.

Phytopathogenic bacteria have a significant global economic cost, and are the cause of multiple food security issues [151]. The use of antibiotics in plant agriculture is controversial because its contribution to the development of antibiotic resistance in human pathogens is undetermined [152] Although its impact may be small, ideally, an alternative strategy to control phytopathogenic bacteria will be developed. As such, the use of endolysins to protect plants from bacterial diseases has been proposed [16]. Widespread implementation of these endolysins will, however, be a significant challenge because of the vast number of crops that would require treatment, and the presence of beneficial soil-borne bacteria. A proposed strategy is the development of transgenic crops that express endolysins, providing protection against the pathogenic bacteria. The potential of this approach has been demonstrated by Düring et al. [153], who produced transgenic potato plants expressing T4 lysozyme. These engineered plants displayed resistance to *Pectobacterium carotovora* (formerly *Erwinia carotovora*) species, which are the cause soft rot [153,154]. Wittmann et al. [155] produced transgenic tomato plants expressing the endolysins from bacteriophage CMP1 in an attempt to prevent *Clavibacter michiganensis* subsp. *michiganensis* infection, the causative agent of bacterial wilt and canker [156]. No symptoms of bacterial infection were observed in the transgenic plants; however, the bacteria were not completely eliminated [155]. A key limitation of this research was that the bacterial infection model may not be representative of natural infection, and therefore the efficacy of these transgenic tomato plants should to be evaluated under more natural conditions. It is also unknown whether *C. michiganensis* subsp. *michiganensis* could acquire resistance to these transgenic plants. However, this is a promising advancement in the development of transgenic plants. As new endolysins are characterized, more opportunities for bioengineering to optimize the activity of the protection mechanism will be possible.

Xanthmonas oryzae pv. *oryzae* causes bacterial leaf blight in rice [157], with a number of antibiotic resistant strains having been isolated [158]. In 2006, Lee et al. [159] identified Lys411 from ΦXo411, which exhibited strong lytic activity against *Xanthmonas*. Additionally, it displayed activity against the multidrug-resistant bacterium *Stenotrophomonas maltophilia* [159], which has growing clinical significance with regards to nosocomial infections and immunocompromised patients [160]. However, no follow-up studies investigating Lys411 have been published, which means the potential of this enzyme for medical or agriculture applications is unknown.

Attai et al. [161] recently characterized an endolysin from bacteriophages Atu_ph02 and Atu_ph03 for the biocontrol of *Agrobacterium tumefaciens. A. tumefaciens* is a Gram-negative soil-borne bacterium that is the etiologic agent of crown gall disease in a variety of orchard and vineyard crops [162].

Its severity and widespread impact has contributed to it to becoming the subject of many recent studies [163]. The lytic protein displayed interesting properties, with the ability to not only rapidly lyse the cell, but to also block cell division, ensuring potent antimicrobial activity [161]. Therefore, the enzyme is a candidate for biocontrol of *A. tumefaciens*; however, the method of implementation needs to be researched before a viable strategy for crop protection can be developed. The practicalities of implementing these endolysins on a global scale for individual phytobacteria may be a significant challenge, and may contribute to the limited information currently available on the use of endolysins for treatment of plant bacterial diseases. However, the cost to society of plant bacterial disease as current strategies become ineffective means that endolysin research should be an important focus.

5. Challenges of Endolysin Development and Engineering

The potential for endolysins to supplement, or replace antibiotics is exciting. However, this field is still emerging, with very few clinical trials on endolysin-based drugs being conducted. There are a number of challenges and considerations which researchers still face to bring these to market. As highlighted by several studies, the immunogenicity of endolysins must be considered and fully assessed. Undesirable immune responses to these foreign proteins could result in decreased efficacy of the enzymes, or possibly anaphylaxis and autoimmunity [164,165]. While there are studies that have reported on the immunogenicity in the application of endolysins [40–42,45,60,63] assessing the degree of immunogenicity in humans using traditional animal models has so far proven unreliable [166]. This was highlighted in studies of SAL200, which showed varying degrees of antibody production between rats, dogs, monkeys and humans [40–42]. Although the efficacy of the endolysin may not be observably impacted in vitro or in vivo, the clinical effects may be more significant. Until more human and animal-specific (for animal husbandry applications) clinical trials are conducted, the immunogenic nature of endolysins will remain unpredicatable.

In light of the current antibiotic resistance crisis, new antibacterials should be rigorously assessed for their potential susceptibility to developed resistance by bacteria. Promisingly, bacterial mutants resistant to endolysins are very infrequent [28–30,167]. Fischetti [167] proposed the lack of developed resistance to endolysins has resulted from the evolution of the interactions between bacteriophage and bacteria. The endolysins have evolved to target essential, immutable, molecules within the cell wall, thereby reducing the likelihood of the bacteria developing resistance mechanisms [167]. However, there are also reports of resistance to other peptidoglycan-cleaving enzymes including lysozymes and lysostaphin [168–173]. In the event of bacterial adaptation, enzyme engineering may prove useful to combat changes to bacteria in order to maintain the efficacy of endolysins.

The potential for endolysin bioengineering are seemingly endless, including optimizing or changing the catalytic abilities, modifying the lytic spectrum, improving its ability to permeate outer membranes and increasing stability (Table 1). Designing new enzymes requires an understanding of the structure and function of individual domains, the interactions between domains, the placement and composition of linker regions, and elucidation of key residues involved in catalysis. Bioinformatics and structural characterization studies are integral in this process [174]. Often, structural characterization can be achieved by X-ray crystallography, a powerful and effective technique for elucidating high resolution 3-D structures of proteins [11]; however, the limited ability to crystallize endolysins is a major challenge. The majority of endolysin crystal structures published to date are of single domains, with very few full-length endolysin crystal structures having been solved. This is attributed to the short flexible linker regions between domains [8], as protein flexibility is a common hindrance of crystal formation [175]. It is important to study full-length proteins to get a better understanding of the potential synergistic/antagonistic interactions between domains. Because of the difficulties in obtaining endolysin crystals, alternative structural characterization strategies need to be considered. These include the fusion of endolysins to proteins to decrease their flexibility, thereby allowing for crystallography, and other structural elucidation techniques such as nuclear magnetic resonance [176,177] and cryo-electron microscopy [178]. Although these approaches also have limitations, such as physiological relevance or size restrictions, exploration of these techniques may advance the structural characterization of endolysins.

6. Conclusions

The field of endolysin research is dynamic, with many potential applications being investigated in the medical, veterinary, and food sectors. The current global crisis of antimicrobial resistance is driving much of this work, with endolysins showing great promise to replace or supplement antibiotics. Engineering endolysins with optimized or new properties provides an opportunity to create even more effective tools. As more bacteriophage endolysins are biochemically and structurally characterized, the ability to design new enzymes improves, therefore expanding the arsenal of lytic tools. However, there are still many challenges that need to be addressed before this technology can be widely adopted by practitioners and industry. While many researchers have described the isolation and in vitro characterization of endolysins, establishing the in vivo efficacy and operating parameters of endolysins for human clinical use, food protection and supplementation, animal husbandry and welfare, and in the environment will be of great importance over the coming years. New technology to cost-effectively scale up endolysin production is also required, as this is currently a significant barrier to implementation. Finally, regulatory pathways need to be established for the use of endolysins in each of the various fields of application, and this can only be achieved by early and effective dialogue with the relevant authorities.

Acknowledgments: This article is supported by ESR SSIF funding. M.L. is supported by an UC Connect scholarship. R.C.J.D. acknowledges the following for funding support, in part: (1) the New Zealand Royal Society Marsden Fund (UOC1506); (2) a Ministry of Business, Innovation and Employment Smart Ideas grant (UOCX1706) the Biomolecular Interactions Centre, University of Canterbury.

Author Contributions: M.J.L., D.B., R.C.J.D. and C.B. wrote the paper.

Conflicts of Interest: The authors declare no conflict of interest.

References

1. World Health Organization. *Antimicrobial Resistance: Global Report on Surveillance*; World Health Organization: Geneva, Switzerland, 2014.
2. World Health Organization. *Global Framework for Development & Stewardship to Combat Antimicrobial Resistance—Draft Roadmap*; World Health Organization: Geneva, Switzerland, 2017.
3. Mwangi, W.; de Figueiredo, P.; Criscitiello, M.F. One health: Addressing global challenges at the nexus of human, animal, and environmental health. *PLoS Pathog.* **2016**, *12*, e1005731. [CrossRef] [PubMed]
4. Young, R. Bacteriophage lysis: Mechanism and regulation. *Microbiol. Rev.* **1992**, *56*, 430–481. [PubMed]
5. Loessner, M.J. Bacteriophage endolysins—Current state of research and applications. *Curr. Opin. Microbiol.* **2005**, *8*, 480–487. [CrossRef] [PubMed]
6. Fischetti, V.A. Bacteriophage endolysins: A novel anti-infective to control Gram-positive pathogens. *Int. J. Med. Microbiol.* **2010**, *300*, 357–362. [CrossRef] [PubMed]
7. Borysowski, J.; Weber-Dabrowska, B.; Gorski, A. Bacteriophage endolysins as a novel class of antibacterial agents. *Exp. Biol. Med. (Maywood)* **2006**, *231*, 366–377. [CrossRef] [PubMed]
8. Schmelcher, M.; Donovan, D.M.; Loessner, M.J. Bacteriophage endolysins as novel antimicrobials. *Future Microbiol.* **2012**, *7*, 1147–1171. [CrossRef] [PubMed]
9. Nelson, D.C.; Schmelcher, M.; Rodriguez-Rubio, L.; Klumpp, J.; Pritchard, D.G. Endolysins as antimicrobials. *Adv. Virus Res.* **2012**, *83*, 299–365. [PubMed]
10. Vollmer, W.; Bertsche, U. Murein (peptidoglycan) structure, architecture and biosynthesis in *Escherichia coli. Biochim. Biophys. Acta. Biomembr.* **2008**, *1778*, 1714–1734. [CrossRef] [PubMed]
11. Donovan, D.M.; Lardeo, M.; Foster-Frey, J. Lysis of Staphylococcal mastitis pathogens by bacteriophage Phi11 endolysin. *FEMS Microbiol. Lett.* **2006**, *265*, 133–139. [CrossRef] [PubMed]

12. Korndorfer, I.P.; Danzer, J.; Schmelcher, M.; Zimmer, M.; Skerra, A.; Loessner, M.J. The crystal structure of the bacteriophage PSA endolysin reveals a unique fold responsible for specific recognition of *Listeria* cell walls. *J. Mol. Biol.* **2006**, *364*, 678–689. [CrossRef] [PubMed]
13. Sass, P.; Bierbaum, G. Lytic activity of recombinant bacteriophage phi11 and phi12 endolysins on whole cells and biofilms of *Staphylococcus aureus*. *Appl. Environ. Microbiol.* **2007**, *73*, 347–352. [CrossRef] [PubMed]
14. Low, L.Y.; Yang, C.; Perego, M.; Osterman, A.; Liddington, R.C. Structure and lytic activity of a *Bacillus anthracis* prophage endolysin. *J. Biol. Chem.* **2005**, *280*, 35433–35439. [CrossRef] [PubMed]
15. Mayer, M.J.; Garefalaki, V.; Spoerl, R.; Narbad, A.; Meijers, R. Structure-based modification of a *Clostridium difficile*-targeting endolysin affects activity and host range. *J. Bacteriol.* **2011**, *193*, 5477–5486. [CrossRef] [PubMed]
16. Gaeng, S.; Scherer, S.; Neve, H.; Loessner, M.J. Gene cloning and expression and secretion of *Listeria monocytogenes* bacteriophage-lytic enzymes in *Lactococcus lactis*. *Appl. Environ. Microbiol.* **2000**, *66*, 2951–2958. [CrossRef] [PubMed]
17. Schmelcher, M.; Tchang, V.S.; Loessner, M.J. Domain shuffling and module engineering of *Listeria* phage endolysins for enhanced lytic activity and binding affinity. *Microb. Biotechnol.* **2011**, *4*, 651–662. [CrossRef] [PubMed]
18. Gerstmans, H.; Criel, B.; Briers, Y. Synthetic biology of modular endolysins. *Biotechnol. Adv.* **2017**, in press. [CrossRef] [PubMed]
19. Horgan, M.; O'Flynn, G.; Garry, J.; Cooney, J.; Coffey, A.; Fitzgerald, G.F.; Ross, R.P.; McAuliffe, O. Phage lysin Lysk can be truncated to its chap domain and retain lytic activity against live antibiotic-resistant staphylococci. *Appl. Environ. Microbiol.* **2009**, *75*, 872–874. [CrossRef] [PubMed]
20. Daniel, A.; Euler, C.; Collin, M.; Chahales, P.; Gorelick, K.J.; Fischetti, V.A. Synergism between a novel chimeric lysin and oxacillin protects against infection by methicillin-resistant *Staphylococcus aureus*. *Antimicrob. Agents Chemother.* **2010**, *54*, 1603–1612. [CrossRef] [PubMed]
21. Schmelcher, M.; Powell, A.M.; Becker, S.C.; Camp, M.J.; Donovan, D.M. Chimeric phage lysins act synergistically with lysostaphin to kill mastitis-causing *Staphylococcus aureus* in murine mammary glands. *Appl. Environ. Microbiol.* **2012**, *78*, 2297–2305. [CrossRef] [PubMed]
22. Swift, S.; Seal, B.; Garrish, J.; Oakley, B.; Hiett, K.; Yeh, H.-Y.; Woolsey, R.; Schegg, K.; Line, J.; Donovan, D. A thermophilic phage endolysin fusion to a *Clostridium perfringens*-specific cell wall binding domain creates an anti-clostridium antimicrobial with improved thermostability. *Viruses* **2015**, *7*, 3019–3034. [CrossRef] [PubMed]
23. Proença, D.; Leandro, C.; Garcia, M.; Pimentel, M.; São-José, C. EC300: A phage-based, bacteriolysin-like protein with enhanced antibacterial activity against *Enterococcus faecalis*. *Appl. Microbiol. Biotechnol.* **2015**, *99*, 5137–5149. [CrossRef] [PubMed]
24. Briers, Y.; Walmagh, M.; Van Puyenbroeck, V.; Cornelissen, A.; Cenens, W.; Aertsen, A.; Oliveira, H.; Azeredo, J.; Verween, G.; Pirnay, J.-P.; et al. Engineered endolysin-based "Artilysins" to combat multidrug-resistant Gram-negative pathogens. *mBio* **2014**, *5*. [CrossRef] [PubMed]
25. Briers, Y.; Walmagh, M.; Grymonprez, B.; Biebl, M.; Pirnay, J.-P.; Defraine, V.; Michiels, J.; Cenens, W.; Aertsen, A.; Miller, S.; et al. Art-175 is a highly efficient antibacterial against multidrug-resistant strains and persisters of *Pseudomonas aeruginosa*. *Antimicrob. Agents Chemother.* **2014**, *58*, 3774–3784. [CrossRef] [PubMed]
26. Peng, S.-Y.; You, R.-I.; Lai, M.-J.; Lin, N.-T.; Chen, L.-K.; Chang, K.-C. Highly potent antimicrobial modified peptides derived from the *Acinetobacter baumannii* phage endolysin LysAB2. *Sci. Rep.* **2017**, *7*, 11477. [CrossRef] [PubMed]
27. O'Flaherty, S.; Coffey, A.; Meaney, W.; Fitzgerald, G.F.; Ross, R.P. The recombinant phage lysin LysK has a broad spectrum of lytic activity against clinically relevant staphylococci, including methicillin-resistant *Staphylococcus aureus*. *J. Bacteriol.* **2005**, *187*, 7161–7164. [CrossRef] [PubMed]
28. Loeffler, J.M.; Nelson, D.; Fischetti, V.A. Rapid killing of *Streptococcus pneumoniae* with a bacteriophage cell wall hydrolase. *Science* **2001**, *294*, 2170–2172. [CrossRef] [PubMed]
29. Schuch, R.; Nelson, D.; Fischetti, V.A. A bacteriolytic agent that detects and kills *Bacillus anthracis*. *Nature* **2002**, *418*, 884–889. [CrossRef] [PubMed]
30. Pastagia, M.; Euler, C.; Chahales, P.; Fuentes-Duculan, J.; Krueger, J.G.; Fischetti, V.A. A novel chimeric lysin shows superiority to mupirocin for skin decolonization of methicillin-resistant and -sensitive *Staphylococcus aureus* strains. *Antimicrob. Agents Chemother.* **2011**, *55*, 738–744. [CrossRef] [PubMed]

31. Kretzer, J.W.; Lehmann, R.; Schmelcher, M.; Banz, M.; Kim, K.P.; Korn, C.; Loessner, M.J. Use of high-affinity cell wall-binding domains of bacteriophage endolysins for immobilization and separation of bacterial cells. *Appl. Environ. Microbiol.* **2007**, *73*, 1992–2000. [CrossRef] [PubMed]
32. Enright, M.C.; Robinson, D.A.; Randle, G.; Feil, E.J.; Grundmann, H.; Spratt, B.G. The evolutionary history of methicillin-resistant *Staphylococcus aureus* (MRSA). *Proc. Natl. Acad. Sci. USA* **2002**, *99*, 7687–7692. [CrossRef] [PubMed]
33. The White House. *National Action Plan for Combating Antibiotic-Resistant Bacteria. Interagency Task Force for Combating Antibiotic-Resistant Bacteria*; U.S. Office of the Press Secretary: Washington, DC, USA, 2015.
34. Filatova, L.Y.; Becker, S.C.; Donovan, D.M.; Gladilin, A.K.; Klyachko, N.L. Lysk, the enzyme lysing *Staphylococcus aureus* cells: Specific kinetic features and approaches towards stabilization. *Biochimie* **2010**, *92*, 507–513. [CrossRef] [PubMed]
35. Becker, S.C.; Foster-Frey, J.; Donovan, D.M. The phage K lytic enzyme LysK and lysostaphin act synergistically to kill MRSA. *FEMS Micriobiol. Lett.* **2008**, *287*, 185–191. [CrossRef] [PubMed]
36. Fenton, M.; Casey, P.G.; Hill, C.; Gahan, C.G.M.; Ross, R.P.; McAuliffe, O.; O'Mahony, J.; Maher, F.; Coffey, A. The truncated phage lysin CHAP(k) eliminates *Staphylococcus aureus* in the nares of mice. *Bioeng. Bugs* **2010**, *1*, 404–407. [CrossRef] [PubMed]
37. Fenton, M.; Ross, R.P.; McAuliffe, O.; O'Mahony, J.; Coffey, A. Characterization of the staphylococcal bacteriophage lysin CHAP(k). *J. Appl. Microbiol.* **2011**, *111*, 1025–1035. [CrossRef] [PubMed]
38. Jun, S.Y.; Jung, G.M.; Son, J.-S.; Yoon, S.J.; Choi, Y.-J.; Kang, S.H. Comparison of the antibacterial properties of phage endolysins SAL-1 and Lysk. *Antimicrob. Agents Chemother.* **2011**, *55*, 1764–1767. [CrossRef] [PubMed]
39. Jun, S.Y.; Jung, G.M.; Yoon, S.J.; Oh, M.-D.; Choi, Y.-J.; Lee, W.J.; Kong, J.-C.; Seol, J.G.; Kang, S.H. Antibacterial properties of a pre-formulated recombinant phage endolysin, SAL-1. *Int. J. Antimicrob. Agents* **2013**, *41*, 156–161. [CrossRef] [PubMed]
40. Jun, S.Y.; Jung, G.M.; Yoon, S.J.; Choi, Y.-J.; Koh, W.S.; Moon, K.S.; Kang, S.H. Preclinical safety evaluation of intravenously administered SAL200 containing the recombinant phage endolysin SAL-1 as a pharmaceutical ingredient. *Antimicrob. Agents Chemother.* **2014**, *58*, 2084–2088. [CrossRef] [PubMed]
41. Jun, S.Y.; Jung, G.M.; Yoon, S.J.; Youm, S.Y.; Han, H.-Y.; Lee, J.-H.; Kang, S.H. Pharmacokinetics of the phage endolysin-based candidate drug SAL200 in monkeys and its appropriate intravenous dosing period. *Clin. Exp. Pharmacol. Physiol.* **2016**, *43*, 1013–1016. [CrossRef] [PubMed]
42. Jun, S.Y.; Jang, I.J.; Yoon, S.; Jang, K.; Yu, K.S.; Cho, J.Y.; Seong, M.W.; Jung, G.M.; Yoon, S.J.; Kang, S.H. Pharmacokinetics and tolerance of the phage endolysin-based candidate drug SAL200 after a single intravenous administration among healthy volunteers. *Antimicrob. Agents Chemother.* **2017**, *61*. [CrossRef] [PubMed]
43. Bryers, J.D. Medical biofilms. *Biotechnol. Bioeng.* **2008**, *100*, 1–18. [CrossRef] [PubMed]
44. Linden, S.B.; Zhang, H.; Heselpoth, R.D.; Shen, Y.; Schmelcher, M.; Eichenseher, F.; Nelson, D.C. Biochemical and biophysical characterization of PlyGRCS, a bacteriophage endolysin active against methicillin-resistant *Staphylococcus aureus*. *Appl. Microbiol. Biotechnol.* **2015**, *99*, 741–752. [CrossRef] [PubMed]
45. Rashel, M.; Uchiyama, J.; Ujihara, T.; Uehara, Y.; Kuramoto, S.; Sugihara, S.; Yagyu, K.-I.; Muraoka, A.; Sugai, M.; Hiramatsu, K.; et al. Efficient elimination of multidrug-resistant *Staphylococcus aureus* by cloned lysin derived from bacteriophage ϕMR11. *J. Infect. Dis.* **2007**, *196*, 1237–1247. [CrossRef] [PubMed]
46. Schuch, R.; Lee, H.M.; Schneider, B.C.; Sauve, K.L.; Law, C.; Khan, B.K.; Rotolo, J.A.; Horiuchi, Y.; Couto, D.E.; Raz, A.; et al. Combination therapy with lysin CF-301 and antibiotic is superior to antibiotic alone for treating methicillin-resistant *Staphylococcus aureus*—induced murine bacteremia. *J. Infect. Dis.* **2014**, *209*, 1469–1478. [CrossRef] [PubMed]
47. Schuch, R.; Khan, B.K.; Raz, A.; Rotolo, J.A.; Wittekind, M. Bacteriophage lysin CF-301, a potent antistaphylococcal biofilm agent. *Antimicrob. Agents Chemother.* **2017**, *61*. [CrossRef] [PubMed]
48. Altoparlak, U.; Erol, S.; Akcay, M.N.; Celebi, F.; Kadanali, A. The time-related changes of antimicrobial resistance patterns and predominant bacterial profiles of burn wounds and body flora of burned patients. *Burns* **2004**, *30*, 660–664. [CrossRef] [PubMed]
49. Chopra, S.; Harjai, K.; Chhibber, S. Potential of combination therapy of endolysin MR-10 and minocycline in treating MRSA induced systemic and localized burn wound infections in mice. *Int. J. Med. Microbiol.* **2016**, *306*, 707–716. [CrossRef] [PubMed]

50. Herpers, B.; Badoux, P.; Pietersma, F.; Eichenseher, F.; Loessner, M. Specific lysis of methicillin susceptible and resistant *Staphylococcus aureus* by the endolysin staphefekt SA. 100. In Proceedings of the 24th European Congress of Clinical Microbiology and Infectious Diseases (ECCMID), Barcelona, Spain, 10–13 May 2014.
51. Totté, J.E.E.; van Doorn, M.B.; Pasmans, S.G.M.A. Successful treatment of chronic *Staphylococcus aureus*-related dermatoses with the topical endolysin staphefekt sa.100: A report of 3 cases. *Case Rep. Dermatol.* **2017**, *9*, 19–25. [CrossRef] [PubMed]
52. Rafii, F.; Sutherland, J.B.; Cerniglia, C.E. Effects of treatment with antimicrobial agents on the human colonic microflora. *Ther. Clin. Risk Manag.* **2008**, *4*, 1343–1358. [CrossRef] [PubMed]
53. Murray, B.E. The life and times of the Enterococcus. *Clin. Micriobiol. Rev.* **1990**, *3*, 46–65. [CrossRef]
54. Hammerum, A.M. Enterococci of animal origin and their significance for public health. *Clin. Microbiol. Infect.* **2012**, *18*, 619–625. [CrossRef] [PubMed]
55. Courvalin, P. Vancomycin resistance in Gram-positive cocci. *Clin. Infect. Dis.* **2006**, *42*, S25–S34. [CrossRef] [PubMed]
56. Boneca, I.G.; Chiosis, G. Vancomycin resistance: Occurrence, mechanisms and strategies to combat it. *Expert Opin. Ther. Targets* **2003**, *7*, 311–328. [CrossRef] [PubMed]
57. Yoong, P.; Schuch, R.; Nelson, D.; Fischetti, V.A. Identification of a broadly active phage lytic enzyme with lethal activity against antibiotic-resistant *Enterococcus faecalis* and *Enterococcus faecium*. *J. Bacteriol.* **2004**, *186*, 4808–4812. [CrossRef] [PubMed]
58. Son, J.S.; Jun, S.Y.; Kim, E.B.; Park, J.E.; Paik, H.R.; Yoon, S.J.; Kang, S.H.; Choi, Y.J. Complete genome sequence of a newly isolated lytic bacteriophage, EFAP-1 of *Enterococcus faecalis*, and antibacterial activity of its endolysin EFAL-1. *J. Appl. Microbiol.* **2010**, *108*, 1769–1779. [CrossRef] [PubMed]
59. Zhang, W.; Mi, Z.; Yin, X.; Fan, H.; An, X.; Zhang, Z.; Chen, J.; Tong, Y. Characterization of *Enterococcus faecalis* phage IME-EF1 and its endolysin. *PLoS ONE* **2013**, *8*, e80435. [CrossRef] [PubMed]
60. Cheng, M.; Zhang, Y.; Li, X.; Liang, J.; Hu, L.; Gong, P.; Zhang, L.; Cai, R.; Zhang, H.; Ge, J.; et al. Endolysin LysEF-P10 shows potential as an alternative treatment strategy for multidrug-resistant *Enterococcus faecalis* infections. *Sci. Rep.* **2017**, *7*, 10164. [CrossRef] [PubMed]
61. Jedrzejas, M.J. Pneumococcal virulence factors: Structure and function. *Microbiol. Mol. Biol. Rev.* **2001**, *65*, 187–207. [CrossRef] [PubMed]
62. Cunningham, M.W. Pathogenesis of group a streptococcal infections. *Clin. Microbiol. Rev.* **2000**, *13*, 470–511. [CrossRef] [PubMed]
63. Loeffler, J.M.; Djurkovic, S.; Fischetti, V.A. Phage lytic enzyme Cpl-1 as a novel antimicrobial for pneumococcal bacteremia. *Infect. Immun.* **2003**, *71*, 6199–6204. [CrossRef] [PubMed]
64. Garcia, J.L.; Garcia, E.; Arraras, A.; Garcia, P.; Ronda, C.; Lopez, R. Cloning, purification, and biochemical characterization of the pneumococcal bacteriophage Cp-1 lysin. *J. Virol.* **1987**, *61*, 2573–2580. [PubMed]
65. Djurkovic, S.; Loeffler, J.M.; Fischetti, V.A. Synergistic killing of *Streptococcus pneumoniae* with the bacteriophage lytic enzyme Cpl-1 and penicillin or gentamicin depends on the level of penicillin resistance. *Antimicrob. Agents Chemother.* **2005**, *49*, 1225–1228. [CrossRef] [PubMed]
66. Jado, I.; López, R.; García, E.; Fenoll, A.; Casal, J.; García, P. Phage lytic enzymes as therapy for antibiotic-resistant *Streptococcus pneumoniae* infection in a murine sepsis model. *J. Antimicrob. Chemother.* **2003**, *52*, 967–973. [CrossRef] [PubMed]
67. Loeffler, J.M.; Fischetti, V.A. Synergistic lethal effect of a combination of phage lytic enzymes with different activities on penicillin-sensitive and -resistant *Streptococcus pneumoniae* strains. *Antimicrob. Agents Chemother.* **2003**, *47*, 375–377. [CrossRef] [PubMed]
68. Entenza, J.; Loeffler, J.; Grandgirard, D.; Fischetti, V.; Moreillon, P. Therapeutic effects of bacteriophage Cpl-1 lysin against *Streptococcus pneumoniae* endocarditis in rats. *Antimicrob. Agents Chemother.* **2005**, *49*, 4789–4792. [CrossRef] [PubMed]
69. Grandgirard, D.; Loeffler, J.M.; Fischetti, V.A.; Leib, S.L. Phage lytic enzyme Cpl-1 for antibacterial therapy in experimental pneumococcal meningitis. *J. Infect. Dis.* **2008**, *197*, 1519–1522. [CrossRef] [PubMed]
70. Doehn, J.M.; Fischer, K.; Reppe, K.; Gutbier, B.; Tschernig, T.; Hocke, A.C.; Fischetti, V.A.; Löffler, J.; Suttorp, N.; Hippenstiel, S.; et al. Delivery of the endolysin Cpl-1 by inhalation rescues mice with fatal pneumococcal pneumonia. *J. Antimicrob. Chemother.* **2013**, *68*, 2111–2117. [CrossRef] [PubMed]

71. Witzenrath, M.; Schmeck, B.; Doehn, J.M.; Tschernig, T.; Zahlten, J.; Loeffler, J.M.; Zemlin, M.; Müller, H.; Gutbier, B.; Schütte, H. Systemic use of the endolysin Cpl-1 rescues mice with fatal pneumococcal pneumonia. *Crit. Care Med.* **2009**, *37*, 642–649. [CrossRef] [PubMed]
72. Resch, G.; Moreillon, P.; Fischetti, V.A. A stable phage lysin (Cpl-1) dimer with increased antipneumococcal activity and decreased plasma clearance. *Int. J. Antimicrob. Agents* **2011**, *38*, 516–521. [CrossRef] [PubMed]
73. Usobiaga, P.; Medrano, F.J.; Gasset, M.; García, J.L.; Saiz, J.L.; Rivas, G.; Laynez, J.; Menéndez, M. Structural organization of the major autolysin from streptococcus pneumoniae. *J. Biol. Chem.* **1996**, *271*, 6832–6838. [CrossRef] [PubMed]
74. Beveridge, T.J. Structures of Gram-negative cell walls and their derived membrane vesicles. *J. Bacteriol.* **1999**, *181*, 4725–4733. [PubMed]
75. Peleg, A.Y.; Seifert, H.; Paterson, D.L. *Acinetobacter baumannii*: Emergence of a successful pathogen. *Clin. Microbial. Rev.* **2008**, *21*, 538–582. [CrossRef] [PubMed]
76. Dijkshoorn, L.; Nemec, A.; Seifert, H. An increasing threat in hospitals: Multidrug-resistant *Acinetobacter baumannii*. *Nat. Rev. Microbiol.* **2007**, *5*, 939–951. [CrossRef] [PubMed]
77. Bodey, G.P.; Bolivar, R.; Fainstein, V.; Jadeja, L. Infections caused by *Pseudomonas aeruginosa*. *Rev. Infect. Dis.* **1983**, *5*, 279–313. [CrossRef] [PubMed]
78. Lister, P.D.; Wolter, D.J.; Hanson, N.D. Antibacterial-resistant *Pseudomonas aeruginosa*: Clinical impact and complex regulation of chromosomally encoded resistance mechanisms. *Clin. Microbial. Rev.* **2009**, *22*, 582–610. [CrossRef] [PubMed]
79. Lai, M.-J.; Lin, N.-T.; Hu, A.; Soo, P.-C.; Chen, L.-K.; Chen, L.-H.; Chang, K.-C. Antibacterial activity of *Acinetobacter baumannii* phage ϕAB2 endolysin (LysAB2) against both Gram-positive and Gram-negative bacteria. *Appl. Microbiol. Biotechnol.* **2011**, *90*, 529–539. [CrossRef] [PubMed]
80. Lood, R.; Winer, B.Y.; Pelzek, A.J.; Diez-Martinez, R.; Thandar, M.; Euler, C.W.; Schuch, R.; Fischetti, V.A. Novel phage lysin capable of killing the multidrug-resistant Gram-negative bacterium *Acinetobacter baumannii* in a mouse bacteremia model. *Antimicrob. Agents Chemother.* **2015**, *59*, 1983–1991. [CrossRef] [PubMed]
81. Walmagh, M.; Briers, Y.; dos Santos, S.B.; Azeredo, J.; Lavigne, R. Characterization of modular bacteriophage endolysins from *Myoviridae* phages OBP, 201φ2-1 and PVP-SE1. *PLoS ONE* **2012**, *7*, e36991. [CrossRef] [PubMed]
82. Sato, H.; Feix, J.B. Peptide–membrane interactions and mechanisms of membrane destruction by amphipathic α-helical antimicrobial peptides. *Biochim. Biophys. Acta. Biomembr.* **2006**, *1758*, 1245–1256. [CrossRef] [PubMed]
83. Guo, M.; Feng, C.; Ren, J.; Zhuang, X.; Zhang, Y.; Zhu, Y.; Dong, K.; He, P.; Guo, X.; Qin, J. A novel antimicrobial endolysin, LysPA26, against *Pseudomonas aeruginosa*. *Front. Microbiol.* **2017**, *8*. [CrossRef] [PubMed]
84. Thummeepak, R.; Kitti, T.; Kunthalert, D.; Sitthisak, S. Enhanced antibacterial activity of acinetobacter baumannii bacteriophage øABP-01 endolysin (LysABP-01) in combination with colistin. *Front. Microbiol.* **2016**, *7*. [CrossRef] [PubMed]
85. Briers, Y.; Walmagh, M.; Lavigne, R. Use of bacteriophage endolysin EL188 and outer membrane permeabilizers against *Pseudomonas aeruginosa*. *J. Appl. Microbiol.* **2011**, *110*, 778–785. [CrossRef] [PubMed]
86. Triantaphyllopoulos, D.C.; Quick, A.J.; Greenwalt, T.J. Action of disodium ethylenediamine tetracetate on blood coagulation; evidence of the development of heparinoid activity during incubation or aeration of plasma. *Blood* **1955**, *10*, 534–544. [PubMed]
87. Briers, Y.; Lavigne, R. Breaking barriers: Expansion of the use of endolysins as novel antibacterials against Gram-negative bacteria. *Future Microbiol.* **2015**, *10*, 377–390. [CrossRef] [PubMed]
88. Zasloff, M. Antimicrobial peptides of multicellular organisms. *Nature* **2002**, *415*, 389. [CrossRef] [PubMed]
89. Dawson, R.M.; Liu, C.-Q. Cathelicidin peptide SMAP-29: Comprehensive review of its properties and potential as a novel class of antibiotics. *Drug Dev. Res.* **2009**, *70*, 481–498. [CrossRef]
90. Timbermont, L.; Haesebrouck, F.; Ducatelle, R.; Van Immerseel, F. Necrotic enteritis in broilers: An updated review on the pathogenesis. *Avian Pathol.* **2011**, *40*, 341–347. [CrossRef] [PubMed]
91. Gervasi, T.; Lo Curto, R.; Minniti, E.; Narbad, A.; Mayer, M.J. Application of *Lactobacillus johnsonii* expressing phage endolysin for control of *Clostridium perfringens*. *Lett. Appl. Microbiol.* **2014**, *59*, 355–361. [CrossRef] [PubMed]

92. Gervasi, T.; Horn, N.; Wegmann, U.; Dugo, G.; Narbad, A.; Mayer, M.J. Expression and delivery of an endolysin to combat *Clostridium perfringens*. *Appl. Microbiol. Biotechnol.* **2014**, *98*, 2495–2505. [CrossRef] [PubMed]
93. Tamai, E.; Yoshida, H.; Sekiya, H.; Nariya, H.; Miyata, S.; Okabe, A.; Kuwahara, T.; Maki, J.; Kamitori, S. X-ray structure of a novel endolysin encoded by episomal phage phiSM101 of *Clostridium perfringens*. *Mol. Microbiol.* **2014**, *92*, 326–337. [CrossRef] [PubMed]
94. Nariya, H.; Miyata, S.; Tamai, E.; Sekiya, H.; Maki, J.; Okabe, A. Identification and characterization of a putative endolysin encoded by episomal phage phiSM101 of *Clostridium perfringens*. *Appl. Microbiol. Biotechnol.* **2011**, *90*, 1973–1979. [CrossRef] [PubMed]
95. Wilson, D.J.; Gonzalez, R.N.; Das, H.H. Bovine mastitis pathogens in New York and Pennsylvania: Prevalence and effects on somatic cell count and milk production. *J. Dairy Sci.* **1997**, *80*, 2592–2598. [CrossRef]
96. Donovan, D.M.; Dong, S.; Garrett, W.; Rousseau, G.M.; Moineau, S.; Pritchard, D.G. Peptidoglycan hydrolase fusions maintain their parental specificities. *Appl. Environ. Microbiol.* **2006**, *72*, 2988–2996. [CrossRef] [PubMed]
97. Schmelcher, M.; Powell, A.M.; Camp, M.J.; Pohl, C.S.; Donovan, D.M. Synergistic streptococcal phage λSA2 and B30 endolysins kill streptococci in cow milk and in a mouse model of mastitis. *Appl. Microbiol. Biotechnol.* **2015**, *99*, 8475–8486. [CrossRef] [PubMed]
98. Fan, J.; Zeng, Z.; Mai, K.; Yang, Y.; Feng, J.; Bai, Y.; Sun, B.; Xie, Q.; Tong, Y.; Ma, J. Preliminary treatment of bovine mastitis caused by *Staphylococcus aureus*, with Trx-SA1, recombinant endolysin of *S. aureus* bacteriophage IME-SA1. *Vet. Microbiol.* **2016**, *191*, 65–71. [CrossRef] [PubMed]
99. Fasanella, A.; Galante, D.; Garofolo, G.; Jones, M.H. Anthrax undervalued zoonosis. *Vet. Microbiol.* **2010**, *140*, 318–331. [CrossRef] [PubMed]
100. Toole, T.O.; Henderson, D.; Bartlett, J.G.; Ascher, M.S.; Eitzen, E.; Friedlander, A.M.; Gerberding, J.; Hauer, J.; Hughes, J.; McDade, J.; et al. Anthrax as a biological weapon, 2002: Updated recommendations for management. *J. Am. Med. Assoc.* **2002**, *287*, 2236–2253.
101. Sykes, J.E.; Hartmann, K.; Lunn, K.F.; Moore, G.E.; Stoddard, R.; Goldstein, R.E. ACVIM Consensus Statement. *J. Vet. Intern. Med.* **2011**, *19*, 1–13. [CrossRef] [PubMed]
102. Hoopes, J.T.; Stark, C.J.; Kim, H.A.; Sussman, D.J.; Donovan, D.M.; Nelson, D.C. Use of a bacteriophage lysin, PlyC, as an enzyme disinfectant against *Streptococcus equi*. *Appl. Environ. Microbiol.* **2009**, *75*, 1388–1394. [CrossRef] [PubMed]
103. Junjappa, R.P.; Desai, S.N.; Roy, P.; Narasimhaswamy, N.; Raj, J.R.M.; Durgaiah, M.; Vipra, A.; Bhat, U.R.; Satyanarayana, S.K.; Shankara, N.; et al. Efficacy of anti-staphylococcal protein P128 for the treatment of canine pyoderma: Potential applications. *Vet. Res. Commun.* **2013**, *37*, 217–228. [CrossRef] [PubMed]
104. Vipra, A.A.; Desai, S.N.; Roy, P.; Patil, R.; Raj, J.M.; Narasimhaswamy, N.; Paul, V.D.; Chikkamadaiah, R.; Sriram, B. Antistaphylococcal activity of bacteriophage derived chimeric protein P128. *BMC Microbiol.* **2012**, *12*, 41. [CrossRef] [PubMed]
105. Wang, Y.; Sun, J.H.; Lu, C.P. Purified recombinant phage lysin LySMP: An extensive spectrum of lytic activity for swine streptococci. *Curr. Microbiol.* **2009**, *58*, 609–615. [CrossRef] [PubMed]
106. Bates, J.; Jordens, J.; Griffith, D.T. Farm animals as putative reservoir for vancomycin resistant enterococcal infections in man. *J. Antiomicrob. Chemother.* **1994**, *34*, 507–516. [CrossRef]
107. Coque, T.M.; Tomayko, J.F.; Ricke, S.C.; Okhyusen, P.C.; Murray, B.E. Vancomycin-resistant enterococci from nosocomial, community, and animal sources in the United States. *Antimicrob. Agents Chemother.* **1996**, *40*, 2605–2609. [PubMed]
108. Madewell, B.R.; Tang, Y.J.; Jang, S.; Madigan, J.E.; Hirsh, D.C.; Gumerlock, P.H.; Silva, J., Jr. Apparent outbreaks of *Clostridium difficile*-associated diarrhea in horses in a veterinary medical teaching hospital. *J. Vet. Diagn. Invest.* **1995**, *7*, 343–346. [CrossRef] [PubMed]
109. Debast, S.B.; Van Leengoed, L.A.M.G.; Goorhuis, A.; Harmanus, C.; Kuijper, E.J.; Bergwerff, A.A. *Clostridium difficile* PCR ribotype 078 toxinotype V found in diarrhoeal pigs identical to isolates from affected humans. *Environ. Microbiol.* **2009**, *11*, 505–511. [CrossRef] [PubMed]
110. Kelly, C.P.; Pothoulakis, C.; LaMont, J.T. *Clostridium difficile* Colitis. *N. Engl. J. Med.* **1994**, *330*, 257–262. [CrossRef] [PubMed]
111. Mayer, M.J.; Narbad, A.; Gasson, M.J. Molecular characterization of a *Clostridium difficile* bacteriophage and its cloned biologically active endolysin. *J. Bacteriol.* **2008**, *190*, 6734–6740. [CrossRef] [PubMed]

112. Oliveira, A.; Leite, M.; Kluskens, L.D.; Santos, S.B.; Melo, L.D.R.; Azeredo, J. The first *Paenibacillus* larvae bacteriophage endolysin (PlyPl23) with high potential to control American Foulbrood. *PLoS ONE* **2015**, *10*, e0132095.
113. Zhang, H.; Bao, H.; Billington, C.; Hudson, J.A.; Wang, R. Isolation and lytic activity of the *Listeria* bacteriophage endolysin LysZ5 against *Listeria monocytogenes* in soya milk. *Food Microbiol.* **2012**, *31*, 133–136. [CrossRef] [PubMed]
114. van Nassau, T.J.; Lenz, C.A.; Scherzinger, A.S.; Vogel, R.F. Combination of endolysins and high pressure to inactivate *Listeria monocytogenes*. *Food Microbiol.* **2017**, *68*, 81–88. [CrossRef] [PubMed]
115. Turner, M.S.; Waldherr, F.; Loessner, M.J.; Giffard, P.M. Antimicrobial activity of lysostaphin and a *Listeria monocytogenes* bacteriophage endolysin produced and secreted by lactic acid bacteria. *Syst. Appl. Microbiol.* **2007**, *30*, 58–67. [CrossRef] [PubMed]
116. van Tassell, M.L.; Angela Daum, M.; Kim, J.S.; Miller, M.J. Creative lysins: *Listeria* and the engineering of antimicrobial enzymes. *Curr. Opin. Biotechnol.* **2016**, *37*, 88–96. [CrossRef] [PubMed]
117. Schmelcher, M.; Waldherr, F.; Loessner, M.J. *Listeria* bacteriophage peptidoglycan hydrolases feature high thermoresistance and reveal increased activity after divalent metal cation substitution. *Appl. Microbiol. Biotechnol.* **2012**, *93*, 633–643. [CrossRef] [PubMed]
118. Hennekinne, J.A.; De Buyser, M.L.; Dragacci, S. *Staphylococcus aureus* and its food poisoning toxins: Characterization and outbreak investigation. *FEMS. Microbiol. Rev.* **2012**, *36*, 815–836. [CrossRef] [PubMed]
119. Chang, Y.; Kim, M.; Ryu, S. Characterization of a novel endolysin LysSA11 and its utility as a potent biocontrol agent against *Staphylococcus aureus* on food and utensils. *Food. Microbiol.* **2017**, *68*, 112–120. [CrossRef] [PubMed]
120. Chang, Y.; Yoon, H.; Kang, D.H.; Chang, P.S.; Ryu, S. Endolysin LysSA97 is synergistic with carvacrol in controlling *Staphylococcus aureus* in foods. *Int. J. Food Microbiol.* **2017**, *244*, 19–26. [CrossRef] [PubMed]
121. Obeso, J.M.; Martinez, B.; Rodriguez, A.; Garcia, P. Lytic activity of the recombinant staphylococcal bacteriophage ΦH5 endolysin active against *Staphylococcus aureus* in milk. *Int. J. Food Microbiol.* **2008**, *128*, 212–218. [CrossRef] [PubMed]
122. Rodriguez-Rubio, L.; Martinez, B.; Donovan, D.M.; Garcia, P.; Rodriguez, A. Potential of the virion-associated peptidoglycan hydrolase HydH5 and its derivative fusion proteins in milk biopreservation. *PLoS ONE* **2013**, *8*, e54828. [CrossRef] [PubMed]
123. World Health Organization. *WHO Estimates of the Global Burden of Foodborne Diseases: Foodborne Disease Burden Epidemiology Reference Group 2007–2015*; World Health Organization: Geneva, Switzerland, 2015.
124. Interagency Food Safety Analytics Collaboration (IFSAC) Project. Foodborne Illness Source Attribution Estimates for *Salmonella*, *Escherichia coli* O157 (*E. coli* O157), *Listeria monocytogenes* (lm) and *Campylobacter* Using Outbreak Surveillance Data. 2014. Available online: https://www.cdc.gov/foodsafety/pdfs/IFSAC-2013FoodborneillnessSourceEstimates-508.pdf (accessed on 13 March 2015).
125. Lim, J.A.; Shin, H.; Kang, D.H.; Ryu, S. Characterization of endolysin from a *Salmonella* Typhimurium-infecting bacteriophage SPN1S. *Res. Microbiol.* **2012**, *163*, 233–241. [CrossRef] [PubMed]
126. Oliveira, H.; Thiagarajan, V.; Walmagh, M.; Sillankorva, S.; Lavigne, R.; Neves-Petersen, M.T.; Kluskens, L.D.; Azeredo, J. A thermostable *Salmonella* phage endolysin, Lys68, with broad bactericidal properties against Gram-negative pathogens in presence of weak acids. *PLoS ONE* **2014**, *9*, e108376. [CrossRef] [PubMed]
127. Rodríguez-Rubio, L.; Gerstmans, H.; Thorpe, S.; Mesnage, S.; Lavigne, R.; Briers, Y. DUF3380 domain from a *Salmonella* phage endolysin shows potent *N*-acetylmuramidase activity. *Appl. Environ. Microbiol.* **2016**, *82*, 4975–4981. [CrossRef] [PubMed]
128. Drudy, D.; Mullane, N.R.; Quinn, T.; Wall, P.G.; Fanning, S. *Enterobacter sakazakii*: An emerging pathogen in powdered infant formula. *Clin. Infect. Dis.* **2006**, *42*, 996–1002. [CrossRef] [PubMed]
129. Endersen, L.; Guinane, C.M.; Johnston, C.; Neve, H.; Coffey, A.; Ross, R.P.; McAuliffe, O.; O'Mahony, J. Genome analysis of *Cronobacter* phage vB_CsaP_Ss1 reveals an endolysin with potential for biocontrol of Gram-negative bacterial pathogens. *J. Gen. Virol.* **2015**, *96*, 463–477. [CrossRef] [PubMed]
130. Endersen, L.; Coffey, A.; Ross, R.P.; McAuliffe, O.; Hill, C.; O'Mahony, J. Characterisation of the antibacterial properties of a bacterial derived peptidoglycan hydrolase (LysCs4), active against *C. sakazakii* and other Gram-negative food-related pathogens. *Int. J. Food Microbiol.* **2015**, *215*, 79–85. [CrossRef] [PubMed]
131. Son, B.; Yun, J.; Lim, J.-A.; Shin, H.; Heu, S.; Ryu, S. Characterization of LysB4, an endolysin from the *Bacillus cereus*-infecting bacteriophage B4. *BMC Microbiol.* **2012**, *12*, 33. [CrossRef] [PubMed]

132. Loessner, M.J.; Maier, S.K.; Daubek-Puza, H.; Wendlinger, G.; Scherer, S. Three *Bacillus cereus* bacteriophage endolysins are unrelated but reveal high homology to cell wall hydrolases from different bacilli. *J. Bacteriol.* **1997**, *179*, 2845–2851. [CrossRef] [PubMed]
133. Park, J.; Yun, J.; Lim, J.A.; Kang, D.H.; Ryu, S. Characterization of an endolysin, LysBPS13, from a *Bacillus cereus* bacteriophage. *FEMS Microbiol. Lett.* **2012**, *332*, 76–83. [CrossRef] [PubMed]
134. Mayer, M.J.; Payne, J.; Gasson, M.J.; Narbad, A. Genomic sequence and characterization of the virulent bacteriophage *ϕ*CTP1 from *Clostridium tyrobutyricum* and heterologous expression of its endolysin. *Appl. Envrion. Microbiol.* **2010**, *76*, 5415–5422. [CrossRef] [PubMed]
135. Mayer, M.J.; Gasson, M.J.; Narbad, A. Genomic sequence of bacteriophage ATCC 8074-B1 and activity of its endolysin and engineered variants against *Clostridium sporogenes*. *Appl. Environ. Microbiol.* **2012**, *78*, 3685–3692. [CrossRef] [PubMed]
136. Vasala, A.; Valkkila, M.; Caldentey, J.; Alatossava, T. Genetic and biochemical characterization of the *Lactobacillus delbrueckii* subsp. *Lactis* bacteriophage LL-H lysin. *Appl. Environ. Microbiol.* **1995**, *61*, 4004–4011. [PubMed]
137. Deutsch, S.-M.; Guezenec, S.; Piot, M.; Foster, S.; Lortal, S. Mur-LH, the broad-spectrum endolysin of *Lactobacillus helveticus* temperate bacteriophage φ-0303. *Appl. Environ. Microbiol.* **2004**, *70*, 96–103. [CrossRef] [PubMed]
138. Kashige, N.; Nakashima, Y.; Miake, F.; Watanabe, K. Cloning, sequence analysis, and expression of *Lactobacillus casei* phage PL-1 lysis genes. *Arch. Virol.* **2000**, *145*, 1521–1534. [CrossRef] [PubMed]
139. Labrie, S.; Vukov, N.; Loessner, M.J.; Moineau, S. Distribution and composition of the lysis cassette of *Lactococcus lactis* phages and functional analysis of bacteriophage Ul36 holin. *FEMS Microbiol. Lett.* **2004**, *233*, 37–43. [CrossRef] [PubMed]
140. Han, F.; Li, M.; Lin, H.; Wang, J.; Cao, L.; Khan, M.N. The novel *Shewanella putrefaciens*-infecting bacteriophage Spp001: Genome sequence and lytic enzymes. *J. Ind. Microbiol. Biotechnol.* **2014**, *41*, 1017–1026. [CrossRef] [PubMed]
141. Lu, Z.; Altermann, E.; Breidt, F.; Kozyavkin, S. Sequence analysis of *Leuconostoc mesenteroides* bacteriophage Phi1-A4 isolated from an industrial vegetable fermentation. *Appl. Environ. Microbiol.* **2010**, *76*, 1955–1966. [CrossRef] [PubMed]
142. Yoon, S.S.; Kim, J.W.; Breidt, F.; Fleming, H.P. Characterization of a lytic *Lactobacillus plantarum* bacteriophage and molecular cloning of a lysin gene in *Escherichia coli*. *Int. J. Food Microbiol.* **2001**, *65*, 63–74. [CrossRef]
143. Kim, W.-S.; Salm, H.; Geider, K. Expression of bacteriophage φEa1h lysozyme in *Escherichia coli* and its activity in growth inhibition of *Erwinia amylovora*. *Microbiology* **2004**, *150*, 2707–2714. [CrossRef] [PubMed]
144. Nakimbugwe, D.; Masschalck, B.; Anim, G.; Michiels, C.W. Inactivation of Gram-negative bacteria in milk and banana juice by hen egg white and lambda lysozyme under high hydrostatic pressure. *Int. J. Food Microbiol.* **2006**, *112*, 19–25. [CrossRef] [PubMed]
145. Wang, W.; Li, M.; Lin, H.; Wang, J.; Mao, X. The *Vibrio parahaemolyticus*-infecting bacteriophage qdvp001: Genome sequence and endolysin with a modular structure. *Arch. Virol.* **2016**, *161*, 2645–2652. [CrossRef] [PubMed]
146. Solanki, K.; Grover, N.; Downs, P.; Paskaleva, E.E.; Mehta, K.K.; Lillian, L.; Schadler, L.S.; Kane, R.S.; Dordick, J.S. Enzyme-based Listericidal nanocomposites. *Sci. Rep.* **2013**, *3*, 1584. [CrossRef] [PubMed]
147. Misiou, O.; van Nassau, T.J.; Lenz, C.A.; Vogel, R.F. The preservation of listeria-critical foods by a combination of endolysin and high hydrostatic pressure. *Int. J. Food Microbiol.* **2017**, in press. [CrossRef] [PubMed]
148. Chmielewski, R.A.N.; Frank, J.F. Biofilm formation and control in food processing facilities. *Compr. Rev. Food Sci. Food Saf.* **2003**, *2*, 22–32. [CrossRef]
149. Gutiérrez, D.; Fernández, L.; Martínez, B.; Ruas-Madiedo, P.; García, P.; Rodríguez, A. Real-time assessment of *Staphylococcus aureus* biofilm disruption by phage-derived proteins. *Front. Microbiol.* **2017**, *8*. [CrossRef] [PubMed]
150. Simmons, M.; Morales, C.A.; Oakley, B.B.; Seal, B.S. Recombinant expression of a putative amidase cloned from the genome of *Listeria monocytogenes* that lyses the bacterium and its monolayer in conjunction with a protease. *Probiotics Antimicrob. Proteins* **2012**, *4*, 1–10. [CrossRef] [PubMed]
151. Strange, R.N.; Scott, P.R. Plant disease: A threat to global food security. *Annu. Rev. Phytopathol.* **2005**, *43*, 83–116. [CrossRef] [PubMed]

152. McManus, P.S.; Stockwell, V.O.; Sundin, G.W.; Jones, A.L. Antibiotic use in plant agriculture. *Annu. Rev. Phytopathol.* **2002**, *40*, 443–465. [CrossRef] [PubMed]
153. Düring, K.; Porsch, P.; Fladung, M.; Lörz, H. Transgenic potato plants resistant to the phytopathogenic bacterium *Erwinia carotovora*. *Plant J.* **1993**, *3*, 587–598. [CrossRef]
154. De Vries, J.; Harms, K.; Broer, I.; Kriete, G.; Mahn, A.; Düring, K.; Wackernagel, W. The bacteriolytic activity in transgenic potatoes expressing a chimeric T4 lysozyme gene and the effect of T4 lysozyme on soil- and phytopathogenic bacteria. *Syst. Appl. Microbiol.* **1999**, *22*, 280–286. [CrossRef]
155. Wittmann, J.; Brancato, C.; Berendzen, K.W.; Dreiseikelmann, B. Development of a tomato plant resistant to *Clavibacter michiganensis* using the endolysin gene of bacteriophage CMP1 as a transgene. *Plant Pathol.* **2016**, *65*, 496–502. [CrossRef]
156. Hausbeck, M.K.; Bell, J.; Medina-Mora, C.; Podolsky, R.; Fulbright, D.W. Effect of bactericides on population sizes and spread of *Clavibacter michiganensis* subsp. *Michiganensis* on tomatoes in the greenhouse and on disease development and crop yield in the field. *Phytopathology* **2000**, *90*, 38–44. [PubMed]
157. Tang, J.L.; Feng, J.X.; Li, Q.Q.; Wen, H.X.; Zhou, D.L.; Wilson, T.J.; Dow, J.M.; Ma, Q.S.; Daniels, M.J. Cloning and characterization of the rpfc gene of *Xanthomonas oryzae* pv. *Oryzae*: Involvement in exopolysaccharide production and virulence to rice. *Mol. Plant Microbe. Interact.* **1996**, *9*, 664–666. [PubMed]
158. Xu, Y.; Luo, Q.-q.; Zhou, M.-g. Identification and characterization of integron-mediated antibiotic resistance in the phytopathogen *Xanthomonas oryzae* pv. *Oryzae*. *PLoS ONE* **2013**, *8*, e55962. [CrossRef] [PubMed]
159. Lee, C.-N.; Lin, J.-W.; Chow, T.-Y.; Tseng, Y.-H.; Weng, S.-F. A novel lysozyme from *Xanthomonas oryzae* phage ϕxo411 active against *Xanthomonas* and *Stenotrophomonas*. *Protein Expr. Purif.* **2006**, *50*, 229–237. [CrossRef] [PubMed]
160. Brooke, J.S. *Stenotrophomonas maltophilia*: An emerging global opportunistic pathogen. *Clin. Microbiol. Rev.* **2012**, *25*, 2–41. [CrossRef] [PubMed]
161. Attai, H.; Rimbey, J.; Smith, G.P.; Brown, P.J.B. Expression of a peptidoglycan hydrolase from lytic bacteriophages Atu_ph02 and Atu_ph03 triggers lysis of *Agrobacterium tumefaciens*. *Appl. Environ. Microbiol.* **2017**, *83*, 17. [CrossRef] [PubMed]
162. Pulawska, J. Crown gall of stone fruits and nuts, economic significance and diversity of its causal agents: Tumorigenic *Agrobacterium* spp. *J. Plant Pathol.* **2010**, *92*, S87–S98.
163. Mansfield, J.; Genin, S.; Magori, S.; Citovsky, V.; Sriariyanum, M.; Ronald, P.; Dow, M.; Verdier, V.; Beer, S.V.; Machado, M.A.; et al. Top 10 plant pathogenic bacteria in molecular plant pathology. *Mol. Plant Pathol.* **2012**, *13*, 614–629. [CrossRef] [PubMed]
164. Rosenberg, A.S. Immunogenicity of biological therapeutics: A hierarchy of concerns. *Dev. Biol.* **2003**, *112*, 15–21.
165. De Groot, A.S.; Scott, D.W. Immunogenicity of protein therapeutics. *Trends Immunol.* **2007**, *28*, 482–490. [CrossRef] [PubMed]
166. Baker, M.P.; Reynolds, H.M.; Lumicisi, B.; Bryson, C.J. Immunogenicity of protein therapeutics: The key causes, consequences and challenges. *Self Nonself* **2010**, *1*, 314–322. [CrossRef] [PubMed]
167. Fischetti, V.A. Bacteriophage lytic enzymes: Novel anti-infectives. *Trends Microbiol.* **2005**, *13*, 491–496. [CrossRef] [PubMed]
168. DeHart, H.P.; Heath, H.E.; Heath, L.S.; LeBlanc, P.A.; Sloan, G.L. The lysostaphin endopeptidase resistance gene (epr) specifies modification of peptidoglycan cross bridges in *Staphylococcus simulans* and *Staphylococcus aureus*. *Appl. Environ. Microbiol.* **1995**, *61*, 1475–1479. [PubMed]
169. Sugai, M.; Fujiwara, T.; Ohta, K.; Komatsuzawa, H.; Ohara, M.; Suginaka, H. Epr, which encodes glycylglycine endopeptidase resistance, is homologous to femab and affects serine content of peptidoglycan cross bridges in *Staphylococcus capitis* and *Staphylococcus aureus*. *J. Bacteriol.* **1997**, *179*, 4311–4318. [CrossRef] [PubMed]
170. Gründling, A.; Missiakas, D.M.; Schneewind, O. *Staphylococcus aureus* mutants with increased lysostaphin resistance. *J. Bacteriol.* **2006**, *188*, 6286–6297. [CrossRef] [PubMed]
171. Vollmer, W. Structural variation in the glycan strands of bacterial peptidoglycan. *FEMS. Microbiol. Rev.* **2008**, *32*, 287–306. [CrossRef] [PubMed]
172. Guariglia-Oropeza, V.; Helmann, J.D. *Bacillus subtilis* σ(v) confers lysozyme resistance by activation of two cell wall modification pathways, peptidoglycan o-acetylation and d-alanylation of teichoic acids. *J. Bacteriol.* **2011**, *193*, 6223–6232. [CrossRef] [PubMed]

173. Davis, K.M.; Weiser, J.N. Modifications to the peptidoglycan backbone help bacteria to establish infection. *Infect. Immun.* **2011**, *79*, 562–570. [CrossRef] [PubMed]
174. Schmelcher, M.; Shabarova, T.; Eugster, M.R.; Eichenseher, F.; Tchang, V.S.; Banz, M.; Loessner, M.J. Rapid multiplex detection and differentiation of *Listeria* cells by use of fluorescent phage endolysin cell wall binding domains. *Appl. Environ. Microbiol.* **2010**, *76*, 5745–5756. [CrossRef] [PubMed]
175. Buck, M. Crystallography: Embracing conformational flexibility in proteins. *Structure* **2003**, *11*, 735–736. [CrossRef]
176. Kashyap, M.; Jagga, Z.; Das, B.K.; Arockiasamy, A.; Bhavesh, N.S. H-1, C-13 and N-15 NMR assignments of inactive form of P1 endolysin Lyz. *Biomol. NMR Assign.* **2012**, *6*, 87–89. [CrossRef] [PubMed]
177. Kutyshenko, V.P.; Mikoulinskaia, G.V.; Molochkov, N.V.; Prokhorov, D.A.; Taran, S.A.; Uversky, V.N. Structure and dynamics of the retro-form of the bacteriophage T5 endolysin. *Biochim. Biophys. Acta Proteins Proteom.* **2016**, *1864*, 1281–1291. [CrossRef] [PubMed]
178. Topf, M.; Lasker, K.; Webb, B.; Wolfson, H.; Chiu, W.; Sali, A. Protein structure fitting and refinement guided by Cryo-EM density. *Structure* **2008**, *16*, 295–307. [CrossRef] [PubMed]

antibiotics

MDPI

Article

Characterization of a Lytic Bacteriophage as an Antimicrobial Agent for Biocontrol of Shiga Toxin-Producing *Escherichia coli* O145 Strains

Yen-Te Liao [1], Alexandra Salvador [1], Leslie A. Harden [1], Fang Liu [1,3], Valerie M. Lavenburg [1], Robert W. Li [2] and Vivian C. H. Wu [1,*]

1 Produce Safety and Microbiology Research Unit, Department of Agriculture (USDA), Agricultural Research Service (ARS), Western Regional Research Center (WRRC), Albany, CA 94710, USA; yen-te.liao@ars.usda.gov (Y.-T.L.); alexandra.salvador@ars.usda.gov (A.S.); leslie.harden@ars.usda.gov (L.A.H.); fang.liu@ars.usda.gov (F.L.); Valerie.lavenburg@ars.usda.gov (V.M.L.)

2 Animal Genomics and Improvement Laboratory, Department of Agriculture (USDA), Agricultural Research Service (ARS), Beltsville, MD 20705, USA; Robert.Li@ars.usda.gov

3 College of Food Science and Engineering, Ocean University of China, Qingdao 266100, China

* Correspondence: vivian.wu@ars.usda.gov; Tel.: +1-510-559-5829

Received: 20 April 2019; Accepted: 31 May 2019; Published: 5 June 2019

Abstract: Shiga toxin-producing *Escherichia coli* (STEC) O145 is one of the most prevalent non-O157 serogroups associated with foodborne outbreaks. Lytic phages are a potential alternative to antibiotics in combatting bacterial pathogens. In this study, we characterized a *Siphoviridae* phage lytic against STEC O145 strains as a novel antimicrobial agent. *Escherichia* phage vB_EcoS-Ro145clw (Ro145clw) was isolated and purified prior to physiological and genomic characterization. Then, in vitro antimicrobial activity against an outbreak strain, *E. coli* O145:H28, was evaluated. Ro145clw is a double-stranded DNA phage with a genome 42,031 bp in length. Of the 67 genes identified in the genome, 21 were annotated with functional proteins, none of which were *stx* genes. Ro145clw had a latent period of 21 min and a burst size of 192 phages per infected cell. The phage could sustain a wide range of pH (pH 3 to pH 10) and temperatures (−80 °C to −73 °C). Ro145clw was able to reduce *E. coli* O145:H28 in lysogeny broth by approximately 5 log at 37 °C in four hours. These findings indicate that the Ro145clw phage is a promising antimicrobial agent that can be used to control *E. coli* O145 in adverse pH and temperature conditions.

Keywords: STEC-specific bacteriophage; whole genome sequencing; STEC O145 strains; antimicrobial agent

1. Introduction

Shiga toxin-producing *Escherichia coli* (STEC) is a notorious foodborne pathogen that can cause severe illness, such as hemolytic uremic syndrome (HUS), which has a high mortality rate among young children and the elderly [1]. The first known STEC outbreak occurred in 1982 and was associated with the consumption of undercooked hamburger patties contaminated with *E. coli* O157:H7 strains [2]. Since then, the number of STEC-related infections, including the serogroups of O157 and non-O157 with O26, O45, O103, O111, O121, and O145 in particular, has increased every year, with an estimated 176,000 cases, 2400 hospitalizations, and 20 deaths annually in the U.S. [3]. In recent years, foodborne illnesses associated with consuming contaminated produce have increased [4]. STEC is one of the most frequently-occurring bacterial pathogens, responsible for about 18% of produce-associated outbreaks in the United States [5]. Next to STEC O157, STEC O145 is the most widespread pathogen among the top six non-O157 STEC serogroups associated with human infections in the U.S. [6,7]. In 2010,

STEC-O145-contaminated romaine lettuce led to a serious foodborne outbreak in multiple states [8]. Infections caused by STEC O145 strains have been reported around the world [9,10].

To trace the contamination source, Carter et al. compared various STEC O145 strains isolated from different environmental sources that had previously been implicated in produce-associated outbreaks including animal feces (cattle and pigs), sediment, surface water around produce-growing regions, and other environmental sources [11]. The authors found that although these strains belonged to the same serogroup (O145), they had significant phenotypic variation that could be associated with natural selection as a result of exposure to different environmental stresses. The changing phenotypic characteristics of these pathogens and/or the development of antimicrobial resistance could challenge existing antimicrobial interventions. A previous study revealed that several weak acids that were effective against *E. coli* O157:H7 were insufficient to control non-O157 STEC strains [12]. Therefore, alternative approaches are needed to prevent the spread of these pathogens.

Bacteriophages (or phages) are some of the most highly diverse and abundant entities in the biosphere, being approximately 10 times more prevalent than bacteria [13]. Phages may have different associations with their bacterial hosts primarily due to two different infection cycles: lytic and lysogenic [14]. Due to concerns surrounding antibiotic resistance as well as advantages of host specificity of lytic phages against the bacterial hosts, interest in isolating and characterizing various lytic phages is growing to use phages as an alternative to antibiotics in controlling bacterial pathogens [15]. Another study evaluated the effectiveness of seven phages isolated from different sources on the reduction of *E. coli* O157:H7 in a post-harvest setting [16]. The authors found that the most promising phage, with strong lytic effects, was isolated from municipal wastewater; it resulted in approximately 2.5 and 3.5 log colony forming unit (CFU)/g reduction in *E. coli* O157:H7 on cut green pepper and spinach leaves after phage treatment, respectively. Amarillas et al. isolated and characterized a phage from horse feces that was capable of infecting multidrug-resistant *E. coli* O157:H7 strains and some *Salmonella* strains [17]. Another study isolated two T5-like phages from food samples; these phages were able to infect different species of pathogens including STEC O152 and O103, *Shigella sonnei*, *Salmonella*, multi-drug resistant *E. coli*, and generic *E. coli* strains [18]. Tolen et al. evaluated the effectiveness of current available prototype bacteriophage intervention on the reduction of STEC O157 and non-O157 strains inoculated on cattle hide [19]. However, none of these commercial phage products targets STEC O145 strains.

The studies discussed above primarily focused on the antimicrobial activities of phages targeting either *E. coli* O157:H7 or other antibiotic-resistant *E. coli* strains; similar information regarding phages that are lytic against non-O157 STEC is relatively scarce. Therefore, the objective of this study was to characterize a bacteriophage isolated from non-fecal compost and to examine its potential as a novel biocontrol agent for STEC O145 strains.

2. Results

2.1. Genomic Analyses of Ro145clw

In this study, *Escherichia* phage vB_EcoS-Ro145clw (also known as Ro145clw) was isolated from non-fecal compost samples using *E. coli* O145 (RM10808) as the primary host strain. After purification, the extracted phage DNA was sequenced; the assembled phage genome had 4350× coverage. Phage Ro145clw has a 42,031 bp double-stranded DNA and an average G + C content of 50.6%. The BLASTn results showed that both *Escherichia* phage K1G (GenBank accession #GU196277) and *Escherichia* phage P AB-2017 (GenBank accession #KY295898), belonging to the family *Siphoviridae*, had the highest nucleotide similarity to phage Ro145clw. Further JSpeciesWS analysis revealed that Ro145clw shared an 84.05% and 83.58% average nucleotide identity based on BLAST (ANIb) with K1G and P AB-2017 phages, respectively. These results indicated that the Ro145clw genome belongs to the *Siphoviridae* family.

Genome annotation predicted 67 putative open reading frames (ORFs), of which 21 encoded functional proteins that were associated with phage DNA replication, packaging, structural proteins, and host cell lysis (Supplementary, Table S1). The six annotated ORFs in the Ro145clw genome associated with DNA replication included putative thermostable DNA polymerase I, putative helicase, transcriptional repressor DicA, putative helicase-primase, putative PD-(D/E)XK nuclease superfamily protein, and putative calcineurin-like phosphoesterase superfamily domain protein. At least nine predicted ORFs in Ro145clw were annotated as phage structural proteins in Ro145clw, including tail protein, tail assembly chaperone protein, tail fiber, head protein, major capsid protein, decoration protein, and tape measure protein. Three consecutive ORFs in Ro145clw (ORF_51, ORF_52, and ORF_53) coded for putative holin-like class II, holin-like class I, and endolysin, respectively, were associated with forming a holin-dependent host cell lysis system [20]. Both ORF_51 and ORF_52 shared 92% and 95% average nucleotide identity, respectively, with their counterparts in phage G AB-2017, whereas ORF_53 shared 86% average nucleotide identity to the gene encoding lysozyme in *Escherichia* phage P AB-2017 [21]. Two spanin proteins encoded by ORF_66 and ORF_67 were located downstream of the cell lysis system in the Ro145clw genome and were associated with the final step of cell lysis by breaking the structure of the outer membrane of the host cell to release the phage progenies [22]. No virulence genes (such as *stx*, antibiotic-resistance genes, or tRNAs) were found in Ro145clw. PhageTerm analysis predicted that phage Ro145clw had a headful DNA packaging mechanism with a preferred packaging (*pac*) site [23].

Comparative analysis showed that phages Ro145clw and K1G shared similar ORF content (Figure 1); however, one ORF-encoding tailspike protein that was present in K1G was absent in Ro145clw. One putative tail fiber protein encoded by ORF_21 and four hypothetical proteins encoded by ORF_10, ORF_11, ORF_45, and ORF_46 in the Ro145clw genome were absent in the K1G phage genome (Figure 1). Phylogenetic analysis of terminase showed that Ro145clw was closely related to the phage K1ind2 (Figure 2A), indicating a similar headful packaging strategy. The analysis of both genes encoding tail and endolysin indicated that Ro145clw had a close evolutionary relationship with phage VB EcoS-Golestan (Figure 2B,E). The genes encoding tape measure protein and holin-like class I in Ro145clw showed maximum similarity with the counterparts in phage L AB-2017 (Figure 2C) and phage ST2 (Figure 2D), respectively.

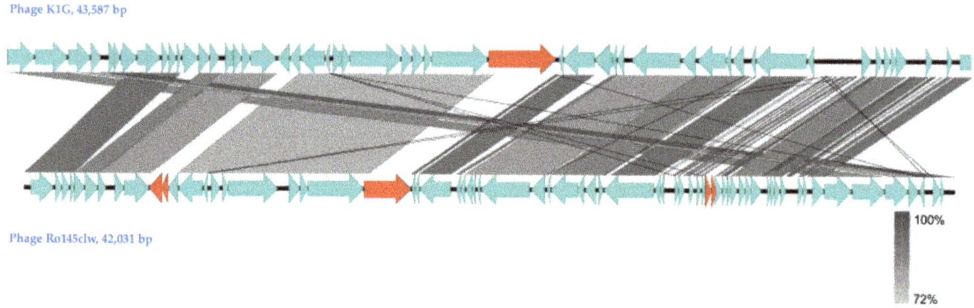

Figure 1. Genome comparison of Ro145clw and its reference phage K1G, using BLASTn and visualization with EasyFig. Genome maps of phages K1G and Ro145clw are presented as turquoise blue arrows, which indicate the order of annotated open reading frames (ORFs) from left to right along the phage genomes. Regions of sequence similarity are connected by a gray-scale shaded area, and the unshared ORFs are highlighted in red. Capital letters indicate the ORFs associated with (A) putative tail assembly chaperone, (B) tape measure, (C) putative structural, (D) tail, and (E) major capsid protein observed on the SDS-PAGE gel (refer to Figure 6).

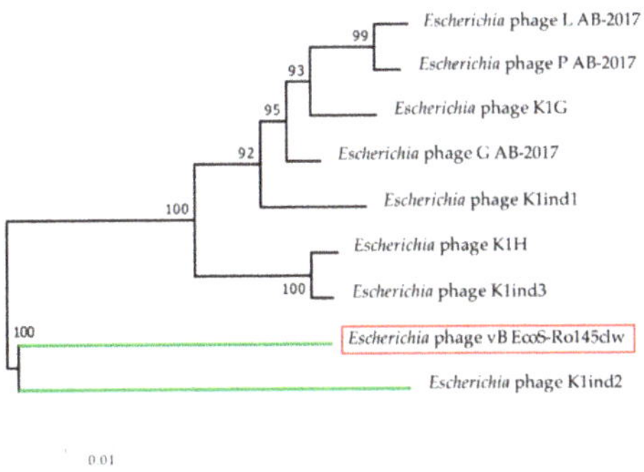

(A)

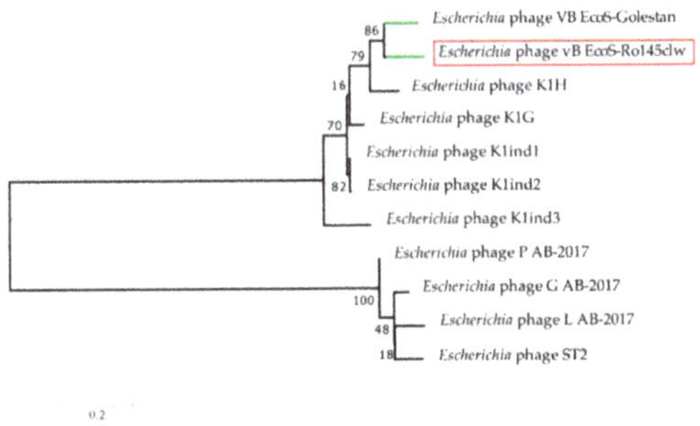

(B)

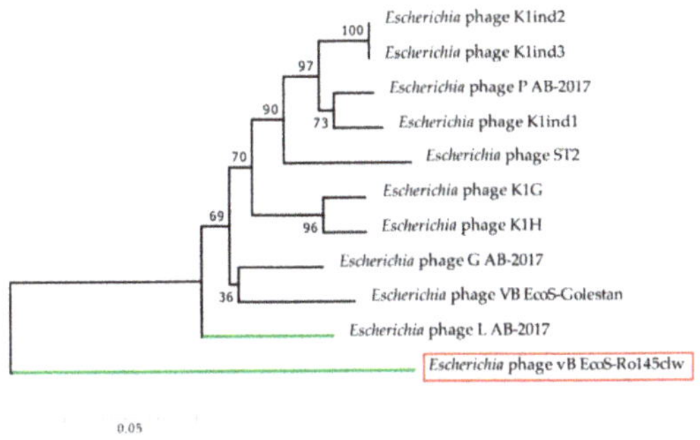

(C)

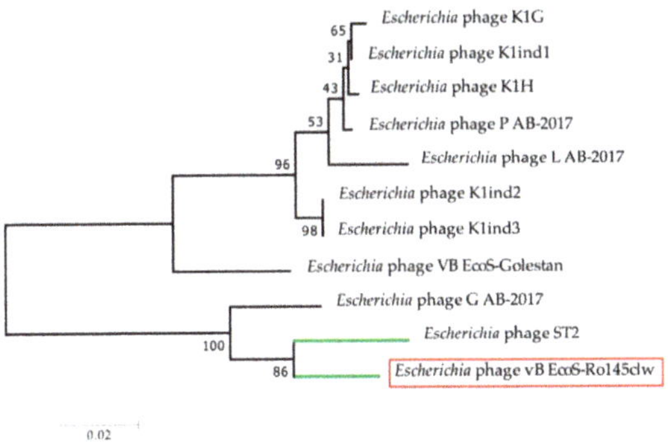

(D)

Figure 2. *Cont.*

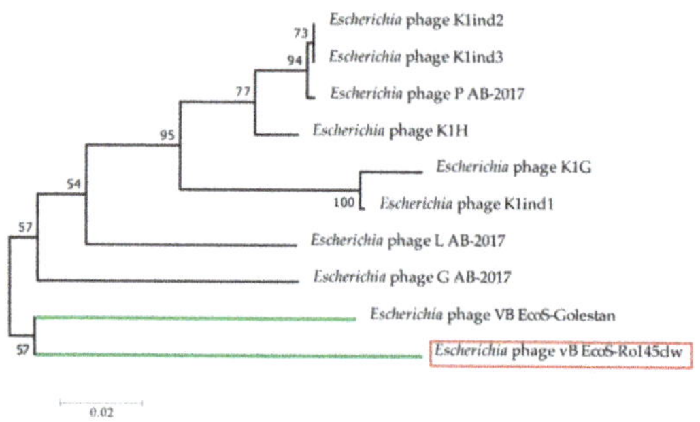

(**E**)

Figure 2. Neighbor-joining phylogenetic tree of phage vB-EcoS-Ro145clw (highlighted with a red box) and the closely-related K1glikevirus reference genomes based on the Clustal Omega alignment of the sequences of (**A**) terminase, (**B**) tail protein, (**C**) tape measure, (**D**) holin-like class I, and (**E**) endolysin. Numbers next to the branches are bootstrap values (500 replicates). The scale represents the homology percentage. Green lines are used to indicate the closest evolutionary relationship between the reference phages and phage vB-EcoS-Ro145clw.

2.2. Morphology and Host Range of Ro145clw

Phage Ro145clw displayed a morphology containing an icosahedral head approximately 58–62 nm in diameter and a long non-contractile tail 122.7 ± 2.5 nm in length, which is typical of phages belonging to the *Siphoviridae* family (Figure 3). Ro145clw contained a base plate structure that resembled a rosette with three to four leaves, with an estimated diameter of 27.3 ± 2.5 nm.

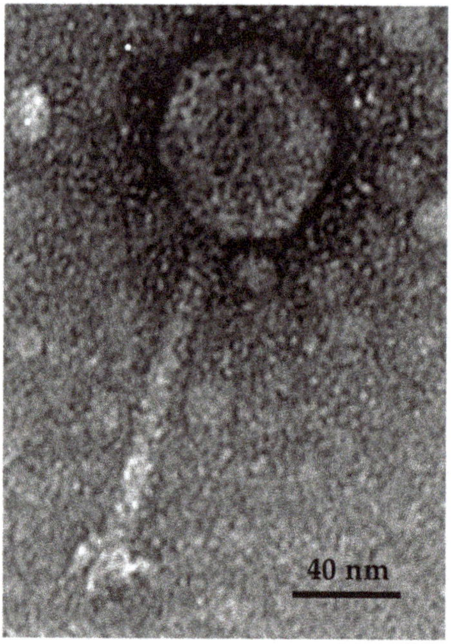

Figure 3. Transmission electron microscopy image of phage Ro145clw with a long and non-contractile tail, showing *Siphoviridae* morphology.

The results of the spot test assay indicated that phage Ro145clw is able to produce a lysis zone on the selected STEC O145 strains, with complete lysis against all the environmental STEC O145 strains, but incomplete lysis against the outbreak strains including RM13514, RM13516, RM12581, and RM12761 (Table 1). Efficiency of plating (EOP) was used to determine the productive infection of the test strains in comparison with the primary strain used for phage isolation. The results showed that the *E. coli* O145:NM strain had a high phage-producing efficiency similar to that of the primary host strain, and four *E. coli* O145 strains (RM8732, RM11691, RM9872, and RM12367) had high phage-producing efficiency (EOP > 0.5) after phage Ro145clw infection (Table 1). One *E. coli* O145:H-strain had medium phage-producing efficiency, and four *E. coli* O145:H28 strains resulted in inefficient phage production (EOP < 0.001).

Table 1. Host range and efficiency of plating (EOP) of phage Ro145clw against different serogroups of Shiga toxin-producing *Escherichia coli* (STEC) and *Salmonella* strains.

Strains	**Strain Ref. No.**	**EOP** $^{\alpha}$
STEC O26	*E. coli* O26:H18 (RM17857), *E. coli* O26:H- (RM18118) *E. coli* O26:H- (RM18132), *E. coli* O26:H- (RM17133)	R *
STEC O103	*E. coli* O103:H2 (RM12551), *E. coli* O103:H2 (RM13322) *E. coli* O103:H- (RM8356), *E. coli* O103:H- (RM10744)	R
STEC O121	*E. coli* O121:H19 (RM10046), *E. coli* O121:H19 (RM10068) *E. coli* O121:H- (RM8082), *E. coli* O121:H- (RM12997)	R
STEC O111	*E. coli* O111:H2 (RM13483), *E. coli* O111:H- (RM13789) *E. coli* O111:H- (RM11765), *E. coli* O111:H8 (RM14488)	R
STEC O145	*E. coli* O145:H+ (RM8732)	0.73
	E. coli O145:H+ (RM11691)	0.64
	E. coli O145:H+ (RM12367)	0.67
	E. coli O145:H- (RM10808)	H ^
	E. coli O145:H28 (RM9872)	0.59
	E. coli O145:H28 (RM13514)	<0.001
	E. coli O145:H28 (RM13516)	<0.001
	E. coli O145:H28 (RM12761)	<0.001
	E. coli O145:H28 (RM12581)	<0.001
	E. coli O145:NM (SJ23)	1.05
	E. coli O145:H- (94-0491)	0.29
STEC O45	*E. coli* O45:H- (RM10729), *E. coli* O45:H- (RM13726) *E. coli* O45:H- (RM13745), *E. coli* O45:H- (RM13752)	R
STEC O157	*E. coli* O157:H7 (RM18959), *E. coli* O157:H7 (RM18961) *E. coli* O157:H7 (RM18972), *E. coli* O157:H7 (RM18974) *E. coli* O157:H7 (ATCC 43888)	R
Salmonella	*Salmonella* Montevideo 51, *Salmonella* Newport H1073 *Salmonella* Heidelberg 45955, *Salmonella* Enteritidis PT30 *Salmonella* Typhimurium 14028	R

$^{\alpha}$ EOP was conducted on spot test-positive strains and is presented with a value that was calculated by the ratio of phage titer on test bacterium relative to the phage titer on the primary bacterium used for isolation. High production efficiency is EOP ≥ 0.5, medium production efficiency is 0.5 > EOP ≥ 0.1, low production efficiency is 0.1 > EOP > 0.001, and inefficiency of phage production is EOP ≤ 0.001. * R denotes no lysis in the spot test assay. ^ H was the primary bacterial strain used for isolation.

2.3. *One-Step Growth Curve*

The growth factors of phage Ro145clw, including the eclipse period, latent period, and burst size, were evaluated. The results demonstrated that Ro145clw phage had an approximately 14-min-long eclipse period and a 21-min-long latent period (Figure 4). An average burst size of 192 phages per infected cell was observed at approximately 35 min of incubation at 37 °C (Figure 4).

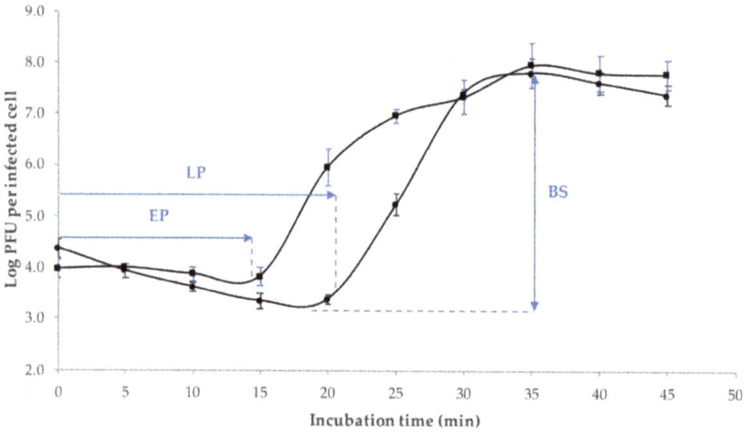

Figure 4. One-step growth curve of the phage Ro145clw using *E. coli* O145 strain (RM10808). The growth parameters of the phage indicate an eclipse period (EP) of 14 min, a latent period (LP) of 21 min, and an average burst size (BS) of 192 phages per infected cell. Closed circles indicate non-chloroform-treated samples; closed squares indicate chloroform-treated samples. The error bars present the standard error of the mean for each time point of the one-step growth curve.

2.4. *Phage pH and Temperature Stability*

Regardless of the different initial phage concentrations for treatment, Ro145clw was stable at 65 °C and 73 °C during the one-hour investigation, with only 0.1 and 0.3 log plaque-forming unit (PFU)/mL reductions in phage titers, respectively (Figure 5A). The phage stock with an initial concentration of 1×10^{10} PFU/mL in 25% glycerol remained at a similar titer when stored at −80 °C for five months (Supplementary, Table S2). Regarding pH stability, the results showed that phage Ro145clw maintained similar titers ($p > 0.05$) in a range of final pH from 3.1 to 10.5 after incubation at 37 °C for 24 h (Figure 5B). However, the phage titer was significantly reduced at pH 3.1 by 2.2 log PFU/mL compared to other pH treatments ($p < 0.05$). The results indicated that phage Ro145clw was able to be sustained in a wide pH range.

2.5. *Analysis of Phage Structural Proteins*

The separation of phage proteins by sodium dodecyl sulfate-polyacrylamide gel (SDS-PAGE) revealed five bands related to phage structural proteins, with molecular weights ranging from approximately 37 to 100 kDa (Figure 6). The identified structural proteins included putative tail assembly chaperone, tape measure protein, phage structural protein, tail protein, and major capsid protein, with coverage of amino acid sequences ranging from 11 to 62% by mass spectrometry (Table 2). Five out of eight structural proteins predicted from the genomic data were identified by mass spectrometry. None of the proteins associated with DNA replication or host lysis were detected in the SDS-PAGE gel.

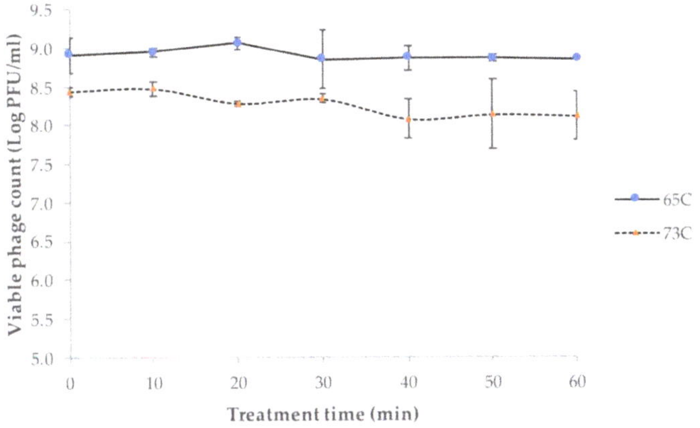

(A)

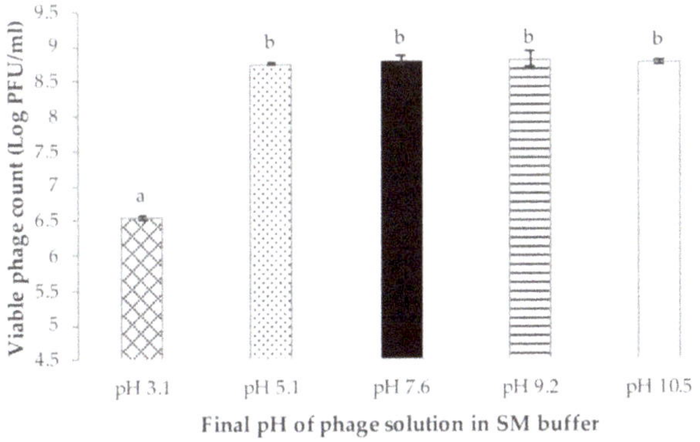

(B)

Figure 5. Stability of phage Ro145clw at (**A**) different temperatures (65 °C and 73 °C) for one hour and (**B**) various final pH (pH 3.1, pH 5.1, pH 7.6, pH 9.2, and pH 10.5) for 24 h. No statistical differences were observed between each time point of the thermal stability ($p > 0.05$). For the pH stability test, means of phage titers from different pH treatments that lack common letters (a and b) differ ($p < 0.05$). SM buffers with the initial pH of 2.2, 4.5, 7.5, 10, and 12 were used for the pH test. The error bars show the standard error of the mean (SEM).

Table 2. Structural proteins of phage Ro145clw, identified by matrix assisted laser desorption ionization-time of flight (MALDI-TOF) mass spectrometry.

Gel Band	ORF	Putative Function	Sequence Coverage (%)	No. of Peptides	Predicted Mass (kDa)
A	20	Putative tail assembly chaperone	11	11	95.5
B	17	Tape measure protein	29	16	80.8
C	62	Putative structural protein	22	8	54.7
D	9	Tail protein	25	5	40.9
E	1	Major capsid protein	62	12	37.8

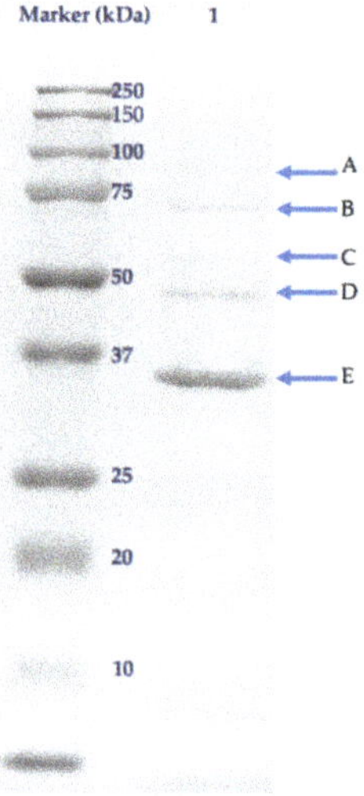

Figure 6. The structural proteins of phage Ro145clw (lane 1) on a 12% SDS-PAGE gel, visualized by Coomassie brilliant blue R-250. A = Putative tail assembly chaperone; B = Tape measure protein; C = Putative structural protein; D = Tail protein; E = Major capsid protein.

2.6. Antimicrobial Activity against E. coli O145:H28 in Lysogeny Broth (LB)

The effect of the multiplicity of infection (MOI) of phage Ro145clw on the growth of the *E. coli* O145:H28 strain was monitored in 96-well plates using a spectrophotometer prior to the in vitro bacterial reduction study in LB. Regardless of MOI, the bacteria of all treatment groups started to grow in a similar pattern to the control group before four hours of incubation at 37 °C (Supplementary, Figure S1). However, in contrast to the control, bacterial growth was suppressed (not a prompt decrease) by the treatment of phage Ro145clw; regardless of the MOIs, this was first observed at approximately five hours' incubation, and continued with minimum growth throughout the experiment period (Supplementary, Figure S1). Due to similar bacterial suppression between the treatments of MOI 10 and MOI 100, a MOI of 100 was selected for the in vitro antimicrobial study in LB.

The in vitro antimicrobial effects of Ro145clw against the *E. coli* O145:H28 strain at different temperatures are illustrated in Figure 7. At 37 °C, the culture of *E. coli* O145:H28 without phage treatment (control) increased 0.8 log after two hours of inoculation and reached 9 log CFU/mL after 24 h of incubation. In the treatment group using phage Ro145clw, *E. coli* O145:H28 levels showed significant decreases of 2.87, 5.07, and 3.20 log ($p < 0.05$) in comparison to the control at two, four, and six hours, respectively (Figure 7A). Although the phage-treated culture commenced growing after 4 h of incubation, viable *E. coli* O145:H28 cells were still reduced by 1.25 log ($p < 0.05$) less than the control group after 24 h (Figure 7A). The treated *E. coli* O145:H28 was reduced by 3.51 and 1.05 log after six hours and 24 h of incubation at 25 °C, respectively (Supplementary, Table S3). At 8 °C, the phage treatment resulted in reductions in *E. coli* O145:H28 by 0.82, 1.15, 1.17, and 1.74 log compared to the control group at two, four, six, and 24 h, respectively (Figure 7B). The phage was able to reduce the *E. coli* O145:NM strain with high EOP by approximately 2.7 log more than the reduction of the low EOP strain, *E. coli* O145:H28, after six hours incubation at 25 °C (Supplementary, Table S3). Additionally, the bacteriophage-insensitive mutant (BIM) frequency was $8.70 \pm 1.22 \times 10^{-2}$ for *E. coli* O145:H28 strain and $3.52 \pm 2.27 \times 10^{-5}$ for *E. coli* O145: NM strain. As expected, phage Ro145clw was more common against environmental STEC O145 strain than the outbreak strain.

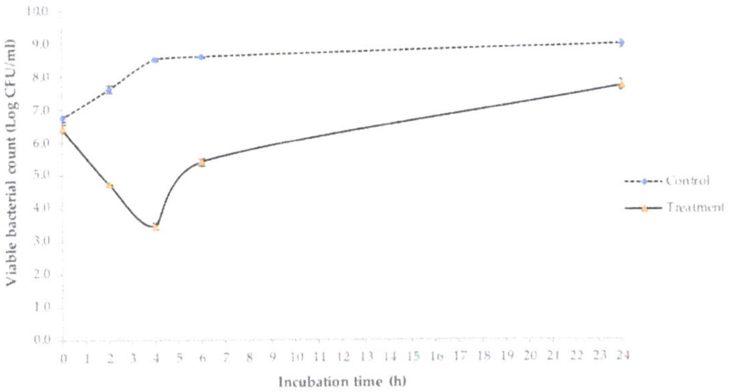

(A)

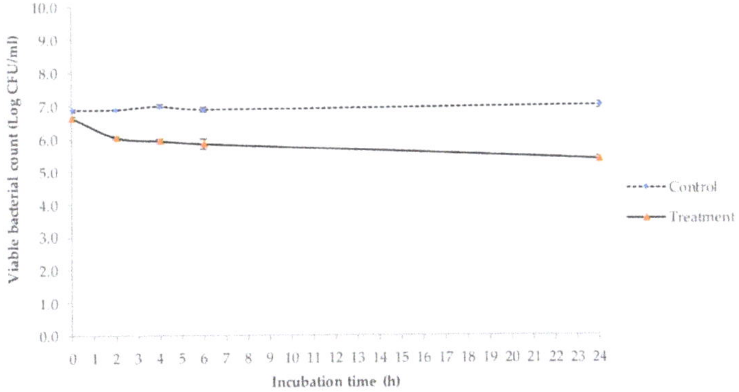

(B)

Figure 7. Antimicrobial effects of phage Ro145clw (multiplicity of infection (MOI) of 100) on *E. coli* O145:H28 (RM13514) in LB at (**A**) 37 °C and (**B**) 8 °C for 24 h. The control group contained bacterial culture without phages (dashed line) and the treatment group contained bacterial culture treated with phage Ro145clw (solid line). The error bars present the SEM for each time point of the treatment.

3. Discussion

Increasing numbers of foodborne outbreaks have been associated with food contaminated by non-O157 STEC. Although phages have been studied for their potential to control bacterial pathogens, the number of commercially-available phages and published studies on phages that are lytic against non-O157 STEC are relatively few in comparison with those on STEC O157. In the current study, a new phage, vB_EcoS-Ro145clw (or Ro145clw), which produces lytic activity against STEC O145 strains, was isolated from a non-fecal environment, unlike the majority of STEC-specific phages that have been isolated from animal-associated environments such as feedlots or fecal samples [24,25]. Though the prevalence of STEC-specific phages is highly related to the presence of their STEC hosts, which are commonly found in ruminant feces, our previous study demonstrated that various STEC-specific phages could also be isolated from produce-growing areas in Salinas Valley, CA, U.S. with STEC O145-specific phages being the most prevalent [26].

For biocontrol applications to be developed, knowledge of the genome sequence of lytic phages is critical for ensuring that no lysogenic factors, virulence-related genes, or antibiotic-resistance genes are encoded [27]. The phage Ro145clw contains a 42,031-bp double-stranded DNA genome with an average G + C content of 50.6%, but no tRNA. There are no unwanted genes such as lysogenic genes, *stx*, *eae*, or antibiotic resistance genes present in the Ro145clw genome, which makes the phage suitable for biocontrol uses. Comparative analysis showed that the phage Ro145clw is closely related to a group of phages belonging to the genus *K1gvirus*, with 84.05% BLAST-based ANI over 64% of the aligned K1G phage sequences. No *K1gvirus* phages encode tRNA, and they all have average genome sizes ranging from 42 to 46 kb, with G + C contents over 50%. This evidence suggests that phage Ro145clw might belong to the genus *K1gviru*, according to the Bacterial and Archaeal Viruses Subcommittee description [28]. Phylogenetic analysis indicates that different core genes in the Ro145clw genome have a close evolutionary relationship with different *K1gviru* phages, most likely due to horizontal genetic transfer during the evolutionary process [29,30].

Transmission electron microscopy (TEM) showed that the phage Ro145clw morphology belongs to the family *Siphoviridae*; this result is consistent with the taxonomic classification based on the BLASTn search of the genome. The phage contains a unique rosette-like base plate structure, which is not commonly seen in siphophages, but has been previously described in Rtp [31] and KLPN1 [32] phages. However, the actual function of the structure has not been well elaborated. The host range of Ro145clw is narrow and STEC-O145-specific. The results of this study indicate that environmental STEC O145 strains are more susceptible to phage Ro145clw infection than outbreak strains. The phage has a latent period of 21 min and a burst size of 192 phages per infected cell, which is considered to be a large burst size in comparison with other coliphages such as phiC119 (210 phages per infected cell) and vB_EcoS-B2 (224 phages per infected cell) that are lytic against different *E. coli* strains [17,33]. A previous study demonstrated that a phage with a large burst size is favorable as a biocontrol agent because the phage population could substantially multiply within a short period of time to control the target bacterial cells [34]. A complete cycle of Ro145clw infection against the STEC O145 strain used for isolation took approximately 45 min, which was approximately 20 min and 30 min shorter than those of phage phiC119 [17] and phage PE37 [35], respectively, against *E. coli* O157:H7 strains.

These findings suggest that phage Ro145clw has a relatively short infection cycle against its host strain. Wang et al. isolated a phage, phage AYO145A, that was lytic against STEC O145 strains from bovine feces in Canada [24]. The phage belongs to the family *Myoviridae* and has a narrow host range. Although whole genome sequencing was used to unveil the genomic characters of AYO145A [36], information regarding further physiological characterization, such as the one-step growth curve and the antimicrobial activity of the phage, is lacking. Another study evaluated a number of phages that were lytic against different serogroups of non-O157 STEC that had been isolated from cattle feces in the United States [25]. Their results showed that four myophages that were lytic against STEC O145 had different burst sizes, ranging from 23 to 195 phages per infected cell, and were also able to infect STEC O111 strains. However, a genomic characterization of these phages was not conducted and thus the presence of unwanted genes such as *stx* could not be determined. To the best of our knowledge, our current study is the first to provide both physiological and genomic information regarding a phage specific to STEC O145 strains.

Phage stability is critical and closely associated with the effectiveness of controlling pathogens during application. Previous studies indicate that each phage responds differently to exposure to external stress [25,37]. The results of this study indicate that phage Ro145clw is resistant to a pH change from 5.1 to 10.5, but a reduction of approximately 2.2 log was observed at pH 3.1. Ro145clw is able to withstand pasteurization temperatures (63 and 73 °C), and the thermal stability feature of the phage is likely to be associated with the gene encoding tape measure protein [38]. Son et al. found that a STEC O157-specific phage (isolated from bovine intestine) was susceptible to pH changes, with 1 log PFU/mL reductions at either pH 6 or 8 [35]. Litt et al. found that most phages were able to survive in a pH range from 5 to 9, but dropped off significantly when the pH went higher or lower than this range [25]. The findings of this study indicate the potential utility of phage Ro145clw in a wider range of conditions.

The genomic and biological characterization of phage Ro145clw indicates that the phage has a biocontrol potential for STEC O145 strains. The antimicrobial activities of the phage against an outbreak strain (*E. coli* O145:H28) were evaluated in LB in this study. The results showed that the strain was reduced by 5.07 log after 4 h of incubation at 37 °C and commenced growing for the remaining incubation time. Though the difference was not significant, phage Ro145clw was more active in reducing *E. coli* O145:H28 at 25 °C than at 37 °C during the first quarter of incubation period. However, at 8 °C, the antimicrobial activity of phage Ro145clw decreased. Son et al. reported a similar trend, showing that the phage reduced higher levels of *E. coli* O157:H7 at 25 °C than at 8 °C in LB [35]. The authors also found that phage-treated *E. coli* O157:H7 culture started to grow after 6 h of incubation at room temperature. This trend has been observed in other studies, as the time required for the treated bacterial culture to grow varies from strain to strain [39,40]. However, the mechanisms associated with the phage-treated bacteria's regrowth require further investigation. The antimicrobial activities of phage Ro145clw resulted in a greater reduction in environmental STEC O145 strains than the outbreak STEC O145 strain (RM13514) at room temperature (25 °C). These findings suggest that phage Ro145clw is a suitable biocontrol agent for STEC O145 in pre-harvest environments, such as those present in produce farms.

4. Materials and Methods

4.1. Bacterial Strains Preparation

Three non-pathogenic *E. coli* strains and 14 STEC strains were selected as host strains for the isolation of STEC-specific phages in this study. The *E. coli* strains included ATCC13706, ATCC43888, and DH5α, and the STEC strains included two strains per serogroup of O26, O45, O103, O111, O121, O145, and O157 STEC. These were obtained from the culture collection of the Produce Safety and Microbiology (PSM) Research Unit at the US Department of Agriculture (USDA), Agricultural Research Service (ARS), Western Regional Research Center, Albany, CA, USA. Seven additional STEC O145 strains, including four outbreak strains and five *Salmonella* strains (also obtained from the PSM culture collection), were used to test the antimicrobial activity of the phage (Supplementary, Table S4). All STEC strains were previously isolated from different environmental samples, such as water and animal feces, and were further confirmed with the presence of *eae* and *stx* genes. Fresh cultures of the strains were prepared by inoculating a sterile 5 mL tryptic soy broth (TSB; Difco, Becton Dickinson, Sparks, MD USA) with a loopful of the individual strain, and were incubated overnight at 37 °C prior to use.

4.2. Bacteriophage Isolation

Non-fecal compost samples, derived from composted food scraps and yard trimmings, were collected from a composting operation. For phage isolation, the samples were enriched with a 14-strain STEC cocktail and a 3-strain non-pathogenic *E. coli* cocktail in TSB, supplemented with 10% calcium chloride solution at 37 °C overnight. After centrifugation at 8000× *g*, the supernatant was obtained and used to confirm the presence of STEC-specific phages using a spot test assay against each individual strain of the 14 STEC strains (Supplementary, Table S4). The fresh overnight culture of each STEC strain (0.1 mL) was mixed with 12 mL molten tryptic soy agar (TSA, Difco, Becton Dickinson, Sparks, MD, USA) and poured into a sterile Petri plate. After the strain-mixed agar solidified, 10 μL of the supernatant (obtained from the enrichment) were spotted on the TSA plate and incubated at 37 °C for 24 h. As a result, the supernatant was positive against the STEC O145 strain (RM10808) and further subjected to the phage purification process using a single-layer plaque assay as previously described, with minor modifications [26]. The single-layer plaque assay was conducted immediately after picking a plaque from the previous plaque assay plate for at least three runs, until the plaques were a similar size on the plate. After purification, the phage was propagated with the fresh overnight culture of the STEC O145 host (RM10808) in 40 mL TSB, supplemented with 10 mM of $CaCl_2$ at 37 °C for 24 h. The propagated phage was centrifuged at 8000× *g* for 10 min and filtered through a 0.22-μm filter membrane to remove bacterial debris. The purified phages were subsequently concentrated using a 100 kDa cut-off Amicon Ultra-15 Centrifugal Filter Unit (Merck Millipore, Billerica, MA, USA) prior to downstream analyses such as transmission electron microscopy (TEM) and DNA extraction.

4.3. *Whole-Genome Sequencing and Genomic Analysis*

Phage DNA was extracted using a phage DNA extraction kit from Norgen Biotek (Thorold, ON, Canada). The DNA library was prepared using a TruSeq Nano DNA Library Prep Kit (Illumina, San Deigo, CA, USA), and the final amplified libraries were quantified by a bioanalyzer (Agilent, Santa Clara, CA, USA) before sequencing. Approximately 6 million 2 × 250 bp pair-end sequence reads were generated using a MisSq Reagent Kit v3 (600-cycle) on the MiSeq platform (Illumina, San Deigo, CA, USA). The quality of raw sequence reads was first checked using FASTQC. The poor sequence reads (below Q30) were then trimmed using Trimmomatic (Galaxy Version 0.36.5, with the setting of average quality required for 30 (= Q30) to trim poor quality reads). A de novo assembly of the resulting quality reads was conducted using Unicycler Galaxy v0.4.6.0 (SPAdes) with default parameters. The final contig was annotated by Prokka (v.1.12.0) with default parameters, followed by manual characterization with PHASTER Webserver [41] and BLASTn against the viral nucleotide sequences obtained from National Center for Biotechnology Information (NCBI) with Geneious (v11.0.4).

The annotated functions of the putative ORFs were confirmed using BLASTp. The prediction of tRNA in the phage genome was accomplished using tRNAscan-SE Search Server [42]. The phage termini and the possible packaging mechanism were predicted according to the in silico determination method proposed in PhageTerm [43]. The new phage sequence was subjected to a BLASTn search to obtain reference phage genomes with high nucleotide similarity from the NCBI database. These reference phage genomes were subjected to analysis with JSpeciesWS [44] to facilitate the taxonomic classification of the newly-isolated phage based on the degree of nucleotide sequence similarity [28]. The comparison of genome maps between phage Ro145clw and its reference genomes was visualized with the EasyFig visualization tool [45]. The presence of antibiotic-resistance genes in the phage genome was identified using the ResFinder (version 3.0) database [46]. Core gene analysis was conducted with CoreGenes3.5 Webserver, and genes with scores higher than 75 were considered core genes [47]. Comparative analysis of the core genes was conducted using the ClustalW algorithm for sequence alignment [48]. The phylogenetic tree was performed with MEGA 7 with the maximum composite likelihood method [49]. The reference phage genomes used in this study of *Escherichia* phage K1G (GenBank accession #GU196277), *Escherichia* phages ST2 (GenBank accession #MF153391), *Escherichia* phage G AB-2017 (GenBank accession #KY295895), and *Escherichia* phage P AB-2017 (GenBank accession #KY295898) were obtained from the NCBI database.

4.4. *Biological Characteristics*

4.4.1. One-Step Growth Curve

The experiment of a one-step growth curve was performed following the procedures described in Amarillas et al., with subtle modification [17]. The STEC O145 strain was inoculated in TSB and incubated at 37 °C overnight, then sub-cultured in 20 mL TSB at 37 °C until optical density at 600 nm (OD_{600}) was 0.5. Subsequently, phage Ro145clw was added to a bacterial suspension at a MOI of 0.01, and the phage-bacteria mixture was kept at room temperature for 2 min for phage adsorption onto the bacterial cells. After the 2-min adsorption period, the mixture was centrifuged at 10,000× *g* for one minute at 4 °C. The supernatant was removed, and the phage titers were obtained to calculate the residual titers of the phages. The bacterial pellet containing infected strains was gently re-suspended in 20 mL TSB and incubated at 37 °C for 1 h, with sampling at 5-min intervals. At each sampling point, an aliquot of 1 mL of sample was obtained and centrifuged at 10,000× *g* for 30 s at 4 °C, followed by filtration through a 0.22-μm pore-size membrane filter, then a single-layer plaque assay, as described above in duplication. Simultaneously, an additional aliquot of 1 mL phage-infected culture was collected at each time point and treated with $CHCl_3$. After homogenization for 2 min and centrifugation at 10,000× *g* for 2 min, the supernatant was obtained and subjected to serial dilutions prior to the single-layer plaque assay to determine the eclipse period. This entire experiment was conducted in three replications to estimate the burst size and latent period. The latent period was determined as the time that elapsed between the end of adsorption and the first release of phage progeny. The eclipse period was determined as the time period that elapsed between the end of adsorption and the appearance of phage particles within the bacterial cell. Burst size was calculated as the ratio of total number of phage particles produced to the initial number of infected bacterial cells during the latent period [50].

4.4.2. Transmission Electronic Microscopy

The concentrated phage was used to examine the phage morphology with a transmission electron microscope (FEI Tecnai G_2). An aliquot of 6 μL was placed on copper mesh PLECO grids (Ted Pella Inc., Redding, CA, USA) and left to set for 1 min at room temperature. Whatman filter paper was used to remove excessive phage lysate, followed by negative staining with an added 8 μL of 0.75% uranyl acetate (Sigma-Aldrich, Darmstadt, Germany) for 30 s staining at room temperature.

4.4.3. Phage Stability

To examine pH susceptibility, 100 μL of the phage lysate was added to 900 μL of SM buffer with the final pH levels of 3.1, 5.1, 7.6, 9.2, and 10.5. Samples were incubated at 37 °C for 24 h. Viable phage particles were enumerated using the plaque assay. For the temperature test, phage lysate was added to SM buffer at a volume ratio of 1:9, and 1 mL of the phage solution was dispensed in sterile tubes prior to thermal treatments (65 and 73 °C). The phage titers were determined by the single-layer plaque assay every 10 min for 60 min. These temperatures were selected to evaluate the stability of Ro145clw in high-temperature processes such as pasteurization. The course of thermal treatment time (60 min) was selected to evaluate the short-term thermal stability of phage Ro145clw. For the stability of frozen storage, an aliquot of 500 μL phage lysate was mixed with glycerol at a final concentration of 25% and stored at −80 °C.

4.5. Structural Protein Analysis

The purified and concentrated phage lysate was subjected to sodium dodecyl sulfate polyacrylamide gel electrophoresis (SDS-PAGE) using a 1D Biorad 12% TGX gel with Precision Plus MW standard marker (2 μL; Biorad, Hercules, CA, USA). Electrophoresis was performed at 100 V for 90 min. The gel was stained using an Imperial^{TM} Protein Stain (ThermoFisher, Waltham, MA, USA) containing a formulation of Coomassie brilliant blue R-250. In-gel digestions were conducted with Trypsin (Promega, Madison, WI, USA) using a Digest Pro digestion robot (Intavis, Köln, Germany). The robot was programmed to perform the in-gel digestion following methods that have been previously published [51,52]. Digested samples were subjected to nanoflow reversed-phase chromatography with an Eksigent NanoLC (Sciex, Framingham, MA, USA) using Picochip 105 mm columns packed with REPROSIL-Pur C18-AQ, 3 μM, 120A packing (New Objectives, Woburn, MA, USA). A 10 μL portion of each digested sample was injected with 2% acetonitrile in water with 0.1% formic acid, with a flow rate of 400 nL/min for 1 h. Elution solvents A and B were 2% acetonitrile in water and acetonitrile, respectively (each containing 0.1% formic acid). Sample elution began with 3% B, ramping up to 10% B at 10 min, then to 25% B at 40 min, then to 40% B at 58 min, returning to 3% B at 60 min. Mass spectral analyses were performed with an Orbitrap Elite (Thermo Fisher Scientific, Waltham, MA, USA), operated in positive ion mode using a top three data-dependent data acquisition method. Survey scans were collected at 60 K resolution in the Orbitrap detector. The top three most intense ions above the threshold of 30 K counts were subjected to collision-induced fragmentation (CID) with normalized collision energy set to 30. The resulting fragment ions were detected in the instrument's linear trap. Dynamic exclusion of precursor ions was set to 6 s. Mascot software (Matrix Science,

Boston, MA, USA) was used to match the tandem mass spectrometry (MS–MS) data to amino acid sequences derived from the nucleotide sequences that were obtained from the phage isolates.

4.6. Antimicrobial Activities

4.6.1. Host Range and Efficiency of Plating

After phage purification, the phage was subjected to the host range test against three non-pathogenic *E. coli*, 28 STEC and five *Salmonella* strains using the spot test assay as described above. For the spot test-positive strains, efficiency of plating (EOP) was used to determine productive infection by using phage particles produced against each susceptible strain in comparison to the phage particles produced against the primary host strain [53]. Fresh overnight cultures of the test strains and of the primary host strain were prepared in TSB at 37 °C for 18 h. After serial dilution, the phage lysates with four dilution factors (10^{-3} to 10^{-7}) were subjected to the single-layer plaque assay separately against all test strains and the primary host strain. The plates were then incubated at 37 °C overnight. The experiment was conducted in three replications. EOP was calculated based on the average of plaque-forming units (PFU) against each test bacterium divided by the average of PFU against the primary bacterium used for isolation. Generally, if the EOP was 0.5 or more, it was classified as having a high phage-producing efficiency. An EOP above 0.1 but below 0.5 indicated a medium-producing efficiency; an EOP between 0.001 and 0.1 indicated a low-producing efficiency; any value under 0.001 represented inefficient phage production.

4.6.2. Bacterial Challenge Assay

The bacterial challenge assay was conducted using a spectrophotometer to monitor bacterial growth treated with different concentrations of phage lysate as previously described, with minor modifications [54]. Prior to the experiment, a fresh overnight culture of *E. coli* O145:H28 was prepared in 5 mL of TSB and incubated at 37 °C for 18 h. Subsequently, the culture was pelleted down at 4000× *g* centrifugation and washed twice with the same volume of fresh TSB. After resuspension in TSB, the bacterial culture was diluted down to 1×10^5 CFU/mL and further dispensed into a 96-well plate, with 200 µL per well; then, phage Ro145clw was added at MOIs of 1, 10, and 100. The plate was monitored in a plate reader with the temperature set at 37 °C. The OD_{600} reading was recorded every 30 min for 18 h.

4.6.3. Determination of Bacteriophage-Insensitive Mutant (BIM) Frequency

The emergence frequency of bacteriophage-insensitive mutant (BIM) was conducted by mixing appropriate volume of overnight cultures of STEC strains [*E. coli* O145:H28 (RM13514) and *E. coli* O145:NM (SJ23)] with phage Ro145clw at MOI of 100. The mixture was added with $CaCl_2$ (10 mM) and $MgSO_4$ (10mM) and then incubated at 37 °C for 10 min. After serial dilutions, the diluted bacterium-phage mixture was plated on MacConkey agar with a top thin layer of TSA (5 mL) and incubated at 37 °C overnight. BIM frequency was determined by dividing the number of surviving bacterial cells by the initial bacterial concentration. The experiments were conducted in 3 replications.

4.6.4. Antimicrobial Activity Test in LB

A fresh overnight culture of *E. coli* O145:H28 was prepared in 10 mL TSB at 37 °C for 18 h. An aliquot of 0.2 mL overnight culture was added to 18.8 mL LB (Invitrogen, Carlsbad, CA, USA) to obtain the final concentration at 1×10^7 CFU/mL. One tube of the bacterial suspension was treated with phage lysates at 1×10^9 PFU/mL (MOI 100). A control group was also prepared by adding the same volume of SM buffer to 20 mL of bacterial suspension. Both the control and treatment were incubated at 8 and 37 °C. At 0, 2, 4, 6, and 24 h of incubation, samples were serially diluted using sterile 0.1% peptone water, and an aliquot of 0.1 mL diluted sample was spread-plated on MacConkey agar plates (BD, Franklin Lakes, NJ, USA). The plates were incubated at 37 °C overnight, and colonies of bacteria were counted. The antimicrobial effects of the phage against *E. coli* O145:H28 (representing a low EOP) and *E. coli* O145: NM (high EOP) were compared using the same method as described at 25 °C.

4.6.5. Statistical Analysis

Experiments were performed with at least three individual repetitions. Bacterial colony counts and phage titers were calculated as CFU/mL or PFU/mL and logarithmically transformed for statistical analysis. Least squares mean (LSM) was performed to compare the means of phage titers using JMP® (Version 12.0.1, SAS Institute Inc., Cary, NC, USA). One-way analysis of variance (ANOVA) with the statistical significance at 5% level was used to evaluate the effects of different pH on the recovery of phage titers. The Student's *t*-test was used to evaluate the viable bacterial count between the control group and treatment group with the phage at each time point.

4.7. Nucleotide Sequence Accession Number

The genome sequence of *Escherichia* phage vB_EcoS-Ro145clw was deposited in GenBank under accession number MG852086. The raw sequence reads were submitted to the NCBI sequence read archive (SRA) with accession number PRJNA525899.

5. Conclusions

In this study, phage Ro145clw, with strong lytic infection against environmental STEC O145 strains, was shown to have a relatively short latent period (21 min) and a large burst size (192 phages per infected cell). The genomic data indicated the absence of unwanted genes including virulence genes, antibiotic-resistance genes, and lysogenic genes in the Ro145clw genome. Phage Ro145clw is resistant to adverse pH and temperature conditions; these features might be associated with the environment from which the phage was isolated. The antimicrobial effects of the phage against environmental STEC O145 strains were more prominent than against outbreak strains. These findings substantiate the potential biocontrol alternative of the phage Ro145clw to prevent the spread of STEC O145 in the pre-harvest environment. The genomic information for phage Ro145clw provides valuable insights into the diversity of the specific lytic phages against STEC strains. Future studies in using phage cocktails may be undertaken to improve the biocontrol effectiveness against outbreak STEC strains.

Supplementary Materials: The following are available online at http://www.mdpi.com/2079-6382/8/2/74/s1: Figure S1: Bacterial challenge assay of the phage Ro145clw against outbreak strain *E. coli* O145:H28; Table S1: List of annotated ORFs with the size, location, and predicted functions in the genome of Ro145clw; Table S2: Storage evaluation of phage Ro145clw in 25% glycerol at −80 °C for five months; Table S3: Comparison of the antimicrobial activities of phage Ro145clw (MOI of 100) against different EOP bacterial strains (initial concentration = 7 log CFU/mL) in LB at 25 °C; Table S4: Bacterial strains including *E. coli* and *Salmonella* were used in the present study for either phage isolation or the host range test of the isolated phage.

Author Contributions: Conceptualization, V.C.H.W.; Methodology, Y.-T.L.; Software, R.W.L. and Y.-T.L.; Formal Analysis, R.W.L. and Y.-T.L.; Resources, V.C.H.W.; Data Curation, R.W.L., Y.-T.L., L.A.H., V.M.L. and F.L.; Writing-Original Draft Preparation, Y.-T.L.; Investigation, Y.-T.L., A.S., L.A.H., F.L. and V.M.L.; Writing-Review & Editing, V.C.H.W. and Y.-T.L.; Supervision, V.C.H.W.; Project Administration, V.C.H.W.; Funding Acquisition, V.C.H.W.

Funding: This research was funded by the USDA-ARS CRIS projects 2030-42000-050-00D.

Acknowledgments: The authors would like to thank Anne Bates for her assistance in generating part of the host range test and EOP data.

Conflicts of Interest: The authors declare no conflict of interest.

References

1. Tarr, P.I.; Gordon, C.A.; Chandler, W.L. Shiga-toxin-producing *Escherichia coli* and haemolytic uraemic syndrome. *Lancet* **2005**, *365*, 1073–1086. [CrossRef]

2. Riley, L.W.; Remis, R.S.; Helgerson, S.D.; McGee, H.B.; Wells, J.G.; Davis, B.R.; Hebert, R.J.; Olcott, E.S.; Johnson, L.M.; Hargrett, N.T.; et al. Hemorrhagic colitis associated with a rare *Escherichia coli* serotype. *N. Engl. J. Med.* **1983**, *308*, 681–685. [CrossRef] [PubMed]
3. Scallan, E.; Hoekstra, R.M.; Angulo, F.J.; Tauxe, R.V.; Widdowson, M.-A.; Roy, S.L.; Jones, J.L.; Griffin, P.M. Foodborne illness acquired in the United States–Major pathogens. *Emerg. Infect. Dis.* **2011**, *17*, 7–15. [CrossRef] [PubMed]
4. Karmali, M.A. Emerging Public Health Challenges of Shiga Toxin-Producing *Escherichia coli* Related to Changes in the Pathogen, the Population, and the Environment. *Clin. Infect. Dis.* **2017**, *64*, 371–376. [CrossRef] [PubMed]
5. Herman, K.M.; Hall, A.J.; Gould, L.H. Outbreaks attributed to fresh leafy vegetables, United States, 1973–2012. *Epidemiol. Infect.* **2015**, *143*, 3011–3021. [CrossRef] [PubMed]
6. Brooks, J.T.; Sowers, E.G.; Wells, J.G.; Greene, K.D.; Griffin, P.M.; Hoekstra, R.M.; Strockbine, N.A. Non-O157 Shiga toxin-producing *Escherichia coli* infections in the United States, 1983–2002. *J. Infect. Dis.* **2005**, *192*, 1422–1429. [CrossRef] [PubMed]
7. Luna-Gierke, R.E.; Griffin, P.M.; Gould, L.H.; Herman, K.; Bopp, C.A.; Strockbine, N.; Mody, R.K. Outbreaks of non-O157 Shiga toxin-producing *Escherichia coli* infection: USA. *Epidemiol. Infect.* **2014**, *142*, 2270–2280. [CrossRef] [PubMed]
8. Taylor, E.V.; Nguyen, T.A.; Machesky, K.D.; Koch, E.; Sotir, M.J.; Bohm, S.R.; Folster, J.P.; Bokanyi, R.; Kupper, A.; Bidol, S.A.; et al. Multistate outbreak of *Escherichia coli* O145 infections associated with romaine lettuce consumption, 2010. *J. Food Prot.* **2013**, *76*, 939–944. [CrossRef] [PubMed]
9. Rivero, M.A.; Passucci, J.A.; Rodriguez, E.M.; Parma, A.E. Role and clinical course of verotoxigenic *Escherichia coli* infections in childhood acute diarrhoea in Argentina. *J. Med. Microbiol.* **2010**, *59*, 345–352. [CrossRef]
10. De Schrijver, K.; Buvens, G.; Posse, B.; Van den Branden, D.; Oosterlynck, O.; De Zutter, L.; Eilers, K.; Pierard, D.; Dierick, K.; Van Damme-Lombaerts, R.; et al. Outbreak of verocytotoxin-producing *E. coli* O145 and O26 infections associated with the consumption of ice cream produced at a farm, Belgium, 2007. *Euro Surveill.* **2008**, *13*, 8041. [CrossRef]
11. Carter, M.Q.; Quinones, B.; He, X.; Zhong, W.; Louie, J.W.; Lee, B.G.; Yambao, J.C.; Mandrell, R.E.; Cooley, M.B. An Environmental Shiga Toxin-Producing *Escherichia coli* O145 Clonal Population Exhibits High-Level Phenotypic Variation That Includes Virulence Traits. *Appl. Environ. Microbiol.* **2015**, *82*, 1090–1101. [CrossRef] [PubMed]
12. Liao, Y.T.; Brooks, J.C.; Martin, J.N.; Echeverry, A.; Loneragan, G.H.; Brashears, M.M. Antimicrobial interventions for O157:H7 and non-O157 Shiga toxin-producing *Escherichia coli* on beef subprimal and mechanically tenderized steaks. *J. Food Prot.* **2015**, *78*, 511–517. [CrossRef] [PubMed]
13. Hatfull, G.F. Bacteriophage genomics. *Curr. Opin. Microbiol.* **2008**, *11*, 447–453. [CrossRef] [PubMed]
14. Clokie, M.R.J.; Millard, A.D.; Letarov, A.V.; Heaphy, S. Phages in nature. *Bacteriophage* **2011**, *1*, 31–45. [CrossRef] [PubMed]
15. Hagens, S.; Loessner, M.J. Bacteriophage for biocontrol of foodborne pathogens: Calculations and considerations. *Curr. Pharm. Biotechnol.* **2010**, *11*, 58–68. [CrossRef] [PubMed]
16. Snyder, A.B.; Perry, J.J.; Yousef, A.E. Developing and optimizing bacteriophage treatment to control enterohemorrhagic *Escherichia coli* on fresh produce. *Int. J. Food Microbiol.* **2016**, *236*, 90–97. [CrossRef] [PubMed]
17. Amarillas, L.; Chaidez, C.; Gonzalez-Robles, A.; Lugo-Melchor, Y.; Leon-Felix, J. Characterization of novel bacteriophage phiC119 capable of lysing multidrug-resistant Shiga toxin-producing *Escherichia coli* O157:H7. *PeerJ* **2016**, 13. [CrossRef]
18. Svab, D.; Falgenhauer, L.; Rohde, M.; Szabo, J.; Chakraborty, T.; Toth, I. Identification and Characterization of T5-Like Bacteriophages Representing Two Novel Subgroups from Food Products. *Front. Microbiol.* **2018**, *9*, 202. [CrossRef] [PubMed]
19. Tolen, T.N.; Xie, Y.; Hairgrove, T.B.; Gill, J.J.; Taylor, T.M. Evaluation of Commercial Prototype Bacteriophage Intervention Designed for Reducing O157 and Non-O157 Shiga-Toxigenic *Escherichia coli* (STEC) on Beef Cattle Hide. *Foods* **2018**, *7*, 114. [CrossRef] [PubMed]
20. Catalao, M.J.; Gil, F.; Moniz-Pereira, J.; Sao-Jose, C.; Pimentel, M. Diversity in bacterial lysis systems: Bacteriophages show the way. *FEMS Microbiol. Rev.* **2013**, *37*, 554–571. [CrossRef] [PubMed]

21. Baig, A.; Colom, J.; Barrow, P.; Schouler, C.; Moodley, A.; Lavigne, R.; Atterbury, R. Biology and Genomics of an Historic Therapeutic *Escherichia coli* Bacteriophage Collection. *Front. Microbiol.* **2017**, *8*, 1652. [CrossRef] [PubMed]
22. Summer, E.J.; Berry, J.; Tran, T.A.T.; Niu, L.; Struck, D.K.; Young, R. Rz/Rz1 lysis gene equivalents in phages of Gram-negative hosts. *J. Mol. Biol.* **2007**, *373*, 1098–1112. [CrossRef] [PubMed]
23. Oliveira, L.; Tavares, P.; Alonso, J.C. Headful DNA packaging: Bacteriophage SPP1 as a model system. *Virus Res.* **2013**, *173*, 247–259. [CrossRef] [PubMed]
24. Wang, J.; Niu, Y.D.; Chen, J.; Anany, H.; Ackermann, H.W.; Johnson, R.P.; Ateba, C.N.; Stanford, K.; McAllister, T.A. Feces of feedlot cattle contain a diversity of bacteriophages that lyse non-O157 Shiga toxin-producing *Escherichia coli*. *Can. J. Microbiol.* **2015**, *61*, 467–475. [CrossRef] [PubMed]
25. Litt, P.K.; Saha, J.; Jaroni, D. Characterization of Bacteriophages Targeting Non-O157 Shiga Toxigenic *Escherichia coli*. *J. Food Prot.* **2018**, *81*, 785–794. [CrossRef] [PubMed]
26. Liao, Y.-T.; Quintela, I.A.; Nguyen, K.; Salvador, A.; Cooley, M.B.; Wu, V.C.H. Investigation of prevalence of free Shiga toxin-producing *Escherichia coli* (STEC)-specific bacteriophages and its correlation with STEC bacterial hosts in a produce-growing area in Salinas, California. *PLoS ONE* **2018**, *13*, e0190534. [CrossRef]
27. Endersen, L.; Guinane, C.M.; Johnston, C.; Neve, H.; Coffey, A.; Ross, R.P.; McAuliffe, O.; O'Mahony, J. Genome analysis of *Cronobacter* phage vB_CsaP_Ss1 reveals an endolysin with potential for biocontrol of Gram-negative bacterial pathogens. *J. Gen. Virol.* **2015**, *96*, 463–477. [CrossRef]
28. Adriaenssens, E.M.; Brister, J.R. How to Name and Classify Your Phage: An Informal Guide. *Viruses* **2017**, *9*, 70. [CrossRef]
29. Hatfull, G.F.; Hendrix, R.W. Bacteriophages and their genomes. *Curr. Opin. Virol.* **2011**, *1*, 298–303. [CrossRef]
30. Peng, Q.; Yuan, Y. Characterization of a newly isolated phage infecting pathogenic *Escherichia coli* and analysis of its mosaic structural genes. *Sci. Rep.* **2018**, *8*, 8086. [CrossRef]
31. Wietzorrek, A.; Schwarz, H.; Herrmann, C.; Braun, V. The genome of the novel phage Rtp, with a rosette-like tail tip, is homologous to the genome of phage T1. *J. Bacteriol.* **2006**, *188*, 1419–1436. [CrossRef] [PubMed]
32. Hoyles, L.; Murphy, J.; Neve, H.; Heller, K.J.; Turton, J.F.; Mahony, J.; Sanderson, J.D.; Hudspith, B.; Gibson, G.R.; McCartney, A.L.; et al. *Klebsiella pneumoniae* subsp. *pneumoniae*-bacteriophage combination from the caecal effluent of a healthy woman. *PeerJ* **2015**, *3*, e1061. [CrossRef] [PubMed]
33. Xu, Y.; Yu, X.; Gu, Y.; Huang, X.; Liu, G.; Liu, X. Characterization and Genomic Study of Phage vB_EcoS-B2 Infecting Multidrug-Resistant *Escherichia coli*. *Front. Microbiol.* **2018**, *9*, 793. [CrossRef] [PubMed]
34. Nilsson, A.S. Phage therapy–Constraints and possibilities. *Upsala J. Med. Sci.* **2014**, *119*, 192–198. [CrossRef] [PubMed]
35. Son, H.M.; Duc, H.M.; Masuda, Y.; Honjoh, K.I.; Miyamoto, T. Application of bacteriophages in simultaneously controlling *Escherichia coli* O157:H7 and extended-spectrum beta-lactamase producing *Escherichia coli*. *Appl. Microbiol. Biotechnol.* **2018**, *102*, 10259–10271. [CrossRef] [PubMed]
36. Wang, J.; Niu, Y.D.; Chen, J.; McAllister, T.A.; Stanford, K. Complete Genome Sequence of *Escherichia coli* O145:NM Bacteriophage vB_EcoM_AYO145A, a New Member of O1-Like Phages. *Genome Announc.* **2015**, *3*, e00539-15. [CrossRef]
37. Merabishvili, M.; Vervaet, C.; Pirnay, J.-P.; De Vos, D.; Verbeken, G.; Mast, J.; Chanishvili, N.; Vaneechoutte, M. Stability of *Staphylococcus aureus* phage ISP after freeze-drying (lyophilization). *PLoS ONE* **2013**, *8*, e68797. [CrossRef]
38. Geagea, H.; Labrie, S.J.; Subirade, M.; Moineau, S. The Tape Measure Protein Is Involved in the Heat Stability of *Lactococcus lactis* Phages. *Appl. Environ. Microbiol.* **2018**, *84*, e02082-17. [CrossRef]
39. Hudson, J.A.; Billington, C.; Cornelius, A.J.; Wilson, T.; On, S.L.W.; Premaratne, A.; King, N.J. Use of a bacteriophage to inactivate *Escherichia coli* O157:H7 on beef. *Food Microbiol.* **2013**, *36*, 14–21. [CrossRef]
40. Tomat, D.; Migliore, L.; Aquili, V.; Quiberoni, A.; Balagué, C. Phage biocontrol of enteropathogenic and shiga toxin-producing *Escherichia coli* in meat products. *Front. Cell. Infect. Microbiol.* **2013**, *3*, 20. [CrossRef]
41. Arndt, D.; Grant, J.R.; Marcu, A.; Sajed, T.; Pon, A.; Liang, Y.; Wishart, D.S. PHASTER: A better, faster version of the PHAST phage search tool. *Nucleic Acids Res.* **2016**, *44*, W16–W21. [CrossRef] [PubMed]
42. Lowe, T.M.; Chan, P.P. tRNAscan-SE On-line: Integrating search and context for analysis of transfer RNA genes. *Nucleic Acids Res.* **2016**, *44*, W54–W57. [CrossRef] [PubMed]

43. Garneau, J.R.; Depardieu, F.; Fortier, L.C.; Bikard, D.; Monot, M. PhageTerm: A tool for fast and accurate determination of phage termini and packaging mechanism using next-generation sequencing data. *Sci. Rep.* **2017**, *7*, 8292. [CrossRef] [PubMed]
44. Richter, M.; Rossello-Mora, R.; Oliver Glockner, F.; Peplies, J. JSpeciesWS: A web server for prokaryotic species circumscription based on pairwise genome comparison. *Bioinformatics* **2016**, *32*, 929–931. [CrossRef] [PubMed]
45. Sullivan, M.J.; Petty, N.K.; Beatson, S.A. Easyfig: A genome comparison visualizer. *Bioinformatics* **2011**, *27*, 1009–1010. [CrossRef] [PubMed]
46. Zankari, E.; Cosentino, S.; Vestergaard, M.; Rasmussen, S.; Lund, O.; Aarestrup, F.M.; Larsen, M.V. Identification of acquired antimicrobial resistance genes. *J. Antimicrob. Chemother.* **2012**, *67*, 2640–2644. [CrossRef] [PubMed]
47. Mahadevan, P.; King, J.F.; Seto, D. CGUG: In silico proteome and genome parsing tool for the determination of "core" and unique genes in the analysis of genomes up to ca. 1.9 Mb. *BMC Res. Methods* **2009**, *2*, 168. [CrossRef] [PubMed]
48. McWilliam, H.; Li, W.; Uludag, M.; Squizzato, S.; Park, Y.M.; Buso, N.; Cowley, A.P.; Lopez, R. Analysis Tool Web Services from the EMBL-EBI. *Nucleic Acids Res.* **2013**, *41*, W597–W600. [CrossRef] [PubMed]
49. Tamura, K.; Stecher, G.; Peterson, D.; Filipski, A.; Kumar, S. MEGA6: Molecular Evolutionary Genetics Analysis version 6.0. *Mol. Biol. Evol.* **2013**, *30*, 2725–2729. [CrossRef]
50. Adams, M.H. *Bacteriophage*; Interscience Publishers, Inc.: New York, NY, USA, 1959.
51. Shevchenko, A.; Tomas, H.; Havlis, J.; Olsen, J.V.; Mann, M. In-gel digestion for mass spectrometric characterization of proteins and proteomes. *Nat. Protoc.* **2006**, *1*, 2856–2860. [CrossRef]
52. Shevchenko, A.; Wilm, M.; Vorm, O.; Mann, M. Mass spectrometric sequencing of proteins silver-stained polyacrylamide gels. *Anal. Chem.* **1996**, *68*, 850–858. [CrossRef] [PubMed]
53. Mirzaei, M.K.; Nilsson, A.S. Correction: Isolation of phages for phage therapy: A comparison of spot tests and efficiency of plating analyses for determination of host range and efficacy. *PLoS ONE* **2015**, *10*, e0118557. [CrossRef] [PubMed]
54. Fong, K.; LaBossiere, B.; Switt, A.I.M.; Delaquis, P.; Goodridge, L.; Levesque, R.C.; Danyluk, M.D.; Wang, S. Characterization of Four Novel Bacteriophages Isolated from British Columbia for Control of Non-typhoidal *Salmonella in Vitro* and on Sprouting Alfalfa Seeds. *Front. Microbiol.* **2017**, *8*, 2193. [CrossRef] [PubMed]

antibiotics

MDPI

Article

Comparison of *Staphylococcus* Phage K with Close Phage Relatives Commonly Employed in Phage Therapeutics

Jude Ajuebor [1], Colin Buttimer [1], Sara Arroyo-Moreno [1], Nina Chanishvili [2], Emma M. Gabriel [1], Jim O'Mahony [1], Olivia McAuliffe [3], Horst Neve [4], Charles Franz [4] and Aidan Coffey [1,5,*]

1 Department of Biological Sciences, Cork Institute of Technology, Bishopstown, Cork T12 P928, Ireland; jude.ajuebor@mycit.ie (J.A.); colin.buttimer@mycit.ie (C.B.); sara.arroyo-moreno@mycit.ie (S.A.-M.); emma.gabriel@mycit.ie (E.M.G.); Jim.OMahony@cit.ie (J.O.)
2 Eliava Institute of Bacteriophages, Microbiology and Virology, Tbilisi 0160, Georgia; nina.chanishvili@pha.ge
3 Teagasc, Moorepark Food Research Centre, Fermoy, Cork P61 C996, Ireland; Olivia.McAuliffe@teagasc.ie
4 Department of Microbiology and Biotechnology, Max Rubner-Institut, DE-24103 Kiel, Germany; horst.neve@mri.bund.de (H.N.); charles.franz@mri.bund.de (C.F.)
5 Alimentary Pharmabiotic Centre, University College, Cork T12 YT20, Ireland
* Correspondence: aidan.coffey@cit.ie; Tel.: +353-214-335-486

Received: 9 March 2018; Accepted: 19 April 2018; Published: 25 April 2018

Abstract: The increase in antibiotic resistance in pathogenic bacteria is a public health danger requiring alternative treatment options, and this has led to renewed interest in phage therapy. In this respect, we describe the distinct host ranges of *Staphylococcus* phage K, and two other K-like phages against 23 isolates, including 21 methicillin-resistant *S. aureus* (MRSA) representative sequence types representing the Irish National MRSA Reference Laboratory collection. The two K-like phages were isolated from the *Fersisi* therapeutic phage mix from the Tbilisi Eliava Institute, and were designated B1 (vB_SauM_B1) and JA1 (vB_SauM_JA1). The sequence relatedness of B1 and JA1 to phage K was observed to be 95% and 94% respectively. In terms of host range on the 23 *Staphylococcus* isolates, B1 and JA1 infected 73.9% and 78.2% respectively, whereas K infected only 43.5%. Eleven open reading frames (ORFs) present in both phages B1 and JA1 but absent in phage K were identified by comparative genomic analysis. These ORFs were also found to be present in the genomes of phages (Team 1, vB_SauM-fRuSau02, Sb_1 and ISP) that are components of several commercial phage mixtures with reported wide host ranges. This is the first comparative study of therapeutic staphylococcal phages within the recently described genus *Kayvirus*.

Keywords: phage isolation; bacteriophage; phage resistance; MRSA; *Staphylococcus*; *Kayvirus*

1. Introduction

Staphylococcus aureus (*S. aureus*) is an opportunistic and important pathogen in clinical and health-care settings causing a wide variety of diseases commonly involving the skin, soft tissue, bone, and joints [1]. It is also a well-known causative agent of prosthetic joint infections (PJI), cardiac device infections, and intravascular catheter infections [1]. *S. aureus* pathogenicity is due, in part, to its ability to acquire and express a wide array of virulence factors, as well as antimicrobial resistance determinants [2], an example of which involves the acquisition of the staphylococcal cassette chromosome (SCCmec) leading to the development of methicillin resistance in *S. aureus* [3]. Methicillin-resistant *S. aureus* (MRSA) was first reported in 1961 [4], and has since been observed to cause serious infections in hospitals worldwide. Reports of MRSA clones resistant to the majority of antibiotics are a growing concern [5]. As such, new treatment options are needed.

Bacteriophages (phages) are biological entities composed of either DNA or RNA enclosed within a protein coat [6]. They are highly specific, with most phages capable of infecting only a single bacterial species [6,7], and studies on these viruses have been performed since the late 19th century [8]. The phage infection process usually begins with the recognition of the receptor on the bacterial cell surface by its receptor binding protein [9]. In natural environments bacterial hosts have evolved many mechanisms to protect themselves from phage attack to include; adsorption blocking, DNA injection blocking, restriction-modification system (R/M), abortive infection, and the clustered regularly interspaced short palindromic repeats (CRISPR)-Cas systems [10,11]. In turn, phages have evolved several strategies for overcoming these systems to ensure their survival in the phage-host co-evolutionary race [12–14].

The use of phages as therapeutics to eliminate pathogenic bacteria dates back to experiments conducted by Felix d'Herelle in 1919 at a French hospital to treat dysentery [15]. Since then, a wide range of phage therapy trials have been undertaken, many with very promising results [15,16]. Pyophage and Intesti-phage are among the commercial phage mixtures currently produced at the Eliava Institute. Metagenomic studies on these phage mixtures have been reported [17,18] and the staphylococcal phages Sb-1 and ISP are key components of Pyophage [19,20]. Other phages isolated from these commercial phages mixes have also been reported [21–24]. Phages like vB_SauM-fRuSau02 was isolated from a phage mix produced by Microgen (Moscow, Russia) [21] and Team 1 was isolated from PhageBioDerm, a wound healing preparation consisting of a biodegradable polymer impregnated with an antibiotic and lytic phages [22–24]. These phages all possess a wide host range against a number of clinically relevant *S. aureus* isolates, demonstrating the efficacy of such commercial phage mixtures in treating a range of bacterial infections [19–24].

In this paper, we employed another phage mixture from the Eliava Institute, namely the Fersisi phage mix. Fersisi is a relatively new combination developed approximately 15–20 years ago on the basis of Pyophage, although with fewer phage components. Two phages from this mix were designated B1 (vB_SauM_B1) and JA1 (vB_SauM_JA1). Phage K, on the other hand, is a well-known phage being the type phage of the recently designated genus *Kayvirus* of the subfamily *Spounavirinae* [25]. The exact origin of phage K is unknown, but descriptions of the phage are made as far back as 1949 [26,27]. An initial host range study involving this phage reported it to be ineffective against many MRSA strains [26]. Thus, phages B1 and JA1 were compared (on the basis of their host range) to phage K to explore possible host range differences and it was observed that both phages had broader host ranges. A comparative study was performed on their genomes and the genomes of similar phages from other commercial phage mixtures (Team 1, vB_SauM-fRuSau02, Sb_1 and ISP) with reported wide host ranges, to provide molecular insight into the differences in host range encountered in this study.

2. Results and Discussion

2.1. Origin of Phages B1 and JA1

Phages B1 and JA1 were isolated from the Fersisi commercial phage mixtures; batch 010112 (B1) and F-062015 (JA1). This product is used in the treatment of staphylococcal and streptococcal infections. For the isolation of B1, phage enrichment was carried out using staphylococcal host cultured from the sonicate fluid of a hospital patient suffering from PJI. DPC5246 was subsequently used as propagating host for B1, as a prophage was encountered in the PJI strain. Phage enrichment in the isolation of JA1 was done using the Cork Institute of Technology (CIT) collection strain *S. aureus* CIT281189. Both the PJI strain and CIT281189 were insensitive to phage K.

2.2. Morphology and Host Range of Phages K, B1 and JA1

Phages B1 and JA1 exhibited typical characteristics of phages belonging to the *Myoviridae* family, similar to the reported morphology of phage K [26]. All three phages possessed an A1 morphology [28], displaying an icosahedral head as well as a long contractile tail. They also contained a structure previously described as knob-like appendages by O'Flaherty et al. [26], extending from their base plates (likely "clumped/aggregated" base plate appendices) and clearly visible in Figure 1. Estimations were made on the dimensions of these phages (Table 1). Capsid heights were estimated as 92.9 ± 4.0 nm (B1), 87.0 ± 2.1 nm (JA1) and 92.9 ± 3.8 nm (K). Tail dimension were also estimated as 233.0 ± 4.4 × 23.4 ± 1.2 nm (B1), 231.5 ± 4.7 × 22.7 ± 0.9 nm (JA1), and 227.5 ± 5.5 × 23.8 ± 1.0 nm (K), and base plates/knobs complexes were estimated as 30.1 ± 1.8 × 47.2 ± 3.7 nm (B1), 32.5 ± 7.9 × 45.8 ± 1.4 nm (JA1), and 36.6 ± 5.1 × 41.7 ± 2.6 nm (K). Owing to the similar morphology of all three phages, a host range study was conducted to explore possible differences in host spectra across a number of hospital isolates. Twenty-one of these isolates represented the entire collection of MRSA sequence-types identified in Ireland by the National MRSA Reference Laboratory (Dublin, Ireland), and includes the commonly encountered ST22-MRSA-IV, which has been predominant in Irish hospitals since the late 1990s [29]. The other two *S. aureus* strains used in this study were included as additional phage propagation strains. Host range was assessed by plaque assay technique on lawns of various MRSA strains listed in Table 2. The efficiency of plaquing (EOP) was used to represent the degree to which each of the phages studied infected all 23 staphylococcal strains. Phage JA1 had the broadest host range, forming plaques on 18 out of the 23 staphylococcal strains examined. B1 also had a broad host range and was capable of forming plaques on 17 isolates (with some in common with the 18 lysed by phage JA1). Phage K had the narrowest host range, forming plaques on only 10 of the isolates (including its propagating strain DPC5246). All 23 staphylococcal strains were effectively lysed by at least one of the three phages, with the exception of E1139 (IV) ST45 and E1185 (IV) ST12, whose EOP were significantly low at 3.88×10^{-6} and 1.16×10^{-6} respectively; as well as 3488 (VV) ST8, which was resistant to all three phages. Plaque size ranged from 0.5 mm to 1.5 mm, with a halo occurring in some instances (Table 3 and Supplementary Materials, Figure S1). The wide host range encountered in this study is common among K-like phages and has been reported for other staphylococcal K-like phages, such as JD007, which infected 95% of *S. aureus* isolates obtained from several hospitals in Shanghai, China [30].

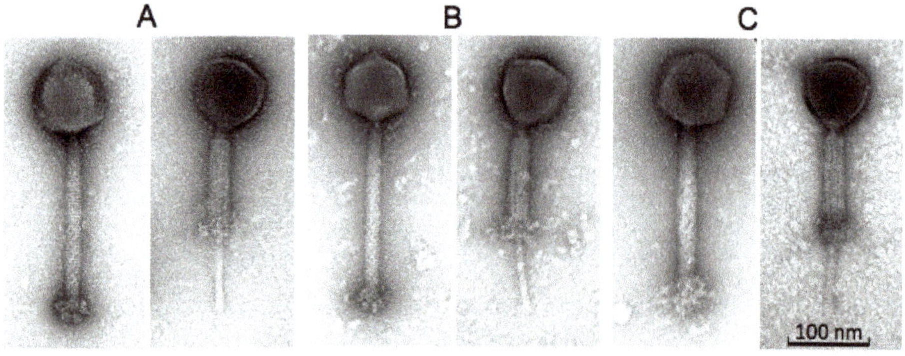

Figure 1. Transmission electron micrographs of phages B1 (**A**), JA1 (**B**), and K (**C**) showing their icosahedral capsid and their long contractile tail (both extended and contracted).

Table 1. Dimensions of staphylococcal phages B1, JA1, and K derived from micrographs obtained from transmission electron microscopy.

Phages	Head (nm)	Tail Length (nm) (incl. "knob")	Tail Width (nm)	Baseplate "knob" Length (nm)	Baseplate "knob" Width (nm)
B1	92.9 ± 4.0 (n = 11)	233.0 ± 4.4 (n = 12)	23.4 ± 1.2 (n = 12)	30.1 ± 1.8 (n = 12)	47.2 ± 3.7 (n = 10)
JA1	87.0 ± 2.1 (n = 9)	231.5 ± 4.7 (n = 9)	22.7 ± 0.9 (n = 9)	32.5 ± 7.9 (n = 9)	45.8 ± 1.4 (n = 9)
K	92.9 ± 3.8 (n = 16)	227.5 ± 5.5 (n = 16)	23.8 ± 1.0 (n = 16)	36.6 ± 5.1 (n = 16)	41.7 ± 2.6 (n = 16)

Table 2. Host ranges of staphylococcal phages B1, JA1, and K against methicillin-resistant *Staphylococcus aureus* (MRSA) strains from the Irish National Reference Laboratory (St. James's Hospital Dublin, Ireland) including the efficiency of plaquing (EOP) of these strains.

S. aureus Strain	Phage K	Phage B1	Phage JA1
DPC5246*	1.00 ± 0.0	1.00 ± 0.0	$8.98 \times 10^{-1} \pm 0.8$
CIT281189*	No infection	No infection	1.00 ± 0.0
0.0066 (IIIV) ST239	No infection	No infection	2.59 ± 2.5
0.1206 (IV) ST250	No infection	$3.89 \times 10^{-1} \pm 0.3$	1.35 ± 1.2
0.1239 (III) ST239	No infection	$1.46 \times 10^{-1} \pm 0.1$	$4.17 \times 10^{-2} \pm 0.0$
0.1345 (II) ST5	No infection	No infection	$2.08 \times 10^{-1} \pm 0.1$
0073 (III) ST239	No infection	$3.21 \times 10^{-1} \pm 0.2$	No infection
0104 (III) ST239	No infection	$3.95 \times 10^{-1} \pm 0.2$	1.82 ± 1.6
0220 (II) ST5	$3.03 \times 10^{-1} \pm 0.1$	$2.17 \times 10^{-1} \pm 0.2$	$2.38 \times 10^{-1} \pm 0.2$
0242 (IV) ST30	$4.43 \times 10^{-1} \pm 0.1$	$5.23 \times 10^{-1} \pm 0.5$	$4.90 \times 10^{-1} \pm 0.3$
0308 (IA) ST247	1.40 ± 0.2	1.36 ± 1.3	1.71 ± 1.6
3045 (IIV) ST8	No infection	$4.93 \times 10^{-2} \pm 0.0$	1.69 ± 0.7
3144 (IIV) ST8	No infection	1.21 ± 1.0	2.17 ± 1.2
3488 (VV) ST8	No infection	No infection	No infection
3581 (IA) ST247	No infection	No infection	$9.26 \times 10^{-1} \pm 0.7$
3594 (II) ST36	$4.38 \times 10^{-1} \pm 0.1$	$8.67 \times 10^{-1} \pm 0.4$	1.06 ± 0.7
3596 (IIV) ST8	$2.49 \times 10^{-4} \pm 0.0$	1.29 ± 0.9	3.59 ± 2.7
E1038 (IIV) ST8	$1.27 \times 10^{-4} \pm 0.0$	$2.02 \times 10^{-1} \pm 0.2$	1.89 ± 1.4
E1139 (IV) ST45	No infection	$3.88 \times 10^{-6} \pm 0.0$	No infection
E1174 (IV) ST22	$7.03 \times 10^{-1} \pm 0.7$	$3.11 \times 10^{-1} \pm 0.2$	No infection
E1185 (IV) ST12	$1.16 \times 10^{-6} \pm 0.0$	No infection	No infection
E1202 (II) ST496	No infection	$4.79 \times 10^{-1} \pm 0.2$	$9.49 \times 10^{-1} \pm 0.8$
M03/0073 (III) ST239	1.76 ± 0.5	1.51 ± 0.8	2.30 ± 0.7

* *S. aureus* strains for phage propagation; data is represented as means ± standard deviations based on triplicate measurements.

Table 3. Zone sizes and morphologies of B1, JA1, and K plaques formed on MRSA strains collected from the Irish National MRSA Reference Laboratory (St. James's Hospital Dublin, Ireland).

S. aureus Strain	Phage K	Phage B1	Phage JA1
DPC5246	2 mm	1 mm with halo to 2 mm	1 mm with halo to 2 mm
CIT281189	No plaques	No plaques	1.5 mm
0.0066 (IIV) ST239	No plaques	No plaques	1 mm
0.1206 (IV) ST250	No plaques	2 mm	0.5 mm with halo to 1 mm
0.1239 (III) ST239	No plaques	0.5 mm, faint plaques	1 mm
0.1345 (II) ST5	No plaques	No plaques	1 mm
0073 (III) ST239	No plaques	0.5 mm	No plaques
0104 (III) ST239	No plaques	0.5 mm	1 mm
0220 (II) ST5	0.5 mm	1 mm	1 mm
0242 (IV) ST30	1 mm	1.5 mm	1.5 mm
0308 (IA) ST247	1 mm	1 mm	0.5 mm, faint plaques
3045 (IIV) ST8	No plaques	1 mm	1 mm

Table 3. *Cont.*

S. aureus Strain	Phage K	Phage B1	Phage JA1
3144 (IIV) ST8	No plaques	1.5 mm, faint plaques	1 mm
3488 (VV) ST8	No plaques	0.5 mm, faint plaques	0.5 mm with halo to 1 mm
3581 (IA) ST247	No plaques	No plaques	1 mm
3594 (II) ST36	1.5 mm	1 mm	1.5 mm
3596 (IIV) ST8	0.5 mm	0.5 mm with halo to 1.5 mm	0.5 mm with halo to 1.5 mm
E1038 (IIV) ST8	0.5 mm, faint plaques	0.5 mm, faint plaques	1.5 mm
E1139 (IV) ST45	No plaques	0.5 mm, faint plaques	No plaques
E1174 (IV) ST22	0.5 mm, faint plaques	0.5 mm	No plaques
E1185 (IV) ST12	0.5 mm, faint plaques	No plaques	No plaques
E1202 (II) ST496	No plaques	1 mm	0.5 mm
M03/0073 (III) ST239	2 mm	0.5 mm with halo to 1.5 mm	0.5 mm with halo to 1.5 mm

2.3. Phage Adsorption on Phage Resistant Isolates

While some level of phage insensitivity was encountered against all three phages, phage K was the frequently insensitive virion to the *S. aureus* strains tested, and thus, was chosen to evaluate whether or not adsorption inhibition played a role in its insensitivity. Phage K was able to adsorb to all phage-insensitive strains to approximately the same extent as the propagating strain DPC5246. This rules out the possibility of adsorption inhibition playing a role in the narrow host range encountered with phage K in comparison to both phages B1 and JA1 (Supplementary Materials, Figure S2). Additionally, adsorption studies with phages B1 and JA1 indicated that adsorption did not play a role in the differences observed.

2.4. Genome Comparison between Phages B1, JA1, and K

The genome of phage K is 139,831 bp in size with long terminal repeats (LTRs) of 8486 bp [31]. Genomes of similar sizes were obtained for phages B1 and JA1, these being 140,808 bp and 139,484 bp, respectively. Examination of sequence reads allowed the identification of LTRs for these phages, due to the identification of a region within their genomes with roughly double the average number of reads, these regions being 8076 bp and 7651 bp in size for phages B1 and JA1, respectively. This approach to the determination of terminal repeats has been utilized for a number of phages [32–34]. The sequences of all three phages, when analyzed, contained the 12 bp inverted repeat sequences 5′-TAAGTACCTGGG-3′ and 5′-CCCAGGTACTTA-3′, which separates the LTRs from the non-redundant part of the phage DNA, and are characteristic of K-like phages [22,35]. Thus, the entire packaged genome sizes are 148,884 bp (B1), 147,135 bp (JA1), and 148,317 bp (K). Phage K possessed 212 ORFs in its genome [31,36], whereas phages B1 and JA1 possessed 219 (Supplementary Materials, Table S1) and 215 ORFs (Supplementary Materials, Table S2) respectively.

Nucleotide pairwise sequence alignment based on BLASTN revealed phages B1 and JA1 (including their LTRs) to be 99% identical to each other, thus can be considered different isolates of the same phage species [37]. On the other hand, phages B1 and JA1 (including their LTRs) showed 95% and 94% identity (respectively) to phage K, placing these phages on the boundary of speciation.

The examination of 100 bp sequences upstream of each ORFs on the non-redundant genome of these phages, using MEME [38], identified 44 and 43 RpoD-like promoters for phages B1 and JA1, respectively. It was observed that these promoters where heavily concentrated in regions with ORFs encoding short hypothetical proteins and those with functions associated with nucleotide metabolism and DNA replication, rather than those associated with virion structure (Supplementary Materials, Tables S3 and S4). A similar finding was also reported with K-like phage vB_SauM-fRuSau02 [21]. Additionally, 30 Rho-independent terminators were identified on the non-redundant genomes for both B1 and JA1 (Supplementary Materials, Tables S5 and S6).

Four ORFs present in phage B1 were observed to be absent in JA1 (Table 4). These ORFs encoded two putative terminal repeat-encoded proteins (PhageB1_009, 016) and two proteins of unknown function (phageB1_202, 203). Although both B1 and JA1 had similar content of ORFs with 1% difference between their genomes, both phages varied in their host range on the *S. aureus* strains they infected. This variation is likely attributed to the difference encountered in their genome. Additionally, multiple ORFs present in phage K but absent in both B1 and JA1 were encountered (Figure 2, Table 5). Furthermore, ORFs present in both phages B1 and JA1 but absent in K were also encountered (Figure 2, Table 6). These ORFs are discussed below.

Table 4. List of missing ORFs present in phage B1 but absent in phage JA1.

ORFs	Amino Acid Numbers	Protein Size (kDa)	Predicted Function
PhageB1_009	112	13.5	Terminal repeat encoded protein
PhageB1_016	107	12.4	Terminal repeat encoded protein
PhageB1_202	32	3.5	Unknown
PhageB1_203	104	11.6	Unknown

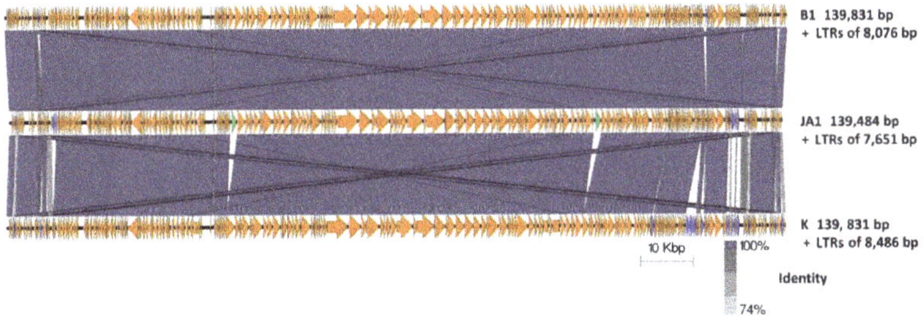

Figure 2. Genome comparison of phages B1, JA1, and K (including their long terminal repeats) using currently available annotations employing BLASTN and visualized with Easyfig. Regions of sequence similarity are connected by the shaded area, using a grey scale; genome maps consisting of orange arrows indicating the location of ORFs along the phage genomes, with unshared ORFs highlighted in blue with those indicating unshared homing endonuclease highlighted in green.

Table 5. List of missing ORFs and their predicted putative functions absent in both phages B1 and JA1 but present in phage K.

ORFs	Amino Acid Number	Protein Size (kDa)	Predicted Function
PhageK_004	108	12.7	Unknown
PhageK_016*	107	12.4	Unknown
PhageK_019	57	4.7	Unknown
PhageK_020	89	10.2	Unknown
PhageK_168	185	21.7	Predicted to contain a transmembrane region based on InterProScan
PhageK_187	101	11.7	Unknown
PhageK_188	123	13.8	Predicted to contain a transmembrane region based on InterProScan
PhageK_189	78	9.2	Unknown
PhageK_190	175	20.6	Predicted as a putative metallophoshatase
PhageK_191	106	12.9	Unknown
PhageK_192	76	8.9	Predicted to contain a transmembrane region based on InterProScan
PhageK_196	226	25.8	Unknown
PhageK_205	83	9.7	Unknown
PhageK_206	98	11.2	Unknown
PhageK_208	99	11.6	Unknown
PhageK_209	75	8.9	Unknown
PhageK_211	117	13.9	Predicted to possess a transmembrane region based on InterProScan
PhageK_212	128	15.6	Unknown

* ORF that phage JA1 does not share with phage K.

Table 6. List of missing ORFs and their predicted function absent in phage K but present in phages B1 and JA1.

ORFs	Amino Acid Number	Protein Size (kDa)	Predicted Function
PhageJA1_003 (PhageB1_003)	96	11.3	Unknown
PhageJA1_020 (PhageB1_022)	161	19.1	Unknown
PhageJA1_021 (PhageB1_023)	135	16.5	Unknown
PhageJA1_084 (PhageB1_087)	323	39.6	Predicted as a putative endonuclease interrupting the terminase large subunit [PhageJA1_083 (PhageB1_086) and PhageJA1_085 (PhageB1_088)]
PhageJA1_152 (PhageB1_155)	322	38.3	Predicted as a putative endonuclease containing a LAGLIDADG-like domain and an Intein splicing domain and interrupts the DNA repair protein [PhageJA1_151 (PhageB1_154) and PhageJA1_153 (PhageB1_156)]
PhageJA1_206 (PhageB1_212)	73	8.9	Unknown
PhageJA1_208 (PhageB1_214)	169	20.3	HHpred indicates homology to cell wall hydrolases
PhageJA1_209 (PhageB1_215)	109	12.6	Unknown
PhageJA1_211 (PhageB1_217)	104	12.0	Unknown
PhageJA1_212 (PhageB1_218)	55	6.5	Unknown
PhageJA1_213 (PhageB1_219)	33	3.7	Predicted to possess a transmembrane region based on InterProScan

2.4.1. Characteristic Features of Phage K ORFs Absent in Both JA1 and B1

Seventeen ORFs present in phage K were absent in both phages B1 and JA1, with one additional ORF found not to be shared between JA1 and K. These ORFs are listed in Table 5. No function could be assigned to these with the exception of phageK_190, which based on NCBI conserved domain search possessed a metallophosphatase-like domain (cd07390; E value; 3.94×10^{-30}) and is a member of the metallophosphatase (MPP) superfamily. Families within this superfamily of enzymes are functionally diverse, involved in the cleavage of phosphoester bonds, and include Mre11/SbcD-like exonucleases, Dbr1-like RNA lariat debranching enzymes, YfcE-like phosphodiesterases, purple acid phosphatases (PAPs), YbbF-like UDP-2,3-diacylglucosamine hydrolases, and acid sphingomyelinases (ASMases) [39].

2.4.2. Characteristic Features of Phages B1 and JA1 ORFs Absent in Phage K

Eleven ORFs present in both phages B1 and JA1 were absent in phage K (Table 6). No putative function could be assigned to the majority of these ORFs based on BLASTP, InterProScan or HHpred analysis, with the exception of phageJA1_084 (phageB1_087) and phageJA1_152 (phageB1_155), which encoded homing endonucleases interrupting both the terminase large subunit and the DNA repair protein, respectively. These homing endonucleases are site-specific DNA endonucleases capable of initiating DNA breaks leading to repair and recombination event that results in the integration of this endonuclease ORF into a gene that was previously lacking it [40]. The presence of these mobile genetic elements is common among known staphylococcal phages of the subfamily *Spounavirinae*, and these endonucleases ORFs are known to insert themselves into essential phage genes [21,41]. Additionally, HHpred analysis indicated ORFs PhageJA1_208 and PhageB1_214 to possess remote homology to cell-degrading proteins. The majority of these ORFs were found to be located next to the genome termini of JA1 and B1, with genes located in this region having been previously reported in similar phages to be expressed early in phage development [35]. Such proteins are usually involved in subversion of the host's machinery to aid phage takeover [42,43].

2.4.3. Comparison of Phages K, B1, and JA1 with other Similar Therapeutic Phages (Team1, vB_SauM-fRuSau02, Sb-1 and ISP)

Four additional staphylococcal phages that originate in commercial phage therapeutic mixtures are Team1, vB_SauM-fRuSau02, Sb-1 and ISP, as discussed earlier [19–24]. These phages were also reported to possess wide host ranges towards a number of clinically relevant *S. aureus* strains. Although similar, these phages have several feature differences from each other and from phages B1 and JA1. Comparison of nucleotide identities (BLASTN) with phage K shows that they belong to the genus *Kayvirus* (Supplementary Materials, Table S7) possessing genomes of similar sizes, apart from Sb-1, being smaller than would be expected, suggesting the genome submission may have been incomplete (Figure 3). Additionally, the arrangement of ORFs is quite similar. Furthermore, tRNA genes of these phages were also examined. All seven phages were found to possess the same four tRNA genes for methionine, tryptophan, phenylalanine, and aspartic acid (Supplementary Materials, Table S8). The eleven ORFs which were present in B1 and JA1 but absent in K (Table 6, Supplementary Materials, Figure S3) were similarly present in Team 1, vB_SauM-fRuSau02, Sb-1 and ISP. And likewise, the ORFs present in K, but absent in both B1 and JA1, were also missing in these phages. However, vB_SauM-fRuSau02 possesses a much shorter putative tail protein (RS_159) of 73 amino acids compared to the phage K counterpart (PhageK_151) of 170 amino acids. Non-hypothetical proteins that differed between these phages were a membrane protein (Phage B1_180, PhageJA1_177, and Phage_170) and an ATPase-like protein (Protein id: CCA65911.1 for phage ISP). Other ORFs that differed among these phages were mostly hypothetical proteins.

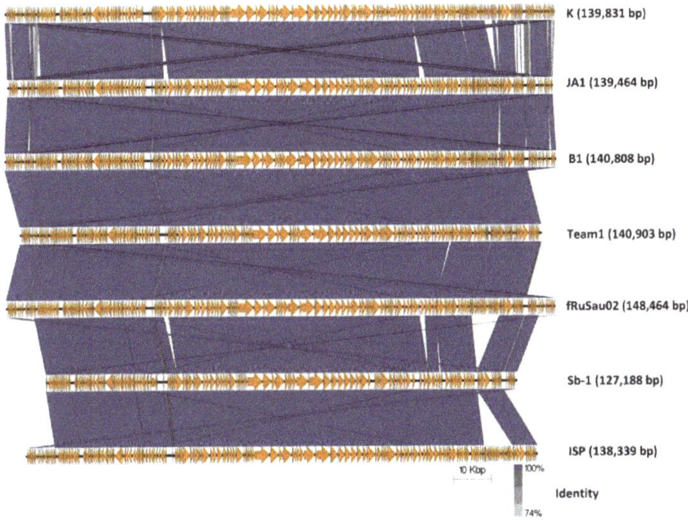

Figure 3. Genome comparison of phage K with the six staphylococcal phages employed in commercial phages mixture (B1, JA1, Team 1 [22–24], vB_SauM-fRuSau02 [21], Sb-1 [19] and ISP [20]) using currently available annotations employing BLASTN and visualized with Easyfig.

S. aureus employ several defense strategies against viral attack [10,44] and these, such as restriction modification systems [45] and CRISPR-Cas systems [46], may vary from strain to strain. These defenses along with several variations encountered at the genetic level across phages B1, JA1, and K may explain the differences in host ranges observed in this study.

3. Materials and Methods

3.1. Bacterial Strains, Phage and Growth Requirement

Phages B1 and JA1 were isolated from a commercial phage cocktail purchased from the George Eliava Institute of Bacteriophage, Microbiology and Virology, Tbilisi, Georgia. The MRSA strains utilized in this study were all acquired from the Irish National MRSA Reference Laboratory, Dublin, Ireland [2] with the exception of DPC5246 and CIT281189, which are routine propagation strains utilized in our laboratory [26,36]. These strains were routinely cultured in Brain Heart Infusion broth (BHI; Sigma, St. Louis, MO, USA) at 37 °C with shaking or on BHI plates containing 1.5% (*w/v*) bacteriological agar (Sigma). All strains were stocked in BHI containing 40% glycerol and stored at −80 °C.

3.2. CsCl Gradient Purification

Isopycnic centrifugation through CsCl gradients was performed as previously described [47], with a number of modifications. A high titer phage lysate (>1 × 10^9 plaque forming units [PFU] mL^{-1}), was precipitated using polyethylene glycol (15% *w/v* PEG8000, 1 M NaCl) at 4 °C overnight and centrifuged, after which the pellet was resuspended in TMN buffer (10 mM Tris-HCl pH 7.4, 10 mM $MgSO_4 \cdot 7H_2O$, 0.5 M NaCl). The resulting phage preparation was placed onto a CsCl step gradient composed of 1.3, 1.5, and 1.7 g/mL layers and spun in a 100 Ti rotor (Beckman Coulter, Brea, CA, USA) at 200,480 *g* for 3 h at 4 °C. The resulting phage preparations were dialyzed in Tris-HCl buffer (10 mM, pH 7.5) at 4 °C.

3.3. Phage Host Range and Adsorption Study

Host range assay was performed for phages B1, JA1, and K using the plaque assay plating technique (Tables 2 and 3). This was done in triplicate for three independent experiments. The efficiency of plaquing (EOP) was determined by dividing the phage titer on each test strain by the phage titer of the reference strain (*S. aureus* DPC5246, in the case of phages B1 and K, and *S. aureus* CIT281189 for phage JA1) [48]. An adsorption assay was performed according to the protocol previously described elsewhere with some modification [49]. Briefly, MRSA strains were grown to an optical density (OD) of 0.2 at 600 nm (estimated cell count at 10^8 colony forming unit (cfu) mL^{-1}) and 100 µL of cells were mixed with 100 µL of respective phage titered at approximately 1 × 10^7 PFU/mL for a multiplicity of infection (MOI) of 0.1. The resulting mixtures were incubated at room temperature for 5 min to allow for phage adsorption. The bound phages were separated from the free phages by centrifugation at 14,000 rpm for 5 min. Adsorption of the phage on each strain was determined by subtracting the number of unbound phage (per mL) from the total input PFU/mL. Adsorption efficiency was expressed as a percentage relative to the propagating strain DPC5246.

3.4. Transmission Electron Microscopy

Electron microscopic analysis was performed following negative staining of the CsCl gradient prepared phages on freshly prepared carbon films with 2% (*w/v*) uranyl acetate. Electron micrographs were taken using a Tecnai 10 transmission electron microscope (FEI Thermo Fisher, Eindhoven, the Netherlands) at an acceleration voltage of 80 kV with a MegaView G2 CDD camera (EMSIS, Muenster, Germany).

3.5. Phage DNA Isolation

Phage DNA extraction was performed on CsCl purified high titer phages. These were initially treated with $MgCl_2$ followed by pre-treatment with DNase and RNase for 60 min at 37 °C. Following that subsequent treatment with SDS, EDTA and proteinase K with further incubation for 60 min at 55 °C. DNA extractions were then performed on the pre-treated samples with phenol/chloroform/isoamyl alcohol (25:24:1 *v/v/v*) and chloroform/isoamyl alcohol (24:1 *v/v*). DNA precipitation was achieved using sodium acetate and 95% ethanol. DNA quality and quantity were estimated using a Nanodrop (ND-1000) and visualized following agarose gel electrophoresis

3.6. Phage DNA Sequencing

DNA sequencing was performed with a high throughput Illumina HiSeq system sequencing (GATC Biotech, Konstanz, Germany). Library preparation was performed by DNA fragmentation together with adapter ligation. The libraries were then measured and quantified on a Fragment Analyzer and then sequenced to generate 2 × 300 bp paired-end reads. *De novo* assembly was performed using CLC Bio Genomics Workbench v8.0 (Aarhus, Denmark).

3.7. Bioinformatic Analysis

Open reading frames (ORFs) for the sequenced phages were predicted with Glimmer [50] and GenemarkS [51]. Putative functions were assigned to these ORFs using BLASTP (https://blast.ncbi.nlm.nih.gov/Blast.cgi?PAGE=Proteins), HHpred (https://toolkit.tuebingen.mpg.de/#/tools/hhpred; [52]) and InterProscan (http://www.ebi.ac.uk/interpro/search/sequence-search; [53]). Transfer RNA was predicted using tRNAscan-SE (http://lowelab.ucsc.edu/tRNAscan-SE/; [54]) and ARAGORN (http://130.235.46.10/ARAGORN/; [55]). Potential promoters were predicted using MEME (Multiple Em for Motif Elicitation) (http://meme-suite.org/tools/meme; [38]), followed by manual curation. Potential Rho-independent terminators were identified using ARNold (http://rna.igmors.u-psud.fr/toolbox/arnold; [56]) with Mfold QuikFold (http://unafold.rna.albany.edu/?q=DINAMelt/Quickfold; [57]) using RNA energy rules 3.0 to verify predictions. Artemis Comparison Tool (ACT) was used for the identification of feature variations between the genomes of phages, with homology being assessed with BLASTN [58] Genome comparison maps between phages were visualized using the Easyfig visualization tool [59]. K-like *Staphylococcus* phages used in comparative studies were K (KF766114), Team 1 (KC012913), vB_SauM-fRuSau02 (MF398190), Sb-1 (HQ163896) and ISP (FR852584).

3.8. Nucleotide Sequence Accession Number

The genome sequence for phages B1 and JA1 were deposited into GenBank under the accession numbers MG656408 and MF405094, respectively.

4. Conclusions

Host range of three highly similar phages was performed in this study, and it was identified that phages B1 and JA1 from the Fersisi commercial phage mix had a much broader host range in comparison to phage K on a representative Irish bank of clinical MRSA sequence type isolates. Comparisons of their genomes lead to the identification of several ORFs absent in phage K, but present in both phages B1 and JA1. These ORFs were also identified in several other staphylococcal phages sourced from commercial phage mixtures (B1, JA1, Team 1 [22–24], vB_SauM-fRuSau02 [21], Sb-1 [19] and ISP [20]), also with a reported wide host range. The exact role of these ORFs is currently unknown. However, these ORFs along with several variations encountered at the genetic level between these phages may, in part, explain their different host range. Unfortunately, information is lacking on the influences of various phage resistance systems, which may be active in *Staphylococcus aureus*. Phage research also needs to focus more on elucidation of the functions of hypothetical proteins to allow greater understanding of how phages overcome such systems.

Supplementary Materials: The following are available online at http://www.mdpi.com/2079-6382/7/2/37/s1, Figure S1: Plaque morphologies of phages B1, JA1 and K with common morphology types encountered in their host range study to include plaques sizes of 2mm (A), 0.5mm (B) and 1.0mm (C), Figure S2: *Staphylococcus* phage K adsorption to strains of *Staphylococcus aureus* resistant to infection by, in comparison host strain DPC5246, Figure S3: Comparison of regions within the genome of phage K to closely related staphylococcal phages, Table S1: Annotation of the staphylococcal phage vB_SauM_B1, Table S2: Annotation of the staphylococcal phage vB_SauM_JA1, Table S3: Predicted Rho-like promoters of *Staphylococcus* phage B1 genome found using MEME, Table S4: Predicted Rho-like promoters of *Staphylococcus* phage JA1 genome found using MEME, Table S5: High ΔG rho-independent terminators predicted in the genome *Staphylococcus* phage B1 identified using ARNold and QuikFold, Table S6: High ΔG rho-independent terminators predicted in the genome *Staphylococcus* phage JA1 identified using ARNold and QuikFold, Table S7: Percentage similarity based on BLASTN of broad host range *Staphylococcus* phages that form commercial phage cocktails to that of *Staphylococcus* phage K, Table S8: tRNA genes of phages B1, JA1, K, vB_SauM-fRuSau02, ISP, Sb-1, Team 1.

Author Contributions: J.A. conducted the majority of lab work and wrote the manuscript, C.B. assisted with bioinformatics, writing, and critically read the manuscript; S.A.M. assisted with the phage host-range work; N.C. provided the Fersisi therapeutic phages and their information, and critically read the manuscript, E.M.G. isolated and helped characterize phage B1, J.OM and O.M. were co-supervisors of J.A. and both critically read the manuscript, H.N. and C.F. were responsible for electron microscopy and phage measurements, and A.C. conceived, funded, guided the study and critically read the manuscript.

Acknowledgments: This work was supported by Science Foundation Ireland, project reference 12/R1/2335. We thank undergraduate students Aoife Keating, Ceile Berkery and Adonai Djankah for their technical assistance with the host range studies and Angela Back for assistance with the electron microscope preparations.

Conflicts of Interest: The authors declare no conflict of interest.

References

1. Tong, S.Y.C.; Davis, J.S.; Eichenberger, E.; Holland, T.L.; Fowler, V.G. *Staphylococcus aureus* infections: Epidemiology, pathophysiology, clinical manifestations, and management. *Clin. Microbiol. Rev.* **2015**, *28*, 603–661. [CrossRef] [PubMed]
2. Shore, A.C.; Rossney, A.S.; O'Connell, B.; Herra, C.M.; Sullivan, D.J.; Humphreys, H.; Coleman, D.C. Detection of staphylococcal cassette chromosome mec-associated DNA segments in multiresistant methicillin-susceptible *Staphylococcus aureus* (MSSA) and identification of *Staphylococcus epidermidis* ccrAB4 in both methicillin-resistant *S. aureus* and MSSA. *Antimicrob. Agents Chemother.* **2008**, *52*, 4407–4419. [CrossRef] [PubMed]
3. Hiramatsu, K.; Cui, L.; Kuroda, M.; Ito, T. The emergence and evolution of methicillin-resistant *Staphylococcus aureus*. *Trends Microbiol.* **2001**, *9*, 486–493. [CrossRef]
4. Jevons, M.P. "Celbenin"—Resistant Staphylococci. *BMJ* **1961**, *1*, 124–125. [CrossRef]
5. Klein, E.; Smith, D.L.; Laxminarayan, R. Hospitalizations and Deaths Caused by Methicillin-Resistant *Staphylococcus aureus*, United States, 1999–2005. *Emerg. Infect. Dis.* **2007**, *13*, 1840–1846. [CrossRef] [PubMed]
6. O'Flaherty, S.; Ross, R.P.; Coffey, A. Bacteriophage and their lysins for elimination of infectious bacteria. *FEMS Microbiol. Rev.* **2009**, *33*, 801–819. [CrossRef] [PubMed]
7. Schmelcher, M.; Loessner, M.J. Application of bacteriophages for detection of foodborne pathogens. *Bacteriophage* **2014**, *4*, e28137. [CrossRef] [PubMed]
8. Wittebole, X.; De Roock, S.; Opal, S.M. A historical overview of bacteriophage therapy as an alternative to antibiotics for the treatment of bacterial pathogens. *Virulence* **2014**, *5*, 226–235. [CrossRef] [PubMed]
9. Bertozzi Silva, J.; Storms, Z.; Sauvageau, D. Host receptors for bacteriophage adsorption. *FEMS Microbiol. Lett.* **2016**, *363*, 1–11. [CrossRef] [PubMed]
10. Hyman, P.; Abedon, S.T. Bacteriophage Host Range and Bacterial Resistance. In *Advances in Applied Microbiology*, 1st ed.; Laskin, A.I., Sarislani, S., Gadd, G.M., Eds.; Elsevier Inc.: Cambridge, MA, USA, 2010; Volume 70, pp. 217–248, ISBN 9780123809919.
11. Labrie, S.J.; Samson, J.E.; Moineau, S. Bacteriophage resistance mechanisms. *Nat. Rev. Microbiol.* **2010**, *8*, 317–327. [CrossRef] [PubMed]
12. Hall, A.R.; Scanlan, P.D.; Buckling, A. Bacteria-Phage Coevolution and the Emergence of Generalist Pathogens. *Am. Nat.* **2011**, *177*, 44–53. [CrossRef] [PubMed]
13. Hall, J.P.J.; Harrison, E.; Brockhurst, M.A. Viral host-adaptation: Insights from evolution experiments with phages. *Curr. Opin. Virol.* **2013**, *3*, 572–577. [CrossRef] [PubMed]

14. Samson, J.E.; Magadán, A.H.; Sabri, M.; Moineau, S. Revenge of the phages: Defeating bacterial defences. *Nat. Rev. Microbiol.* **2013**, *11*, 675–687. [CrossRef] [PubMed]
15. Sulakvelidze, A.; Alavidze, Z.; Morris, J.G. Bacteriophage therapy. *Antimicrob. Agents Chemother.* **2001**, *45*, 649–659. [CrossRef] [PubMed]
16. Abedon, S.T.; Kuhl, S.J.; Blasdel, B.G.; Kutter, E.M. Phage treatment of human infections. *Bacteriophage* **2011**, *1*, 66–85. [CrossRef] [PubMed]
17. Villarroel, J.; Larsen, M.V.; Kilstrup, M.; Nielsen, M. Metagenomic analysis of therapeutic PYO phage cocktails from 1997 to 2014. *Viruses* **2017**, *9*. [CrossRef] [PubMed]
18. Zschach, H.; Joensen, K.G.; Lindhard, B.; Lund, O.; Goderdzishvili, M.; Chkonia, I.; Jgenti, G.; Kvatadze, N.; Alavidze, Z.; Kutter, E.M.; et al. What can we learn from a metagenomic analysis of a georgian bacteriophage cocktail? *Viruses* **2015**, *7*, 6570–6589. [CrossRef] [PubMed]
19. Kvachadze, L.; Balarjishvili, N.; Meskhi, T.; Tevdoradze, E.; Skhirtladze, N.; Pataridze, T.; Adamia, R.; Topuria, T.; Kutter, E.; Rohde, C.; et al. Evaluation of lytic activity of staphylococcal bacteriophage Sb-1 against freshly isolated clinical pathogens. *Microb. Biotechnol.* **2011**, *4*, 643–650. [CrossRef] [PubMed]
20. Vandersteegen, K.; Mattheus, W.; Ceyssens, P.J.; Bilocq, F.; de Vos, D.; Pirnay, J.P.; Noben, J.P.; Merabishvili, M.; Lipinska, U.; Hermans, K.; et al. Microbiological and molecular assessment of bacteriophage ISP for the control of *Staphylococcus aureus*. *PLoS ONE* **2011**, *6*, e24418. [CrossRef] [PubMed]
21. Leskinen, K.; Tuomala, H.; Wicklund, A.; Horsma-Heikkinen, J.; Kuusela, P.; Skurnik, M.; Kiljunen, S. Characterization of vB_SauM-fRuSau02, a Twort-Like Bacteriophage Isolated from a Therapeutic Phage Cocktail. *Viruses* **2017**, *9*, 258. [CrossRef] [PubMed]
22. El Haddad, L.; Abdallah, N.B.; Plante, P.L.; Dumaresq, J.; Katsarava, R.; Labrie, S.; Corbeil, J.; St-Gelais, D.; Moineau, S. Improving the safety of *Staphylococcus aureus* polyvalent phages by their production on a *Staphylococcus xylosus* strain. *PLoS ONE* **2014**, *9*, e102600. [CrossRef] [PubMed]
23. Markoishvili, K.; Tsitlanadze, G.; Katsarava, R.; Morris, J.G.; Sulakvelidze, A. A novel sustained-release matrix based on biodegradable poly(ester amide)s and impregnated with bacteriophages and an antibiotic shows promise in management of infected venous stasis ulcers and other poorly healing wounds. *Int. J. Dermatol.* **2002**, *41*, 453–458. [CrossRef] [PubMed]
24. Jikia, D.; Chkhaidze, N.; Imedashvili, E.; Mgaloblishvili, I.; Tsitlanadze, G.; Katsarava, R.; Morris, J.G.; Sulakvelidze, A. The use of a novel biodegradable preparation capable of the sustained release of bacteriophages and ciprofloxacin, in the complex treatment of multidrug-resistant *Staphylococcus aureus*-infected local radiation injuries caused by exposure to Sr90. *Clin. Exp. Dermatol.* **2005**, *30*, 23–26. [CrossRef] [PubMed]
25. Adriaenssens, E.M.; Clokie, C.M.R.; Sullivan, M.B.; Gillis, A.; Jens Kuhn, B.H.; Kropinski, A.M. Taxonomy of prokaryotic viruses: 2016 update from the ICTV bacterial and archaeal viruses subcommittee. *Arch. Virol.* **2017**, *162*, 1153–1157. [CrossRef] [PubMed]
26. O'Flaherty, S.; Ross, R.P.; Meaney, W.; Fitzgerald, G.F.; Elbreki, M.F.; Coffey, A. Potential of the Polyvalent Anti-Staphylococcus Bacteriophage K for Control of Antibiotic-Resistant Staphylococci from Hospitals. *Appl. Environ. Microbiol.* **2005**, *71*, 1836–1842. [CrossRef] [PubMed]
27. Rountree, P.M. The serological differentiation of staphylococcal bacteriophages. *J. Gen. Microbiol.* **1949**, *3*, 164–173. [CrossRef] [PubMed]
28. Ackermann, H.W. Frequency of morphological phage descriptions in the year 2000. *Arch. Virol.* **2001**, *146*, 843–857. [CrossRef] [PubMed]
29. Rossney, A.S.; Lawrence, M.J.; Morgan, P.M.; Fitzgibbon, M.M.; Shore, A.; Coleman, D.C.; Keane, C.T.; O'Connell, B. Epidemiological typing of MRSA isolates from blood cultures taken in Irish hospitals participating in the European Antimicrobial Resistance Surveillance System (1999–2003). *Eur. J. Clin. Microbiol. Infect. Dis.* **2006**, *25*, 79–89. [CrossRef] [PubMed]
30. Cui, Z.; Feng, T.; Gu, F.; Li, Q.; Dong, K.; Zhang, Y.; Zhu, Y.; Han, L.; Qin, J.; Guo, X. Characterization and complete genome of the virulent Myoviridae phage JD007 active against a variety of *Staphylococcus aureus* isolates from different hospitals in Shanghai, China. *Virol. J.* **2017**, *14*, 26. [CrossRef] [PubMed]
31. Gill, J.J. Revised Genome Sequence of *Staphylococcus aureus* Bacteriophage K. *Genome Announc.* **2014**, *2*, 12–13. [CrossRef] [PubMed]
32. Buttimer, C.; Hendrix, H.; Oliveira, H.; Casey, A.; Neve, H.; McAuliffe, O.; Paul Ross, R.; Hill, C.; Noben, J.P.; O'Mahony, J.; et al. Things are getting hairy: Enterobacteria bacteriophage vB_PcaM_CBB. *Front. Microbiol.* **2017**, *8*, 1–16. [CrossRef] [PubMed]

33. Li, S.; Fan, H.; An, X.; Fan, H.; Jiang, H.; Chen, Y.; Tong, Y. Scrutinizing virus genome termini by high-throughput sequencing. *PLoS ONE* **2014**, *9*, e85806. [CrossRef] [PubMed]
34. Fouts, D.E.; Klumpp, J.; Bishop-Lilly, K.A.; Rajavel, M.; Willner, K.M.; Butani, A.; Henry, M.; Biswas, B.; Li, M.; Albert, M.J.; et al. Whole genome sequencing and comparative genomic analyses of two *Vibrio cholerae* O139 Bengal-specific Podoviruses to other N4-like phages reveal extensive genetic diversity. *Virol. J.* **2013**, *10*, 165. [CrossRef] [PubMed]
35. Łobocka, M.; Hejnowicz, M.S.; Dąbrowski, K.; Gozdek, A.; Kosakowski, J.; Witkowska, M.; Ulatowska, M.I.; Weber-Dąbrowska, B.; Kwiatek, M.; Parasion, S.; et al. Genomics of staphylococcal Twort-like phages—Potential therapeutics of the post-antibiotic era. *Adv. Virus Res.* **2012**, *83*, 143–216. [CrossRef] [PubMed]
36. O'Flaherty, S.; Coffey, A.; Edwards, R.; Meaney, W.; Fitzgerald, G.F.; Ross, R.P. Genome of Staphylococcal Phage K: A New Lineage of Myoviridae Infecting Gram-Positive Bacteria with a Low G+C Content. *J. Bacteriol.* **2004**, *186*, 2862–2871. [CrossRef] [PubMed]
37. Adriaenssens, E.M.; Rodney Brister, J. How to name and classify your phage: An informal guide. *Viruses* **2017**, *9*, 70. [CrossRef] [PubMed]
38. Bailey, T.L.; Boden, M.; Buske, F.A.; Frith, M.; Grant, C.E.; Clementi, L.; Ren, J.; Li, W.W.; Noble, W.S. MEME SUITE: Tools for motif discovery and searching. *Nucleic Acids Res.* **2009**, *37*, W202–W208. [CrossRef] [PubMed]
39. Matange, N.; Podobnik, M.; Visweswariah, S.S. Metallophosphoesterases: structural fidelity with functional promiscuity. *Biochem. J.* **2015**, *467*, 201–216. [CrossRef] [PubMed]
40. Gogarten, J.P.; Hilario, E. Inteins, introns, and homing endonucleases: recent revelations about the life cycle of parasitic genetic elements. *BMC Evol. Biol.* **2006**, *6*, 94. [CrossRef] [PubMed]
41. Vandersteegen, K.; Kropinski, A.M.; Nash, J.H.E.; Noben, J.-P.; Hermans, K.; Lavigne, R. Romulus and Remus, two phage isolates representing a distinct clade within the Twortlikevirus genus, display suitable properties for phage therapy applications. *J. Virol.* **2013**, *87*, 3237–3247. [CrossRef] [PubMed]
42. Wei, P.; Stewart, C.R. A Cytotoxic Early Gene of *Bacillus subtilis* Bacteriophage SPOl. *J. Bacteriol.* **1993**, *175*, 7887–7900. [CrossRef] [PubMed]
43. Stewart, C.R.; Gaslightwala, I.; Hinata, K.; Krolikowski, K.A.; Needleman, D.S.; Peng, A.S.Y.; Peterman, M.A.; Tobias, A.; Wei, P. Genes and regulatory sites of the "host-takeover module" in the terminal redundancy of *Bacillus subtilis* bacteriophage SPO1. *Virology* **1998**, *246*, 329–340. [CrossRef] [PubMed]
44. Seed, K.D. Battling Phages: How Bacteria Defend against Viral Attack. *PLoS Pathog.* **2015**, *11*, e1004847. [CrossRef] [PubMed]
45. Roberts, G.A.; Houston, P.J.; White, J.H.; Chen, K.; Stephanou, A.S.; Cooper, L.P.; Dryden, D.T.F.; Lindsay, J.A. Impact of target site distribution for Type i restriction enzymes on the evolution of methicillin-resistant *Staphylococcus aureus* (MRSA) populations. *Nucleic Acids Res.* **2013**, *41*, 7472–7484. [CrossRef] [PubMed]
46. Cao, L.; Gao, C.H.; Zhu, J.; Zhao, L.; Wu, Q.; Li, M.; Sun, B. Identification and functional study of type III—A CRISPR-Cas systems in clinical isolates of *Staphylococcus aureus*. *Int. J. Med. Microbiol.* **2016**, *306*, 686–696. [CrossRef] [PubMed]
47. Sambrook, J.; Russell, D.W. Purification of bacteriophage lamda particles by isopycnic centrifugation through CsCl gradients. In *Molecular Cloning: A Laboratory Manual*; Cold Spring Harbor Laboratory Press: New York, NY, USA, 2001; Volume 1, p. 247, ISBN 0879695773.
48. Gutiérrez, D.; Vandenheuvel, D.; Martínez, B.; Rodríguez, A.; Lavigne, R.; García, P. Two Phages, phiIPLA-RODI and phiIPLA-C1C, Lyse Mono- and Dual-Species Staphylococcal Biofilms. *Appl. Environ. Microbiol.* **2015**, *81*, 3336–3348. [CrossRef] [PubMed]
49. Li, X.; Koç, C.; Kühner, P.; Stierhof, Y.-D.; Krismer, B.; Enright, M.C.; Penadés, J.R.; Wolz, C.; Stehle, T.; Cambillau, C.; et al. An essential role for the baseplate protein Gp45 in phage adsorption to *Staphylococcus aureus*. *Sci. Rep.* **2016**, *6*, 26455. [CrossRef] [PubMed]
50. Delcher, A. Improved microbial gene identification with GLIMMER. *Nucleic Acids Res.* **1999**, *27*, 4636–4641. [CrossRef] [PubMed]
51. Besemer, J.; Lomsadze, A.; Borodovsky, M. GeneMarkS: A self-training method for prediction of gene starts in microbial genomes. Implications for finding sequence motifs in regulatory regions. *Nucleic Acids Res.* **2001**, *29*, 2607–2618. [CrossRef] [PubMed]
52. Söding, J.; Biegert, A.; Lupas, A.N. The HHpred interactive server for protein homology detection and structure prediction. *Nucleic Acids Res.* **2005**, *33*, W244–W248. [CrossRef] [PubMed]

53. Mitchell, A.; Chang, H.-Y.; Daugherty, L.; Fraser, M.; Hunter, S.; Lopez, R.; McAnulla, C.; McMenamin, C.; Nuka, G.; Pesseat, S.; et al. The InterPro protein families database: The classification resource after 15 years. *Nucleic Acids Res.* **2015**, *43*, D213–D221. [CrossRef] [PubMed]
54. Lowe, T.M.; Eddy, S.R. tRNAscan-SE: A program for improved detection of transfer RNA genes in genomic sequence. *Nucleic Acids Res.* **1997**, *25*, 955–964. [CrossRef] [PubMed]
55. Laslett, D.; Canback, B. ARAGORN, a program to detect tRNA genes and tmRNA genes in nucleotide sequences. *Nucleic Acids Res.* **2004**, *32*, 11–16. [CrossRef] [PubMed]
56. Naville, M.; Ghuillot-Gaudeffroy, A.; Marchais, A.; Gautheret, D. ARNold: A web tool for the prediction of rho-independent transcription terminators. *RNA Biol.* **2011**, *8*, 11–13. [CrossRef] [PubMed]
57. Zuker, M. Mfold web server for nucleic acid folding and hybridization prediction. *Nucleic Acids Res.* **2003**, *31*, 3406–3415. [CrossRef] [PubMed]
58. Carver, T.J.; Rutherford, K.M.; Berriman, M.; Rajandream, M.-A.; Barrell, B.G.; Parkhill, J. ACT: The Artemis comparison tool. *Bioinformatics* **2005**, *21*, 3422–3423. [CrossRef] [PubMed]
59. Sullivan, M.J.; Petty, N.K.; Beatson, S.A. Easyfig: A genome comparison visualizer. *Bioinformatics* **2011**, *27*, 1009–1010. [CrossRef] [PubMed]

antibiotics

MDPI

Article

Use of a Regression Model to Study Host-Genomic Determinants of Phage Susceptibility in MRSA

Henrike Zschach [1,*], Mette V. Larsen [2], Henrik Hasman [3], Henrik Westh [4,5], Morten Nielsen [1,6,*], Ryszard Międzybrodzki [7,8], Ewa Jończyk-Matysiak [7], Beata Weber-Dąbrowska [7] and Andrzej Górski [7,8]

1 Department of Bio and Health Informatics, Technical University of Denmark, 2800 Kgs. Lyngby, Denmark
2 GoSeqIt ApS, Ved Klaedebo 9, 2970 Hoersholm, Denmark; MVL@goseqit.com
3 Department of Bacteria, Fungi and Parasites, Statens Serum Institut, 2300 Copenhagen S, Denmark, henh@ssi.dk
4 Department of Clinical Microbiology, MRSA Knowledge Center, Hvidovre Hospital, 2650 Hvidovre, Denmark; Henrik.torkil.westh@regionh.dk
5 Faculty of Health and Medical Sciences, Institute of Clinical Medicine, University of Copenhagen, 2200 Copenhagen, Denmark
6 Instituto de Investigaciones Biotecnológicas, Universidad Nacional de San Martín, San Martín, B 1650 HMP, Buenos Aires, Argentina
7 Bacteriophage Laboratory, Hirszfeld Institute of Immunology and Experimental Therapy, Polish Academy of Sciences, 53-114 Wroclaw, Poland; mbrodzki@iitd.pan.wroc.pl (R.M.); ewa.jonczyk@iitd.pan.wroc.pl (E.J.-M.); weber@iitd.pan.wroc.pl (B.W.-D.); agorski@ikp.pl (A.G.)
8 Department of Clinical Immunology, Transplantation Institute, Medical University of Warsaw, 02-006 Warsaw, Poland
* Correspondence: henrike@bioinformatics.dtu.dk (H.Z.); mniel@bioinformatics.dtu.dk (M.N.); Tel.: +45-45-25-24-25 (M.N.)

Received: 15 November 2017; Accepted: 24 January 2018; Published: 29 January 2018

Abstract: *Staphylococcus aureus* is a major agent of nosocomial infections. Especially in methicillin-resistant strains, conventional treatment options are limited and expensive, which has fueled a growing interest in phage therapy approaches. We have tested the susceptibility of 207 clinical *S. aureus* strains to 12 (nine monovalent) different therapeutic phage preparations and subsequently employed linear regression models to estimate the influence of individual host gene families on resistance to phages. Specifically, we used a two-step regression model setup with a preselection step based on gene family enrichment. We show that our models are robust and capture the data's underlying signal by comparing their performance to that of models build on randomized data. In doing so, we have identified 167 gene families that govern phage resistance in our strain set and performed functional analysis on them. This revealed genes of possible prophage or mobile genetic element origin, along with genes involved in restriction-modification and transcription regulators, though the majority were genes of unknown function. This study is a step in the direction of understanding the intricate host-phage relationship in this important pathogen with the outlook to targeted phage therapy applications.

Keywords: phage therapy; bacterial phage resistance; regression modeling; MRSA

1. Introduction

Methicillin-resistant *Staphylococcus aureus* (MRSA) is a growing health concern. It is the agent of many chronic bacterial infections in hospitals as well as in the community. Its resistance to beta-lactamases severely limits treatment options, drives up the price for therapy, increases unwanted side effects, and leads in many cases to worse clinical outcomes [1]. MRSA has been classified as a high-priority pathogen on the 2017 list of antibiotic-resistant priority pathogens published by the World Health Organization [2]. Pathogens on this list are considered to pose the greatest threat to human health and to require urgently discovery and development of new antibiotics.

Phage therapy has been proposed as a promising substitute for conventional antibiotics or a co-treatment in the treatment of multi-resistant bacterial pathogens [3–7]. Of the *S. aureus* phage known to date, most are temperate phages and belong to the Siphoviridae family [8]. Strictly lytic staphylococcal phages, as are typically required for therapy, are almost exclusively found in the Podoviridae and Myoviridae families [8].

The Hirszfeld Institute of Immunology and Experimental Therapy of the Polish Academy of Science in Wroclaw (HI) has been producing staphylococcal phages for therapeutic purposes since the 1970s [9]. At present, its collection consists of nine monovalent staphylococcal phages (see Materials and Methods) [10]. Those phages are used at the Phage Therapy Unit in Wrocław under the rules of a therapeutic experiment to conduct treatment of patients with chronic bacterial infections resistant to antibiotic therapy. The result have been encouraging, as a good response has been observed in one third of patients [6].

However, in order for phage therapy to be efficient, it is necessary to have a good understanding of the specific interaction between phage and host. There are many strategies by which bacteria aim to evade predation by phages, which is a significant fitness factor and therefore under high evolutionary pressure. *S. aureus* is known to be deficient in CRISPR, one of the major phage defense mechanisms [11]. Instead, its principle defense against invading DNAs are extensive restriction-modification (RM) systems [12]. RM systems are two-part system composed of a methylase and a nuclease. The methylase introduces specific modifications on the organism's DNA, thereby marking it is as self. DNA lacking those modifications, i.e., DNA of foreign origin, will be cleaved by the nuclease. All four types of RM systems known to date are present in *S. aureus* [12]. Another, highly specialized phage defense mechanism is present in the form of staphylococcal pathogenicity islands (SaPIs) [13]. These mobile genetic elements interfere with the packaging of phage DNA in the late phase of infection, instead packaging and thereby disseminating copies of themselves. However, a small percentage of phage particles are still produced normally, leading to a reduced load of phage progeny instead of a total block. It has been implied that this may be an advantage to *S. aureus* as a species as it facilitates gene transfer [14]. Akin to abortive infection mechanisms, phage resistance by SaPI includes the lysis of the infected cell [13].

S. aureus is known to have a rather large accessory genome that can make up as much as 25% of total genome size [8]. We therefore hypothesize in this study that *S. aureus* may be carrying accessory genes that encode various mechanisms that are geared toward phage resistance. The presence of such mechanisms may hamper the efficacy of phage therapy, and it is therefore important to study these in order to perform optimization of phages used for treatment. With the advent of affordable high-throughput sequencing methods, it is now becoming possible to determine the whole genome sequences of the infecting strain in a clinical setting, making them accessible to this kind of investigation.

The relationship between *S. aureus* and its phages is intricate. A large proportion of *S. aureus* virulence factors are phage-encoded [8], and phages are the major agents of horizontal gene transfer in this species [11]. Furthermore, *S. aureus* is known to harbor prophages with a very high frequency, as detailed in a review by Lindsay in 2010 that states that all *S. aureus* sequenced up to that point contained at least one prophage [15]. In accordance with that, there is a sizeable body of research into staphylococcal phages, their genomes, their influence on their host's evolution, and their contribution to *S. aureus*' virulence (see for example [8,14,16]). Furthermore, phage susceptibility patterns have been used to classify *S. aureus* before the advent of molecular typing methods [17]. Despite that, there is a distinct lack of studies investigating the genetic basis for phage susceptibility and resistance in *S. aureus* from the host perspective, in particular with regard to whole genome approaches as opposed to studies focusing on single loci.

In this study, we seek to elucidate the interplay between *S. aureus* and therapeutic phage preparations. To do so, we have tested the susceptibility of a collection of clinical MRSA isolates towards a collection of staphylococcal phage preparations from HI. Both the bacterial and phage collections we used are of great relevance to the phage therapy efforts, since the phages are either already in use or under consideration for experimental therapy in accordance with European Union (EU) rules concerning compassionate use. Furthermore, the bacterial isolates were provided by Hvidovre Hospital in Hvidovre, Denmark and were obtained from patients showing complicated nosocomial MRSA infections. This strain set represents the most prevalent clonal complexes observed in Denmark. MRSA is predominantly imported, making the collection very diverse [18]. However, it is not representative of MRSA in all localities. The genomes of the bacterial strains were determined by whole genome sequencing and through employing a number of bioinformatics tools and machine-learning methods. We attempted to shed light on the genes of MRSA that play a role in determining the susceptibility or resistance towards phages. A similar approach but with different methodology was proposed by Allen et al., who tested for associations between phage and antibiotic resistance profiles with phylogenetic similarity in *E. coli* [19].

In this way, we aim to contribute to the development of predictive tools of phage susceptibility in the phage therapy–targeted bacteria and ultimately to devising strategies for the prevention, delay, or circumvention of phage resistance in a phage therapy setting.

2. Results

2.1. General Results of the Susceptibility Testing

A total of 207 MRSA strains were successfully tested for susceptibility to 12 phage preparations. The ratio of susceptible to resistant strains differed between the preparations. Note that phage preparations were standardized to routine test dilution (RTD). The percentage of susceptible strains ranged from 19% to 68%, as can be seen in Table 1. We have chosen to regard both weakly susceptible and resistant reactions as negatives for the modelling. We did not observe a large difference in efficacy between single phage preparations and mixtures.

Table 1. Wet lab results of susceptibility testing. All phage preparations were tested at RTD, see Methods. MS-1, OP_MS-1 and OP_MS-1_TOP are mixtures of P4/6409, A5/80 and 676/Z.

Phage Preparation	Percent Sensitive	Percent Resistant
1N/80	31.9%	68.1%
676/F	50.7%	49.3%
676/T	68.1%	31.9%
676/Z	40.6%	59.4%
A3/R	18.8%	81.2%
A5/L	47.3%	52.7%
A5/80	55.1%	44.9%
P4/6409	37.7%	62.3%
phi200/6409	44.0%	56.0%
MS-1	33.8%	66.2%
OP_MS-1	38.6%	61.4%
OP_MS-1 TOP	39.6%	60.4%

2.2. Genetic Diversity of the Strain Collection

Genetic distance between the MRSA strains was measured as 1-orthoANI (see Methods), and the result is depicted in form of a heatmap in Figure 1. This figure reveals a clear clustering of strains into groups with high identity, which follows the established clonal complexes and sequence types of *S. aureus* [20]. Based on this clustering, the strains were split into five partitions by visual inspection.

Partition 1 is substantially larger than the other four. This is due to the fact that the strains belonging to clonal complexes CC1, CC5, CC8, and CC80 have a high degree of identity to each other, compare large blue area in the upper left corner. Partitions 2 and 3 are well defined, encompassing CC22 and CC30, respectively. Partition 4 is made up of CC45 and CC398. CC398 is known for its prevalence in swine and cattle. Those strains are genetically distant from the rest of the strains, though there is some degree of similarity to CC30. Partition 5 is composed of two clusters of related strains, as indicated in Figure 1. It contains a number of rarer CCs that also show a comparatively high distance in terms of orthoANI to the rest of the data set.

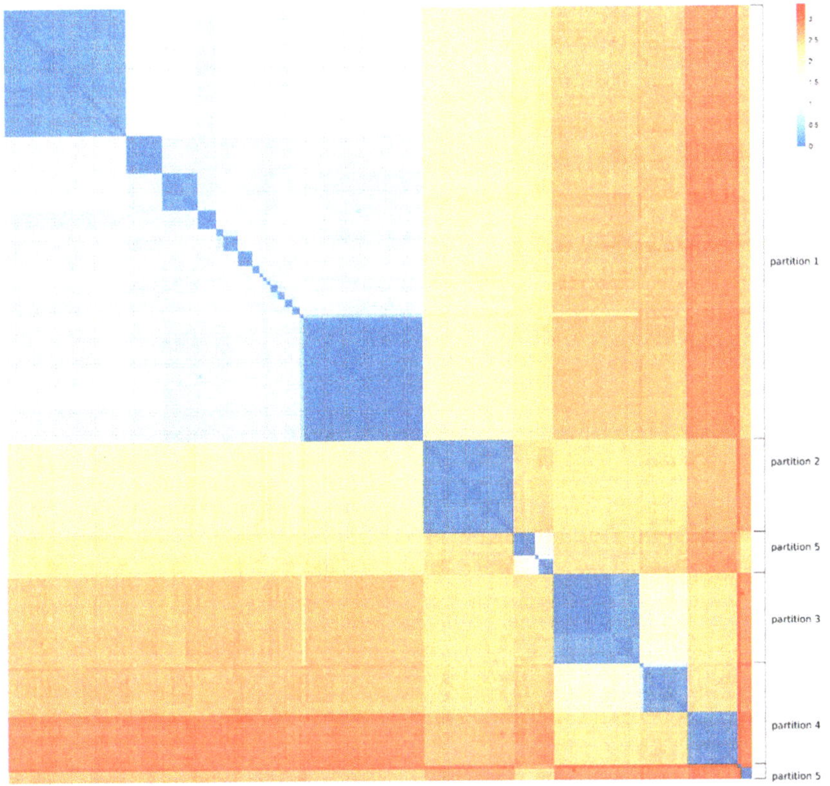

Figure 1. All-against-all matrix of the genetic distance between the 207 methicillin-resistant *Staphylococcus aureus* (MRSA) strains used for this study. Distance is calculated as 1-orthoANI and represented as color, where blue corresponds to lower and red corresponds to greater distance. The assignment of strains to partitions is marked on the right margin.

2.3. Identification of Gene Families

When predicting and clustering genes, we identified a total of 6419 gene families in the MRSA strain dataset. The distribution of these gene families across the 207 MRSA strains can be seen in Figure 2, which shows a histogram of abundances of the gene families. Here, 1777 gene families were identified in all 207 strains. These are the housekeeping genes. Furthermore, there is a heavy tail of gene families that were only observed in few strains (left side of the histogram).

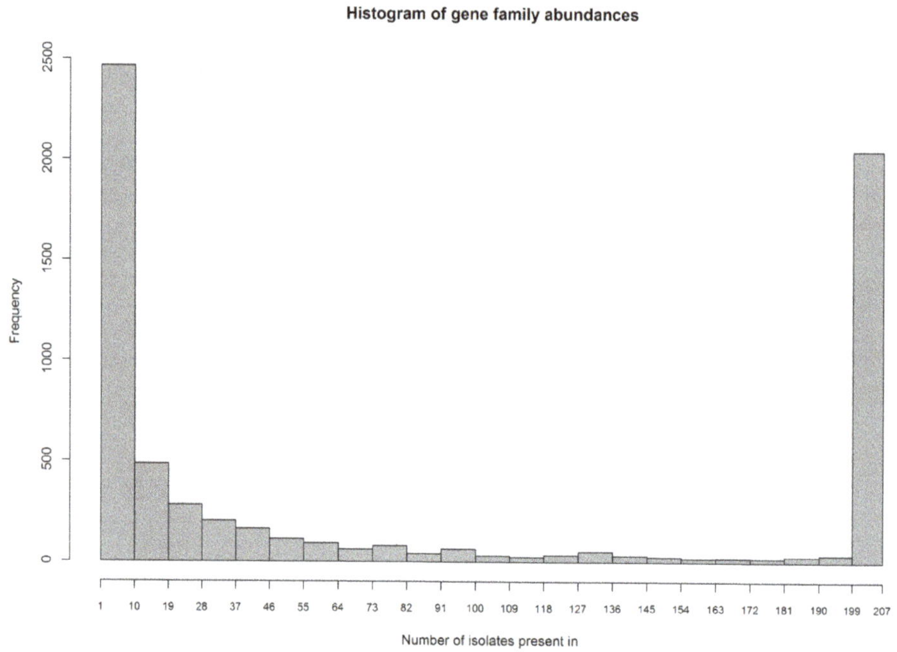

Figure 2. Abundance of gene families in the 207 strains. The peak depicted in the histogram is slightly higher than the number of housekeeping genes, 1.777, since the bin is wider than 1.

2.3.1. *p*-Value Distribution from Association Tests

Models are set up in five-fold cross validation frameworks (see Methods Section 4.5). For each cross validation fold, each gene family was assigned a *p*-value calculated from its corresponding contingency table estimated once from the original data and once from permuted data. We chose here to illustrate results for phage P4/6409 as it was representative of the other phage preparations.

When plotting the distributions of these *p*-values, see Figure 3, we can make several observations.

(a) In most phage interactions, there is a small tail of gene families with very low *p*-values, while the majority of gene families have non-significant *p*-values.

(b) In the permuted data, this tail vanishes, as was to be expected. We also observed that the *p*-value distributions of phages 1N/80, A3/R and cocktail MS-1 resemble those of the permuted data much more than those of the real data (see Supplementary Figure S1). This indicates there were not enough positive examples of lysed strains to produce a signal that is distinguishable from random.

Based on these observations, a *p*-value threshold of 0.01 or lower was implemented to admit gene families to the first step model. As seen in Table 2, the number of gene families picked by enrichment varied both by fold as well as by phage. In preparations 1N/80, A3/R, and mix MS-1, the number of gene families picked was very low. Further, as expected, we find that no or only very few gene families are selected when analyzing the permuted data.

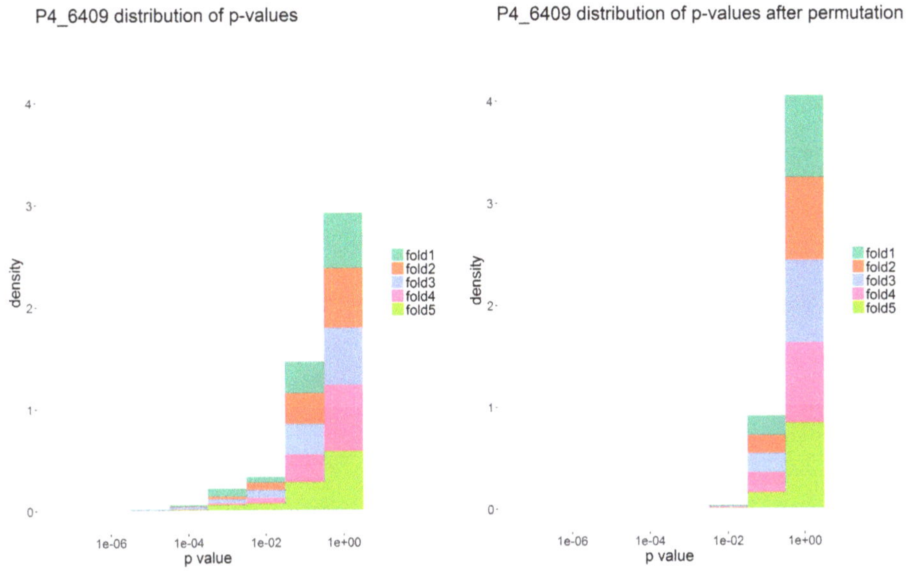

Figure 3. Stacked histogram of p-value distributions across the five folds for the interaction with phage P4/6409. The density is shown instead of counts to account for fold 1 having a 100 times less p-values compared to the other folds, since it does not include partition 1 and therefore did not need to be subsampled. **Left**: Real data. **Right**: Permuted data.

2.3.2. Refinement Based on Regression Models

In the second step of feature selection, we employed linear regression models fitted using Ridge regression. An internal cross validation was used to identify the optimal parameter for the Ridge penalty lambda. The optimal lambda penalty value across the different folds in the cross validation were comparable, indicating that the models are robust, though the size of the feature space varies (see Supplementary Figure S2).

We next required that a gene family should have absolute regression weights greater than 0.01 in at least three of the five partitions to have passed a second selection step. The number of gene families selected in this manner is listed per phage on the right side of Table 2. We term this the set of significant gene families for a certain phage. The number of significant gene families in interaction with phages 1N/80, A3/R, and mix MS-1 was too small to train a final model. For the remaining phages, the amount of significant gene families varied between the different phages, though the sets were comparable in size, with the smallest comprising 13 and the largest 80 gene families (see Table 2). In total, there were 167 significant gene families. When performing the same procedure on permuted data, significant gene families could only be identified in four phages, and a final model could only be trained for two.

Table 2. Summary of the modelling results for real and permuted data. The "First Model" section reports the results of the first filtering procedure based of association analyses. The "Final Model" section gives the result of the second filtering procedure based on regression model fitting combined with consistency constraints. The area under the curve (AUC) is used as performance measure of the final model. The number of gene families selected given in the left part of the table is calculated as the average ± standard deviation across the five folds. If less than two gene families were selected based on regression weights, a final model could not be trained and the associated AUC is reported as NA (not applicable).

	First Model		Final Model			
	Real Data	Permuted Data	Real Data		Permuted Data	
Phage Preparation	No. of Gene Families Selected by Enrichment	No. of Gene Families Selected by Enrichment	No. of Gene Families Selected on Regression Weights	AUC	No. of Gene Families Selected on Regression Weights	AUC
1N/80	10 ± 16	0	2	NA	0	NA
676/F	222 ± 144	0	45	0.78	0	NA
676/T	361 ± 243	12 ± 11	79	0.87	3	0.63
676/Z	112 ± 87	11 ± 14	31	0.72	4	0.61
A3/R	13 ± 26	0	1	NA	0	NA
A5/L	184 ± 124	0	37	0.8	0	NA
A5/80	265 ± 148	0	80	0.78	0	NA
P4/6409	200 ± 137	2 ± 4	61	0.79	0	NA
phi200/6409	160 ± 138	0	56	0.79	0	NA
MS-1	6 ± 10	0	0	NA	0	NA
OP_MS-1	86 ± 78	0	29	0.65	0	NA
OP_MS-1_TOP	54 ± 52	1 ± 1	13	0.67	0	NA

2.3.3. Final Model

Final models were next retrained including only the significant gene families passing both steps of feature selection (low association *p*-values and high regression weights) as input features. Plots of the regression weights assigned by those final models showed the direction of weights to be consistent across folds, i.e., gene families are consistently found to have either positive or negative weights across all of the five partitions. This is depicted for the example of phage P4/6409 in Figure 4. Results for other phage preparations were comparable.

Out of all the 167 gene families, a total of 97 increased phage resistance, 62 increased phage susceptibility, and eight were ambiguous, meaning that they increased resistance to some phages but susceptibility to others. This further shows that the vast majority of significant gene families identified were consistent in their direction of influence across all 12 tested phage preparations.

Figure 4. Heat map of the regression weights for the final model of phage P4/6409. Columns are gene families, rows are cross validation folds. The color indicates the value and direction of each weight, with blue being strongly positive and red being strongly negative. Weights with low values are white. Results were comparable for other phages with the exception of 1N/80, A3/R, and mix MS-1 (see Table 2).

2.4. Functional Annotation of the Significant Genes

We further sought to characterize the function of the identified significant gene families by comparing them to the eggNOG database. The distribution of functional annotation terms identified for the full set of significant genes is shown in Figure 5 and shows that it was possible to identify a match in eggNOG for only 60% of gene families. Most genes had either no hit in the eggNOG database or a hit to a NOG of unknown function.

Case-by-case inspection of the functional annotation terms retrieved from both RAST and eggNOG for the 167 significant gene families identified 13 gene families that have terms directly related to phages, while another 18 were related either to other mobile genetic elements such as genomic islands and transposons or to processes associated to them such as transposase activity. Of these, three gene families have homologs found in SaPIs, which are a phage defense system of *S. aureus* [13]. Four additional gene families appeared to be part of restriction-modification systems and six had hits to transcriptional regulators.

Out of these groups, the gene families related to restriction-modification systems and SaPIs were found to consistently be associated with resistance to phage infection (as measured by the sign of the weights in the final model described earlier), as can be seen in Supplementary Table S1. Of the gene families associated with transcriptional regulators, five were found to increase phage resistance, while one was found to increase susceptibility. The gene families related to phages and mobile elements encompass both gene families promoting resistance and families promoting susceptibility, further pointing to the complexity of the host–phage interaction. The full list of annotation terms for all significant gene families can be found in the Supplementary Table S1, together with the gene family's average regression weight across the five cross validation folds per phage.

We have estimated cumulative density functions (CDF) for each eggNOG category from the full gene set and next evaluated which functional categories in the significant gene set were enriched or depleted. With a threshold of $p = 0.05$, we found that categories "No hit" and "Replication, recombination, and repair" were enriched, while "Post-translational modification, protein turnover, and chaperones" and "Inorganic ion transport and metabolism" were depleted (see Supplementary Table S2).

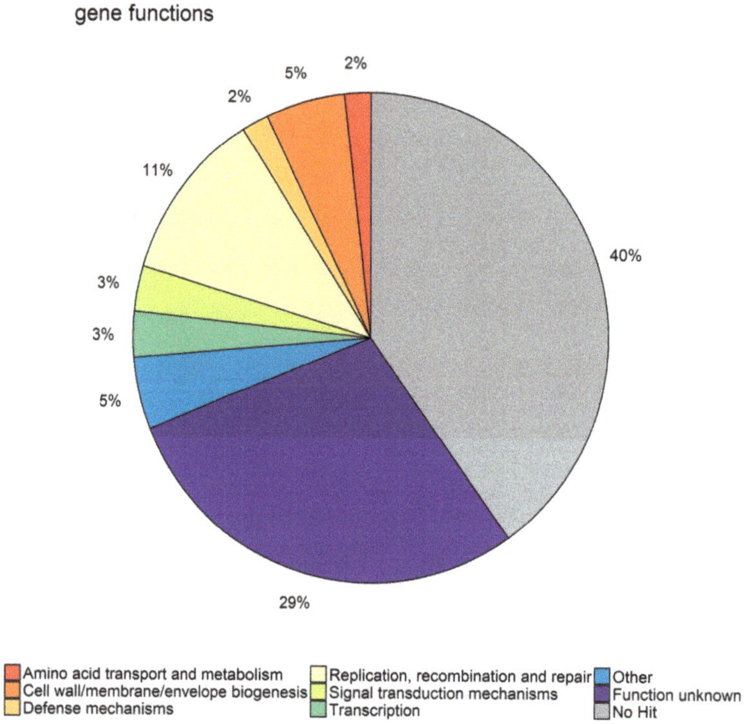

Figure 5. Functional annotation categories of the eggNOGs matching to the set of significant genes across all nine phages.

2.5. Overlap of Significant Gene Family Sets

We further analyzed the overlap between the significant gene family sets found for each phage model. Figure 6 shows a histogram of the number of phage models where a given gene family was identified as significant. It clearly presents that very few significant gene families are shared by many phage models, and only one is shared by all nine. The majority of significant gene families have been observed in interaction with only one or two different phages. This in turn means that each of the phages we tested has a distinct and specific interaction with our bacterial strain set, since different genes in the bacterial host dictate whether infection will be successful.

Further, the significant gene families of the three cocktails are not a linear combination of the sets identified for their component phages, though there is a sizeable overlap (data not shown).

There were four gene families found significant in at least eight phage models. They are listed in Table 3, along with their direction of influence and the annotation and category of their matching eggNOG, if any. Out of the four, three increase resistance to phage, while one was ambiguous in its direction of influence. Two gene families had no hit in the eggNOG database and one was categorized as being of "unknown function". We were therefore unable to deduce a possible function for them though they appear to be of great importance for phage susceptibility. One, cluster 3112, appears to be involved in regulation of transcription and signal transduction that may play a role in host takeover. There were no direct indications for how exactly those gene families effect their influence biologically, but it is evident from the models that they do.

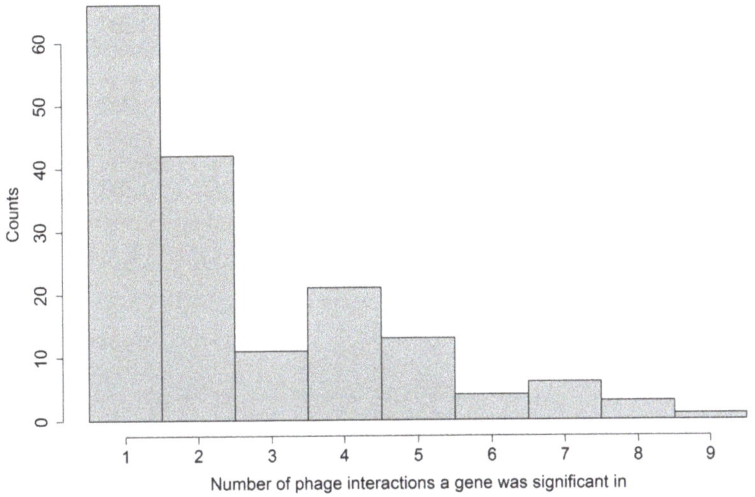

Figure 6. Histogram depicting the number of phage models where a given gene family was identified significant.

Table 3. Predicted functions of the gene families found significant in interaction with eight or more phages.

Gene Family ID	Times Observed	Increases	eggNOG Annotation	eggNOG Category
cluster_1791	9	Resistance	-	No Hit
cluster_389	8	Resistance	-	Function unknown
cluster_3112	8	Resistance	Transcriptional regulator	Transcription
cluster_3992	8	Ambiguous *	-	No Hit

* This gene family always confers phage resistance except in one interaction in which it confers susceptibility.

3. Discussion

In this study, we sought to model the host-genetic determinants of MRSA phage susceptibility with a two-step logistic regression model fitted via ridge regression. We succeeded in building models of acceptable performance for nine of the 12 tested phage preparations with AUCs ranging from 0.65 to 0.87. By doing so, we identified 167 host gene families that influence *S. aureus'* interaction with those nine phages.

Our dataset is, with 207 observations, rather small for this type of analysis, since there are many more covariates—i.e., gene families—than observations. We have addressed this by building a two-step model and including a filtering step based on *p*-values, thereby greatly reducing the number of covariates going into the analysis. As biological entities are shaped by evolution, the strains share some degree of relatedness, and the testing results are not completely independent observations. We have partitioned the data according to phylogeny in a way that ensures highly similar strains are located to the same partition. Doing that ensures that the observations we are aiming to predict are more independent from the ones we feed into the model during training. The partitioning was maintained at all steps, ensuring that data from highly similar strains was never used to predict the outcome. Furthermore, there was an uneven partitioning of the data due to a high percentage of strains from two very related sequence types, which may lead to bias. The challenge of uneven partitions was addressed by subsampling the oversized partition 1 so we could obtain a realistic distribution of *p*-values for the association of all genes to the observed phenotype. Finally, our set of strains with its composition of clonal complexes is specific to Denmark [18]. It is not necessarily representative of *S. aureus* populations observed in different settings.

It should further be noted that our approach can only identify gene families that are part of the accessory genome, since the first selection step is based on differential abundance of those gene families in susceptible vs. resistant strains. Furthermore, this analysis does not consider point mutations as far wild type and mutant version of a gene are more than 90% identical, since we have clustered genes into families with that threshold.

Regarding the electronic gene family annotation, we were able to identify four gene families related to restriction-modification systems and three related the genes found in SaPIs, all of which increased the resistance to phage as expected. Further, six of the significant gene families were related to transcriptional regulation, which fits well with the fact that phages try to shut down host transcription during takeover.

A multitude of gene families found appear to be mobile elements of some kind. Interestingly, Ram et al. have stated that "Most genes involved in phage resistance are carried by plasmids and other mobile genetic elements, including bacteriophages and their relatives" [14], though this statement is quite possibly related to SaPIs and phage-inducible chromosomal islands (PICIs) in general. Those mobile element related gene families had varying direction of influence. They may be related to the interplay of integrated prophages and external phages, which can either complement each other or oppose each other. An integrated prophage may for example protect from further infection via a principle known as superinfection-exclusion [21]. For a large proportion of the significant gene families, however, no hit could be found in the eggNOG database, and of those that had a hit, the most common category was "Function unknown". This may be due to the fact *S. aureus* has a large accessory genome that is made up mostly of different types of mobile genetic elements, among them prophages, that are highly diverse and not well characterized [8]. We have not determined whether either the gene families with hits to phage related proteins or those without hits or with hits to proteins of unknown functions are parts of integrated prophages. Identification of the prophages present in our strain set could add to the interpretation of the analysis; however, it is out of the scope of this study.

We also found that there is only a minor overlap between the sets of significant gene families identified for different phages. This means that each phage had a different and specific interaction with the set of bacterial strains.

Further, we found that generally more gene families promoted resistance than susceptibility. Among the four gene families that were found significant in interaction with at least eight different phages, three promote resistance, and one was ambiguous (see Table 3). This overrepresentation of gene families promoting resistance was expected, since in our set-up resistance to phage can more easily be explained by a gain of function model, meaning the gaining of a defense mechanism of which there are plenty found in nature. We were unfortunately unable to identify the nature of the defense mechanism in most resistance promoting gene families from electronic annotation alone.

Conversely, a gain in susceptibility linked to the presence of a certain gene family is more difficult to explain. The most ready interpretation is that these gene families somehow improve conditions for the phage. The observation can also be explained by integrated prophages that may become activated upon infection or stress caused by the adsorption of an external phage and then lyse their host after completing the lytic cycle. Since the products of the bacterial lysis by the phages were not sequenced, we cannot say whether the external, therapeutic phage or an integrated prophage is the agent of the lysis. Intriguingly, evidence of an interplay between virulence and phage resistance has also been shown. Laanto et al. report that after co-cultivation with lytic phage, strains of the fish pathogen *Flavobacterium columnare* that have acquired phage-resistance have also lost their virulence compared to phage-sensitive paternal strains [22]. Similar observations have been made for *S. aureus* by Capparelli et al. [23], who show that phage-resistance is associated with reduced fitness. Accordingly, the opposite correlation may hold as well, meaning that genes associated with higher virulence and host fitness may at the same time effect higher susceptibility to phages. As our strain set was isolated from patients displaying severe *S. aureus* infections, it is conceivable that these strains are both very virulent and of high fitness.

In conclusion, we have shown that while our methodology does not have predictive power, it allows for the association of the observed phenotype with the genetic background, thereby producing interpretable results that can be used for gene function discovery. This type of analysis, which combines phenotypic and whole genome sequencing (WGS) data, can be used to identify genetic determinants of observed bacterial phenotypes in other settings as well and is expected to be a useful tool in future analyses of phage-host relationships

4. Materials and Methods

4.1. Collection of Clinical MRSA Strains Used for Susceptibility Testing

The collection of 207 MRSA strains tested in this project as well as their whole genome sequences (WGS) were obtained from the Clinical Microbiology Department of Hvidovre Hospital, Hvidovre, Denmark. The strains originate from patient samples. They were selected to represent a broad genetic diversity of the more than 5000 WGS MRSA from Hvidovre Hospital. The fasta sequences of the 207 selected strains have been submitted to the European Nucleotide Archive (Hinxton, Cambridgeshire, UK) [24] with the accession numbers ERZ485118–ERZ485325. They can be viewed under the link: http://www.ebi.ac.uk/ena/data/view/<AccessionNumbers>.

Although no methicillin-sensitive (MSSA) strains were included in the study, we nonetheless chose MRSA strains of the spa-types that are common in MSSA infections [25]. Spa-typing is a single-locus classification scheme for *S. aureus* based on the polymorphic region in protein A [26]. We included MRSA strains positive for PVL and containing *mecC*. All inclusion criteria are listed Supplementary Section 1 'List of inclusion criteria for MRSA strains' and the properties of selected isolates can be found in the Supplementary Table S3.

4.2. Collection of Phages Used for Susceptibility Testing

A total of 12 therapeutic staphylococcal phage preparations were used for susceptibility testing. They contain phages which are part of the proprietary collection of therapeutic phages used by the phage therapy unit of the Hirszfeld Institute of Immunology and Experimental Therapy of the Polish Academy of Science in Wroclaw (HI) [27]. Nine of the preparations are monovalent phage lysates: 1N/80, 676/F, 676/T, 676/Z, A3/R, A5/L, A5/80, P4/6409, and phi200/6409. Crude phage lysates were prepared according to the modified method of Ślopek et al. [9]. Six of those phages (1N/80, 676/Z, A3/R, A5/80, P4/6409, and phi200/6409) were sequenced and confirmed to be obligatory lytic and belonging to a Twortlikevirus genus of a Spounavirinae subfamily of Myoviruses. A detailed report on characteristics of these six phages can be found in Łobocka et al. [28]. All monovalent phage preparations were standardized to routine test dilution (RTD) and had a titer between 10^6 and 10^8. RTD is the highest dilution that still gives confluent lysis on the designated propagating strain of *S. aureus* [17] and the standardization method of choice at HI.

MS.1, OP_MS.1, and OP_MS.1_TOP were equal mixtures of A5/80, P4/6409, and 676/Z phages prepared at the Institute of Biotechnology, Sera and Vaccines BIOMED S.A. in Cracow, Poland. MS-1 phage cocktail lysate contained each component phage in a titer no less than 5×10^5 pfu/mL, OP_MS-1_TOP cocktail of purified phages was suspended in phosphate buffered saline containing each phage at no less than 10^9 pfu/mL [29], and OP_MS-1 phage cocktail had similar characteristics as OP_MS-1_TOP but contained up to 10% of saccharose as a phage stabilizer.

4.3. Susceptibility Testing Procedure

Testing for phage susceptibility was performed as described by Ślopek et al. [30]. In short, 50 μL of phage preparation was applied onto a fresh bacterial lawn from day culture and the results were assessed the next day following 6 h incubation at 37 °C.

Results were assessed according to a 7-point scale as described by Ślopek et al. [30] and shortly summarized in the supplement Section 9 'Details on susceptibility testing as described by Ślopek et al.'

Results were further discretized into two levels: "susceptible" and "resistant". The "susceptible" label was applied to the two strongest reactions, resulting in confluent or semi confluent lysis. According to standards applied at the Bacteriophage Laboratory of the HI, those two levels enable the phage procurement for therapeutic phage preparation. All other weak reactions as well as a negative reaction and opaque lysis were regarded as "resistant". Susceptibility testing results in these two levels, as used for the modelling, can be found in Table 1, while Supplementary Table S4 details results in three levels: resistant, weakly susceptible and strongly susceptible.

The full set of 207 strains was challenged with each of the 12 phage preparations. We call the result of susceptibility testing to a preparation the "interaction" of our strain set with said phage.

4.4. Data Partitioning

For the purpose of modelling the phage response from the genomic composition of the bacterial strains, the 207 MRSA strains were divided into five partitions. This division was based on the orthogonal average nucleotide identity (orthoANI) as described by Lee et al. [31]. OrthoANI is suitable for creating a distance matrix, because it is a symmetric measure of distance, unlike the traditional ANI. Calculations were performed on all pairs of strains with the standalone tool OAT by Lee et al. Distances were subsequently calculated as 1-orthoANI, and a heat map was generated that can be found in Figure 1.

The resulting heat map showed very clear clusters of closely related sequences. Partitioning was therefore done by visual inspection.

The partitions thus obtained were then used in a five-fold cross validation framework, i.e., four of them were combined into the training set, and one was left out for testing. This process was repeated five times so that each partition was in turn the testing set.

4.5. Model Framework

We sought to model a binary outcome (resistant/susceptible) based on weighted binary features (absence/presence of gene families). Logistic regression models were chosen for this task and set-up inside a five-fold cross validation. Each cross validation fold was trained using a Ridge regression to avoid overfitting. A nested cross validation was used to identify the optimal parameter for the Ridge penalty lambda.

Due to challenges posed by the large feature space, the modelling was further split into a two-step process: a first-step model in which we performed feature selection by association testing, and a second-step model whose features were selected based on the regression weights obtained from the first model. The following sections describe details of each modelling step.

4.6. Feature Selection by Association Testing

The genetic background of the MRSA strains was established by first predicting genes and performing functional annotation through the RAST service [32] for all 207 strains. The predicted genes were then clustered with cd-hit [33] using a cutoff of 90% on global sequence identity, word size 5 and the -g 1 option to cluster with the best match instead of the first match. This resulted in a total of 6.419 gene families in the 207 MRSA strains.

Next, the feature space, i.e., the number of gene families included in the model, was reduced by removing gene families with limited power for distinguishing susceptible from non-susceptible bacterial strains. This was done by constructing 2×2 contingency tables as illustrated in Supplementary Table S5, and from these tables calculating a p-value to each gene family in each phage interaction using Fischer-Boschloo's exact unconditional test. We then imposed a threshold of 0.01 on the p-value for the gene family to be admitted to the second step of modelling.

As can be seen in Figure 1, one of the partitions was significantly larger than the other four. This obliged us to employ bootstrapping in every fold that included partition 1 so as to not bias the feature selection on partition size. Details to this can be found in the Supplementary Section 10 'Details on Feature selection by association testing'.

4.7. Feature Selection by Regression Weights

Due to the five-fold cross validation setup, each gene family was assigned five regression weights for interaction with each phage preparation. These may be NA (not applicable) if the gene family was not chosen by association testing for that fold. Weights can be either positive or negative. As we chose to model susceptibility as the positive outcome and resistance as the negative outcome, this means that positive weights point towards increased susceptibility, while negative weights point towards increased resistance.

We hypothesized that gene families with a high weight across many folds drive the response to this particular phage. Therefore, we next trained and tested a second five-fold cross validated regression model with only the genes that (1) were significant according to the Fischer-Boschloo's test ($p \leq 0.01$) and (2) had absolute regression weights greater or equal to 0.01 in at least three folds in the first regression model. We term the gene families selected in this fashion the set of significant gene families. They are the main focus of this study as they are thought to be driving the response to the tested phage preparations.

In order to verify that the set of gene families we identified were indeed descriptive of the phage susceptibility and not an artifact of overfitting, we repeated the model construction and feature selection with shuffled target values. That is, we randomly associated susceptibility outcomes and bacterial genomes while keeping the ratio between susceptible and resistant as in the original data. We then re-ran the modelling and evaluated the predictive performance and the number of predictive gene-families identified.

4.8. Assignment of eggNOGs

We compared each selected gene family to the eggNOG database [34] by using the eggNog-mapper available on their webpage. eggNOG is a database of non-supervised orthologous groups (NOG) of proteins based on the clustering of the 9.6 million proteins from 2031 genomes. Each NOG has only one annotation term compiled from the integrated and summarized functional annotation of its group members, as well as being part of a broader functional category. EggNOG was chosen primarily because of this functional category assignment that allows a broad overview of the functions present in a set of genes.

To estimate whether the observed distribution of functional categories in the set of significant gene families was different from what could be expected by chance, we employed the cumulative density function (CDF). We first drew 10,000 random subsamples of the same size as the full set of significant genes families from the total set of 6419 gene families. From these data, we established an estimated cumulative density function (eCDF) for each functional category. We could then calculate likelihoods for each category of obtaining the actual observed frequency or lower or, conversely, the actual observed frequency or higher.

Supplementary Materials: The following are available online at http://www.mdpi.com/xxx/s1: Figure S1: *p*-Value distributions of gene enrichment analysis on phage preparations 1N_80, A3_R and cocktail MS-1. Figure S2: Plot of the cumulative mean square error of the inner cross validation vs. strength of the ridge penalty. Table S1: List of all significant gene families along with their functional annotation terms. Table S2: Probabilities of observing a given prevalence per functional category based on the cumulative density function. Table S3: List of MRSA strains included in the test set and their properties. Table S4: Detailed phage typing results. Table S5: Layout of the contingency tables. The supplement further contains sections one the following: Details of inclusion criteria for MRSA strains. Details on susceptibility testing as described by Ślopek et al. Details on Feature selection by association testing.

Acknowledgments: This work was supported financially by a full PhD scholarship granted by the Technical University of Denmark (DTU).

Author Contributions: Mette V. Larsen and Ryszard Międzybrodzki conceived and designed the overall project idea. Morten Nielsen coordinated the modeling part. Mette V. Larsen and Morten Nielsen coordinated the gene functional analysis. Ryszard Międzybrodzki and Ewa Jończyk-Matysiak coordinated the experimental part. Ewa Jończyk-Matysiak and Henrike Zschach conducted the laboratory work. Beata Weber-Dąbrowska supplied the phage preparations. Henrik Westh supplied the bacterial strains and advised on the strain selection criteria. Henrik Hasman, Henrik Westh, and Andrzej Górski provided feedback on the biological relevance of the findings. Henrike Zschach and Ryszard Międzybrodzki wrote the paper. Mette V. Larsen, Morten Nielsen, and Andrzej Górski advised the paper writing and performed edits. All authors contributed to the final proof read.

Conflicts of Interest: The authors declare no conflict of interest. The founding sponsors had no role in the design of the study; in the collection, analyses, or interpretation of data; in the writing of the manuscript, and in the decision to publish the results.

References

1. World Health Organization (WHO). Antimicrobial Resistance Fact Sheet. 2016. Available online: http://www.who.int/mediacentre/factsheets/fs194/en/ (accessed on 5 September 2017).
2. World Health Organization (WHO). WHO Global Priority List of Antibiotic-Resistant Bacteria. 2017. Available online: http://www.who.int/medicines/publications/WHO-PPL-Short_Summary_25Feb-ET_NM_WHO.pdf (accessed on 27 February 2017).
3. Chhibber, S.; Kaur, T.; Kaur, S.S.; Wilson, B.; Cheung, A. Co-Therapy Using Lytic Bacteriophage and Linezolid: Effective Treatment in Eliminating Methicillin Resistant *Staphylococcus aureus* (MRSA) from Diabetic Foot Infections. *PLoS ONE* **2013**, *8*, e56022. [CrossRef] [PubMed]
4. Abedon, S.T.; Kuhl, S.J.; Blasdel, B.G.; Kutter, E.M. Phage treatment of human infections. *Bacteriophage* **2011**, *1*, 66–85. [CrossRef] [PubMed]
5. Pincus, N.B.; Reckhow, J.D.; Saleem, D.; Jammeh, M.L.; Datta, S.K.; Myles, I.A. Strain specific phage treatment for *Staphylococcus aureus* infection is influenced by host immunity and site of infection. *PLoS ONE* **2015**, *10*, e0124280. [CrossRef] [PubMed]
6. Miedzybrodzki, R.; Borysowski, J.; Weber-Dabrowska, B.; Fortuna, W.; Letkiewicz, S.; Szufnarowski, K.; Pawelczyk, Z.; Rogóz, P.; Klak, M.; Wojtasik, E.; et al. Clinical aspects of phage therapy. *Adv. Virus Res.* **2012**, *83*, 73–121. [PubMed]
7. Borysowski, J.; Łobocka, M.; Międzybrodzki, R.; Weber-Dabrowska, B.; Górski, A. Potential of Bacteriophages and Their Lysins in the Treatment of MRSA. *BioDrugs* **2011**, *25*, 347–355. [CrossRef] [PubMed]
8. Deghorain, M.; van Melderen, L. The Staphylococci Phages Family: An Overview. *Viruses* **2012**, *4*, 3316–3335. [CrossRef] [PubMed]
9. Ślopek, S.; Durlakowa, I.; Weber-Dąbrowska, B.; Kucharewicz-Krukowska, A.; Dąbrowski, M.; Bisikiewicz, R. Results of bacteriophage treatment of suppurative bacterial infections. I. General evaluation of the results. *Arch. Immunol. Ther. Exp.* **1983**, *31*, 267–291.
10. Weber-Dąbrowska, B.; Jończyk-Matysiak, E.; Żaczek, M.; Łobocka, M.; Łusiak-Szelachowska, M.; Górski, A. Bacteriophage Procurement for Therapeutic Purposes. *Front. Microbiol.* **2016**, *7*, 1177. [CrossRef] [PubMed]
11. Sadykov, M.R. Restriction-Modification Systems as a Barrier for Genetic Manipulation of *Staphylococcus aureus*. In *The Genetic Manipulation of Staphylococci. Methods in Molecular Biology*; Bose, J., Ed.; Humana Press: New York, NY, USA, 2016; Volume 1373, pp. 9–23.
12. Seed, K.D. Battling Phages: How Bacteria Defend against Viral Attack. *PLoS Pathog.* **2015**, *11*, e1004847. [CrossRef] [PubMed]
13. Ram, G.; Chen, J.; Ross, H.F.; Novick, R.P. Precisely modulated pathogenicity island interference with late phage gene transcription. *Proc. Natl. Acad. Sci. USA* **2014**, *111*, 14536–14541. [CrossRef] [PubMed]
14. Xia, G.; Wolz, C. Phages of *Staphylococcus aureus* and their impact on host evolution. *Infect. Genet. Evol.* **2014**, *21*, 593–601. [CrossRef] [PubMed]
15. Lindsay, J.A. Genomic variation and evolution of *Staphylococcus aureus*. *Int. J. Med. Microbiol.* **2010**, *300*, 98–103. [CrossRef] [PubMed]
16. Goerke, C.; Pantucek, R.; Holtfreter, S.; Schulte, B.; Zink, M.; Grumann, D.; Bröker, B.M.; Doskar, J.; Wolz, C. Diversity of prophages in dominant *Staphylococcus aureus* clonal lineages. *J. Bacteriol.* **2009**, *191*, 3462–3468. [CrossRef] [PubMed]
17. Blair, J.E.; Williams, R.E.O. Phage typing of staphylococci. *Bull World Heal. Organ.* **1961**, *24*, 771–784.

18. Bartels, M.D.; Larner-Svensson, H.; Meiniche, H.; Kristoffersen, K.; Schonning, K.; Nielsen, J.B.; Rohde, S.M.; Christensen, L.B.; Skibsted, A.W.; Jarlov, J.O.; et al. Monitoring meticillin resistant *Staphylococcus aureus* and its spread in Copenhagen, Denmark, 2013, through routine whole genome sequencing. *Eurosurveillance* **2015**, *20*, 21112. [CrossRef] [PubMed]
19. Allen, R.C.; Pfrunder-Cardozo, K.R.; Meinel, D.; Egli, A.; Hall, A.R. Associations among Antibiotic and Phage Resistance Phenotypes in Natural and Clinical *Escherichia coli* Isolates. *MBio* **2017**, *8*, e01341-17. [CrossRef] [PubMed]
20. Monecke, S.; Coombs, G.; Shore, A.C.; Coleman, D.C.; Akpaka, P.; Borg, M.; Chow, H.; Ip, M.; Jatzwauk, L.; Jonas, D.; et al. A field guide to pandemic, epidemic and sporadic clones of methicillin-resistant *Staphylococcus aureus*. *PLoS ONE* **2011**, *6*, e17936. [CrossRef] [PubMed]
21. Hofer, B.; Ruge, M.; Dreiseikelmann, B. The superinfection exclusion gene (sieA) of bacteriophage P22: Identification and overexpression of the gene and localization of the gene product. *J. Bacteriol.* **1995**, *177*, 3080–3086. [CrossRef] [PubMed]
22. Laanto, E.; Bamford, J.K.H.; Laakso, J.; Sundberg, L.R. Phage-Driven Loss of Virulence in a Fish Pathogenic Bacterium. *PLoS ONE* **2012**, *7*, e53157. [CrossRef] [PubMed]
23. Capparelli, R.; Nocerino, N.; Lanzetta, R.; Silipo, A.; Amoresano, A.; Giangrande, C.; Becker, K.; Blaiotta, G.; Evidente, A.; Cimmino, A.; et al. Bacteriophage-resistant *Staphylococcus aureus* mutant confers broad immunity against staphylococcal infection in mice. *PLoS ONE* **2010**, *5*, e11720. [CrossRef] [PubMed]
24. Leinonen, R.; Akhtar, R.; Birney, E.; Bower, L.; Cerdeno-Tárraga, A.; Cheng, Y.; Cleland, I.; Faruque, N.; Goodgame, N.; Gibson, R.; et al. The European Nucleotide Archive. *Nucleic Acids Res.* **2011**, *39*, D28–D31. [CrossRef] [PubMed]
25. Aanensen, D.M.; Feil, E.J.; Holden, M.T.; Dordel, J.; Yeats, C.A.; Fedosejev, A.; Goater, R.; Castillo-Ramírez, S.; Corander, J.; Colijn, C.; et al. Whole-Genome Sequencing for Routine Pathogen Surveillance in Public Health: A Population Snapshot of Invasive *Staphylococcus aureus* in Europe. *MBio* **2016**, *7*, e00444-16. [CrossRef] [PubMed]
26. Shopsin, B.; Gomez, M.; Montgomery, S.O.; Smith, D.H.; Waddington, M.; Dodge, D.E.; Bost, D.A.; Riehman, M.; Naidich, S.; Kreiswirth, B.N. Evaluation of protein A gene polymorphic region DNA sequencing for typing of *Staphylococcus aureus* strains. *J. Clin. Microbiol.* **1999**, *37*, 3556–3563. [PubMed]
27. Weber-Dąbrowska, B.; Mulczyk, M.; Górski, A.; Boratyński, J.; Łusiak-Szelachowska, M.; Syper, D. Methods of Polyvalent Bacteriophage Preparation for the Treatment of Bacterial Infections. U.S. Patent US7232564 B2, 2002.
28. Łobocka, M.; Hejnowicz, M.S.; Dąbrowski, K.; Gozdek, A.; Kosakowski, J.; Witkowska, M.; Ulatowska, M.I.; Weber-Dąbrowska, B.; Kwiatek, M.; Parasion, S.; et al. Genomics of Staphylococcal Twort-like Phages—Potential Therapeutics of the Post-Antibiotic Era. *Adv. Virus Res.* **2012**, *83*, 143–216. [PubMed]
29. Górski, A.; Weber-Dąbrowska, B.; Miedzybrodzki, R.; Stefański, G.; Dechnik, K.; Olchawa, E. A Method for Obtaining Bacteriophage Purified Preparations. Polish Patent No. PL 212811 B1, 2012.
30. Slopek, S.; Durlakowa, I.; Kucharewicz-Krukowska, A.; Krzywy, T.; Slopek, A.; Weber, B. Phage typing of Shigella flexneri. *Arch. Immunol. Ther. Exp.* **1972**, *20*, 1–60.
31. Lee, I.; Kim, Y.O.; Park, S.C.; Chun, J. OrthoANI: An improved algorithm and software for calculating average nucleotide identity. *Int. J. Syst. Evol. Microbiol.* **2016**, *66*, 1100–1103. [CrossRef] [PubMed]
32. Overbeek, R.; Olson, R.; Pusch, G.D.; Olsen, G.J.; Davis, J.J.; Disz, T.; Edwards, R.A.; Gerdes, S.; Parrello, B.; Shukla, M.; et al. The SEED and the Rapid Annotation of microbial genomes using Subsystems Technology (RAST). *Nucleic Acids Res.* **2014**, *42*, D206–D214. [CrossRef] [PubMed]
33. Fu, L.; Niu, B.; Zhu, Z.; Wu, S.; Li, W. CD-HIT: Accelerated for clustering the next-generation sequencing data. *Bioinformatics* **2012**, *28*, 3150–3152. [CrossRef] [PubMed]
34. Huerta-Cepas, J.; Szklarczyk, D.; Forslund, K.; Cook, H.; Heller, D.; Walter, M.C.; Rattei, T.; Mende, D.R.; Sunagawa, S.; Kuhn, M.; et al. eggNOG 4.5: A hierarchical orthology framework with improved functional annotations for eukaryotic, prokaryotic and viral sequences. *Nucleic Acids Res.* **2016**, *44*, D286–D293. [PubMed]

antibiotics

MDPI

Article

Directed Evolution of a Mycobacteriophage

María Cebriá-Mendoza [1], Rafael Sanjuán [1,2] and Pilar Domingo-Calap [1,2,*]

[1] Institute for Integrative Systems Biology (I2SysBio), Universitat de València-CSIC, 46980 Paterna, Valencia, Spain; maria.c.cebria@uv.es (M.C.-M.); rafael.sanjuan@uv.es (R.S.)
[2] Department of Genetics, Universitat de València, 46100 Burjassot, Valencia, Spain
* Correspondence: domingocalap@gmail.com; Tel.: +34-963-543-261

Received: 10 April 2019; Accepted: 23 April 2019; Published: 25 April 2019

Abstract: Bacteriophages represent an alternative strategy to combat pathogenic bacteria. Currently, *Mycobacterium tuberculosis* infections constitute a major public health problem due to extensive antibiotic resistance in some strains. Using a non-pathogenic species of the same genus as an experimental model, *Mycobacterium smegmatis*, here we have set up a basic methodology for mycobacteriophage growth and we have explored directed evolution as a tool for increasing phage infectivity and lytic activity. We demonstrate mycobacteriophage adaptation to its host under different conditions. Directed evolution could be used for the development of future phage therapy applications against mycobacteria.

Keywords: phage therapy; *Mycobacterium smegmatis*; mycobacteriophages; directed evolution

1. Introduction

Antimicrobial resistance is a major global health concern with a current estimated cost of around 700,000 deaths annually which, if not controlled, could raise up to 10 million deaths by 2050 [1]. Tuberculosis, caused by the bacillus *Mycobacterium tuberculosis*, is the 9th leading cause of death worldwide, being responsible for 1.3 million deaths in 2016 [2]. Most of this mortality, though, could be prevented with early diagnosis and appropriate treatment. Some *M. tuberculosis* infections are treated successfully with different antibiotics, but the emergence of antibiotic-resistant strains is a growing source of concern. In addition, *M. tuberculosis* can stay in a dormant state inside alveolar macrophages, a population not targeted by most drugs, making it difficult to eradicate the disease [3]. Under this scenario, there is a need to find potential alternative therapies. One possibility is phage therapy, that is, using bacteriophages (phages) to treat bacterial infections. An interesting property of phages is that some can penetrate into macrophages by phagocytosis [4].

Phages are the most abundant biological entity in the planet with an estimated 10^{31} total particles. Phages were discovered independently in 1915 and 1917 by Twort and d'Hérelle, respectively. Soon after their discovery, d'Hérelle employed phages to treat dysentery in France and cholera in India. Until the Second World War, phage therapy was considered the only therapeutic tool and the only treatment for bacterial diseases, but the discovery of antibiotics and their introduction in the 1940s replaced phage therapy in most countries, particularly in the West. In contrast, in the former Soviet Union phage therapy was used and is still in practice in countries such as Georgia, Russia, and Poland. Differences in language, in the setting-up of clinical trials between Eastern and Western countries, and the emergence of antibiotics has slowed down the progress of phage therapy in the USA and Western Europe for the past 50 years [5]. Currently, however, the antibiotic resistance crisis has led to a reappraisal of phage therapy worldwide.

The field of phage therapy focuses essentially on virulent phages [6]. In addition to killing bacteria rapidly, phages have some advantages over antibiotics, such as their high host specificity, which reduces the damage to other bacteria and hence avoids dysbiosis. Furthermore, phages multiply at the site of infection and only in the presence of their specific bacteria [7]. Based on this, it is expected that phage treatment will require relatively low dosages and treatment frequencies to reach the optimal therapeutic effect [8]. Furthermore, one of the most interesting aspects of phage therapy relates to bacterial resistance, since phages can evolve and overcome resistance [8,9]. Also, directed evolution can help improve phage infectivity, and may be used to reduce resistance emergence rates. Although there are currently no phage therapy products approved as antibacterial drugs for human use in the EU or US, there are ongoing or completed clinical trials [10]. Moreover, the food industry accepts several commercial phage preparations used for biocontrol of bacterial pathogens, which are approved by the FDA, like Listex™ and ListShield™ used to protect food from *Listeria* [11,12].

In this context, we sought to establish proof of concept for the application of phage therapy to mycobacteria. Mycobacteriophages were first isolated in the 1940s using *Mycobacterium smegmatis* as host and are mostly double-stranded DNA, tailed phages belonging to the *Caudovirales* group. Currently, there are 10,454 phages described whose host belongs to *Mycobacterium* genus, of which 1751 are sequenced [13]. Most of these mycobacteriophages belong to the family *Siphoviridae* (with long, flexible, non-contractile tails) and, in a lower proportion, to the family *Myoviridae* (with contractile tails) [14]. Mycobacteriophages have provided a wealth of information on the diversity of phages that infect a common bacterial host. In addition, published sequences suggest a mosaic nature for their genomes, with extensive illegitimate recombination and horizontal gene exchange. Mycobacteriophages have been classified in different groups, called clusters, based on sequence similarity [15]. This previous knowledge has provided a variety of tools that can be employed to study mycobacterial genetics and, also, to establish new strategies to control, diagnose and treat diseases caused by mycobacteria [16].

Our main goal here was to demonstrate that evolution under controlled conditions can help us obtain more infective mycobacteriophages. To achieve this aim, we evolved a phage by serial passages under different conditions using *M. smegmatis* as model host. *M. smegmatis* is a non-pathogenic bacterium with a faster life cycle than other *Mycobacterium* species [17], thus offering a good system to explore and set up directed evolution protocols.

2. Results

2.1. Directed Evolution

We evolved six independent lines (evolution replicates) of *M. smegmatis bacteriophage* (American Type Culture Collection, ATCC® 11759-B1™) in *M. smegmatis* for 20 serial transfers (passages) in semi-solidified medium, three (Figure 1A–C) using a large phage inoculum per passage (10^5 plaque forming units, PFU) and three (Figure 1D–F) using a small phage inoculum (10^2 PFU). Independent of passage number, we observed that the phage reached higher titers with a small inoculum ($1.28 \pm 0.53 \times 10^9$ PFU/mL) than with a high inoculum treatment ($1.94 \pm 0.54 \times 10^7$ PFU/mL; *t*-test using log-transformed titers: $p = 0.001$). This probably indicates that the large inoculum exhausted the cell population rapidly, whereas a smaller inoculum allowed cells to proliferate for longer, increasing the number of susceptible host cells and, thus, the total amount of viral progeny produced. For the small-inoculum lines, a linear model using the evolution replicate (Figure 1D–F) as a random factor and passage number as a covariate showed that the log-titer increased significantly with passage number ($F = 9.071$, $p = 0.004$; Figure 1). In contrast, in phages evolved using a large inoculum size, we could not detect an effect of passage number on log-titer ($F = 1.131$, $p = 0.292$; Figure 1A–C).

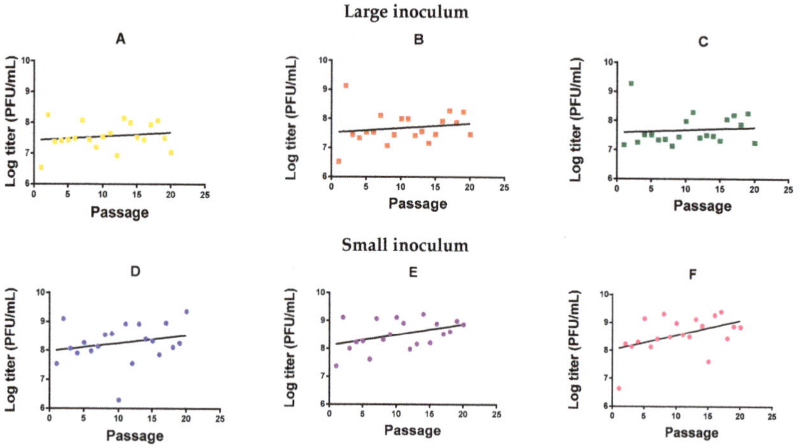

Figure 1. Log-titer reached after each passage of mycobacteriophage evolution. Large inoculum: (**A**) Yellow, (**B**) red, (**C**) green. Small inoculum: (**D**) Blue, (**E**) purple, (**F**) pink. Dots: Experimental points. Line: Linear regression.

2.2. Analysis of Mycobacteriophage Growth Rate and Stability

In order to more directly test whether viral fitness changed significantly after 20 passages, we performed standard growth curves in triplicate, in which we compared the founder phage and each of the evolved phages in the same experimental block. Each line was tested using the same inoculum size employed during the evolution experiment. Significant differences between the founder and the evolved lines were found at intermediate and late points (*t*-tests, $p < 0.05$, Figure 2), suggesting an adaptation of all the evolved lines under our different conditions, although less marked for those lines evolved under large inoculum conditions.

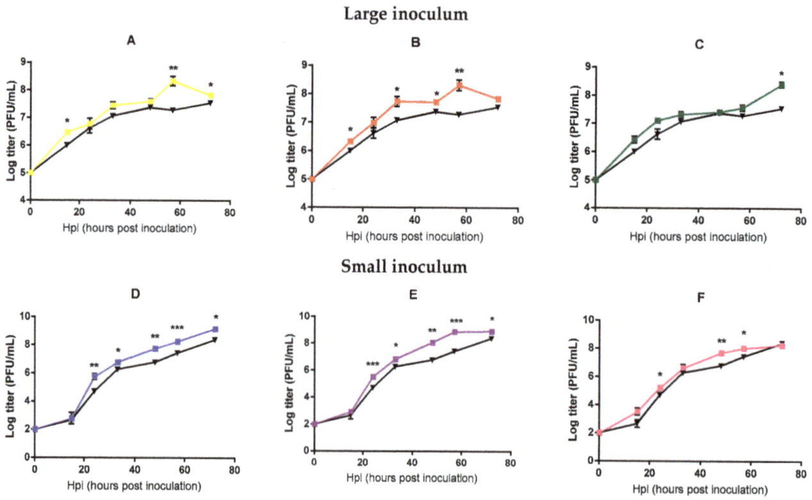

Figure 2. Mycobacteriophages growth curves of the founder and evolved lines. Founder: Black. Large inoculum: (**A**) Yellow, (**B**) red, (**C**) green. Small inoculum: (**D**) Blue, (**E**) purple, (**F**) pink. Dots show the average titer at each time point and error bars indicate the SEM ($n = 3$). *: $p < 0.05$, **: $p < 0.005$, ***: $p < 0.0005$.

In principle, differences in the population growth rate of the phage could be due to the faster infection rate or to the slower degradation rate of the phages. We determined the phage's degradation by measuring the decrease in viral titer as a function of time on the infection medium (in the absence of bacteria). For each lineage (founder and evolved lines), three replicates were done to estimate the degradation rate (Figure 3). Our results showed no differences in degradation rate between the different lines (unpaired *t*-tests, $p > 0.05$). Hence, the observed acceleration of phage growth was driven by a more rapid infection, which in turn could be due to faster adsorption, faster replication, or increase phage yield per cell.

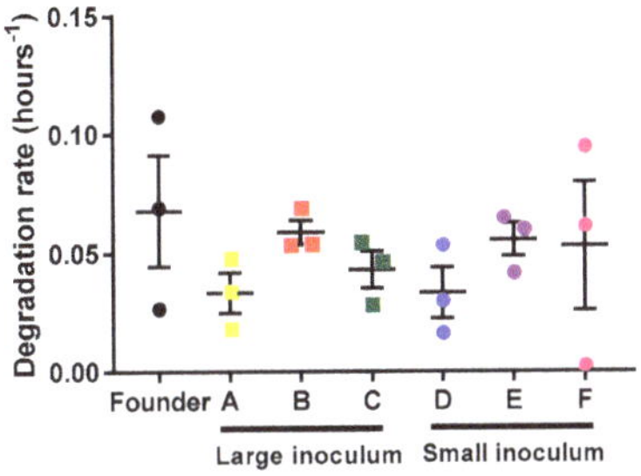

Figure 3. Mycobacteriophages degradation rates of the founder and evolved lines. Founder: Black. Large inoculum: (**A**) Yellow, (**B**) red, (**C**) green. Small inoculum: (**D**) Blue, (**E**) purple, (**F**) pink. Dots show the experimental replicates, and error bars indicate the SEM ($n = 3$).

2.3. *Analysis of Bacterial Lysis Efficiency*

In addition to examining the phage growth rate, we were interested in assessing whether the evolved lines killed *M. smegmatis* more efficiently than the founder phage. Measurements of bacterial densities at the same time points used above for the phage growth curves were done in triplicate. Again, each phage line was tested using the same inoculum size employed during passages. For large inoculum condition, evolved phages were able to lyse bacteria more efficiently than the founder (nested ANOVA: Founder vs. evolved $p = 0.007$, among lines $p = 0.082$; Figure 4). In contrast, these assays were not capable of detecting differences in lysis activity under small-inoculum conditions (nested ANOVA: Founder vs. evolved $p = 0.715$, among lines $p = 0.174$; Figure 4) because small inocula did not produce an appreciable change in bacterial density, independent of the phage used. In other words, with large inocula phages were able to reach the majority of bacterial cells growing in the culture dish, whereas, small inocula gave rise to few, isolated plaques which had little effect on overall bacterial counts.

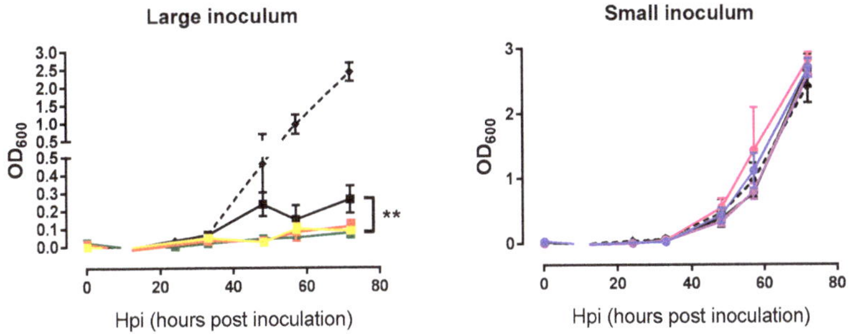

Figure 4. Bacterial growth curves of the founder and evolved lines. Founder: Black. Large inoculum: (**A**) Yellow, (**B**) red, (**C**) green. Small inoculum: (**D**) Blue, (**E**) purple, (**F**) pink. Dots show the average OD_{600} at each time point and error bars indicate the SEM ($n = 3$). Control without phage infection in dashed line. **: $p < 0.005$.

To better examine the killing capacity of phage evolved under the small-inoculum regime, we plated all the evolved lines in triplicate to determine differences in the surface area of the plaques using image analysis. Clear differences between the plaque sizes of lines evolved under the small inoculum regime and the founder were detected (*t*-tests: D: $p < 0.001$; E: $p < 0.001$; F: $p = 0.009$; Figure 5). A tendency towards increased plaque size was also observed for lines evolved under the large inoculum regime, albeit the effect was less evident than for small-inoculum lines and reached significance only for one of the evolved lines (*t*-tests: A: $p = 0.019$; B: $p = 0.408$; C: $p = 0.134$; Figure 5).

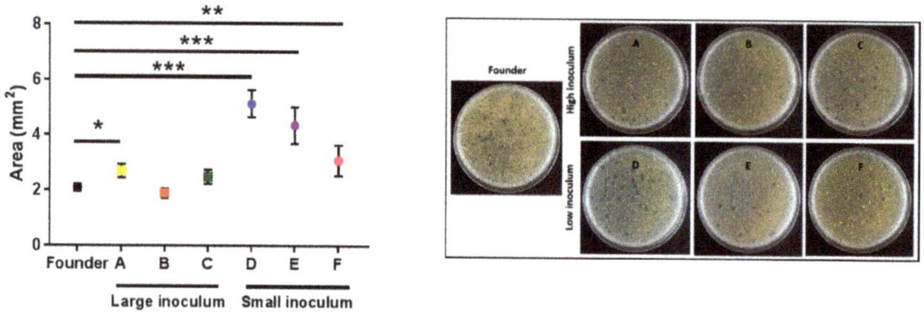

Figure 5. The average area of the plaques of the founder and evolved lines. Founder: Black. Large inoculum: (**A**) Yellow, (**B**) red, (**C**) green. Small inoculum: (**D**) Blue, (**E**) purple, (**F**) pink. Dots show the average area for each line and error bars indicate the SEM ($n = 3$). *: $p < 0.05$, **: $p < 0.005$, ***: $p < 0.0005$. Right panel: Representative images of the plates for each line.

3. Discussion

Directed evolution is a powerful approach for the optimization of different biological processes. The evolution of phage and bacteria has been investigated experimentally in many previous studies [18–22]. However, the experimental evolution of large double-stranded DNA viruses has been less extensively explored, probably because of their lower mutation rates [23], which should slow down evolutionary processes compared to RNA viruses. Yet, some DNA phages have been shown to evolve at rates close to those of RNA phages in the laboratory [18]. We note, though, that there is little or no previous work exploring the evolution of mycobacteriophages under controlled laboratory conditions. Here, we found that phages evolved under our experimental conditions improved fitness as indicated by different analyses. Adaptation has been observed in all the evolved lines (large and small inoculum), demonstrating that the ability to kill bacteria increases after 20 serial passages. In general, higher inoculum sizes achieve higher effective population sizes [19], and increasing the population size typically improves adaptation because this increases allelic diversity and strengthens the efficacy of natural selection relative to random genetic drift [24,25]. However, the dynamics of the host population is also an important factor. If the inoculum is large, the population of host cells can become exhausted rapidly, leaving no resources for further viral replication [26]. This shortens the number of viral generations per passage, potentially slowing adaptation down on an absolute time basis. Clearly, our small-inoculum evolution regime selected for larger plaque sizes than the large-inoculum regime. This increase in plaque size indicates faster spread and more efficient lysis, both of which are interesting properties for phage therapy. Phages evolved under the large-inoculum regime showed increased lysis activity as determined by OD_{600} measurements, but plaque size measurements revealed little or no improvement in spread ability. Based on this and, since in real treatment settings it is likely that the initial ratio of phage to target bacteria is small, we suggest that directed evolution protocols aimed at increasing the ability of a phage to lyse bacterial populations should be preferably performed under small-inoculum conditions.

Directed evolution can be carried out in liquid or in semi-solidified media. Liquid media combined with shaking allows for more efficient propagation of the phage, whilst semi-solidified media impose spread constraints due to the spatial structure. We believe that the latter approach provides a more realistic scenario for phage adaptation and, hence, may select for better phages. Additionally, a liquid culture of mycobacteria is generally difficult because mycolic acids promote cellular aggregation, generating clumps. Adding detergents to the culture can avoid clumps, but these will typically inhibit phage infectivity. For these reasons, we performed our serial transfers in culture dishes instead of liquid-culture tubes. Phage diffusion is limited in semi-solidified media and is dependent on virion morphology and size. In most cases, phage progeny from one cell will infect neighbor cells only, imposing selection pressures that probably differ from those of well-mixed populations. Mycobacteriophages should be able to increase fitness in many different ways, including improvement in the attachment to cell receptors, increased burst size, reduced lysis time, or faster replication, among others, and the outcome may depend on medium viscosity. A better understanding of the molecular mechanisms responsible for adaptation would be obtained by sequencing the evolved lines. Yet, this can be difficult due to the relatively large genome of mycobacteriophages compared to other viruses and the large number of genes with unknown functions.

Bacteria can evolve phage resistance fast by different mechanisms, including loss of the receptor or activation of the CRISPR system [27,28]. Antagonistic coevolution between phage and bacteria implies ongoing selection of host resistance and resistance breaking, which results in changes in phage and host genomes across evolutionary time [29]. Previous works have shown that infection efficiency improves in comparison with a wild-type phage when the coevolution process is performed using preadapted phages (phages which have been subjected to evolutionary passages) [19,20]. This might be an alternative strategy for achieving resistance breaking and for obtaining in the laboratory a phage population capable of preventing the emergence of resistant bacteria during treatment.

Our work provides proof of concept for the use of directed evolution in a mycobacteriophage. Despite the fact that phages are highly specific to their hosts, some mycobacteriophages can infect both *M. tuberculosis* and *M. smegmatis* [30]. Thus, it is possible that our results are applicable to *M. tuberculosis*. However, additional experiments should be done to explore this possibility.

4. Materials and Methods

4.1. Bacterial Strains and Culture Conditions

We used *M. smegmatis* Δ1366 (*M. smegmatis*) [31], kindly provided by Dr. José Luis García (CSIC). This strain is a clone of mc^{2}155 which has the *MSMEG_1366* gene deleted (ABC transporter, ATP-binding protein) and has a plasmid encoding kanamycin resistance and Venus fluorescent protein. *M. smegmatis* was grown in Lysogeny Broth (LB) medium at 37 °C with shaking (250 rpm) in the case of liquid media [16]. LB was supplemented with kanamycin 20 μg·mL^{-1}, as well as with 5 mM $CaCl_2$ to improve phage growth. For growing bacteria in semi-solidified media, 1.5% of agar was added to LB and kanamycin was omitted. In order to preserve bacteria, glycerol 20% (*v*/*v*) was added to bacterial cultures, which were stored at −70 °C. A stock of non-evolved bacteria was done in a first step. For this purpose, a culture of *M. smegmatis* in supplemented LB was incubated until OD_{600} = 0.3 was reached. The culture was divided into 50 mL tubes and chilled on ice. These tubes were centrifuged at 2000× *g* for 15 min at 4 °C. The supernatant was discarded and the pellet was resuspended in ca. 30 mL of LB with 15% of glycerol to obtain a 40× concentrated stock. The culture was split into 250 μL aliquots that were flash-frozen in liquid nitrogen and stored at −70 °C.

4.2. Mycobacteriophage

The *M. smegmatis bacteriophage* was obtained from ATCC (reference ATCC® 11759B1™). Phage buffer (Tris-HCl pH 7.5, 10 mM $MgSO_4$, 5 mM $CaCl_2$ and 68.5 mM NaCl) [13] was used for phage dilutions, phage recovery and phage storage. For phage propagation, *M. smegmatis* was grown to an OD_{600} of 0.3 in supplemented LB. We mixed 100 μL of *M. smegmatis* culture with 100 μL of phage. This mix was incubated without shaking for 15 min at room temperature to allow for phage adsorption. The infection was inoculated in 2 mL of supplemented LB and was incubated for 48 h at 37 °C. After that, cultures were centrifuged at 16,000× *g* for 1 min, the pellet was discarded and the supernatant was aliquoted and stored at −70 °C. For phage titration, serial dilutions of each aliquot were prepared and 100 µL of these dilutions were employed to infect 10^7 CFU of *M. smegmatis*. Titration was done onto LB plates with top agar (LB with 0.7% agar) and plates were incubated for 40 h at 37 °C. After this, plaques were counted to determine titers.

4.3. Evolution

The indicated amount of PFU was used to inoculate 10^7 CFU of non-evolved *M. smegmatis* obtained from our frozen stock (see above) at an initial OD_{600} of 0.3. Two different conditions were tested: Large inoculum (10^5 PFU) and small inoculum (10^2 PFU). Twenty serial passages from plate to plate were performed, controlling the initial inoculum (10^5 PFU or 10^2 PFU) for each replicate at each passage. For this, after each passage, phages were titrated by the plaque assay. For each condition, three evolution replicates (lines) were established. Inoculations were carried with 100 µL of phage suspension onto LB plates with top agar and incubated for 40 h at 37 °C. To collect phages, 5 mL of phage buffer was added to the plates and incubated for 2 h at 37 °C [32]. Then, 2 mL of this phage buffer were collected and cells were removed by centrifugation (16,000× *g* for 1 min). The supernatant was aliquoted and stored at −70 °C. One aliquot of each condition was used for the next passage.

4.4. Growth Curves

Infections were started with 10^5 PFUs in the case of phages evolved using a high inoculum size and with 10^2 PFUs in the case of phages evolved with a low inoculum size, that is, under the same conditions used for evolution. The founder phage was included in all assays. For each time tested, we performed three replicates per line. Infections were performed as above onto LB plates with top agar and incubated at 37 °C for different times (0 hpi, 15 hpi, 24 hpi, 33 hpi, 46 hpi, 57 hpi and 72 hpi). Then, plates were flooded with 5 mL of phage buffer and incubated for 2 h at 37 °C. The buffer (2 mL) was collected and OD_{600} measures were done for each replicate at each time point. Afterward, samples were centrifuged (16,000× *g* for 1 min) to remove cells, and supernatants were aliquoted, stored at −70 °C and titrated.

4.5. Mycobacteriophage Degradation Rates

Founder and evolved lines were used to estimate the mycobacteriophage degradation rate. For this, 10^4 PFUs were added to 3.5 mL of top agar in the absence of bacteria and were plated in semi-solidified agar plates and titrated at 0 hpi, 24 hpi, 48 hpi, and 72 hpi, as described previously.

4.6. Plaque Sizes

Estimates of plaque sizes for the evolved and ancestral lines were done in triplicate. Plating was done by carrying out plaque assays with an estimated 100 PFUs. Plaque size was determined at 72 hpi and was calculated by image analysis using ImageJ software (National Institutes of Health, Bethesda, MD, USA).

5. Conclusions

The present work is a step forward in the use of directed evolution as an optimization tool in mycobacteriophages. We have shown that evolved phages exhibit improved lytic activity compared with the founder phage under our experimental conditions. Our results might help in the development of future treatments against pathogenic and multi-resistant *Mycobacterium tuberculosis* strains, and suggest phage therapy as a potential alternative to the conventional antibiotics.

Author Contributions: Conceptualization, R.S. and P.D.-C.; Data curation, R.S. and P.D.-C.; Formal analysis, R.S. and P.D.-C.; Funding acquisition, R.S.; Investigation, M.C.-M. and P.D.-C.; Methodology, M.C.-M. and P.D.-C.; Resources, P.D.-C.; Supervision, P.D.-C.; Validation, P.D.-C.; Visualization, P.D.-C.; Writing—original draft, M.C.-M.; Writing—review & editing, R.S. and P.D.-C.

Funding: This work was funded by grant BFU2017-84762-R from the Spanish Ministerio de Ciencia, Innovación y Universidades (MCIU) and grant 724519-Vis-à-vis from the European Research Council (ERC) to R.S. P.D.-C. was recipient of a Juan de la Cierva Incorporación contract from MCIU.

Acknowledgments: We thank José Luis García López for the bacterium and María Durán-Moreno for technical assistance.

Conflicts of Interest: The authors declare no conflict of interest.

References

1. Górski, A.; Miedzybrodzki, R.; Weber-Dabrowska, B.; Fortuna, W.; Letkiewicz, S.; Rogóz, P.; Jończyk-Matysiak, E.; Dabrowska, K.; Majewska, J.; Borysowski, J. Phage Therapy: Combating Infections with Potential for Evolving from Merely a Treatment for Complications to Targeting Diseases. *Front. Microbiol.* **2016**, *7*, 1–9. [CrossRef] [PubMed]
2. World Health Organization. Global Tuberculosis Report 2017. Available online: https://www.who.int/tb/publications/global_report/en/ (accessed on 24 April 2019).
3. Samaddar, S.; Grewal, R.K.; Sinha, S.; Ghosh, S.; Roy, S.; Das Gupta, S.K. Dynamics of Mycobacteriophage-Mycobacterial Host Interaction: Evidence for Secondary Mechanisms for Host Lethality. *Appl. Environ. Microbiol.* **2016**, *82*, 124–133. [CrossRef] [PubMed]
4. Jończyk-Matysiak, E.; Weber-Dąbrowska, B.; Owczarek, B.; Międzybrodzki, R.; Łusiak-Szelchowska, M.; Łodej, N.; Górski, A. Phage-Phagocyte Interactions and Their Implications for Phage Application as Therapeutics. *Viruses* **2017**, *9*, 150. [CrossRef] [PubMed]
5. Domingo-Calap, P.; Georgel, P.; Bahram, S. Back to the Future: Bacteriophages as Promising Therapeutic Tools. *HLA* **2016**, *87*, 133–140. [CrossRef] [PubMed]
6. Housby, J.N.; Mann, N.H. Phage Therapy. *Drug Discov. Today* **2009**, *14*, 536–540. [CrossRef]
7. Weber-Dabrowska, B.; Jończyk-Matysiak, E.; Zaczek, M.; Łobocka, M.; Łusiak-Szelachowska, M.; Górski, A. Bacteriophage Procurement for Therapeutic Purposes. *Front Microbiol.* **2016**, 1–14. [CrossRef] [PubMed]
8. Nilsson, A.S. Phage Therapy-Constraints and Possibilities. *Ups. J. Med. Sci.* **2014**, *119*, 192–198. [CrossRef]
9. Ul Haq, I.; Chaudhry, W.N.; Akhtar, M.N.; Andleeb, S.; Qadri, I. Bacteriophages and Their Implications on Future Biotechnology: A Review. *Virol. J.* **2012**, *9*, 9. [CrossRef]
10. Kingwell, K. Bacteriophage Therapies Re-Enter Clinical Trials. *Nat. Rev. Drug Discov.* **2015**, *14*, 515–516. [CrossRef]
11. Lin, D.M.; Koskella, B.; Lin, H.C. Phage Therapy: An Alternative to Antibiotics in the Age of Multi-Drug Resistance. *World J. Gastrointest. Pharmacol. Ther.* **2017**, *8*, 162–173. [CrossRef] [PubMed]
12. Gray, J.A.; Chandry, P.S.; Kaur, M.; Kocharunchitt, C.; Bowman, J.P.; Fox, E.M. Novel Biocontrol Methods for Listeria Monocytogenes Biofilms in Food Production Facilities. *Front. Microbiol.* **2018**, *9*, 1–12. [CrossRef]
13. The Actinobacteriophage Database. Available online: http://phagesdb.org/ (accessed on 24 April 2019).
14. Lima-Junior, J.D.; Viana-Niero, C.; Conde Oliveira, D.V.; Machado, G.E.; Rabello, M.C.D.S.; Martins-Junior, J.; Martins, L.F.; Digiampietri, L.A.; Da Silva, A.M.; Setubal, J.C.; et al. Characterization of Mycobacteria and Mycobacteriophages Isolated from Compost at the São Paulo Zoo Park Foundation in Brazil and Creation of the New Mycobacteriophage Cluster U. *BMC Microbiol.* **2016**, *16*, 1–15. [CrossRef] [PubMed]
15. Hatfull, G.F. Molecular Genetics of Mycobacteriophages. *Microbiol. Spectr.* **2014**, 2. [CrossRef] [PubMed]
16. Jacobs-Sera, D.; Marinelli, L.J.; Bowman, C.; Broussard, G.W.; Guerrero Bustamante, C.; Boyle, M.M.; Petrova, Z.O.; Dedrick, R.M.; Pope, W.H.; Modlin, R.L.; et al. On the Nature of Mycobacteriophage Diversity and Host Preference. *Virology* **2012**, *434*, 187–201. [CrossRef] [PubMed]
17. Klann, A.G.; Belanger, A.E.; Abanes-de Mello, A.; Lee, J.Y.; Hatfull, G.F. Characterization of the DnaG Locus in Mycobacterium smegmatis Reveals Linkage of DNA Replication and Cell Division. *J. Bacteriol.* **1998**, *180*, 65–72.
18. Domingo-Calap, P.; Cuevas, J.M.; Sanjuán, R. The Fitness Effects of Random Mutations in Single-Stranded DNA and RNA Bacteriophages. *PLoS Genet.* **2009**, *5*, 1–7. [CrossRef]
19. Domingo-Calap, P.; Sanjuán, R. Experimental evolution of RNA versus DNA viruses. *Evolution (NY)* **2011**, *65*, 2987–2994. [CrossRef]
20. Pal, C.; Maciá, M.D.; Oliver, A.; Schachar, I.; Buckling, A. Coevolution with Viruses Drives the Evolution of Bacterial Mutation Rates. *Nature* **2007**, *450*, 1079–1981. [CrossRef]

21. Betts, A.; Gray, C.; Zelek, M.; MacLean, R.C.; King, K.C. High Parasite Diversity Accelerates Host Adaptation and Diversification. *Science* **2018**, *360*, 907–911. [CrossRef]
22. Wichman, H.A.; Badgett, M.R.; Scott, L.A.; Boulianne, C.M.; Bull, J.J. Different Trajectories of Parallel Evolution During Viral Adaptation. *Science* **1999**, *285*, 422–424. [CrossRef]
23. Sanjuán, R.; Domingo-Calap, P. Mechanisms of Viral Mutation. *Cell. Mol. Life Sci.* **2016**, *73*, 4433–4448. [CrossRef]
24. Miralles, R.; Gerrish, P.J.; Moya, A.; Elena, S.F. Clonal Interference and the Evolution of RNA Viruses. *Science* **1999**, *285*, 1745–1747. [CrossRef] [PubMed]
25. Arjan, G.J.; de Visser, M.; Zeyl, C.W.; Gerrish, P.J.; Blanchard, J.L.; Lenski, R.E. Diminishing Returns from Mutation Supply Rate in Asexual Populations. *Science* **1999**, *283*, 404–406. [CrossRef]
26. Kick, B.; Hensler, S.; Praetorius, F.; Dietz, H.; Weuster-Botz, D. Specific Growth Rate and Multiplicity of Infection Affect High-Cell-Density Fermentation with Bacteriophage M13 for ssDNA Production. *Biotechnol. Bioeng.* **2017**, *114*, 777–784. [CrossRef]
27. Sun, C.L.; Barrangou, R.; Thomas, B.C.; Horvath, P.; Fremaux, C.; Banfield, J.F. Phage Mutations in Response to CRISPR Diversification in a Bacterial Population. *Environ. Microbiol.* **2012**, *15*, 463–470. [CrossRef]
28. Deveau, H.; Barrangou, R.; Garneau, J.E.; Labonté, J.; Fremaux, C.; Boyaval, P.; Romero, D.A.; Horvath, P.; Moineau, S. Phage Response to CRISPR-Encoded Resistance in Streptococcus Thermophilus. *J. Bacteriol.* **2008**, *190*, 1390–1400. [CrossRef]
29. Scanlan, P.D.; Buckling, A.; Hall, A.R. Experimental Evolution and Bacterial Resistance: (Co)Evolutionary Costs and Trade-Offs as Opportunities in Phage Therapy Research. *Bacteriophage* **2015**, *5*, e1050153. [CrossRef]
30. Broxmeyer, L.; Sosnowska, D.; Miltner, E.; Chacón, O.; Wagner, D.; McGarvey, J.; Barletta, R.G.; Bermudez, L.E. Killing of *Mycobacterium Avium* and *Mycobacterium Tuberculosis* by a Mycobacteriophage Delivered by a Nonvirulent Mycobacterium: A Model for Phage Therapy of Intracellular Bacterial Pathogens. *J. Infect. Dis.* **2002**, *186*, 1155–1160. [CrossRef] [PubMed]
31. García-Fernández, J.; Papavinasasundaram, K.; Galán, B.; Sassetti, C.M.; García, J.L. Unravelling the Pleiotropic Role of the MceG ATPase in Mycobacterium Smegmatis. *Environ. Microbiol.* **2017**, *19*, 2564–2576. [CrossRef]
32. Piuri, M.; Rondón, L.; Urdániz, E.; Hatfull, G.F. Generation of Affinity-Tagged Fluoromycobacteriophages by Mixed Assembly of Phage Capsids. *Appl. Environ. Microbiol.* **2013**, *79*, 5608–5615. [CrossRef] [PubMed]

antibiotics

MDPI

Article

Bystander Phage Therapy: Inducing Host-Associated Bacteria to Produce Antimicrobial Toxins against the Pathogen Using Phages

T. Scott Brady [1], Christopher P. Fajardo [1], Bryan D. Merrill [1], Jared A. Hilton [1], Kiel A. Graves [1], Dennis L. Eggett [2] and Sandra Hope [1,]*

1 Department of Microbiology and Molecular Biology, Brigham Young University, Provo, UT 84602, USA; thomasscottbrady@gmail.com (T.S.B.); christopher.fajardo@gmail.com (C.P.F.); brymerr921@gmail.com (B.D.M.); thehumanjervis@gmail.com (J.A.H.); kielgraves@gmail.com (K.A.G.)

2 Department of Statistics, Brigham Young University, Provo, UT 84602, USA; theegg@byu.edu

* Correspondence: sandrahope2016@gmail.com; Tel.: +1-801-422-1310

Received: 8 November 2018; Accepted: 3 December 2018; Published: 4 December 2018

Abstract: *Brevibacillus laterosporus* is often present in beehives, including presence in hives infected with the causative agent of American Foulbrood (AFB), *Paenibacillus larvae*. In this work, 12 *B. laterosporus* bacteriophages induced bactericidal products in their host. Results demonstrate that *P. larvae* is susceptible to antimicrobials induced from field isolates of the bystander, *B. laterosporus*. Bystander antimicrobial activity was specific against the pathogen and not other bacterial species, indicating that the production was likely due to natural competition between the two bacteria. Three *B. laterosporus* phages were combined in a cocktail to treat AFB. Healthy hives treated with *B. laterosporus* phages experienced no difference in brood generation compared to control hives over 8 weeks. Phage presence in bee larvae after treatment rose to $60.8 \pm 3.6\%$ and dropped to $0 \pm 0.8\%$ after 72 h. In infected hives the recovery rate was 75% when treated, however AFB spores were not susceptible to the antimicrobials as evidenced by recurrence of AFB. We posit that the effectiveness of this treatment is due to the production of the bactericidal products of *B. laterosporus* when infected with phages resulting in bystander-killing of *P. larvae*. Bystander phage therapy may provide a new avenue for antibacterial production and treatment of disease.

Keywords: American Foulbrood; bacteriophage; phage; phage therapy; *Paenibacillus larvae*; *Brevibacillus laterosporus*; treatment; safety; bystander phage therapy

1. Introduction

Brevibacillus laterosporus is a Gram-positive, spore-forming bacterium that can be found in myriad locations including the gut of honeybees [1–5]. While typically found at low levels in healthy honeybees, the population of *B. laterosporus* often increases as a secondary infection when a hive is infected with *Paenibacillus larvae* or *Melissococcus plutonius*, the causative agents of American Foulbrood and European foulbrood, respectively [6]. American Foulbrood (AFB) is the most devastating bacterial infection in honeybees, killing honeybee larvae and spreading easily from hive to hive within an apiary [7–9]. In the wake of antibiotic resistance in *P. larvae*, novel methods for controlling AFB outbreaks are needed, similar to the need for new approaches to treating antibiotic resistant bacterial infections in general.

Strains of *B. laterosporus* produce potent toxins that can kill a wide range of organisms [5,10,11]. *B. laterosporus* has been used as a bio control agent for decreasing the populations of unwanted bacteria and this method yielded modest results in attempts to control American Foulbrood [12,13]. While typically a symbiote to honeybees [14], *B. laterosporus* can produce toxins with insecticidal properties and certain strains of the bacterium are implicated in causing minor disease in honeybee hives after a primary infection [15–18]. The role of *B. laterosporus* as either a beneficial symbiote or as an opportunistic infector is yet to be fully understood.

Prior to this study, phages that specifically infect *B. laterosporus* were isolated from beehives and the genomes of most have been studied and published [19–21]. In this study, isolated phages were tested against strains of *B. laterosporus* to determine the most effective combination of phages to be included in a final cocktail. During isolation and experimentation, we discovered that when *B. laterosporus* was treated with phages, the bacteria began to produce antimicrobials that kill *P. larvae* when undiluted. These findings led us to believe that *B. laterosporus* phages could be used as a biocontrol for AFB by inducing antimicrobial production to kill *P. larvae*.

The studies presented here show: (1) The host range of identified phages; (2) the phages' presence and persistence in the larval gut after treatment; (3) the phages' ability to induce antimicrobial production compared to other forms of induction; (4) the phages' safety to healthy honeybee hives over time; and (5) the phages' ability to control an active AFB infection. We propose a new approach called "bystander phage therapy" as a method for treating pathogenic bacteria.

2. Results

2.1. Phage Characteristics and Host Range

The genome sequences for all of the phages used in these studies, except for Lauren and Fawkes, were previously sequenced and analyzed [19,21]. GenBank accession numbers for the phage genomes are as follows: Jimmer1-KC595515, Jimmer2-KC595514, Emery-KC595516, Abouo-KC595517, Davies-KC595518, Osiris-KT151956, Powder-KT151958, SecTim467-KT151957, Sundance-KT151959, Jenst-KT151955.

Electron microscopy images of Jimmer1, Jimmer2, Emery, Abouo, Davies, Osiris, and Powder were previously published [19,21]. Figure 1 includes electron microscopy images of the two previously unpublished images of phages used in this study, Lauren and Fawkes, from phage lysates. Figure 1C is an image of Fawkes attached to the side of BL2 *B. laterosporus* field isolate.

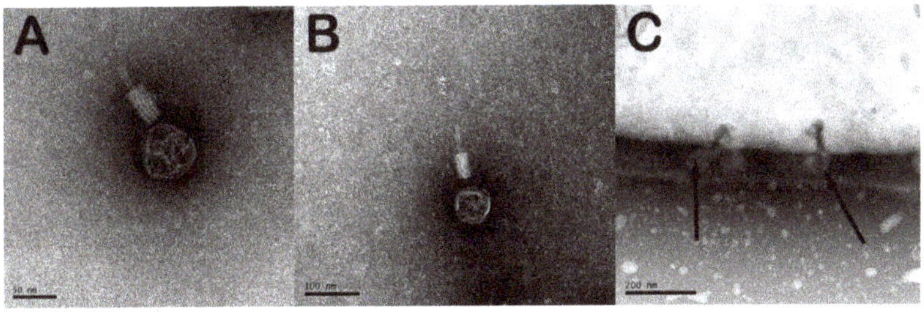

Figure 1. *Brevibacillus laterosporus* phages Lauren and Fawkes. (**A**) Single Lauren phage particle SEM image. (**B**) Single Fawkes phage particle SEM image. (**C**) Fawkes phage particles attached to BL2 bacterium SEM image, arrows point to attached phage particles. Images of the other phages mentioned were previously published by [19] and [21].

Upon isolation, *B. laterosporus* phages were challenged for their ability to infect three field isolates of *B. laterosporus* as well as nine type-strains of *B. laterosporus* from the Bacillus Genetic Stock Center (BGSC) by both spot tests and plaque formation assays. Table 1 indicates bacterial susceptibility to *B. laterosporus* phage infection using *P. larvae* bacteria as a negative control. Emery/Abouo had the largest host range against archived *B. laterosporus* strains, showing infectivity against eight of the 12 strains. Fawkes showed infectivity against seven strains of which three were not covered by Emery/Abouo. None of the tested *B. laterosporus* phages were capable of forming plaques on lawns of 40A4. Furthermore, no plaques formed on *P. larvae* ATCC 9545, a highly phage susceptible strain [22], indicating that the isolated phages are specific to *B. laterosporus* and do not have the ability to cross-infect into *P. larvae*.

Table 1. Host range of *B. laterosporus* phages. Twelve *B. laterosporus* strains and one *P. larvae* strain were challenged with 12 *B. laterosporus* phages. The number of plus signs indicate the level of clearing. A minus sign indicates that no bacterial clearing occurred. BL2–BL14 are our field isolates of *B. laterosporus*, 40A1–40A10 are type strains of *B. laterosporus* from BGSC, and PL ATCC is the type strain of *P. larvae ATCC 9545*.

Phage	BL2	BL6	BL14	40A1	40A2	40A3	40A4	40A5	40A6	40A8	40A9	40A10	PL ATCC
Jimmer1	++++	–	++++	+	–	–	–	–	–	–	–	–	–
Jimmer2	++++	–	++++	+	–	–	–	–	–	–	–	–	–
Osiris	++++	–	++	++	+	+	–	+	++	–	++	+	–
Fawkes	++++	–	++	+++	+	–	–	–	+	–	++++	++	–
Lauren	++++	–	++++	+	–	–	–	–	+	–	+	+	–
Powder/ Sundance	+++	–	+++	+++	–	+	–	–	–	–	+++	–	–
SecTim467	+++	++	+++	++	+	–	–	–	–	–	+++	–	–
Jenst	–	++++	–	+	–	–	–	–	+	–	+++	–	–
Davies	–	++++	–	++++	++	+	–	+++	+++	+++	–	–	–
Emery/ Abouo	–	++++	–	++++	++++	+	–	+++	+++	++	+++	–	–

Underlines designate the bacteria used for phage isolation.

2.2. *Phage Persistence in the Larval Honeybee*

This study aimed to determine whether phages would reach the larval gut and how long the phages would persist in a larval gut. Five hives were previously established in a single apiary and each hives' brood racks (with the worker bees covering the brood) were sprayed with *B. laterosporus* phage lysate suspended in sugar water. One hundred larval specimens were collected from each hive at spaced time points and were tested for the presence of viable phages, see Figure 2. The first samples were collected at time 0 immediately prior to treatment with the phage cocktail to establish a baseline for the presence of naturally occurring phages in honeybee larvae. Phage persistence studies showed that phage presence in bee larvae was 1.5 $\pm$ 0.8% before treatment and rose to 58.8 $\pm$ 3.2% 15 min after treatment, 60.8 $\pm$ 3.6% after 3 hours, 52.2 $\pm$ 1.8% after 24 h, 44.9 $\pm$ 1.8% after 48 h, and 0 $\pm$ 0.8% after 72 h. Phages were found in larvae within 15 min of the treatment and peaked at 3 hours where 60.8 $\pm$ 3.6% of larvae contained detectible, viable phages as determined by spot test. Phage presence in bee larvae remained well above the normal untreated control for 2 days after the treatment was administered. After 3 days, the phage presence returned to the normal nominal levels.

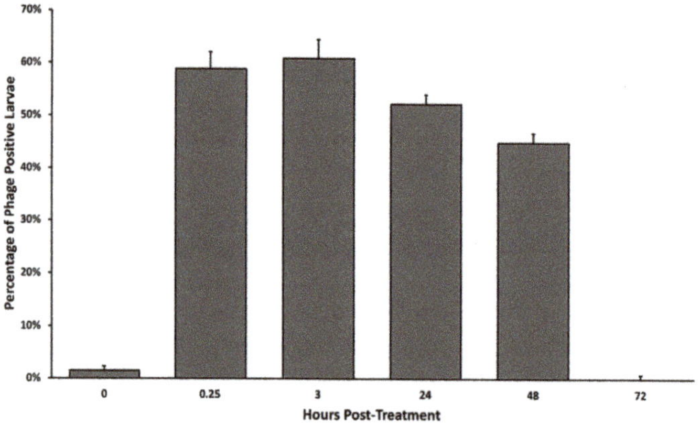

Figure 2. Average presence of phages in larvae samples after treatment. The time 0 sampling was taken just prior to initial treatment to serve as a baseline. Bees and racks were sprayed with phages and larvae were plucked from the racks at each timepoint and tested for the presence of phages.

2.3. Phage Infection Induces B. laterosporus to Produce Antimicrobials

During culture of *B. laterosporus* phages, we observed that bacterial lawns exhibited clearing from phage plaques as well as a diffusion of a bacterial component in the vicinity of a plaque. An experiment was designed to characterize the effects of *B. laterosporus* phage on the production/release of toxins from *B. laterosporus*. Strains BL-2 and BL-6 were infected with the phages Fawkes and Emery/Abouo, respectively in duplicate. The resulting lysates were filtered and three µLs spotted onto lawns of different bacteria. Antimicrobial activity was qualified by the creation of a hole in the bacteria on the plate indicating cell die off distinguished between plaques from phages by observing the shape and size of the clearing (Figure 3, Table 2). Lysates from Fawkes and Emery/Abouo both contained antimicrobial products that were lethal to BL-2, BL-6, *P. larvae* ATCC 9545, and *E. coli* MG1655. Neither lysate type was effective against *Agrobacterium tumefaciens* or *Sinorhizobium meliloti*.

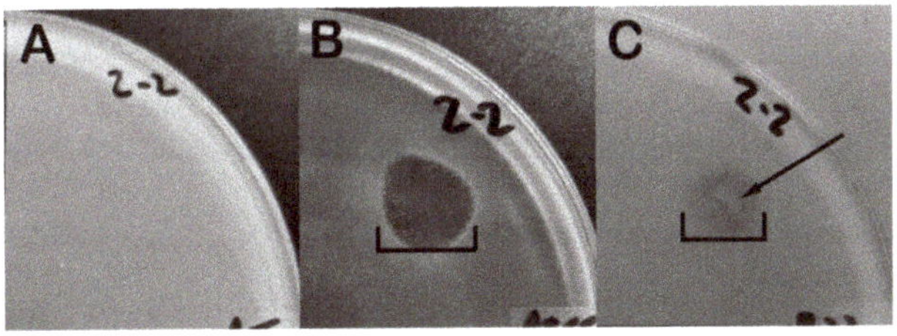

Figure 3. *B. laterosporus* antimicrobial product spot test. Drops of *B. laterosporus* phage lysate were placed and incubated for 24 h onto (**A**) a lawn of *A. tumefaciens* that did not respond to the antimicrobial product or generate plaque clearings, (**B**) a lawn of *P. larvae* that exhibited antimicrobial death, and (**C**) a lawn of *B. laterosporus* strain BL2 that showed antimicrobial death as well as phage infection formation. Brackets indicate antimicrobial clearing, arrow indicated phage plaque formation.

Table 2. Bacterial susceptibility to *B. laterosporus* antimicrobial products. *P. larvae, E. coli, A. tumefaciens, S. Meliloti,* and two strains of *B. laterosporus* were challenged with the supernatant from two phage lysates and the supernatant of live, dead, and mechanically lysed *B. laterosporus*. Antimicrobial-induced death is indicated by plus signs. A minus sign indicates no discernable antimicrobial clearing on the bacterial lawn.

Source Tested	B. Laterosporus (BL-2)	B. laterosporus (BL-6)	P. larvae	E. coli	A. tumefaciens	S. Meliloti
Emery/Abouo Phage lysate (BL-6)	+++	++ *	++++	+	–	–
Fawkes Phage lysate (BL-2)	++ *	+++	++++	+	–	–
Supernatant of live *B. Laterosporus*	–	–	–	–	–	–
Supernatant of UV killed *B. Laterosporus*	–	–	–	–	–	–
Supernatant of mechanically lysed *B. Laterosporus*	–	–	–	–	–	–

* Phage plaques were discernable on the bacterial lawns as well as death from antimicrobial products.

These data indicate the sensitivity of *P. larvae* to the antimicrobial product generated by *B. laterosporus*, and that it has limited killing against other bacteria.

Control samples recreated various stages of the phage life cycle to verify phage-induced antimicrobial production as opposed to products release from other mechanisms. Supernatant from UV killed bacteria was spotted onto lawns of bacteria to identify if bacterial death alone induces antimicrobial production. The supernatant from mechanically lysed bacteria was also tested to determine whether phage lysis releases antimicrobial products present in the bacterial cytoplasm. Supernatant from untreated vegetative *B. laterosporus* was also tested to identify whether unprovoked bacteria releases antimicrobial products. None of the control sample supernatants formed holes in bacterial lawns, indicating that these mechanisms did not result in any production or release as seen in Figure 3. The lack of antimicrobial production via UV killing and mechanical lysis indicates that the bactericidal produced by *B. laterosporus* is not a result of bacterial death or lysis. Phage-induced antimicrobial production may be the result of expression of genes encoded by the bacteria since no known antimicrobial genes reside in the sequenced phage genomes while several have been identified in *B. laterosporus* [5,10]. The fact that more than one genetically unique *B. laterosporus* phage can induce the bacteria to make an identically-acting antimicrobial (Table 3 and [19]), may further suggest that the product arises from the bacterial genome instead of the phage genome.

2.4. Phage-Induced B. laterosporus Antimicrobial Products Shows Inert Characteristics Against Honeybees

This study aimed to determine whether phage treatment for *B. laterosporus* would be problematic for honeybees. Since *B. laterosporus* has been suggested previously to be a commensal to honeybees, this study was conducted to observe if side effects of phage-induced toxin or phage killing of *B. laterosporus* in the bee gut would decrease the overall health of the hives. Twelve hives, six in a test group and six in a mock-treated group, were installed into new boxes with new frames in spring. New queens and approximately 1.1 kg of honeybees were installed into each box and weekly inspections were made to follow the bees' progress by observing the number of bees in the spaces between racks. Hives were allowed to become established for 9 weeks before receiving phage or mock treatments. Populations in all treated and untreated hives stayed below four full racks through early-summer. In mid-summer, the bees began to expand to fill the fourth rack, at which point the phage treatment commenced.

All 12 hives were treated three times at weeks nine and eleven, with 3 days between treatments for each regiment. Our data showed that all hives expanded at approximately the same rate during the study, see Figure 4. There was no statistical difference between the expansion of the bee populations in mock-treated controls versus the phage-treated group. The data was evaluated statistically using the repeated measures, mixed procedure, two-tailed analysis of the number of bee-filled spaces in the treated, and control hives over the 17-week period using an alpha level $\alpha = 0.05$ (p-value of 0.1104). These data indicate that antimicrobial products in the phage lysate treatment was sufficiently low or not active on honeybees. Further, it shows that the hives were either lacking *B. laterosporus* and thus this bacterium is not essential for honeybee health, and/or that any antimicrobials or killing from phage infection of *B. laterosporus* does not adversely affect honeybee expansion.

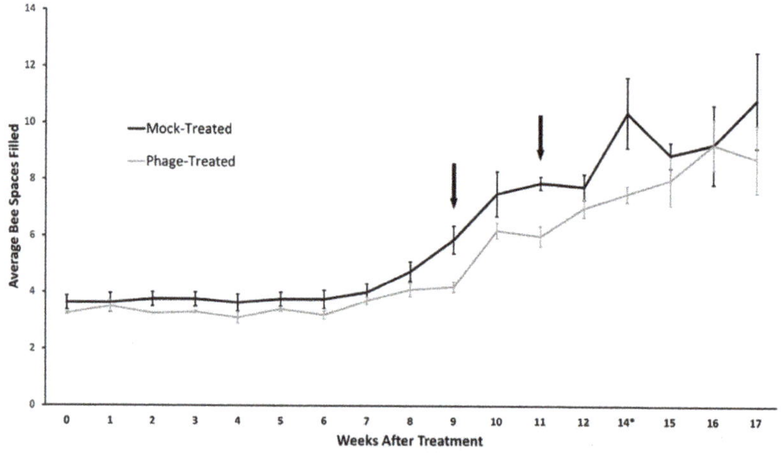

Figure 4. Colony expansion after phage treatment in beehives. New packets of bees with a fertilized queen were allowed to establish in new hives. Arrows indicate when phage treatments were administered to the bees and results demonstrate that healthy hives treated with *B. laterosporus* phage cocktail exhibited no difference in colony expansion when compared to healthy control hives. Bee spaces indicate honeybee population within the hive. Data not collected for week 13.

2.5. *B. laterosporus* Phages Can Effectively Treat an Active AFB Infection

The objective of this experiment was to determine the effectiveness *B. laterosporus* phages in curing honeybee hives of American Foulbrood caused by *P. larvae*. Forty hives of honeybees (*Apis mellifera*) were previously established in one apiary. Of the 40 colonies, 12 presented with American Foulbrood, the remaining 28 colonies were relocated to prevent the spread of the disease to the remaining healthy hives. Government regulation requires immediate treatment or destruction of known American Foulbrood-infected hives and any hives potentially exposed during an outbreak. Beekeepers are allowed to treat sick bees for 2 weeks at which point recovered hives can be kept and any hives with signs of the disease must be burned. Within these regulatory parameters, the 12 sick hives were all assigned to receive phage therapy and were followed for 2 weeks. All 12 sick hives were treated three times (each treatment was given 3 days apart) by spraying each rack on both sides with *B. laterosporus* phages in sugar water. The remaining 28 hives appeared healthy and were treated with antibiotics by the beekeeper. Treatment of the beehives occurred immediately before the onset of winter.

All 40 hives were inspected 2 weeks after the first treatment. Of the 12 infected hives treated with *B. laterosporus* phages, 9 recovered and showed no signs of AFB upon inspection at week two, which indicates a 75% cure rate (see Table 3). Of the 28 originally uninfected hives, two appeared healthy at the beginning of the study but were diagnosed with AFB at the two-week inspection. The two hives had received antibiotic treatments along with the other 26 healthy hives; however, both hives collapsed with severe signs of AFB. It is anticipated that the hives were in very early stages of infection at the time of the outbreak and were missed as infected during hive assessment at the beginning of the study. The complete AFB destruction of these two untreated hives within 2 weeks of the first inspection of the study indicates that the 12 diagnosed-infected hives which received bystander phage therapy from that apiary during that outbreak were infected with a strain of *P. larvae* that was both antibiotic resistant and lethal. Dead-out hives were burned. No further problems were reported with the other 26 hives healthy antibiotic treated hives. From the 12 sick beehives, dead larval samples were taken from the hives before the first phage treatment and healthy larvae were taken at 2 weeks post-phage treatment (healthy larvae were taken post-treatment because no dead larvae were observed). The larval samples were analyzed by PCR for the presence of bacteria as described in the Materials and Methods Section 4.1. Results of PCR confirmed the presence of *P. larvae* and *B. laterosporus* DNA at pre-treatment, and only *B. laterosporus* DNA with no amplification of *P. larvae* DNA at post-treatment from the larval samples. The nine hives that recovered from AFB were followed through winter. In spring, of the nine recovered hives, five survived, four died. No signs of AFB were found in the four dead hives; three of the hives appeared to have frozen to death, and one hive was destroyed by vandals.

Table 3. Survival rate of hives after treatment in fall and after winter. Infected hives received phage cocktail treatments and uninfected hives were prophylactically treated with antibiotics. Survival rates of the hives were evaluated after 2 and 16 weeks.

Hive Status	Total Hives	AFB-Free Post-Treatment	Hive Survival Over Winter
Uninfected hives	28	92.85% *	78.1% †
AFB infected hives	12	75%	62.5%

* Two hives of the 28 uninfected became infected when they were removed from the initial 12 infected hive. † Results excluding the two hives that became infected with AFB.

2.6. *B. Laterosporus* Phages Do not Prevent Reinfection by Latent *P. larvae* Spores

The five recovered, surviving hives were followed for 9 months to investigate the effectiveness of the *B. laterosporus* phage cocktail in the inactivation of latent *P. larvae* spores, see Table 4. Two weeks after phage treatment as well as in the following spring, 16 weeks after the first treatment, all five hives had no signs of AFB infection. At 18 weeks, one of the five surviving hives experienced an AFB infection and that hive recovered after another treatment of *B. laterosporus* phages. At each time of recurrence, all five hives in the apiary were preemptively treated with the phage cocktail. At week 22, the first hive and a second hive experienced symptoms of AFB, which were again treatable with *B. laterosporus*, signs of AFB disappearing within a week of the treatment. By week 26, all five hives presented with AFB symptoms and were treated with the phage cocktail, which again cleared all of the hives of AFB symptoms with 2 weeks. All hives were destroyed mid-summer due to the reoccurring infections. These data indicate that *B. laterosporus* phage treatment could kill active *P. larvae* infections, but could not kill *P. larvae* spores nor prevent future infection.

Table 4. Health of five hives after and AFB outbreak, *B. laterosporus* phage treatment, and overwintering. The surviving hives after *B. laterosporus* phage treatment were monitored for 28 weeks after initial treatment. When hives were seen to relapse, all hives were retreated with phage cocktail.

Hive Status	Week 16	Week 18 *	Week 20	Week 22 *	Week 24	Week 26 *	Week 28
Healthy	5	4	5	3	5	0	5
AFB+	0	1	0	2	0	5	0

* Weeks when phage cocktail was administered.

3. Discussion

The phage cocktail used in these studies was formulated to specifically infect a wide range of *B. laterosporus* field isolate strains. As seen in [23] and [24], phage cocktails designed in this manner (with phages in the cocktail selected according to the ability to kill as many field strains as possible) are effective at reducing the amount of their target bacteria. Here, we observed that the phages selected for a cocktail using laboratory-generated data of phage efficacy was predictive of the efficacy of phages in field tests, as observed by the reduction in the amount of *P. larvae* DNA present in hives and recovery of 75% of the hives with 2 weeks of bystander phage treatment. Furthermore, we studied the antimicrobial-inducing capabilities in the laboratory to observe whether or not the antimicrobials could be effective at reducing *P. larvae* bacteria, and then applied our results to safety and efficacy studies in the field.

Using *B. laterosporus* phages as a biocontrol comes with some inherent risk. We were concerned to know whether, by inducing antimicrobial synthesis and lysing *B. laterosporus*, the phage cocktail could release toxins with insecticidal properties or other adverse effects in honeybees. No such deleterious effects were seen in our studies. Firstly, we observed rapid loss of detectable phages in healthy larvae which indicates that a phage treatment has a relatively short exposure time to the bees. Secondly, we observed no short-term or long-term harm to healthy honeybees treated with multiple doses of phages. These studies add to the expanding literature that indicates that phage cocktails are a safe alternative to traditional antibiotic use [25–31].

The results of our studies further indicate that *B. laterosporus* is not a necessary symbiote for honeybee health, which conclusion is contrary to the postulations of several other researchers [11,12] but supports reports by others [5]. The current field of research surrounding *B. laterosporus* is tempestuous as to its merits and disadvantages. However, the research conducted in this article is uniquely equipped to demonstrate the effects of beehives with and without *B. laterosporus* in vivo and the results indicate that there are no significant differences between hives with or without the bacteria. This study also demonstrates advantages to having the bacteria naturally present and using phages to induce antimicrobials to kill pathogenic bacteria.

One aim of our studies was to determine whether a phage cocktail designed for a co-infecting of commensal bacteria (*B. laterosporus*), could reduce the presence of a pathogenic bacteria (*P. larvae*), during a disease state (AFB infection). Figure 5 depicts this new "bystander phage therapy" as a phage treatment approach compared to the current dogma of phage therapy. Such situations may be more common than just this *B. laterosporus/P. larvae* system because co-existing, non-pathogenic bacteria may evolve to secrete antimicrobials in order to out-compete a pathogenic bacteria. The "bystander" bacteria may be poised to produce antimicrobials under stress as we were able to do using phage infection. It is important to note that none of our *B. laterosporus* phages could infect *P. larvae*. Therefore, any activity of a cocktail of *B. laterosporus* phages against AFB must be either from antimicrobial release to induce bystander killing of AFB or that *B. laterosporus* is responsible for AFB. We do not believe the latter is true. It was already known that *B. laterosporus* can produce antimicrobial toxins [10,32] and results from our laboratory experimentation demonstrate that these compounds are effective against *P. larvae* as well as other unrelated bacteria. In addition, we demonstrated that the phage cocktail can clear an active AFB infection but is not curative as observed by the recurrent infections. We hypothesize that the antimicrobials released by the phage when infecting *B. laterosporus* are effective against the vegetative bacteria that infect the larval brood, but that the antimicrobials are not able to eradicate *P. larvae* spores. *P. larvae* spores resistant to the antimicrobial toxins, unfortunately indicate that in this system bystander phage therapy is not a preventative tool for AFB but a treatment for clearing active infection, which is true of current antibiotic treatments for AFB which also leave *P. larvae* spores intact and viable.

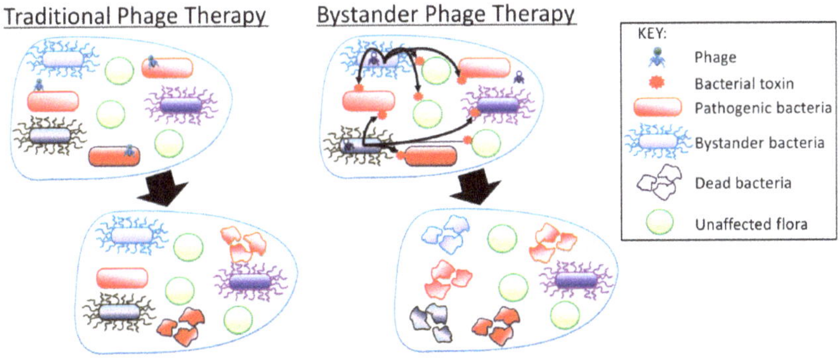

Figure 5. Mechanism of pathogen killing using phage therapy versus bystander phage therapy. In traditional phage therapy, phages against a pathogenic bacterium bind and lyse come bacterial strains, but may leave others unscathed (Left Panel). In Bystander Phage Therapy, phages against a bystander induce the bystander to make a toxin that kills all versions of the pathogenic bacteria while leaving an untouched population of itself that was not infected by phages (Right Panel).

Furthermore, our results show that the antimicrobial from *B. laterosporus* was capable of killing vegetative *B. laterosporus*, albeit at a much lower sensitivity than killing of *P. larvae*. Typically, bacteria are good at defending themselves against their own agents while having potency against others. In this instance, we do see preference of the toxin against the pathogen but still some potential loss of the bystander. This discrepancy may be due to the fact that the toxin does not affect spores. *B. laterosporus* is also a spore-former, so its survival strategy may have evolved to release an antimicrobial that is effective against other bacteria at the risk of killing a few self-bacteria with the confidence that the majority of its own will survive as well as self-survival of its spores. This approach would explain the lack of toxicity to *P. larvae* spores as well. In application of bystander phage therapy to other bacterial commensal/pathogen systems, the lack of complete killing of the pathogen may not be an issue if the commensal bystander is not a spore-former or if the pathogenic bacteria is not a spore former.

Bystander phage therapy has an advantage over typical phage therapy because the range of targets affected by the antimicrobials can be much greater than traditional phage therapy that has limited host range. For instance, bystander phage therapy does not rely on the phage killing all of its targets. Rather, the phages only need to infect and induce enough antimicrobials to kill the pathogen. By this method, a hive could be infected with several strains of *P. larvae* that could include phage resistant *P. larvae* because of the limited host range of the individual phages, but the bacteria could still be killed by the phage-induced *B. laterosporus* antimicrobial product. This bystander effect could occur regardless of whether or not all strains of the non-pathogenic bacteria (*B. laterosporus*) are killed.

An option not to kill all target bacterium is useful and desirable for a phage therapy approach because it means that the cocktail for bystander phage treatment would not need to include phages to kill every possible bacterial strain of its target. This simplifies the cocktail itself, and increases the chances of the treatment being functional since it is not dependent on killing all bystanders, but simply on activating the bystander to kill the pathogen. Bystander phage therapy may be a useful as an added component of phages in a traditional phage cocktail and/or in combination with antibiotics. Others have indicated the need for clinical phage treatments that include co-treatment of both phages and antibiotics at the time to restore antibiotic function as well as attempt complete elimination of all life-threatening bacteria [33]. Co-treatment of traditional phage and antibiotics may well indicate that bystander phages could step into the position of the antibiotic and a traditional/bystander cocktail of phages could be a highly effective therapeutic approach.

Due to the nature of the antimicrobial effects of products made by *B. laterosporus*, bystander phage therapy could function as treatment against other bacterial infections in beehives such as *Melissococcus plutonius*, the causative agent of European Foulbrood. If the phage-induced antimicrobial products are lethal to other pathogens such as *M. plutonius*, then it would be an attractive alternative to standard phage therapies because of its ability to treat various diseases. This approach is especially helpful in the case of misdiagnoses of the pathogen causing foulbrood in a hive. For instance, *B. laterosporus* is often found in the hive regardless of whether foulbrood disease is due to the European or American Foulbrood pathogens (*M. plutonius* or *P. larvae*, respectively). Therefore, a bystander treatment using phages against *B. laterosporus* could be effective in both instances. Furthermore, *M. plutonius* is difficult and expensive to culture due to anaerobic requirements, which presents a barrier to lab work that would otherwise lead to phage isolation for traditional phage therapy against European Foulbrood. This exemplifies a situation where bystander phage therapy is a sensible method to pursue as a phage treatment since the bystander is easy to grow in aerobic conditions and could be used to treat more than one bacterial pathogen whose pathogenic presentation is very similar. Such approaches can be applied to many infectious bacterial systems. By inducing bystander bacteria to produce an antimicrobial, phages can remain a treatment option even for difficult-to-culture bacteria.

4. Materials and Methods

4.1. Gathering B. laterosporus Field Isolates

Samples of honey and hive material were gathered from local apiaries and used for bacterial isolation. Samples were processed as described previously intended for *P. larvae* isolation [19,34] and isolated bacterial colonies were identified as *P. larvae* or *B. laterosporus* by PCR. Specifically, bacteria were initially streaked on *Paenibacillus larvae agar* PLA agar [35] and incubated at 37 °C. Catalase negative [36] and Gram-positive colonies were streaked on LB agar (Becton, Dickinson and Company, Sparks, MD, USA), gathered, archived in 20% glycerol, and stored at −80 °C. Bacteria were confirmed as *B. laterosporus* by PCR amplification of the *B. laterosporus rpo*B gene, see Table 5. Samples were also PCR tested with primers specific for *P. larvae* rpoB and *fts*A to confirm the presence of *P. larvae* [37]. Prior to PCR, bacterial samples were streaked out to single colonies. Template DNA for PCR was extracted by adding part of a colony to 50 μL of water in a PCR tube and incubating it at 100 °C for 10 min. The total PCR reaction volume was 25 μL composed of 22 μL standard PCR reagents (New England Biolabs, Ipswich, MA, USA) plus 3 μL of template DNA. After 30 cycles, PCR products were run in an agarose gel to confirm amplification. Amplicons from the reactions were sequenced using BigDye (Life Technologies, Carlsbad, CA, USA). MEGA6 was used to match sequence results with bacterial genomes.

Table 5. Primer List. Primers used for amplification and sequencing of *rpo*B, *fts*A, and 16S rRNA genes of *B. laterosporus* and *P. larvae*. Results were used to positively identify bacterial isolates from beehives.

Primer	Sequence	Direction	Purpose	Reference
27F 907R	5′-AGAGTTTGATCMTGGCTCAG-3′ 5′-CCGTCAATTCMTTTRAGTTT-3′	Forward Reverse	16S rRNA universal primer	[38]
BLrpoB-F BLrpoB-R	5′-GCAGGTAAACTGGTCCAGAGCG-3′ 5′-CACCTGTTGATTTATCAATCAGCG-3′	Forward Reverse	*B. laterosporus rpo*B	-
KAT1 KAT2	5′-ACAAACACTGGACCCGATCTAC-3′ 5′-CCGCCTTCTTCATATCTCCC-3′	Forward Reverse	*P. larvae* ERIC-1 or ERIC-2	[39]
PLrpoB-F PLrpoB-R	5′-ATAACGCGAGACATTCCTAA-3′ 5′-GAACGGCATATCTTCTTCAG-3′	Forward Reverse	Amplifies *P. larvae rpo*B	[40]
PLftsA-F PLftsA-R	5′-AAATCGGTGAGGAAGACATT-3′ 5′-TGCCAATACGGTTTACTTTA-3′	Forward Reverse	Amplifies *P. larvae fts*A	[40]
ERIC1R ERIC2	5′-ATGTAAGCTCCTGGGGATTCAC-3′ 5′-AAGTAAGTGACTGGGGTGAGCG-3′	Forward Reverse	Generates multiple amplicons to fingerprint the bacteria tested	[41]

4.2. Isolating Phages Specific for *B. laterosporus*

B. laterosporus phages were isolated from bee debris collected near beehives. Bee debris was crushed and added to a flask containing LB broth and a field isolate of *B. laterosporus*. The bee debris and bacteria were incubated overnight at 37 °C. The mixture was spun in a centrifuge and the supernatant was passed through a 0.22 µm filter. A total of 50 µL of the supernatant were incubated at room temperature with 500 µL of *B. laterosporus* bacteria for 30 to 60 min, mixed with LB top agar, plated on LB agar, and incubated at 37 °C overnight. Plaques that appeared were isolated and re-plated a minimum of three times to purify individual phages.

4.3. Host Range and Phage Presence Testing for Isolated Phages

B. laterosporus bacterial strains were tested for phage susceptibility using a plaque formation assay and a spot test assay. For the plaque formation assay, phage lysate was incubated at room temperature with 500 µL of an overnight culture of bacteria for 30 min, plated in 0.8% LB top agar, and incubated overnight at 37 °C. For the spot test assay, 500 µL of an overnight culture of bacteria was plated in 0.8% top agar. After the top agar hardened, 3 µL of phage lysate was placed on the top agar. The plates were incubated agar side facing up overnight at 37 °C.

Phage detection in bee larvae was performed by taking one hundred larval samples at each time point and homogenizing them in 500 µL of LB broth in a 1.7 mL microcentrifuge tube for approximately 1 min. Three µL of the larval homogenate was spotted and incubated on plates *B. laterosporus* strains BL2 and BL6 were plated in top agar as described above.

4.4. Electron Microscopy

Phages were prepared for electron microscopy by incubating carbon-coated copper grids with 50 µL of high-titer lysate for 90 seconds, wicking away moisture, incubating with 50 µL of 2% phosphotungstic acid (pH = 7) for 90 seconds, wicking away moisture, and then allowing the grids to air dry prior to imaging. Electron micrographs were taken by the BYU Microscopy Center, and images were measured using ImageJ [42].

4.5. Creation of Bacterial Lysate to Test for *B. Laterosporus* and Phage Cocktail Treatments

Field isolates of *B. laterosporus*, BL-2 and BL-6, were reconstituted from freezer stock by plating onto Porcine Brain Heart Infusion (PBHI) (Acumedia, Lansing, MI) plates and incubating at 37 °C for 48 h. The resulting colonies were streaked for pure culture and incubated at 37 °C overnight. Fawkes and Emery/Abouo were brought out from freezer stock by streaking onto Porcine Brain-Heart Infusion (PBHI) plates with a lawn of *B. laterosporus* in agar incubated at 37 °C overnight. Picked plaques were grown in liquid culture with overnight growths of *B. laterosporus* to generate a high titer lysate. The lysates were centrifuged at 4000*g* for 30 min to pellet bacterial debris and then filtered (0.45 μm). The controls had no phage added and were processed the same to collect mock lysate.

Overnight cultures of *B. laterosporus* BL-2/BL-6, *P. larvae* ATCC 9545, *Agrobacterium tumefaciens* field isolate, *Sinorhizobium meliloti* field isolate, and *E. coli* MG1655 were plated using top agar onto plates of their respective media. Spot assays were conducted on bacterial lawns using three μL of lysate and incubating overnight. *A. tumefaciens* and *S. meliloti* samples were incubated at 30 °C and all other cultures were incubated at 37 °C.

Phages in the cocktail were generated as described above and then precipitated with polyethylene glycol (PEG) (Spectrum, New Brunswick, NJ) at 10,000*g* for 15 min at 4 °C to obtain a pure phage stock devoid of antimicrobial products. The cocktail was applied to the hives using a spray comprised of phage lysate diluted in a 1:1 sugar/water solution. Control hives received 340 mL of sugar water, while the phage treated hives received 320 mL of sugar water with 50 mL of phages containing a titer of 10^8 mixed into the sugar water.

4.6. Phage Beehive Parameters

In studies beginning with healthy hives, each had a viable laying queen, approximately 40,000 or more adult worker bees, uncapped brood, and no visible signs of American Foulbrood. Sick hives treated in Sections 2.5 and 2.6 were identified by a local beekeeper and experimental treatment was approved through the Utah Department of Food and Agriculture.

Population growth was determined in each of the hives based on the amount of racks the bees occupied. A rack was considered full when the space between the racks was fully crowded. In Section 2.4 the phage treatment started once all 12 of the hives achieved at least four fully occupied racks.

4.7. Statistics

The BYU statistical center analyzed the collected data to generate p-values, standard deviation, standard error, and to determine statistical significance. Statistical analysis included repeated measures, mixed procedure, two-tailed analysis using the Fisher's exact test for 2 × 2 contingency tables with $\alpha = 0.05$.

5. Conclusions

Phage therapies are an attractive alternative to traditional antibiotic use in the face of antibiotic resistance in pathogens. This study presents bystander phage therapy as a new alternative approach for phage therapy. The phages used in this study did not target the pathogen causing the disease that it treated, but rather targeted a known co-infecting bacterium and induced the co-infecting bacteria to produce antimicrobial products to which the pathogen is sensitive.

The properties of phage-induced antimicrobials produced by *B. laterosporus* can be characterized to establish the extent of their host range. This research demonstrated that phages can induce *B. laterosporus* to produce antimicrobial products and demonstrated how phages that kill bystander bacteria can also result in killing of off-target, pathogenic bacteria. This approach could be useful as a single treatment for different diseases caused by different pathogens with overlapping symptoms provided that the phage-induced antimicrobial products can kill both pathogens, and that the loss of the antimicrobial-producing bystander bacteria is not vital to the organism. In this case, *B. laterosporus* is not a vital commensal and treatment of healthy bees with *B. laterosporus* phages did not result in any detectable health consequences in the bees. Use of *B. laterosporus* phages rescued a significant number of sick hives from succumbing to an antibiotic-resistant form of AFB. The use of bystander phage therapy is an exciting and new avenue of study that merits further investigation in the field of phage research.

6. Patents

System and Method for Treating a Disease or Bacterial Infection—Bystander Phage Therapy BYU#2018-037

Author Contributions: Conceptualization, S.H., B.D.M., T.S.B., C.P.F.; methodology, S.H., B.D.M., T.S.B., C.P.F.; software, B.D.M.; validation, S.H., D.L.E.; formal analysis, D.L.E.; investigation, T.S.B., C.P.F., B.D.M., J.A.H., K.A.G.; resources, S.H.; data curation, T.S.B., C.P.F., B.D.M., J.A.H., K.A.G., S.H.; writing—original draft preparation, T.S.B.; writing—review and editing, C.P.F.; visualization, T.S.B., C.P.F.; supervision, S.H.; project administration, S.H.; funding acquisition, S.H.

Funding: This project was funded by the BYU Department of Microbiology & Molecular Biology, the BYU College of Life Sciences, and a Technology Transfer Grant from the BYU Technology Transfer Office. Further funding also came from an undergraduate Office of Research and Creative Activities (ORCA) grant obtained by co-author Bryan Merrill.

Acknowledgments: The authors thank the students and faculty of the Brigham Young University (BYU) Department of Molecular & Microbiology in the Phage Hunters Research Laboratory for their work and assistance. Thanks also to Dr. Dennis Eggett of the BYU Department of Statistics for statistical assistance and Dr. Michael Standing of the BYU Microscopy Center. We also thank local and distant beekeepers as well as Joey Caputo and Stephen Stanko at the Utah Department of Food and Agriculture.

Conflicts of Interest: BYU holds intellectual properties and patent disclosures related to this research.

References

1. Roman-Blanco, C.; Sanz-Gomez, J.J.; Lopez-Diaz, T.M.; Otero, A.; Garcia-Lopez, M.L. Numbers and species of Bacillus during the manufacture and ripening of Castellano cheese. *Milchwissenschaft-Milk Sci. Int.* **1999**, *54*, 385–388.
2. Khan, M.R.; Saha, M.L.; Afroz, H. Microorganisms associated with gemstones. *Bangladesh J. Bot.* **2001**, *30*, 93–96.
3. De Oliveira, E.J.; Rabinovitch, L.; Monnerat, R.G.; Passos, L.K.J.; Zahner, V. Molecular characterization of *Brevibacillus laterosporus* and its potential use in biological control. *Appl. Environ. Microbiol.* **2004**, *70*, 6657–6664. [CrossRef] [PubMed]
4. Suslova, M.Y.; Lipko, I.A.; Mamaeva, E.V.; Parfenova, V.V. Diversity of cultivable bacteria isolated from the water column and bottom sediments of the Kara Sea shelf. *Microbiology* **2012**, *81*, 484–491. [CrossRef]
5. Ruiu, L. *Brevibacillus laterosporus*, a Pathogen of Invertebrates and a Broad-Spectrum Antimicrobial Species. *Insects* **2013**, *4*, 476–492. [CrossRef]
6. Alippi, A.M.; Lopez, A.C.; Aguilar, O.M. Differentiation of *Paenibacillus larvae* subsp larvae, the cause of American foulbrood of honeybees, by using PCR and restriction fragment analysis of genes encoding 16S rRNA. *Appl. Environ. Microbiol.* **2002**, *68*, 3655–3660. [CrossRef]
7. Genersch, E. American Foulbrood in honeybees and its causative agent, *Paenibacillus larvae*. *J. Invertebr. Pathol.* **2010**, *103*, S10–S19. [CrossRef]
8. Pohorecka, K.; Skubida, M.; Bober, A.; Zdanska, D. Screening of Paenibacillus Larvae Spores in Apiaries from Eastern Poland. Nationwide Survey. Part I. *Bull. Vet. Inst. Pulawy* **2012**, *56*, 539–545. [CrossRef]
9. Ebeling, J.; Knispel, H.; Hertlein, G.; Funfhaus, A.; Genersch, E. Biology of *Paenibacillus larvae*, a deadly pathogen of honey bee larvae. *Appl. Microbiol. Biotechnol.* **2016**, *100*, 7387–7395. [CrossRef]
10. Yang, X.; Huang, E.; Yuan, C.H.; Zhang, L.W.; Yousef, A.E. Isolation and Structural Elucidation of Brevibacillin, an Antimicrobial Lipopeptide from *Brevibacillus laterosporus* That Combats Drug-Resistant Gram-Positive Bacteria. *Appl. Environ. Microbiol.* **2016**, *82*, 2763–2772. [CrossRef]
11. Khaled, J.M.; Al-Mekhlafi, F.A.; Mothana, R.A.; Alharbi, N.S.; Alzaharni, K.E.; Sharafaddin, A.H.; Kadaikunnan, S.; Alobaidi, A.S.; Bayaqoob, N.I.; Govindarajan, M.; et al. *Brevibacillus laterosporus* isolated from the digestive tract of honeybees has high antimicrobial activity and promotes growth and productivity of honeybee's colonies. *Environ. Sci. Pollut. Res. Int.* **2017**, *25*, 10447–10455. [CrossRef] [PubMed]
12. Alippi, A.M.; Reynaldi, F.J. Inhibition of the growth of *Paenibacillus larvae*, the causal agent of American foulbrood of honeybees, by selected strains of aerobic spore-forming bacteria isolated from apiarian sources. *J. Invertebr. Pathol.* **2006**, *91*, 141–146. [CrossRef] [PubMed]

13. Saikia, R.; Gogoi, D.K.; Mazumder, S.; Yadav, A.; Sarma, R.K.; Bora, T.C.; Gogoi, B.K. *Brevibacillus laterosporus* strain BPM3, a potential biocontrol agent isolated from a natural hot water spring of Assam, India. *Microbiol. Res.* **2011**, *166*, 216–225. [CrossRef] [PubMed]
14. Marche, M.G.; Mura, M.E.; Ruiu, L. *Brevibacillus laterosporus* inside the insect body: Beneficial resident or pathogenic outsider? *J. Invertebr. Pathol.* **2016**, *137*, 58–61. [CrossRef] [PubMed]
15. Charles, J.F.; Nielsen-LeRoux, C. Mosquitocidal bacterial toxins: Diversity, mode of action and resistance phenomena. *Mem. Inst. Oswaldo Cruz* **2000**, *95*, 201–206. [CrossRef] [PubMed]
16. Ruiu, L.; Satta, A.; Floris, I. Observations on house fly larvae midgut ultrastructure after *Brevibacillus laterosporus* ingestion. *J. Invertebr. Pathol.* **2012**, *111*, 211–216. [CrossRef] [PubMed]
17. Bashir, F.; Aslam, S.; Khan, R.A.; Shahzadi, R. Larvicidal Activity of Bacillus laterosporus Against Mosquitoes. *Pak. J. Zool.* **2016**, *48*, 281–284.
18. Mura, M.E.; Ruiu, L. *Brevibacillus laterosporus* pathogenesis and local immune response regulation in the house fly midgut. *J. Invertebr. Pathol.* **2017**, *145*, 55–61. [CrossRef]
19. Merrill, B.D.; Grose, J.H.; Breakwell, D.P.; Burnett, S.H. Characterization of *Paenibacillus larvae* bacteriophages and their genomic relationships to firmicute bacteriophages. *BMC Genom.* **2014**, *15*, 745. [CrossRef]
20. Merrill, B.D.; Berg, J.A.; Graves, K.A.; Ward, A.T.; Hilton, J.A.; Wake, B.N.; Grose, J.H.; Breakwell, D.P.; Burnett, S.H. Genome Sequences of Five Additional *Brevibacillus laterosporus* Bacteriophages. *Genome Announc.* **2015**, *3*, e01146-15. [CrossRef]
21. Berg, J.A.; Merrill, B.D.; Crockett, J.T.; Esplin, K.P.; Evans, M.R.; Heaton, K.E.; Hilton, J.A.; Hyde, J.R.; McBride, M.S.; Schouten, J.T.; et al. Characterization of Five Novel Brevibacillus Bacteriophages and Genomic Comparison of Brevibacillus Phages. *PLoS ONE* **2016**, *11*, e0156838. [CrossRef]
22. Stamereilers, C.; Fajardo, C.P.; Walker, J.K.; Mendez, K.N.; Castro-Nallar, E.; Grose, J.H.; Hope, S.; Tsourkas, P.K. Genomic Analysis of 48 *Paenibacillus larvae* Bacteriophages. *Viruses* **2018**, *10*, 377. [CrossRef] [PubMed]
23. Brady, T.S.; Merrill, B.D.; Hilton, J.A.; Payne, A.M.; Stephenson, M.B.; Hope, S. Bacteriophages as an alternative to conventional antibiotic use for the prevention or treatment of *Paenibacillus larvae* in honeybee hives. *J. Invertebr. Pathol.* **2017**, *150*, 94–100. [CrossRef] [PubMed]
24. Pereira, S.; Pereira, C.; Santos, L.; Klumpp, J.; Almeida, A. Potential of phage cocktails in the inactivation of *Enterobacter cloacae*–An in vitro study in a buffer solution and in urine samples. *Virus Res.* **2016**, *211*, 199–208. [CrossRef]
25. Mateus, L.; Costa, L.; Silva, Y.J.; Pereira, C.; Cunha, A.; Almeida, A. Efficiency of phage cocktails in the inactivation of Vibrio in aquaculture. *Aquaculture* **2014**, *424*, 167–173. [CrossRef]
26. Yost, D.G.; Tsourkas, P.; Amy, P.S. Experimental bacteriophage treatment of honeybees (*Apis mellifera*) infected with *Paenibacillus larvae*, the causative agent of American Foulbrood Disease. *Bacteriophage* **2016**, *6*, e1122698. [CrossRef] [PubMed]
27. Bruttin, A.; Brussow, H. Human volunteers receiving Escherichia coli phage T4 orally: A safety test of phage therapy. *Antimicrob. Agents Chemother.* **2005**, *49*, 2874–2878. [CrossRef]
28. Endersen, L.; Buttimer, C.; Nevin, E.; Coffey, A.; Neve, H.; Oliveira, H.; Lavigne, R.; O'Mahony, J. Investigating the biocontrol and anti-biofilm potential of a three phage cocktail against *Cronobacter sakazakii* in different brands of infant formula. *Int. J. Food Microbiol.* **2017**, *253*, 1–11. [CrossRef] [PubMed]
29. Mirzaei, M.K.; Nilsson, A.S. Isolation of Phages for Phage Therapy: A Comparison of Spot Tests and Efficiency of Plating Analyses for Determination of Host Range and Efficacy. *PLoS ONE* **2015**, *10*, 13. [CrossRef]
30. Regeimbal, J.M.; Jacobs, A.C.; Corey, B.W.; Henry, M.S.; Thompson, M.G.; Pavlicek, R.L.; Quinones, J.; Hannah, R.M.; Ghebremedhin, M.; Crane, N.J.; et al. Personalized Therapeutic Cocktail of Wild Environmental Phages Rescues Mice from Acinetobacter baumannii Wound Infections. *Antimicrob. Agents Chemother.* **2016**, *60*, 5806–5816. [CrossRef] [PubMed]
31. Kutter, E.; De Vos, D.; Gvasalia, G.; Alavidze, Z.; Gogokhia, L.; Kuhl, S.; Abedon, S.T. Phage Therapy in Clinical Practice: Treatment of Human Infections. *Curr. Pharm. Biotechnol.* **2010**, *11*, 69–86. [CrossRef] [PubMed]
32. Ruiu, L.; Satta, A.; Floris, I. Emerging entomopathogenic bacteria for insect pest management. *Bull. Insectol.* **2013**, *66*, 181–186.

33. Furfaro, L.L.; Payne, M.S.; Chang, B.J. Bacteriophage Therapy: Clinical Trials and Regulatory Hurdles. *Front. Cell. Infect. Microbiol.* **2018**, *8*, 376. [CrossRef] [PubMed]
34. Forsgren, E.; Stevanovic, J.; Fries, I. Variability in germination and in temperature and storage resistance among *Paenibacillus larvae* genotypes. *Vet. Microbiol.* **2008**, *129*, 342–349. [CrossRef]
35. De Graaf, D.C.; Alippi, A.M.; Antunez, K.; Aronstein, K.A.; Budge, G.; De Koker, D.; De Smet, L.; Dingman, D.W.; Evans, J.D.; Foster, L.J.; et al. Standard methods for American Foulbrood research. *J. Apic. Res.* **2013**, *52*, 1–28. [CrossRef]
36. Genersch, E.; Forsgren, E.; Pentikainen, J.; Ashiralieva, A.; Rauch, S.; Kilwinski, J.; Fries, I. Reclassification of *Paenibacillus larvae* subsp. pulvifaciens and *Paenibacillus larvae* subsp. larvae as *Paenibacillus larvae* without subspecies differentiation. *Int. J. Syst. Evol. Microbiol.* **2006**, *56*, 501–511. [CrossRef] [PubMed]
37. Berg, J.A.; Merrill, B.D.; Breakwell, D.P.; Hope, S.; Grose, J.H. A PCR-Based Method for Distinguishing between Two Common Beehive Bacteria, *Paenibacillus larvae* and *Brevibacillus laterosporus*. *J. Appl. Environ. Microbiol.* **2018**, *84*, e01886-18. [CrossRef] [PubMed]
38. Lane, D.J. 16S/23S rRNA sequencing. In *Nucleic Acid Techniques in Bacterial Systematics*; Stackebrandt, E., Goodfellow, M., Eds.; John Wiley & Sons, Ltd.: Chichester, UK, 1991; pp. 115–175.
39. Alippi, A.M.; López, A.C.; Aguilar, O.M. A PCR-based method that permits specific detection of *Paenibacillus larvae* subsp. larvae, the cause of American Foulbrood of honey bees, at the subspecies level. *Lett. Appl. Microbiol.* **2004**, *39*, 25–33. [CrossRef] [PubMed]
40. Morrissey, B.J.; Helgason, T.; Poppinga, L.; Funfhaus, A.; Genersch, E.; Budge, G.E. Biogeography of *Paenibacillus larvae*, the causative agent of American foulbrood, using a new multilocus sequence typing scheme. *Environ. Microbiol.* **2015**, *17*, 1414–1424. [CrossRef] [PubMed]
41. Versalovic, J.; Schneider, M.; De Bruijn, F.J.; Lupski, J.R. Genomic fingerprinting of bacteria using repetitive sequence-based polymerase chain reaction. *Methods Mol. Cell. Biol.* **1994**, *5*, 25–40.
42. Abràmoff, M.D.; Magalhães, P.J.; Ram, S.J. Image processing with ImageJ. *Biophotonics Int.* **2004**, *11*, 36–42.

antibiotics

MDPI

Article

Synergistic Action of Phage and Antibiotics: Parameters to Enhance the Killing Efficacy Against Mono and Dual-Species Biofilms

Ergun Akturk [1], Hugo Oliveira [1], Sílvio B. Santos [1], Susana Costa [1], Suleyman Kuyumcu [2], Luís D. R. Melo [1,*] and Joana Azeredo [1,*]

[1] LIBRO-Laboratório de Investigação em Biofilmes Rosário Oliveira, Centre of Biological Engineering, University of Minho, Campus de Gualtar, 4700-057 Braga, Portugal

[2] Department of Medical Genetics, Medical Faculty, Sifa University, 35535 Izmir, Turkey

* Correspondence: lmelo@deb.uminho.pt (L.D.R.M.); jazeredo@deb.uminho.pt (J.A.)

Received: 13 June 2019; Accepted: 22 July 2019; Published: 25 July 2019

Abstract: *Pseudomonas aeruginosa* and *Staphylococcus aureus* are opportunistic pathogens and are commonly found in polymicrobial biofilm-associated diseases, namely chronic wounds. Their co-existence in a biofilm contributes to an increased tolerance of the biofilm to antibiotics. Combined treatments of bacteriophages and antibiotics have shown a promising antibiofilm activity, due to the profound differences in their mechanisms of action. In this study, 48 h old mono and dual-species biofilms were treated with a newly isolated *P. aeruginosa* infecting phage (EPA1) and seven different antibiotics (gentamicin, kanamycin, tetracycline, chloramphenicol, erythromycin, ciprofloxacin, and meropenem), alone and in simultaneous or sequential combinations. The therapeutic efficacy of the tested antimicrobials was determined. Phage or antibiotics alone had a modest effect in reducing biofilm bacteria. However, when applied simultaneously, a profound improvement in the killing effect was observed. Moreover, an impressive biofilm reduction (below the detection limit) was observed when gentamicin or ciprofloxacin were added sequentially after 6 h of phage treatment. The effect observed does not depend on the type of antibiotic but is influenced by its concentration. Moreover, in dual-species biofilms it was necessary to increase gentamicin concentration to obtain a similar killing effect as occurs in mono-species. Overall, combining phages with antibiotics can be synergistic in reducing the bacterial density in biofilms. However, the concentration of antibiotic and the time of antibiotic application are essential factors that need to be considered in the combined treatments.

Keywords: *Pseudomonas aeruginosa*; *Staphylococcus aureus*; bacteriophage; dual-species; biofilms; antibiotic; synergy; simultaneous; sequential

1. Introduction

Polymicrobial interactions are widespread in many biofilm-associated infections [1], accounting for a significant higher mortality and considerable high costs to the health-care systems [2,3]. Biofilms are communities of microbial cells adhered to biotic or abiotic surfaces and encased in a self-produced extracellular polymeric matrix that confer protection to the community against adverse environmental conditions and antimicrobials including the presence of antibiotics [4]. In an established polymicrobial biofilm, the co-existence of different species is a clear advantage for the overall biofilm population. These biofilms very often exhibit improved capabilities and functions compared to single-species ones [5,6], such as enhanced degradation of organic compounds [7], increased virulence [8], and increased tolerance against antimicrobials [9].

Pseudomonas aeruginosa and *Staphylococcus aureus* are versatile bacterial pathogens and common etiological agents of polymicrobial associated infections. These two opportunistic pathogens exhibit intrinsic and acquired antibiotic resistance [10]. When co-existing in a biofilm this tolerance largely increases, namely due to the decreased metabolic activity, increased bacterial doubling time, and increased level of mutations and upregulation of efflux pumps [11].

Bacteriophages (or phages), viruses that infect bacteria, are natural antibacterial agents that specifically infect and lyse bacteria. Phages are the most abundant biological entities on our planet and can be used as biocontrol agents targeting bacterial cells either in suspension or in biofilms [12,13]. Due to the phage bacterial host specificity and bacteriolytic activity against antibiotic-resistant strains, phage therapy has been suggested as a valuable approach to control numerous pathogenic bacteria. Phages can penetrate the inner layers of the biofilms and infect dormant cells [14], which is a clear advantage of phages compared to antibiotics in killing biofilms. Therefore, it has been proposed that phages may be a useful combination with antibiotic treatment [15–17]. Several studies demonstrated the efficiency of phage and antibiotic combinations in planktonic cultures of *P. aeruginosa* [17,18] and biofilms [19]. Besides, phages and antibiotics use different mechanisms of action, which make them effective against phage/antibiotic resistance pathogens [17]. Consequently, antibiotic and phage resistance have a low chance of evolving at the same time and, besides, bacteria resistant to one agent will be taken by the other agent. However, to our knowledge, this type of studies was never assessed on dual-species biofilms.

In this study, we report the isolation and characterization of a new *Pakpunavirus* phage, named vB_PaM_EPA1, and the use of several phage-antibiotic combinations against mono and dual-species biofilms of *P. aeruginosa* and *S. aureus*.

2. Results

2.1. Isolation and Characterization of a New *P. aeruginosa*-Infecting Phage

Six clinical strains were used for phage enrichment (Table S1), using raw sewage from Sifa Hospital (Izmir, Turkey) as phage sources. A phage vB_PaM_EPA1 (EPA1) was isolated, and its plaque morphology was characterized by clear and small plaques (0.8 mm in diameter) surrounded by halo rings on the host strain Sifa_Pa_1.5 (Table S1). The morphology of EPA1 particles was observed by Transmission Electron Microscopy (TEM). EPA1 has an icosahedral head with 69 nm in diameter and a contractile tail of 145 × 24 nm. According to Ackermann's classification [20], EPA1 belongs to the *Caudovirales* order and *Myoviridae* family (Figure 1).

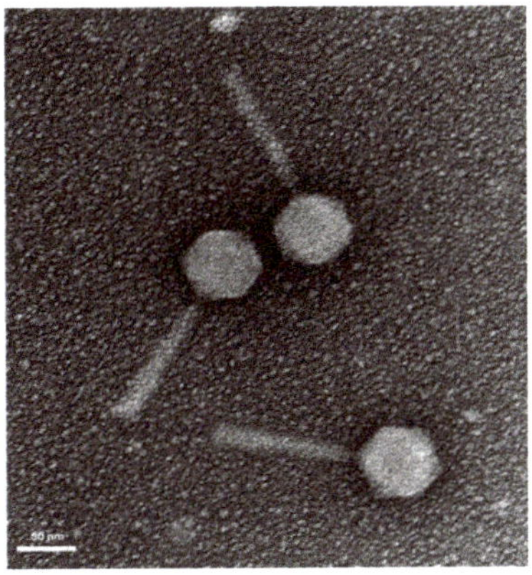

Figure 1. TEM image of *P. aeruginosa* specific phage EPA1 obtained by negative staining with 2% (*w/v*) uranyl acetate. Scale bar represents 50 nm.

2.2. *Host range, Efficiency of Plating and One-Step Growth Curve*

In total, seventeen drug-resistant clinical isolates (Table S1) and three reference *P. aeruginosa* strains were used to determine the host range and the efficiency of plating (EOP) of EPA1 (Table S1). EPA1 has a broad spectrum of activity (within the panel of strains used) and was able to propagate on 70% (14 out of 20) of the *P. aeruginosa* strains with moderate to high EOP. No lysis from without events were observed. Also, no correlation between phage susceptibility and antibiotic resistance was detected (Table S1). Due to the fact that EPA1 propagates better in *P. aeruginosa* PAO1 strain, we have used this strain to produce the phage for further experiments. Nevertheless, we are aware of the fact that PAO1 encloses filamentous phages that could influence phage production, however no filamentous phages were detected by plating methods or TEM [21]. One-step growth curve (OSGC) experiments were performed to examine the infection parameters of EPA1. The latent period of EPA1 was around 10 min, and the burst size was approximately of 34 progeny phages per infected cell (Figure S1).

2.3. *Genome Analysis of EPA1*

EPA1 has a linear double-stranded DNA genome containing 91,394 bp with an average 49.2% GC content. This phage encodes 175 putative CDSs, of which 35 have a putative function, and 140 are considered hypothetical/novel (Table S2). Most predicted gene products exhibit homology to phage known proteins belonging to the *Pakpunavirus* (ICTV 2015.029a-dB ratification) genus, mostly *Pseudomonas* phages JG004 (NC_019450.1), PAK_P4 (NC_022986) and vB_PaeM_C2-10_Ab1 (NC_019918). Moreover, seventeen tRNA genes coding for Arg, Asn, Asp, Cys, Gln, Glu, Gly, Ile, Lys, Leu, Met, Phe, Pro, Ser, Thr, Trp and Tyr were found. Regarding regulatory elements, 16 promoters were identified as well as 14 rho-independent terminators. The general characteristics of the phage genome are summarized in Table 1. Whole-genome comparisons through BLASTN show that EPA1 has a high overall nucleotide identity (>90%) with other *P. aeruginosa* phages, such as JG004 (NC_019450.1), vB_PaeM_SCUT-S2 (MK340761.1) and SRT6 (MH370478.1). EPA1 shares >145 proteins with these phages.

Table 1. General features of EPA1 genome.

Feature	vB_PaM_EPA1
Genome size	91,394 bp
G+C content	49,2%
Number of predicted CDSs	175
Number of proteins with assigned functions	35

2.4. *Characterisation of Mono and Dual-Species Biofilm Models*

In vitro mono and dual-species biofilms were formed in 24-well polystyrene plates for 48 h, and the number of viable bacteria cells were determined by colony forming unit (CFU) counting. It is well documented that *P. aeruginosa* inhibits *S. aureus* proliferation in dual-species biofilms. The reason for the lower density of *S. aureus* population has been attributed to the toxic effect of *P. aeruginosa* exoproducts [22], including LasA protease, 4-hydroxy-2-heptylquinoline-N-oxide (HQNO) [23], the Pel and Psl products [24], and phenazines such as pyocyanin [25]. Therefore, in order to successfully produce dual-species biofilms, biofilm formation has been initiated with an *S. aureus* cell culture, and 24 h later *P. aeruginosa* cells were added on the *S. aureus* biofilm, then incubated for another 24 h. A similar strategy of biofilm formation was also used by DeLeon et al. [26] where biofilms were initiated with *S. aureus*, and 48 h later *P. aeruginosa* cells were added [26]. Our results showed that the number of viable cells of *S. aureus* was 3.77×10^7 CFU/mL and 1.2×10^9 CFU/mL for *P. aeruginosa* in the mono-species biofilms. Regarding the dual-species biofilms, the concentrations were 1.28×10^7 CFU/mL for *S. aureus* and 2×10^8 CFU/mL for *P. aeruginosa*.

2.5. Biofilm Treatments

The selected antibiotics (Table 2) and EPA1 were tested individually or in combinations within intact mono, and dual-species 48 h biofilms and treated for 24 h in total. Phage and antibiotics were simultaneously or sequentially added in combined treatments. Twenty-four hours post-treatment, CFUs were enumerated in order to assess the antibiofilm efficacy and to characterize the possible interactions between antimicrobials.

Table 2. List of the antibiotics, MIC values of *P. aeruginosa* and *S. aureus* planktonic cells and their mechanism of action.

Name of Antibiotics	*P. aeruginosa* MIC Values	*S. aureus* MIC Values	Mechanism of Action	
Gentamicin	4 µg/mL	16 µg/mL	Protein Synthesis Inhibitors	30S ribosomal subunit
Kanamycin	10 µg/mL	*	Protein Synthesis Inhibitors	
Tetracycline	8 µg/mL	*	Protein Synthesis Inhibitors	
Chloramphenicol	32 µg/mL	*	Protein Synthesis Inhibitors	50S ribosomal subunit
Erythromycin	128 µg/mL	*	Protein Synthesis Inhibitors	
Ciprofloxacin	<1 µg/mL	<1 µg/mL	DNA Synthesis Inhibitor	
Meropenem	2 µg/mL	2 µg/mL	Cell wall Synthesis Inhibitor	

* These antibiotics were not tested on *S. aureus* biofilm models.

Further, the effects of phage, gentamicin at MIC and phage-gentamicin at MIC combinations (simultaneous and sequential) in mono and dual-species biofilms were also analyzed by confocal laser microscopy (CLSM). For that assessment, fluorescence probes were designed to specifically target differentially both bacterial species. Generally, microscopy analysis corroborated cell counting results.

2.5.1. Antibiotics and Phages Alone cause a Moderate Killing Effect on Biofilms

Three antibiotics were selected (gentamicin, ciprofloxacin and meropenem), and their anti-biofilm ability was tested against *P. aeruginosa* and *S. aureus* mono-species biofilms. These antibiotics were selected depending on their mechanism of action (Table 2): protein synthesis inhibitor (gentamicin), DNA synthesis inhibitor (ciprofloxacin) and cell wall synthesis inhibitor (meropenem). The killing effect of the antibiotics against *P. aeruginosa* biofilms, used in different concentrations, ranged from 0.8 to 5 orders-of-magnitude (Figure 2).

Regarding *S. aureus*, no significant reduction in the number of viable cells was observed when antibiotics were applied at their MIC (Figure S2). However, when gentamicin was applied with 8xMIC, the number of viable cells was reduced approximately 1.4 orders-of-magnitude (Figure S2).

Additionally, EPA1 was individually tested (at multiplicity of infection, MOI, of 1) on *P. aeruginosa* biofilms for 6 h and 24 h. The observed reductions were 3.4 and 0.5 orders-of-magnitude, respectively (Figure 2, Figure 3b,c). The best reduction was observed at 6 h post-treatment; after that, *P. aeruginosa* cells started to regrow (Figure 2). CLSM images corroborated CFUs results. *P. aeruginosa* biofilms after being challenged for 6 h with EPA1 reduced their thickness from 22.4 µm to 7.2 µm, but after 24 h of phage contact an increase in biofilm thickness to 11.7 µm was observed (Figure 3a–c).

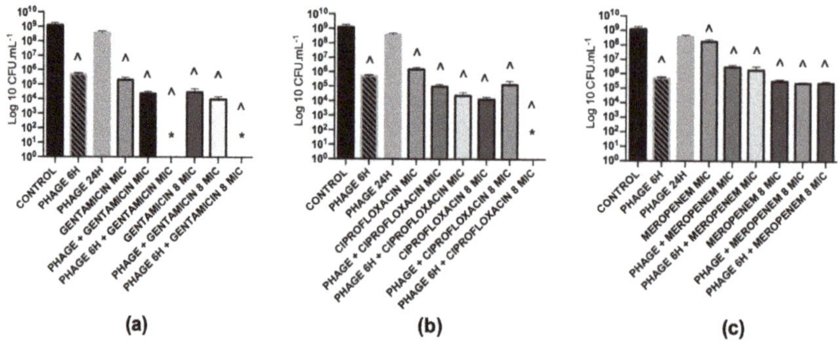

Figure 2. Treatment of *P. aeruginosa* PAO1 48 h biofilms with different antimicrobial agents individually or in combinations; phage EPA1 and (**a**) gentamicin; (**b**) ciprofloxacin; and (**c**) meropenem for 24 h. A prefix PHAGE indicates EPA1 in MOI 1, 6 H and 24 Hindicates treatment time period for 6 and 24 h, MIC indicates the dose of antibiotics with 1-time MIC value of *P. aeruginosa* 8× MIC indicates the dose of antibiotics with 8-times MIC value of *P. aeruginosa*, PHAGE + antibiotic indicates simultaneous treatment, and PHAGE 6 H+ antibiotics indicates phage was added first then antibiotic was added with 6 h delay. * Under detection limit (<10^2). (^) Statistical differences between the control and treated biofilms were determined by two-way repeated-measures analysis of variance (ANOVA) with a Tukey's multiple comparison test.

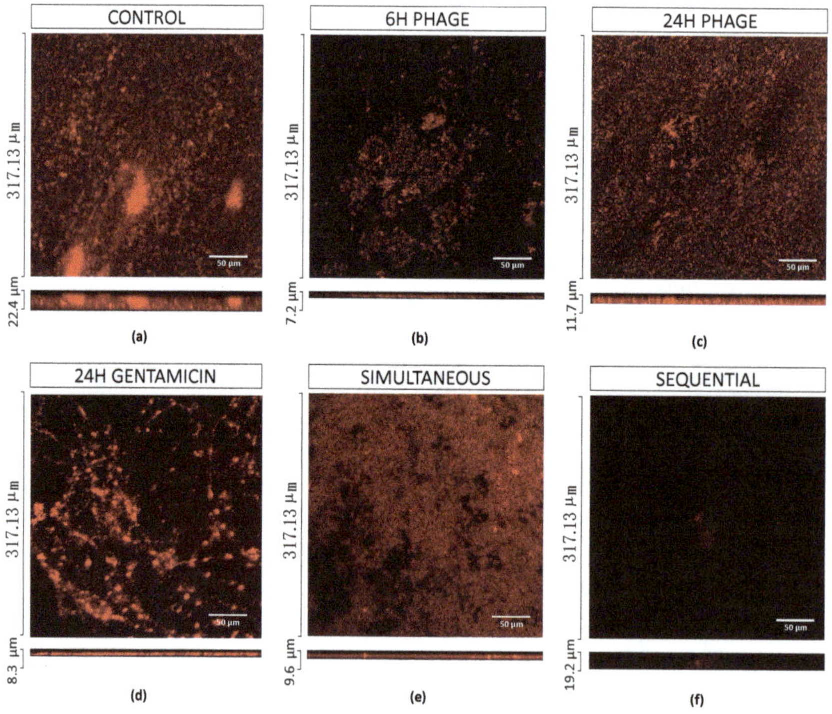

Figure 3. 3D reconstructions of confocal stacks of images of mono-species *P. aeruginosa* biofilms. (**a**) Control, (**b**) 6 h phage treatment, (**c**) 24 h phage treatment, (**d**) 24 h Gentamicin treatment, (**e**) 24 h simultaneous treatment, (**f**) 24 h sequential treatment. All biofilms were stained with EPA1_TFP (with mCherry) recombinant protein. Scale bar represents 50 μm.

2.5.2. Combined Treatments with Simultaneous Application of Phage and Antibiotics have Synergistic Effects for Low Concentrations of Antibiotics

The efficacy of the combinations of phage and antibiotics was also tested on *P. aeruginosa* mono-species biofilms. These maturated intact biofilms were treated in one of two ways; simultaneously (phage and antibiotic were added at the same time) and sequentially (phage was added first, then antibiotic was added with a delay of 6 h).

The combination of phage EPA1 with gentamicin (Figure 3e), ciprofloxacin, or meropenem when applied simultaneously with a lower dose (MIC) resulted in population reductions of 4.7, 4.1 and 2.6 orders-of-magnitudes (Figure 2), respectively. These results show a clear synergistic effect between antimicrobial agents in most cases (Table 3). When the antibiotic concentrations were increased, we were expecting an increase in the killing efficacy of simultaneous combined treatments, however, interestingly 8 × MIC did not increased the overall biofilm killing (in certain cases, we observed an antagonistic effect) (Table 3).

Table 3. General overview of the efficacy of combined treatments in 48 h *P. aeruginosa* mono-species biofilm. Synergistic—the biofilm reduction using phage-antibiotic combinations is greater than the sum of their individual treatments. Additive—the biofilm reduction using phage-antibiotic combination is similar to the sum of their individual treatments. Antagonistic—the biofilm reduction using phage-antibiotic combinations is lower than the sum of their individual treatments.

Treatments	Gentamicin	Ciprofloxacin	Meropenem
Simultaneously MIC	Synergistic	Synergistic	Synergistic
Simultaneously 8 MIC	Additive	Antagonistic	Antagonistic
Sequentially MIC	Synergistic	Synergistic	Synergistic
Sequentially 8 MIC	Synergistic	Synergistic	Antagonistic

2.5.3. Antibiotics that Target Protein and DNA Synthesis Mechanisms Interfere with Phage Replication

In order to understand why increasing the antibiotic concentration did not lead to an increased killing activity, we tested the effect of the antibiotics on phage replication. Phage titer was enumerated after 24 h of simultaneous treatment and compared with the control. Unsurprisingly, the titer of phages, when combined with gentamicin and ciprofloxacin was significantly lower than the titer of phages in control samples (Figure S3). Conversely and as expected, the phage replication was not affected by the presence of meropenem. This antibiotic is affecting bacteria cell wall synthesis and thus does not interfere in phage replication.

2.5.4. Combined Treatments with Sequential Application of Phage and Antibiotics have a better Killing Efficacy than when Applied Simultaneously

The fact that protein and DNA synthesis inhibitors interfere with phage replication, led us to assess a sequential treatment in which the phage was applied first and six hours later the antibiotic. This six hour period was chosen based on previous biofilm/phage interaction studies that refer that after six hours of phage interaction, a regrowth of phage-resistant phenotypes is observed [27]. Our CFU and CLSM results have also corroborated this phenomenon (Figures 2a and 3b,c).

The same phage-drug combinations tested before were applied in sequential treatments. The results showed an almost eradication of the biofilm with gentamicin (MIC, 8× MIC) (Figure 3f) and ciprofloxacin (8xMIC). Besides, other combinations with ciprofloxacin or meropenem (with MIC) also showed an increased killing effect, 4.7 and 2.8 orders-of-magnitudes, respectively. In accordance, increasing the antibiotic concentration of meropenem in sequential treatment (to 8× MIC), resulted in an antagonistic effect (3.7 orders-of-magnitudes), contrarily to what was observed for the other antibiotics (Figure 2). CLSM results also confirmed that almost all biofilm was eradicated except a cluster (Figure 3e). To understand the impact of antimicrobial application order in sequential interaction, the same combinations were applied in the reverse order. Gentamicin was applied first, and then phage was applied six hours after. The collected data showed that killing efficacies of combinations were reduced when gentamicin MIC and 8× MIC were applied first, with reductions of 2.5 and 3.6 orders-of-magnitude, respectively (Figure S4).

The data suggest that biofilm exposure to phage prior to antibiotics is more effective than simultaneous treatment in eliminating biofilm-associated cells. Considering the overall results, when gentamicin was administered at MIC sequentially after six hours of phage addition, it almost eradicated biofilms (Figure 3f).

2.5.5. The Phage Killing Efficacy with the Sequential Treatment of Phage and Gentamicin cannot be Extrapolated to other Protein Synthesis Inhibitors

An impressive biofilm biomass reduction was observed with a protein synthesis inhibitor (gentamicin) at MIC. To understand if this effect can be extrapolated to other antibiotics of the same class, we also tested kanamycin, tetracycline, erythromycin and chloramphenicol (Table 3), in simultaneous and sequential combinations. Contrarily to what we were expecting, the effect observed for gentamicin was not reproduced with the other tested antibiotics (Figure 4). In fact, those antibiotics alone had a low to moderate effect against biofilms, lower than 3 orders-of-magnitude in the overall biomass reduction. Gentamicin alone caused ten times more biomass damage, which could be one of the reasons for the better performance of sequential treatments with gentamicin compared to the other antibiotics.

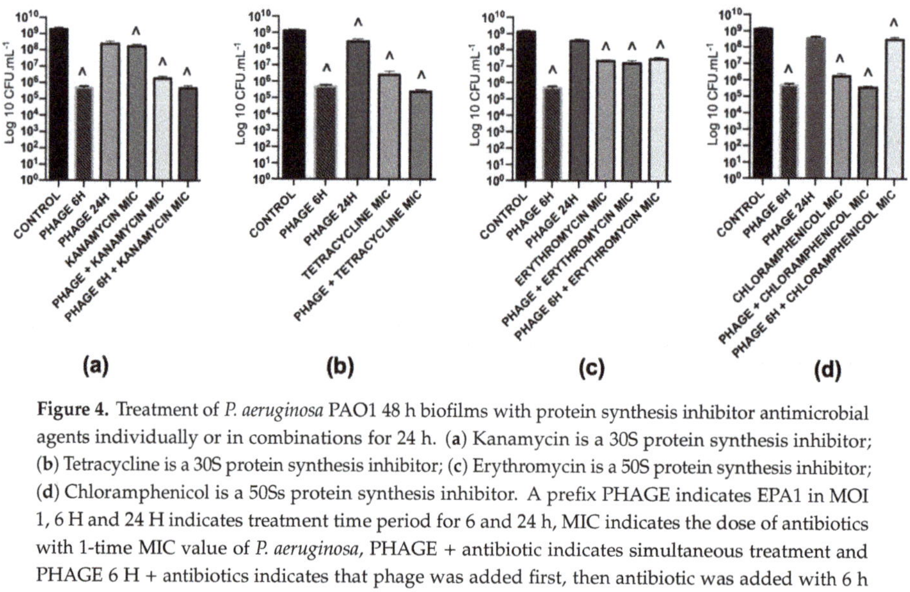

Figure 4. Treatment of *P. aeruginosa* PAO1 48 h biofilms with protein synthesis inhibitor antimicrobial agents individually or in combinations for 24 h. (**a**) Kanamycin is a 30S protein synthesis inhibitor; (**b**) Tetracycline is a 30S protein synthesis inhibitor; (**c**) Erythromycin is a 50S protein synthesis inhibitor; (**d**) Chloramphenicol is a 50Ss protein synthesis inhibitor. A prefix PHAGE indicates EPA1 in MOI 1, 6 H and 24 H indicates treatment time period for 6 and 24 h, MIC indicates the dose of antibiotics with 1-time MIC value of *P. aeruginosa*, PHAGE + antibiotic indicates simultaneous treatment and PHAGE 6 H + antibiotics indicates that phage was added first, then antibiotic was added with 6 h delay. (^) Statistical differences between the control and treated biofilms were determined by two-way repeated-measures analysis of variance (ANOVA) with a Tukey's multiple comparison test.

2.5.6. The Efficacy of Sequential Antibiofilm Treatments is Dependent on the Antibiotic Concentration

We also investigated the effect of the different gentamicin concentrations on the simultaneous and sequential treatment efficacy (Figure 5). A direct correlation was observed between the concentration of gentamicin and the biofilm killing efficacy. An almost complete biofilm eradication (below the detection limits) was observed only when antibiotic concentrations were equal or above the MIC (Table S1).

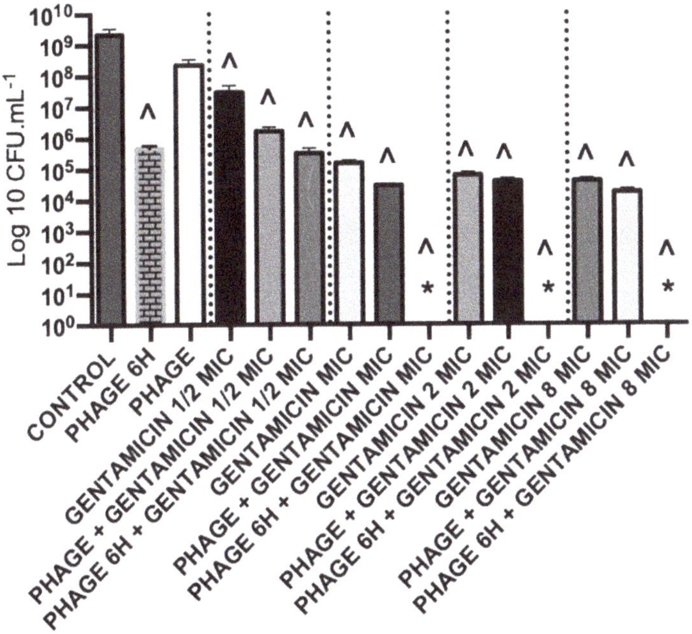

Figure 5. Treatment of *P. aeruginosa* PAO1 48 h biofilms with gentamicin at different concentrations. A prefix PHAGE indicates EPA1 in MOI 1, 1/2 MIC indicates the dose of antibiotics with 1/2× MIC value, MIC indicates the dose of antibiotics with 1× MIC value of *P. aeruginosa*, 2 MIC indicates the dose of antibiotics with 2× MIC value of *P. aeruginosa*, 8 MIC indicates the dose of antibiotics with 8× MIC value of *P. aeruginosa*, PHAGE + antibiotic indicates simultaneous treatment and PHAGE 6 H + antibiotics indicates phage was added first then antibiotic was added with 6 h delay. * Under detection limit ($<10^2$). (^) Statistical differences between the control and treated biofilms were determined by two-way repeated-measures analysis of variance (ANOVA) with a Tukey's multiple comparison test.

2.5.7. Sequential Application of Phages and Gentamicin have a great Antibiofilm Effect in Dual-Species Biofilms

The killing capacity of gentamicin (MIC) and EPA1 was tested individually and in combination (simultaneous and sequential treatments) in a dual-species biofilm model comprising *P. aeruginosa* and *S. aureus* (Figure 6). Intact biofilms were grown for 48 h and treated for 24 h in total. In the control, it was possible to observe a predominance of *P. aeruginosa* (1.4×10^9 CFU/mL), in comparison with *S. aureus* (2.3×10^5 CFU/mL) (Figure 7). Although *S. aureus* was the first colonizer, CLSM images indicate that both species were randomly distributed throughout the biofilm 3D structures (Figure 6).

The individual treatments with gentamicin with MIC and 8× MIC resulted in a significant reduction of approximately 3.3 orders-of-magnitude and 4.6 orders-of-magnitude of *P. aeruginosa* cells, respectively. Phage treatment was less effective than gentamicin, resulting in a reduction of 0.7 orders-of-magnitude of *P. aeruginosa* cells (Figure 8b). None of the individual treatments showed a significant impact on the *S. aureus* population (Figure 7).

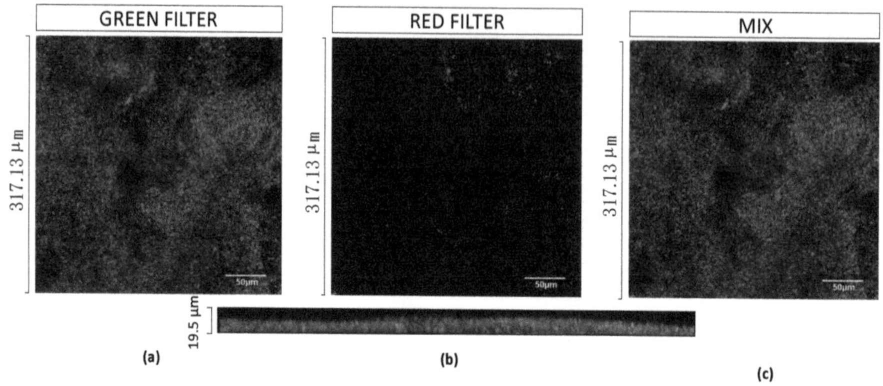

Figure 6. 3D reconstructions of confocal stacks of images of dual-species of *P. aeruginosa* and *S. aureus* biofilms. (**a**) 48 h old intact biofilms were stained by using recombinant proteins, LM12_AMI-SH3 (with GFP) specific for *S. aureus* and (**b**)EPA1_TFP (with mCherry) specific for *P. aeruginosa*. (**c**) 48 h old intact biofilms were stained by using both recombinant proteins. Scale bar represents 50 µm.

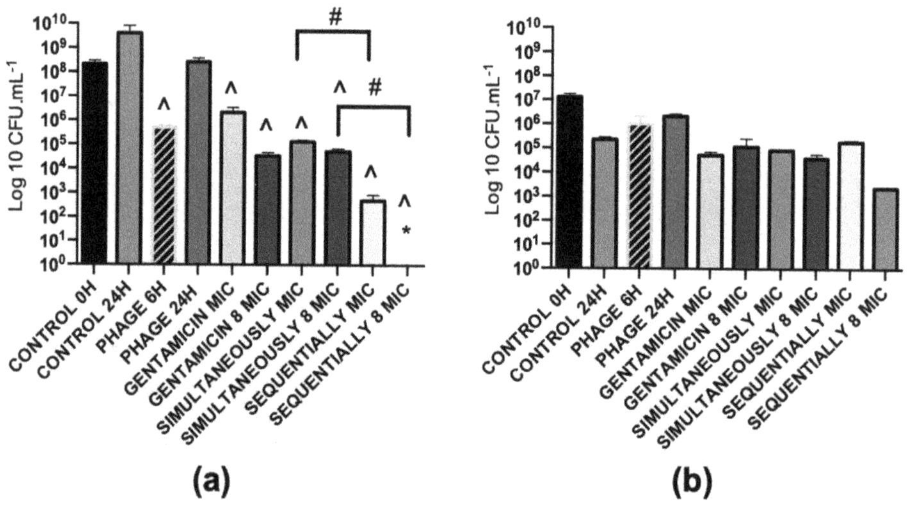

Figure 7. Treatment of 48 h dual-species biofilm. (**a**) *P. aeruginosa* number of viable cells. (**b**) *S. aureus* number of viable cells. A prefix PHAGE indicates EPA1 in MOI 1, 6 H and 24 H indicates treatment time period for 6 and 24 h, MIC indicates the dose of antibiotics with 1× MIC value of *P. aeruginosa*, 8 MIC indicates the dose of antibiotics with 8× MIC value of *P. aeruginosa*, PHAGE + antibiotic indicates simultaneous treatment and PHAGE 6 H + antibiotics indicates phage was added first then antibiotic was added with 6 h delay. * Under detection limit ($<10^2$). (^) Statistical differences between the control and treated biofilms. (#) Statistical differences between the simultaneously and sequentially treated biofilms. Statistical differences were determined by two-way repeated-measures analysis of variance (ANOVA) with a Tukey's multiple comparison test.

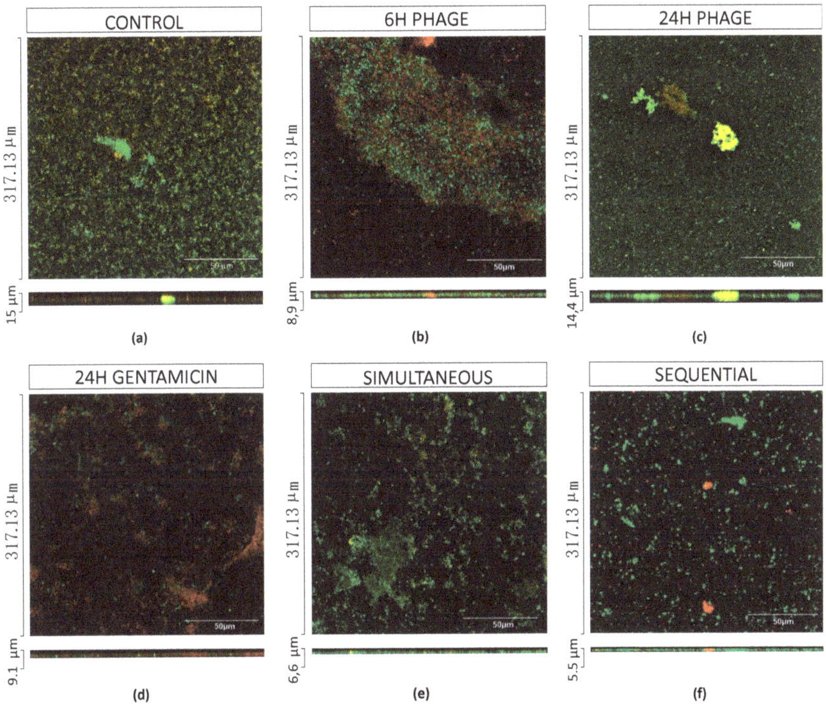

Figure 8. 3D reconstructions of confocal stacks of images of dual-species of *P. aeruginosa* and *S. aureus* biofilms. (**a**) Control, (**b**) 6 h phage treatment, (**c**) 24 h phage treatment, (**d**) 24 h Gentamicin with MIC treatment, (**e**) 24 h simultaneous treatment, (**f**) 24 h sequential treatment. 48 h old intact biofilms were stained by using LM12_AMI-SH3 (with GFP) and EPA1_TFP (with mCherry) recombinant proteins. Scale bar represents 50 µm.

Regarding the simultaneous treatments, no synergy was observed for any combination. On *P. aeruginosa*, phage-gentamicin MIC resulted in 4.1 orders-of-magnitude reduction (additive effect), while the phage-gentamicin 8 × MIC only slightly increased the reduction (4.6 orders-of-magnitude), demonstrating an antagonistic effect (Figure 8d). Both treatments also had a positive effect on *S. aureus* biofilm cell control, significantly reducing CFUs by about 0.4 and 0.8 orders-of-magnitude when using phage-gentamicin MIC and phage-gentamicin 8xMIC treatments, respectively (Figure 7).

Similarly, in mono-species biofilms, a sequential treatment was also tested. Results indicate that a preliminary phage treatment (6 h) before gentamicin application was very effective in biofilm reduction. Phage-gentamicin MIC reduced about 6.3 orders-of-magnitude the *P. aeruginosa* population, while phage-gentamicin 8× MIC almost eradicated *P. aeruginosa* cells (approximately 7orders-of-magnitude reduction) (Figure 8f). Although the phage-gentamicin treatment did not significantly impact the *S. aureus* population, the phage-gentamicin 8× MIC sequential application was also the most efficient treatment in reducing *S. aureus* biofilm cells in about 2 orders-of-magnitude (Figure 7).

In order to infer the impact of combined treatments on the biofilm structure, CLSM observations were performed on dual-species biofilms before and after the treatments. In this case, as in mono-species biofilms, we also observed a reduction of biofilm thickness, from 15 µm to 8.9 µm after 6 h of phage infection, and a biofilm thickness increase to 14.4 µm after 24 h of phage infection. The significant biofilm thickness reductions were observed in simultaneous and sequential combined treatments, after which the remaining biofilms had thicknesses of 6.6 µm and 5.5 µm, respectively.

3. Discussion

EPA1 is a lytic *Pseudomonas* phage, isolated from a hospital sewage, that belongs to the *Myoviridae* family (Figure 1) This phage has no identifiable lysogeny-associated genes or genes coding for toxins (Table S2). Phages from *Pakpunavirus* genus are highly conserved genetically and are described as having a broad host range, being therefore described as good for therapy. Indeed, EPA1 presented similar features and therefore could be considered a good candidate for further phage therapy approaches [28].

Herein, this phage was studied for its potential synergistic activity in combinations (simultaneous and sequential) with different antibiotics (at MIC and 8× MIC) against mono and dual-species biofilms.

Initially, three antibiotics (gentamicin, ciprofloxacin and meropenem) belonging to different classes were tested on mono-species biofilms. The efficacy of antibiotics ranged from 0.8 to 5 orders-of-magnitude (Figure 2) on *P. aeruginosa* biofilms, and no significant reductions were observed for *S. aureus*, except when gentamicin was applied with 8× MIC which reduced the biofilm density by approximately 1.4 orders-of-magnitude (Figure S2). The reductions were of 3.4 and 0.5 orders-of-magnitude when EPA1 was applied individually for 6 h and 24 h on *P. aeruginosa* biofilms, respectively.

To augmenting the effect of antimicrobials, combined therapies were applied. Interestingly, a synergistic effect of phages and antibiotics in simultaneous combinations was only observed when antibiotics were applied at MIC (Figure 2). Surprisingly, the increased concentration of antibiotics (8xMIC) in simultaneous combinations have not resulted in a higher biofilm efficacy (Figure 2). The fact that the increase in antibiotic concentrations did not lead to an increase of the antibiofilm efficacy might be related to phage replication inhibition phenomena [18]. As phage particles are constituted mainly by proteins, protein synthesis inhibitors and DNA synthesis inhibitors might affect the formation of new phage particles and therefore have an antagonistic effect on phage replication [18,29]. Our results (Figure S3) corroborate that phage titers obtained after phage infection in combination with antibiotic treatments were significantly lower than the titer of phage in the control (phage infection without antibiotic treatment).

Phage-antibiotic sequential approaches were already reported as a promising antibiofilm strategy [30]. In this study, an impressive biofilm biomass reduction was obtained when antibiotics were added to biofilms after 6 h of phage treatment. This was the time point where phages caused the maximum biofilm reduction. Similar observations were already reported elsewhere [27]. When phages interact with *P. aeruginosa* biofilms an initial population reduction is observed up to 6 h of phage contact and after that period the biofilm regrows. Biofilm regrowth was attributed to the emergence of phage-resistant variants that were equally well adapted to the biofilm phenotype [27].

On sequential treatments, we observed a synergistic effect with all combinations, when antibiotics were applied at their MICs. However, the antibiotic that showed the best antibiofilm efficacy was gentamicin. The combined treatment with this antibiotic was enough to eliminate all detectable biofilm cells in a concentration-dependent manner. The same phenomenon was not observed when other four protein synthesis inhibitors were used (Figure 4).

The antimicrobial synergy between phages has been in part explained by a more efficient penetration of both antibiotics and phages into the biofilm. In fact, it has been described that phages can degrade the biofilm matrix using depolymerases [31,32], enhancing their penetration into the deep layers of the biofilm. In the case of EPA1, no depolymerase was identified in its genome, therefore we have no evidence that this phenomenon might have been responsible for a synergistic action of the combined therapy. We have previously reported that phages can access the bottom layers of the biofilm migrating through the biofilm void spaces [33]. This might be the case in our study, since both mono and dual-species biofilms appeared to be heterogeneous structures with plenty of void spaces (Figures 3 and 6). This phenomenon leads phages to replicate in the deeper-layer of biofilm, reaching high titers and interrupting the biofilm matrix. The addition of antibiotics following this interruption results in an enhanced bacterial reduction due to the deeper penetration of these agents. When phages are applied prior to the antibiotic, they also avoid the antagonistic effect of antibiotics on phage replication, as was previously described [18]. After determining the best-case scenario to treat *P. aeruginosa* biofilms, it was our intention to study the impact of this treatment on dual-species biofilms. The relevance of studying polymicrobial communities has gained more interest on the last decade, mainly due to the fact that the vast majority of biofilms are formed by more than one species of bacteria and in some case also fungi, protozoa and algae [34]. This reason triggered the design of this study, and to our knowledge, this is the first time that this strategy is reported in dual-species biofilms.

The establishment of dual-species biofilm models was difficult, due to the inhibitory effect of *P. aeruginosa* on *S. aureus* cells, which is widely reported [9,10,35,36]. It is described that polymicrobial biofilms of chronic wounds are first colonized by small numbers of resident Gram-positive aerobic cocci, including *S. aureus*, after which there is a shift on wound microbiome, and Gram-negative bacilli are predominant on this environment [37]. As previously performed by DeLeon et al. [26], in our experiments, dual-species biofilms were established in two-steps. Colonization by *S. aureus* and a further addition of *P. aeruginosa* led to the establishment of stable biofilms.

In general, dual-species biofilms were more tolerant to all treatments than mono-species *P. aeruginosa* biofilms. Nevertheless, the sequential treatment at 8× MIC almost eradicated *P. aeruginosa* biofilm cells, but it did not increase the antimicrobial effect on *S. aureus*.

The presence of external selective pressures, like antibiotics or phages, can stimulate a cell response that can lead to increased tolerance to antibiotics, namely as a result of EPS production by *P. aeruginosa*. Alginate, Pel, and Psl, which are part of the biofilm extracellular matrix have a structural and protective function [38,39]. More specifically, Psl is described as creating a protective barrier against aminoglycosides [40]. Also, quorum-sensing molecules can justify the weaker effect of these approaches on dual-species biofilms [41].

The data collected herein showed that the concentration of gentamicin needs to be increased (8× MIC) to eliminate *P. aeruginosa* cells successfully in dual-species biofilms in comparison with mono-species biofilms. This can result in an overdose of antibiotics that can lead to toxicity in treatments [42].

Overall, from this study, two main conclusions emerge. First, the combined treatment with a sequential application of phages and then antibiotic is the most promising approach to combat infectious biofilms when compared with their individual and simultaneous treatments. Second, the majority of the studies of antibiofilm approaches are conducted in mono-species biofilms, and as demonstrated herein, the treatment outcomes are completely different when a second species is added. So, to achieve success, a phage cocktail comprising different *P. aeruginosa* and *S. aureus* phages targeting different bacterial receptors should be tested prior to antibiotic addition.

4. Material and Methods

4.1. Bacterial Strains and Culture Conditions

The biofilm-forming strains *P. aeruginosa* PAO1 and *S. aureus* (ATCC®25923™) were obtained from LPhage Laboratory in the Centre of Biology (CEB) strain collection (Braga, Portugal). Additional one *P. aeruginosa* clinical isolate and one Spanish Type Culture Collection (CECT) strain were obtained from Lphage Laboratory in CEB strain collection and Sifa Hospital Strain collection (Izmir, Turkey). In total, this accounts for a total of 20 strains used (Table S1) for isolating and characterizing the phage. The antibiogram profile of the clinical strains was previously established by the provider institutes. All strains were grown in Tryptic Soy Broth (TSB, VWR Chemicals, Randor, PA, USA), Tryptic Soy Agar (TSA; VWR Chemicals) or in TSA soft overlays (TSB with 0.6% agar) at 37 °C. In addition, mannitol salt agar (MSA; VWR Chemicals) was used to enumerate *S. aureus* cells in dual-species biofilms, while *P. aeruginosa* cells were counted in non-selective media (TSA).

4.2. *Phage Isolation and Production*

Phage was isolated from effluent samples of raw sewage from Sifa Hospitals in Izmir, Turkey. The phage enrichment method was applied to isolate the phage [43]. Briefly, 100 mL of the effluent were mixed with 100 mL of double-strength TSB and with 10 μL of each of the exponentially grown *P. aeruginosa* strains. The obtained suspensions were incubated at 37 °C and 120 rpm (BIOSAN ES-20/60, Riga, Latvia) overnight. Suspensions were further centrifuged (15 min, 9000× *g*, 4 °C), and the supernatants were filtered through a 0.22 μm polyethersulfone (PES) membrane (ThermoFisher Scientific, Massachusetts, USA). The presence of phages was confirmed by performing spot assays on bacterial lawns. The prepared plates were further incubated overnight at 37 °C, and the presence of inhibition halos observed. When phage plaques appeared, successive rounds of single plaque purification were carried out until purified plaques were observed, reflected by a single plaque morphology.

The purified phage was produced by using the double agar layer method, as described before [43]. Briefly, 100 μL of a phage suspension at 10^8 PFU/mL were spread on *P. aeruginosa* PAO1 lawns for overnight incubation at 37 °C. If full lysis was observed, plates were further incubated at 4 °C for 6 h at 120 rpm (BIOSAN PSU-10i), with 2 mL of SM Buffer (100 mM NaCl, 8 mM MgSO4, 50 mM Tris/HCl, pH 7.5) to resuspend the phage particles. The liquid phase was collected and centrifuged (15 min, 9000× *g*, 4 °C), and the supernatants were filtered through a 0.22 μm PES membrane. Purified phages were stored at 4 °C for further use.

4.3. *Electron Microscopy*

Phage suspension was sedimented by centrifugation (25,000× *g*, 60 min, 4 °C) using a ScanSpeed 1730R centrifuge (Labogene, Lillerød, Denmark). The pellet was further washed in tap water by repeating the centrifugation step. Subsequently, phage suspension was deposited on copper grids with a carbon-coated Formvar carbon film on a 200 square mesh nickel grid, stained with 2% uranyl acetate (pH 4.0) and examined using a Jeol JEM 1400 transmission electron microscope (TEM) (Tokyo, Japan).

4.4. *Phage Host Range and Efficiency of Plating Determination*

Phage host range was determined on the strains listed in Table 3, using the spot test method [43]. Briefly, 100 μL of each host-growing culture were added to 3 mL of TSB-soft agar and poured onto TSB agar plates. The bacterial lawns were spotted with 10 μL of serial 10-fold dilutions of the phage suspension and incubated at 37 °C for overnight and results were analyzed. The EOP was calculated by dividing the titer of the phage (PFU/mL) obtained in each isolate by the titer determined in the propagating host. EOP was recorded as high (>10%), moderate (0.01–9%) or low (<0.01%) [43].

4.5. *Genome Sequencing and in Silico Analysis*

P. aeruginosa EPA1 genomic DNA was extracted according to the standard methods with phenol-chloroform-isoamyl alcohol, as described elsewhere [44]. The genome was sequenced using the genome sequencer FLX Instrument (Roshe Life Science), *de novo* assembled using Geneious R9 and manually inspected. The genome was annotated using MyRAST algorithm [45]. The CDSs putative functions were assigned using BLASTP [46] with tRNAs being predicted with tRNAscan-SE [47] and ARAGORN [48]. HHPRED [49] was used to detect protein homology and structure prediction. N-terminal signal peptides with SignalP 3.0. [50] Transcriptional factors were determined by MEME [51] and ARNold [52] for the promoter and rho-independent terminators, respectively. For comparative studies, genomic comparisons were made using BLASTN and OrthoVenn [53], for DNA and protein sequence similarities.

4.6. Minimal Inhibitory Concentration Determination

Seven different antibiotics were selected to use in the study: gentamicin, kanamycin, tetracycline, chloramphenicol, erythromycin, ciprofloxacin and meropenem. MIC values were determined by the microdilution method for *P. aeruginosa* PAO1 according to the described method [54] (Table 2).

4.7. Establishing Mono and Dual-Species Biofilms

Mono and dual-species biofilm formation was performed in 24 polystyrene well plates (Orange Scientific, Braine-l'Alleud, Belgium). For initiating the biofilms, one bacterial colony (*P. aeruginosa* or *S. aureus)* was incubated in TSB overnight in an orbital shaker (120 rpm, BIOSAN ES-20/60) at 37 °C.

For establishing mono-species biofilms, 10 µL of the starter culture were transferred into 24-well plates containing 990 µL of fresh TSB media. The plates were incubated for 24 h in an orbital shaker incubator (120 rpm, BIOSAN ES-20/60) at 37 °C. After 24 h, half of the growth media (500 µL TSB, 1:1, *v:v*) was replaced with fresh TSB then, incubated for more 24 h.

For dual-species biofilms, the procedure was similar with some differences. *S. aureus* cells were inoculated prior to *P. aeruginosa* addition. Thus, biofilms were initiated with 10 µL of the overnight culture of *S. aureus* (~10^8 CFU/mL) in 990 µL TSB and incubated for 24 h in an orbital shaker (120 rpm) at 37 °C. After that, half of the growth media (500 µL TSB, 1:1, *v:v*) was replaced with TSB including 10 µL of the starter culture of *P. aeruginosa* (~10^8 CFU/mL, 1:49, *v/v*) and incubated for additional 24 h.

In mono and dual-species biofilms, the liquid part was aspirated, and the wells were washed twice with saline solution (0.9% NaCl (*w/v*)) to remove planktonic bacteria. Biofilms were further scraped in saline solution (1 mL) using a micropipette tip, and the number of viable cells was determined using the microdrop method [43]. Three independent experiments were performed in duplicate.

4.8. Biofilm Challenge

Forty-eight hours old mono and dual-species biofilms were treated with the antimicrobials; individually, in simultaneous or sequential combinations for 24 h post-treatment. Briefly, biofilms were washed twice with the saline solution and antibiotics (MIC or 8× MIC) and phage (MOI 1) were applied in TSB, individually. Also, the efficacy of two combinations was tested. In simultaneous combination, one of the selected antibiotics with MIC or 8× MIC combined with phage at MOI 1 in TSB solution was added into biofilm-bearing wells for 24 h post-treatment. In sequential combination, phage at MOI 1 was added into biofilm-bearing wells for 6 h, and then one of the antibiotics (final concentration of MIC or 8× MIC) was added into the well plates for additional 18 h. The reverse sequential combination, gentamicin at MIC and 8× MIC were added into biofilm wells for 6 h, and then phage EPA1 with MOI 1 was added into the well plates for additional 18 h. The number of viable cells was enumerated using the microdrop method [43].

The potential interaction of treatments with biofilms is described as synergistic when the biofilm reduction in combinations is greater than the sum of individual treatments of antimicrobials, as described by Chaudhry et al. [18]. An interaction is described as an additive when the biofilm reduction in combinations is similar/equal to the sum of individual treatments of antimicrobials. An interaction is described as antagonistic when the biofilm reduction in combinations is lower than the sum of individual treatments of antimicrobials. The efficacy of treatment can be defined according to the result of the following equations:

$$\log(\text{AP}) - ((\log(\text{A}) + \log(\text{P}))$$

$$=0, \text{ additive interaction}$$

$$>0, \text{ Synergistic interaction}$$

$$<0, \text{ Antagonistic interaction}$$

In which P is the reduction in the number of viable biofilm cell in individual phage treatment; A the reduction in the number of viable biofilm cell in individual antibiotic treatment; AP the reduction in the number of viable biofilm cell in combined treatment.

4.9. Development of Probes for Biofilm Imaging

To assess the structure of biofilm models, bacteria-specific fluorescent probes were constructed using phage proteins. Given the expertise of our laboratory on phage-based protein construction, this method was selected instead of the use of commercial probes in the biofilm imaging process.

The red fluorescent mCherry gene derived from the DsRed of *Discosoma* sea anemones was inserted into the plasmid pET28a (+) (Novagen, Merck, Darmstady, Germany), between the *Sac*I and *Xho*I restriction sites conserving the plasmid N-terminal hexahistidine (His)-tag sequence and originating the pET_mCherry plasmid. Primers were designed to obtain fragments of the EPA1_gp81 tail fiber C-terminus (further referred to as the EPA1_TFP with mCherry). The fragments were amplified with Phusion DNA Polymerase (ThermoFisher Scientific) with the EPA1 genome as DNA template and digested with the restriction enzymes *Sac*I and *Xho*I. The digested fragments were inserted into pET28a(+) and ligated with the T4 ligase (ThermoFisher Scientific) to obtain the construction (pET_ EPA1_TFP), further used to transform *E. coli* TOP10 competent cells (Invitrogen, California, USA). Colonies were screened through colony PCR and positives used for plasmid extraction and further confirmation through Sanger sequencing. A correct pET_EPA1_TFP plasmid was used to transform competent *E. coli* BL21. Besides, the *S. aureus* phage vB_SauM-LM12 [43] endolysin truncated at its N-terminus and fused with GFP (LM12_AMI-SH3 with GFP) was constructed by our group [55].

Expression of the different peptides was performed as described before [56]. Briefly, the cells harboring recombinant plasmids were grown at 37 °C in Lysogeny Broth (LB) supplemented with 50 μg/mL of kanamycin until reaching an optical density at 620 nm (OD_{620nm}) of 0.6. Recombinant protein expression induced with isopropyl-β-D-thiogalactopyranoside (IPTG; Thermo Fisher Scientific) at 1 mM final concentration was carried overnight at 16 °C, 150 rpm. Cells were collected by centrifugation (9000× *g*, 15 min, 4 °C) and further resuspended in lysis buffer (20 mM NaH_2PO_4, 500 mM sodium chloride, 10 mM imidazole, pH 7.4). Cell disruption was made by thaw-freezing (3 cycles, from −80 °C to room temperature) followed by a 5 min sonication (Cole-Parmer Ultrasonic Processor) for 10 cycles (30 s ON, 30 s OFF), 40% amplitude. Soluble cell-free extracts were separated by centrifugation, filtered, and loaded on a 1 mL HisPur™Ni-NTA Resin (Thermo Fisher Scientific) stacked into a Polypropylene column (Qiagen). After two washing steps with protein-dependent imidazole concentrations (lysis buffer supplemented with 20 mM imidazole in the first wash, and 40 mM imidazole in the second wash), the protein was eluted with 300 mM imidazole. Protein fractions were observed through SDS-PAGE. The purified proteins were quantified using the Pierce™ BCA Protein Assay Kit (Thermo Ficher).

4.10. CLSM Analysis

CLSM was performed as described before [56] with some modifications. Briefly, the 13 mm in diameter Thermanox® Plastic Coverslip (Rochester, New York, USA) were placed in 24-well plates, and mono and dual-biofilms were formed as mentioned before. Coverslips were further washed twice with saline solution, and treatments were applied. After the treatment, the suspension was aspirated, and the wells were washed twice with 0.9% saline solution. The fluorescence probes, EPA1_TFP with mCherry (laser excitation line 635nm and emissions filters BA 655–755, red channel) and LM12_AMI-SH3 with GFP (laser excitation line 488 nm and emissions filters BA 505–605, green channel) were used for detection of cells in biofilms. The coupons were stained with 15 μL of probes in a final concentration of 20 mM for 15 min. After, the images were acquired in a Confocal Scanning Laser Microscope (Olympus B × 61, Model FluoView 1000) with the program FV10-Ver4.1.1.5 (Olympus). For each condition, three independent biofilms were used.

4.11. Statistical Analysis

The results of assays were compared using two-way analysis of variance (ANOVA) by applying the Tukey's multiple comparisons test using Prism 6 (GraphPad, La Jolla, CA, USA). Means and standard deviations (SD) were calculated with the software. Differences among conditions were considered statistically significant when $p < 0.001$.

4.12. Nucleotide Sequence Accession Number

The genome sequence of *Pseudomonas* phage vB_PaM_EPA1 was deposited in the GenBank database under the accession number MN013356.

Supplementary Materials: The following are available online at http://www.mdpi.com/2079-6382/8/3/103/s1, Figure S1. One-step growth curve of phage vB_PaM_EPA1 in *P. aeruginosa* Sifa_Pa_1.5 at 37 °C. Shown are the PFU per infected cell; Figure S2. Treatment of *S. aureus* 48 h biofilms population with different antimicrobial agents (gentamicin, ciprofloxacin and meropenem) for 24 h; Figure S3. The effect of the antibiotics (gentamicin, ciprofloxacin and meropenem) on phage replication for 24 h post-treatment; Figure S4. Treatment of *P. aeruginosa* PAO1 48 h biofilms population with reverse sequential combinations. Table S1. Bacterial strains feature and susceptibility to phage EPA1; Table S2. The genome annotation of phage vB_PaM_EPA1.

Author Contributions: J.A., L.M. and E.A. conceived and designed the experiments; E.A. and S.C. performed the experiments; E.A., L.M., S.S., H.O., S.K. and J.A. analyzed the data; E.A. wrote the paper. The overall editing was performed by L.M. and J.A. All authors read and approved the final manuscript.

Funding: This study was supported by the Portuguese Foundation for Science and Technology (FCT) under the scope of the strategic funding of UID/BIO/04469/2013 unit, COMPETE 2020 (POCI-01-0145-FEDER-006684) and the Project PTDC/BBB-BSS/6471/2014 (POCI-01-0145-FEDER-016678). This work was also supported by BioTecNorte operation (NORTE-01-0145-FEDER-000004) funded by the European Regional Development Fund under the scope of Norte2020 - Programa Operacional Regional do Norte. Ergun Akturk acknowledges FCT for grant PD/BD/135254/2017.

Conflicts of Interest: The authors declare no conflict of interest.

References

1. Røder, H.L.; Sørensen, S.J.; Burmølle, M. Studying Bacterial Multispecies Biofilms: Where to Start? *Trends Microbiol.* **2016**, *24*, 503–513. [CrossRef] [PubMed]
2. Wolcott, R.D.; Rhoads, D.D.; Bennett, M.E.; Wolcott, B.M.; Gogokhia, L.; Costerton, J.W.; Dowd, S.E. Chronic wounds and the medical biofilm paradigm. *J. Wound Care* **2014**, *19*, 45–53. [CrossRef] [PubMed]
3. Römling, U.; Balsalobre, C. Biofilm infections, their resilience to therapy and innovative treatment strategies. *J. Intern. Med.* **2012**, *272*, 541–561. [CrossRef]
4. Hall, C.W.; Mah, T.-F. Molecular mechanisms of biofilm-based antibiotic resistance and tolerance in pathogenic bacteria. *FEMS Microbiol. Rev.* **2017**, *41*, 276–301. [CrossRef] [PubMed]
5. Lopes, S.P.; Ceri, H.; Azevedo, N.F.; Pereira, M.O. Antibiotic resistance of mixed biofilms in cystic fibrosis: Impact of emerging microorganisms on treatment of infection. *Int. J. Antimicrob. Agents* **2012**, *40*, 260–263. [CrossRef] [PubMed]
6. Hotterbeekx, A.; Kumar-Singh, S.; Goossens, H.; Malhotra-Kumar, S. In vivo and In vitro Interactions between *Pseudomonas aeruginosa* and *Staphylococcus* spp. *Front. Cell. Infect. Microbiol.* **2017**, *7*, 1–13. [CrossRef]
7. Yoshida, S.; Ogawa, N.; Fujii, T.; Tsushima, S. Enhanced biofilm formation and 3-chlorobenzoate degrading activity by the bacterial consortium of *Burkholderia sp.* NK8 and *Pseudomonas aeruginosa* PAO1. *J. Appl. Microbiol.* **2009**, *106*, 790–800. [CrossRef]
8. Pastar, I.; Nusbaum, A.G.; Gil, J.; Patel, S.B.; Chen, J.; Valdes, J.; Stojadinovic, O.; Plano, L.R.; Tomic-Canic, M.; Davis, S.C. Interactions of Methicillin Resistant *Staphylococcus aureus* USA300 and *Pseudomonas aeruginosa* in Polymicrobial Wound Infection. *PLoS ONE* **2013**, *8*, e56846. [CrossRef]
9. Kart, D.; Tavernier, S.; Van Acker, H.; Nelis, H.J.; Coenye, T. Activity of disinfectants against multispecies biofilms formed by *Staphylococcus aureus*, *Candida albicans* and *Pseudomonas aeruginosa*. *Biofouling* **2014**, *30*, 377–383. [CrossRef]

10. Radlinski, L.; Rowe, S.E.; Kartchner, L.B.; Maile, R.; Cairns, B.A.; Vitko, N.P.; Gode, C.J.; Lachiewicz, A.M.; Wolfgang, M.C.; Conlon, B.P. *Pseudomonas aeruginosa* exoproducts determine antibiotic efficacy against *Staphylococcus aureus*. *PLoS Biol.* **2017**, *15*, 1–25. [CrossRef]
11. Nguyen, A.T.; Oglesby-Sherrouse, A.G. Interactions between *Pseudomonas aeruginosa* and *Staphylococcus aureus* during co-cultivations and polymicrobial infections. *Appl. Microbiol. Biotechnol.* **2016**, *100*, 6141–6148. [CrossRef] [PubMed]
12. Fong, S.A.; Drilling, A.; Morales, S.; Cornet, M.E.; Woodworth, B.A.; Fokkens, W.J.; Psaltis, A.J.; Vreugde, S.; Wormald, P.-J. Activity of Bacteriophages in Removing Biofilms of *Pseudomonas aeruginosa* Isolates from Chronic Rhinosinusitis Patients. *Front. Cell. Infect. Microbiol.* **2017**, *7*, 418. [CrossRef]
13. Ozkan, I.; Akturk, E.; Yeshenkulov, N.; Atmaca, S.; Rahmanov, N.; Atabay, H.I. Lytic Activity of Various Phage Cocktails on Multidrug-Resistant Bacteria. *Clin. Investig. Med.* **2016**, *39*, S66–S70. [CrossRef]
14. Pires, D.P.; Melo, L.D.R.; Vilas Boas, D.; Sillankorva, S.; Azeredo, J. Phage therapy as an alternative or complementary strategy to prevent and control biofilm-related infections. *Curr. Opin. Microbiol.* **2017**, *39*, 48–56. [CrossRef] [PubMed]
15. Knezevic, P.; Curcin, S.; Aleksic, V.; Petrusic, M.; Vlaski, L. Phage-antibiotic synergism: A possible approach to combatting *Pseudomonas aeruginosa*. *Res. Microbiol.* **2013**, *164*, 55–60. [CrossRef] [PubMed]
16. Torres-Barceló, C.; Arias-Sánchez, F.I.; Vasse, M.; Ramsayer, J.; Kaltz, O.; Hochberg, M.E. A window of opportunity to control the bacterial pathogen *Pseudomonas aeruginosa* combining antibiotics and phages. *PLoS ONE* **2014**, *9*, e106628. [CrossRef] [PubMed]
17. Torres-Barceló, C.; Hochberg, M.E. Evolutionary Rationale for Phages as Complements of Antibiotics. *Trends Microbiol.* **2016**, *24*, 249–256. [CrossRef] [PubMed]
18. Chaudhry, W.N.; Concepcion-Acevedo, J.; Park, T.; Andleeb, S.; Bull, J.J.; Levin, B.R. Synergy and order effects of antibiotics and phages in killing *Pseudomonas aeruginosa* biofilms. *PLoS ONE* **2017**, *12*, e0168615. [CrossRef] [PubMed]
19. Danis-Wlodarczyk, K.; Vandenheuvel, D.; Jang, H.B.; Briers, Y.; Olszak, T.; Arabski, M.; Wasik, S.; Drabik, M.; Higgins, G.; Tyrrell, J.; et al. A proposed integrated approach for the preclinical evaluation of phage therapy in *Pseudomonas infections*. *Sci. Rep.* **2016**, *6*, 28115. [CrossRef] [PubMed]
20. Ackermann, H.W. 5500 Phages examined in the electron microscope. *Arch. Virol.* **2007**, *152*, 227–243. [CrossRef] [PubMed]
21. Knezevic, P.; Voet, M.; Lavigne, R. Prevalence of Pf1-like (pro)phage genetic elements among *Pseudomonas aeruginosa* isolates. *Virology* **2015**, *483*, 64–71. [CrossRef] [PubMed]
22. Palmer, K.L.; Aye, L.M.; Whiteley, M. Nutritional cues control *Pseudomonas aeruginosa* multicellular behavior in cystic fibrosis sputum. *J. Bacteriol.* **2007**, *189*, 8079–8087. [CrossRef] [PubMed]
23. Hoffman, L.R.; Deziel, E.; D'Argenio, D.A.; Lepine, F.; Emerson, J.; McNamara, S.; Gibson, R.L.; Ramsey, B.W.; Miller, S.I. Selection for *Staphylococcus aureus* small-colony variants due to growth in the presence of *Pseudomonas aeruginosa*. *Proc. Natl. Acad. Sci. USA* **2006**, *103*, 19890–19895. [CrossRef] [PubMed]
24. Qin, Z.; Yang, L.; Qu, D.; Molin, S.; Tolker-Nielsen, T. *Pseudomonas aeruginosa* extracellular products inhibit *staphylococcal* growth, and disrupt established biofilms produced by *Staphylococcus epidermidis*. *Microbiology* **2009**, *155*, 2148–2156. [CrossRef] [PubMed]
25. Dietrich, L.E.P.; Price-Whelan, A.; Petersen, A.; Whiteley, M.; Newman, D.K. The phenazine pyocyanin is a terminal signalling factor in the quorum sensing network of *Pseudomonas aeruginosa*. *Mol. Microbiol.* **2006**, *61*, 1308–1321. [CrossRef] [PubMed]
26. DeLeon, S.; Clinton, A.; Fowler, H.; Everett, J.; Horswill, A.R.; Rumbaugh, K.P. Synergistic Interactions of *Pseudomonas aeruginosa* and *Staphylococcus aureus* in an In Vitro Wound Model. *Infect. Immun.* **2014**, *82*, 4718–4728. [CrossRef] [PubMed]
27. Pires, D.P.; Dötsch, A.; Anderson, E.M.; Hao, Y.; Khursigara, C.M.; Lam, J.S.; Sillankorva, S.; Azeredo, J. A Genotypic Analysis of Five *P. aeruginosa* Strains after Biofilm Infection by Phages Targeting Different Cell Surface Receptors. *Front. Microbiol.* **2017**, *8*, 1229. [CrossRef]
28. Essoh, C.; Latino, L.; Midoux, C.; Blouin, Y.; Loukou, G.; Nguetta, S.-P.A.P.A.; Lathro, S.; Cablanmian, A.; Kouassi, A.K.; Vergnaud, G.; et al. Investigation of a Large Collection of *Pseudomonas aeruginosa* Bacteriophages Collected from a Single Environmental Source in Abidjan, Côte d'Ivoire. *PLoS ONE* **2015**, *10*, 1–25. [CrossRef]
29. Sturino, J.M.; Klaenhammer, T.R. Inhibition of bacteriophage replication in *Streptococcus thermophilus* by subunit poisoning of primase. *Microbiology* **2007**, *153*, 3295–3302. [CrossRef]

30. Torres-Barceló, C.; Gurney, J.; Gougat-Barberá, C.; Vasse, M.; Hochberg, M.E. Transient negative effects of antibiotics on phages do not jeopardise the advantages of combination therapies. *FEMS Microbiol. Ecol.* **2018**, *94*, fiy107. [CrossRef]
31. Gutiérrez, D.; Briers, Y.; Rodríguez-Rubio, L.; Martínez, B.; Rodríguez, A.; Lavigne, R.; García, P. Role of the Pre-neck Appendage Protein (Dpo7) from Phage vB_SepiS-phiIPLA7 as an Anti-biofilm Agent in *Staphylococcal* Species. *Front. Microbiol.* **2015**, *6*, 1315. [CrossRef] [PubMed]
32. Lin, D.M.; Koskella, B.; Lin, H.C. Phage therapy: An alternative to antibiotics in the age of multi-drug resistance. *World J. Gastrointest. Pharmacol. Ther.* **2017**, *8*, 162. [CrossRef] [PubMed]
33. Vilas Boas, D.; Almeida, C.; Sillankorva, S.; Nicolau, A.; Azeredo, J.; Azevedo, N.F. Discrimination of bacteriophage infected cells using locked nucleic acid fluorescent in situ hybridization (LNA-FISH). *Biofouling* **2016**, *32*, 179–190. [CrossRef] [PubMed]
34. Burmølle, M.; Ren, D.; Bjarnsholt, T.; Sørensen, S.J. Interactions in multispecies biofilms: Do they actually matter? *Trends Microbiol.* **2014**, *22*, 84–91. [CrossRef] [PubMed]
35. Smith, A.C.; Rice, A.; Sutton, B.; Gabrilska, R.; Wessel, A.K.; Whiteley, M.; Rumbaugh, K.P. Albumin Inhibits *Pseudomonas aeruginosa* Quorum Sensing and Alters Polymicrobial Interactions. *Infect. Immun.* **2017**, *85*, 1–12. [CrossRef] [PubMed]
36. Filkins, L.M.; Graber, J.A.; Olson, D.G.; Dolben, E.L.; Lynd, L.R.; Bhuju, S.; O'Toole, G.A. Coculture of *Staphylococcus aureus* with *Pseudomonas aeruginosa* Drives *S. aureus* towards Fermentative Metabolism and Reduced Viability in a Cystic Fibrosis Model. *J. Bacteriol.* **2015**, *197*, 2252–2264. [CrossRef] [PubMed]
37. Mendes, J.J.; Neves, J. Diabetic Foot Infections: Current Diagnosis and Treatment. *J. Diabet. Foot Complicat.* **2012**, *4*, 26–45.
38. Leid, J.G.; Willson, C.J.; Shirtliff, M.E.; Hassett, D.J.; Parsek, M.R.; Jeffers, A.K. The exopolysaccharide alginate protects *Pseudomonas aeruginosa* biofilm bacteria from IFN-gamma-mediated macrophage killing. *J. Immunol.* **2005**, *175*, 7512–7518. [CrossRef]
39. Ryder, C.; Byrd, M.; Wozniak, D.J. Role of polysaccharides in *Pseudomonas aeruginosa* biofilm development. *Curr. Opin. Microbiol.* **2007**, *10*, 644–648. [CrossRef]
40. Colvin, K.M.; Irie, Y.; Tart, C.S.; Urbano, R.; Whitney, J.C.; Ryder, C.; Howell, P.L.; Wozniak, D.J.; Parsek, M.R. The Pel and Psl polysaccharides provide *Pseudomonas aeruginosa* structural redundancy within the biofilm matrix. *Environ. Microbiol.* **2012**, *14*, 1913–1928. [CrossRef]
41. Rémy, B.; Mion, S.; Plener, L.; Elias, M.; Chabrière, E.; Daudé, D. Interference in Bacterial Quorum Sensing: A Biopharmaceutical Perspective. *Front. Pharmacol.* **2018**, *9*, 203. [CrossRef] [PubMed]
42. Koban, Y.; Genc, S.; Bilgin, G.; Cagatay, H.H.; Ekinci, M.; Gecer, M.; Yazar, Z. Toxic Anterior Segment Syndrome following Phacoemulsification Secondary to Overdose of Intracameral Gentamicin. *Case Rep. Med.* **2014**, *2014*, 143564. [CrossRef] [PubMed]
43. Melo, L.D.R.; Brandão, A.; Akturk, E.; Santos, S.B.; Azeredo, J. Characterization of a new *Staphylococcus aureus* Kayvirus harboring a lysin active against biofilms. *Viruses* **2018**, *10*, 182. [CrossRef] [PubMed]
44. Oliveira, H.; Pinto, G.; Oliveira, A.; Oliveira, C.; Faustino, M.A.; Briers, Y.; Domingues, L.; Azeredo, J. Characterization and genome sequencing of a *Citrobacter freundii* phage CfP1 harboring a lysin active against multidrug-resistant isolates. *Appl. Microbiol. Biotechnol.* **2016**, *100*, 10543–10553. [CrossRef] [PubMed]
45. Aziz, R.K.; Bartels, D.; Best, A.; DeJongh, M.; Disz, T.; Edwards, R.A.; Formsma, K.; Gerdes, S.; Glass, E.M.; Kubal, M.; et al. The RAST Server: Rapid annotations using subsystems technology. *BMC Genom.* **2008**, *9*, 75. [CrossRef] [PubMed]
46. Altschul, S.F.; Gish, W.; Miller, W.; Myers, E.W.; Lipman, D.J. Basic local alignment search tool. *J. Mol. Biol.* **1990**, *215*, 403–410. [CrossRef]
47. Schattner, P.; Brooks, A.N.; Lowe, T.M. The tRNAscan-SE, snoscan and snoGPS web servers for the detection of tRNAs and snoRNAs. *Nucleic Acids Res.* **2005**, *33*, W686–W689. [CrossRef]
48. Laslett, D.; Canback, B. ARAGORN, a program to detect tRNA genes and tmRNA genes in nucleotide sequences. *Nucleic Acids Res.* **2004**, *32*, 11–16. [CrossRef]
49. Söding, J. Protein homology detection by HMM-HMM comparison. *Bioinformatics* **2005**, *21*, 951–960. [CrossRef]
50. Bendtsen, J.D.; Nielsen, H.; Von Heijne, G.; Brunak, S. Improved prediction of signal peptides: SignalP 3.0. *J. Mol. Biol.* **2004**, *340*, 783–795. [CrossRef]

51. Bailey, T.L.; Boden, M.; Buske, F.A.; Frith, M.; Grant, C.E.; Clementi, L.; Ren, J.; Li, W.W.; Noble, W.S. MEME Suite: Tools for motif discovery and searching. *Nucleic Acids Res.* **2009**, *37*, 202–208. [CrossRef] [PubMed]
52. Naville, M.; Ghuillot-Gaudeffroy, A.; Marchais, A.; Gautheret, D. ARNold: A web tool for the prediction of rho-independent transcription terminators. *RNA Biol.* **2011**, *8*, 11–13. [CrossRef] [PubMed]
53. Wang, Y.; Coleman-Derr, D.; Chen, G.; Gu, Y.Q. OrthoVenn: A web server for genome wide comparison and annotation of orthologous clusters across multiple species. *Nucleic Acids Res.* **2015**, *43*, W78–W84. [CrossRef] [PubMed]
54. Cui, H.; Ma, C.; Lin, L. Co-loaded proteinase K/thyme oil liposomes for inactivation of *Escherichia coli* O157:H7 biofilms on cucumber. *Food Funct.* **2016**, *7*, 4030–4040. [CrossRef] [PubMed]
55. Costa, S. Development of a Phage-Based Lab-on-Chip for the Detection of Foodborne Pathogens. Master's Thesis, University of Minho, Braga, Portugal, 2016.
56. Cerca, N.; Gomes, F.; Pereira, S.; Teixeira, P.; Oliveira, R. Confocal laser scanning microscopy analysis of *S. epidermidis* biofilms exposed to farnesol, vancomycin and rifampicin. *BMC Res. Notes* **2012**, *5*, 244. [CrossRef] [PubMed]

antibiotics

MDPI

Article

Phage-Bacterial Dynamics with Spatial Structure: Self Organization around Phage Sinks Can Promote Increased Cell Densities

James J. Bull [1,2,3,*], Kelly A. Christensen [4,5], Carly Scott [4,6], Benjamin R. Jack [7], Cameron J. Crandall [6] and Stephen M. Krone [4,5,8,*]

1 Department of Integrative Biology, University of Texas, Austin, TX 78712, USA
2 The Institute for Cellular and Molecular Biology, University of Texas, Austin, TX 78712, USA
3 Center for Computational Biology and Bioinformatics, University of Texas, Austin, TX 78712, USA
4 Department of Mathematics, University of Idaho, Moscow, ID 83844, USA; chri4898@vandals.uidaho.edu (K.A.C.); scot9278@vandals.uidaho.edu (C.S.)
5 Center for Modeling Complex Interactions, University of Idaho, Moscow, ID 83844, USA
6 Department of Biological Sciences, University of Idaho, Moscow, ID 83844, USA; cjcrandall91@gmail.com
7 The Institute for Cellular and Molecular Biology, University of Texas, Austin, TX 78712, USA; benjamin.r.jack@gmail.com
8 Institute for Bioinformatics and Evolutionary Studies, University of Idaho, Moscow, ID 83844, USA
* Correspondence: bull@utexas.edu (J.J.B.); krone@uidaho.edu (S.M.K.); Tel.: +1-512-471-8266 (J.J.B.)

Received: 27 December 2017; Accepted: 23 January 2018; Published: 29 January 2018

Abstract: Bacteria growing on surfaces appear to be profoundly more resistant to control by lytic bacteriophages than do the same cells grown in liquid. Here, we use simulation models to investigate whether spatial structure per se can account for this increased cell density in the presence of phages. A measure is derived for comparing cell densities between growth in spatially structured environments versus well mixed environments (known as mass action). Maintenance of sensitive cells requires some form of phage death; we invoke death mechanisms that are spatially fixed, as if produced by cells. Spatially structured phage death provides cells with a means of protection that can boost cell densities an order of magnitude above that attained under mass action, although the effect is sometimes in the opposite direction. Phage and bacteria self organize into separate refuges, and spatial structure operates so that the phage progeny from a single burst do not have independent fates (as they do with mass action). Phage incur a high loss when invading protected areas that have high cell densities, resulting in greater protection for the cells. By the same metric, mass action dynamics either show no sustained bacterial elevation or oscillate between states of low and high cell densities and an elevated average. The elevated cell densities observed in models with spatial structure do not approach the empirically observed increased density of cells in structured environments with phages (which can be many orders of magnitude), so the empirical phenomenon likely requires additional mechanisms than those analyzed here.

Keywords: biofilm; phage therapy; resistance; bacteriophage; models; agent based; mass action

1. Introduction

Bacteriophages are ubiquitous predators of bacteria, and they have long been entertained as having possible therapeutic utility in medicine. However, therapeutic utility is typically a matter of controlling the bacterial populations, and population control is not easily inferred from the mere fact that individuals of one species can kill individuals of another species. The difference between killing that achieves population control and killing that has little effect on the population rests on quantitative properties of the killing. Fortunately, phages are easily manipulated in the lab and thus easily studied to address dynamics and the control of bacterial populations.

The history of work on phage-bacterial dynamics has been dominated by liquid cultures in which bacteria are suspended as single cells at uniform density. Such cultures are routinely modeled as ordinary differential equations (ODEs) with assumptions of "mass action." Mass action refers to an environment in which all individuals are "well mixed," as would occur in a chemostat or batch culture, and so collisions occur at random. In such a system (see Equation (7)), interaction terms appear as products of bulk densities and essential parameters are easily estimated. The typical outcome following a lytic phage assault on a dense population of sensitive bacteria in liquid is killing of the bacterial population by many orders of magnitude, followed by a rebound of bacteria genetically resistant to the phage [1,2] with possible long-term coevolutionary arms races [3]. This work has led to many insights about bacterial and bacteriophage biology but has also given rise to a perception that bacterial escape from phages is chiefly through evolution of genetic resistance. However, we now know that many bacteria spend much of their lives in structured environments such as biofilms and aggregates, and bacterial biology in structured environments is fundamentally different than in liquid suspensions [4–6]. Spatially structured bacterial populations are difficult to control—they may persist seemingly indefinitely amid ongoing phage attack (they also survive antibiotic attack), and this persistence does not appear to be from genetic resistance [7–13]. Understanding the nature of this coexistence may be critical to phage therapy. Is it spatial structure itself that allows bacterial escape, or is it an indirect consequence of spatial structure on bacterial habits that allows the escape?

The goal of this study is to use models to understand the maintenance of high densities of sensitive bacteria amid phage attack in spatially structured environments. Our ultimate motivation is to develop phage interventions for controlling bacteria, which requires understanding of how bacteria normally escape. Does spatial structure per se allow for easy persistence, or does escape require cells to behave differently in structured environments than in liquid ones? We use computational models to explore the dynamic nature of the phage-bacterial interaction in spatially structured populations, identifying which mechanisms enable bacterial persistence at high densities. The empirical evidence is that sensitive bacteria easily persist, but identifying a process that may reasonably account for the coexistence is challenging.

2. Empirical Anomalies and Possible Causes

Various observations on bacteria grown under spatial structure suggest that genetically sensitive bacteria can be maintained as the dominant population in the presence of phage, at least in the short term [8,10–12,14]. The environmental contexts in these examples are diverse. The phage typically reduce bacterial numbers 1 or more orders of magnitude, but the remaining population is predominantly sensitive and persists at a much higher density than would occur in liquid. The phage sensitivity of residual populations is sometimes measured directly or is inferred from dynamic principles, such as the continuing high output of phage (which could not grow on genetically resistant cells). In some cases, the surviving bacterial strain is a genetic mutant that is fundamentally sensitive to phage but exhibits reduced adsorption (e.g., mucoidy); the bacteria are merely maintained at higher levels than explicable by basic dynamics principles (e.g., [15]).

As one striking example, Darch et al. [14] grew *Pseudomonas aeruginosa* in a synthetic sputum medium; cell numbers were measured non-destructively with confocal microscopy. The cells grew in aggregates. Addition of phage to an established culture resulted in a less than 1-log drop in bacterial numbers (measured in situ). However, when the bacteria were grown in liquid (albeit in different media), addition of phage resulted in a 7-log drop. In a second example, Lu and Colins [10] grew 24 h *E. coli* biofilms in peg-lid microtiter plates (0.2 mL volumes per well). After media replacement, 24 h treatment with phage T7 led to approximately a 2-log reduction in cell density, but close to 10^5 cells remained (their Fig. 3B). However, treatment with a T7 phage engineered to encode an enzyme that degrades a bacterial matrix component led to another nearly 2-log reduction in cell density. Density of the enzyme-free phage was $\approx 5 \times 10^8$/mL in the surrounding liquid. The fact that the enzyme had such a profound effect indicates that sensitive cells were sequestered from the no-enzyme phage while surrounded with a phage density that should have been more than sufficient to eliminate nearly all of them.

Compared to mass action, the most obvious consequence of spatial structure is local variation in the abundance of bacteria and phage. However, this spatial variation arises, reproduction of phage and bacteria enhances that variation, whereas diffusion diminishes it. Structure leads to expanding concentrations of bacteria (colonies) and to high concentrations of phages near bacterial clusters that have been invaded [16–18]. The spatial variation in abundance will interact with any of several factors that could be contributors to the long-term co-maintenance of sensitive bacteria and lytic phages, as follows.

Resource concentration. Phage growth is known to be reduced on cells that are starved [19,20], a phenomenon easily appreciated from the halting of plaque growth on plates after the bacterial lawn matures. In spatial environments, high concentrations of bacteria will depress resources locally, suppressing phage growth in those zones.

Barriers and gradients. Spatial structure allows the local buildup of substances exuded from cells, such as expolysaccharides (EPS), ions, signalling molecules, and outer membrane vesicles [1,8,21]. These agents may trap phages, drive phages away with electrostatic forces, or alter the concentration of factors necessary for phage adsorption.

Phage-adsorbing debris. The remnants of cells lysed by phages may continue to adsorb phage perhaps irreversibly and thereby reduce the number of phage encountering live cells. Spatial structure will facilitate the buildup of debris around clusters of cells.

Co-infection and superinfection. Phage growth with spatial structure will often concentrate phages around cells, which for many phages will lead to high numbers of phages infecting the same cell [18]. This property will reduce the effective number of phage progeny and may allow cells to reach higher densities than in liquid.

Altered gene expression. Cells may vary gene expression specifically in response to surface attachment or signals received from adjacent cells (e.g., [22]). Changes in gene expression are not necessarily effects of spatially structured dynamics per se, but gene expression changes may themselves enable phage-bacterial co-existence. As an example, non-genetic variation in receptor abundance on cells can lead to high levels of the survival of genetically sensitive bacteria challenged with phages [23–26]. If bacterial growth with spatial structure amplifies variation in gene expression, that variation could enable bacterial escape and subsequent growth, more than in liquid.

3. Perspective: Does Spatial Structure Increase Bacterial Density?

The question addressed here is whether phage and cell dynamics that are spatial in nature allow cells to attain a higher density than if everything is well mixed. As our approach uses mathematical and computational models, this question requires understanding the difference between spatial structure and well-mixed conditions. Phage dynamics have traditionally been modeled under the assumptions of mass action, which assumes cells and phages are fully mixed and that interactions occur at rates determined by population averages. Mass action means that cells and phage have no assigned locations; they just exist. This mathematical convenience allows the process to be studied with ordinary differential equations [1,27–29]. With spatial structure, the locations of cells and phages are tracked over time, and interactions are location dependent. Typically, phages move through diffusion and cells remain in fixed locations (adjacent to parent cells). Thus, high densities of cells or phages can build up in parts of the environment while other parts have few or no individuals. Phage killing is local to the areas of high phage density.

Extensive computational analyses of spatially structured phage-bacterial dynamics have been undertaken in a few previous studies [16–18]. This pioneering work described many properties of dynamics unique to spatial structure, such as strong spatial co-localization of bacteria and phage, as well as spatial structure enabling coexistence over a wider range of parameter values than does mass action (due to greater oscillations with mass action).

Our study uses that foundation to ask a specific question: does spatially structured phage dynamics per se maintain a greater cell density than under mass action? The fact that spatial structure more easily allows coexistence [17] might suggest that spatial structure also increases bacterial density, but the effect of spatial structure was reported to stem from reduced global oscillations rather than an increase in (mean) bacterial density. Reduced oscillations could lead to greater coexistence without affecting mean density.

The reason for using models to study these processes is to develop understanding that cannot feasibly be obtained from empirical studies alone. The models allow control of variables so that effects of single variables can be isolated. From there, one may proceed to empirical studies to test specific processes.

4. Setting the Stage for Evaluating the Effect of Spatial Structure: Biological Consequences of Mass Action Are Well Studied

We use a variety of computational approaches to understand phage-bacterial dynamics in spatially structured environments. Whereas the outcomes of simulations are easy to interpret, understanding the causal parameters can be challenging because of the many environmental details that must simultaneously be specified to model dynamics in space. To help understand simulation results, and especially to motivate the types of analyses done with simulations, we offer a brief review of specific mass action results from previous studies using ordinary differential equations.

1. Mass action does not preclude high cell density. Although the typical pattern of phage-bacterial dynamics under mass action is one in which phages decimate the bacterial populations, there are mass action conditions in which high densities of sensitive bacteria can be maintained, typically with a low adsorption rate [28].
2. Maintenance of phages and bacteria requires some form of phage death. The ODE models typically assume a constant rate of phage death or clearance from the system.
3. Numerical solutions to the equations often exhibit undamped and even accelerating oscillations [17,28,29]. The oscillations complicate comparisons of cell density across systems (see below).

5. Formal Spatial Structure

We use computational simulations to consider the formal dynamics of phage and bacteria with spatial structure. Our simulations were based on a two-dimensional 'grid' of sites and included a mix of stochastic (random) and deterministic processes. In these models, every cell, phage or other agent has a location on the grid; at each time step, infection, reproduction and movement may occur (explained in Methods). These models have many components similar to those in mass action models, but with explicit spatial structure and rates that are locally determined. We are primarily interested in whether and how spatial structure affects the cell density maintained in the presence of phages. The grid models include versions that enforce spatial structure as well as mass action versions, although nearly all trials assumed spatial structure. In the mass action versions of the grid models, each individual gets relocated every time step.

Biologically, there are two general types of bacterial avoidance of phages that may be entertained. One is that bacteria are protected from phages, whether by reduced adsorption rate or by surrounding themselves with anti-phage protection. The second is that cells either produce or associate with phage-killing products but are otherwise intrinsically susceptible when phages encounter them. We focus on the latter here, chiefly because it is non-trivial. It is otherwise clear that fully protected bacteria will be able to grow to the limits permitted by the environment—as is well known from ODE models allowing evolution of genetically resistant bacteria. If spatially structured cell growth combined with phage death does not intrinsically promote higher bacterial densities by several orders of magnitude, then protection of individual bacteria becomes plausible as the main driver.

A challenge in switching from mass action to spatial structure lies in accommodating attachment of phage to bacteria. With mass action models, an adsorption rate coefficient (k) subsumes both the chance encounter of a bacteria with phage and the rate at which the phage sticks to the bacterium given an encounter [1]. With spatial structure, we are forced to separate encounter from attachment because the two processes are operating at different scales in different parts of the environment [16].

Although some types of physiological protection of cells may be imposed by the environment (e.g., temperature, metal ions that affect adsorption, pH), of interest here is how the bacteria can potentially influence the local environment to enact protection by blocking encounter with phages. Excretion of extracellular polysaccharides and other substances may directly slow or block phage access altogether, and some of the extracellular matrix may effectively kill phages by binding them irreversibly. Dead cells and outer membrane vesicles may act as decoys that bind phages and cause them to eject their genomes.

6. Results

The maintenance of sensitive cells amid phage attack depends fundamentally on phage density and thus on phage death mechanisms. In a closed environment with cells and phages, such as a flask, the absence of phage death (or other form of permanent loss/sequestration) will ensure that phage ultimately eliminate all sensitive cells. Once cells are abundant, even phages with poor adsorption rates will ultimately increase to such densities that cells are rapidly eliminated. In the absence of cells being completely protected from phage, some form of phage death is required to prevent the ultimate buildup of phage to the point that all cells are killed. While it is obvious that fully protected cells can grow with impunity in the presence of phages, it is less obvious how the interplay between phage growth and death will collaborate to allow coexistence of sensitive cells and phage. The latter is our focus here—how phage death mechanisms influence the density of cells maintained.

6.1. The Nature of Phage Death Used Here: EPS and Cellular Debris

We will model two phage death mechanisms: adsorption to exopolysaccharides (EPS) and adsorption to dead cells (debris). The main difference in implementation of these two mechanisms is that EPS is treated as a spatially static and permanent mechanism of phage death; debris is also assumed to be spatially static, but its creation waxes and wanes as phage kill more or fewer cells, and it is not permanent, instead having an intrinsic decay rate. The association of debris with phage abundance may lead to substantially different outcomes than with a static phage sink. EPS will be the mechanism employed in all but the last set of studies presented here (for reasons explained below).

We accept that the empirical evidence from liquid cultures does not support a major role of debris in causing phage death (e.g., phage titers in lysed cultures are often stable for months—even when the lysate is not filtered or cleared of bacterial debris—J.J. Bull personal observations). The implementation of death by debris is offered in the spirit of any mechanism that rises and falls with phage attack on cells. Furthermore, if debris is short-lived, it may have an impact but the mechanism be difficult to detect empirically. We note that our mechanisms of phage death do not necessarily obey any empirically established process, mostly for lack of effort to detect such processes. Nonetheless, our assumed processes are seemingly more realistic than the usual assumption of a constant, intrinsic phage death rate, and they fall within the broad realm of mechanisms that cells can use to potentially kill off phages (e.g., outer membrane vesicles). It will be shown that our assumption of a fixed level of permanent EPS is equivalent to a constant phage death rate in mass action models.

Spatial structure will alter the dynamics in several ways [16–18], and indeed, it is likely that different models of spatial structure will do so differently. Most fundamentally, a lack of uniform densities will often result, allowing cells to amplify in zones that are temporarily phage-free. As regards phage death, phage reproduction from individual cells will have progeny phage spatially clustered at least temporarily and thus subject to a common fate. In addition, cells may find refuge and amplify behind materials that bind phage and act as phage sinks.

6.2. A Formal Measure of Whether Cell Density Is Elevated

If spatial structure leads to an elevated cell density above that with complete mixing (mass action), it might seem sufficient to merely observe cell density alone. However, any comparison of cell densities between spatial structure and mass action is not straightforward, in part because there is no single cell density expected under mass action—the cell density, even at equilibrium, depends on many parameters, such as phage burst size, adsorption rate, and death rate, to mention a few. Complicating matters further, mass action processes can themselves lead to a high cell density at equilibrium. Thus, cell density alone cannot tell us whether spatial structure elevates cell density. The effect of spatial structure must be measured via some comparison to cell density in the absence of spatial structure, a comparison that otherwise avoids confounding the many differences between the two types of models.

One such approach is to directly compare cell density when spatial structure is present to that when it is absent in the simulation; abolishing spatial structure can be done by increasing the diffusion rates of phage and cells [16,17]. This approach is free of alternative interpretations, but it has the drawback that bacterial and phage numbers often oscillate with mass action [16,17,28]. Given the limited dynamical range of cell densities afforded by the simulations, bacteria may often go extinct in the simulations even when the equilibrium density is well above extinction (see below).

We adopt a related approach, one that takes advantage of a universal property of equilibrium under mass action, at which phage and bacterial densities are unchanging. Our approach identifies a reproduction number constant that will be used to scale bacterial densities, with a similar use in [29]. Every successful phage infection of a cell will, on average, lead to one new successful infection. This dynamical property of populations in reproductive equilibrium is commonly used in ecology [30]. In the context of phage-bacterial dynamics under mass action, it means that the following equality holds:

$$\frac{\text{rate of productive phage infection}}{\text{all sources of phage loss from the free state}} \times \text{ phage fecundity per infection } = 1. \tag{1}$$

The ratio on the LHS (left hand side) is merely the fraction of all rates leading to phage loss that result in phage reproduction. Since only one phage offspring from an infected cell will go to establish a new successful infection, the product equals unity on average. We denote the ratio on the LHS of Equation (1) as α. Phage fecundity per infection, known as burst size, is represented here as b. We have analytically confirmed that $\alpha b = 1$ at equilibrium in various mass action models (e.g., those in [27,28,31]) and not found any that violate the equality.

For the specific sources of phage loss in the spatial models, we propose

$$\alpha = \frac{k_C C}{k_C C + k_I I + k_D D + k_E E} \tag{2}$$

where C, I, D, E represent the densities of uninfected cells, infected cells, debris, and EPS, and k with appropriate subscripts denotes the various attachment/infection probabilities. The time-variable quantities in Equation (2) are C, I, and D, but not all models here allow infection of I and D; moreover, α is an increasing function of C and a decreasing function of I and D.

In this implementation, α is calculated with the parameters used and values observed in the simulations of spatial structure, but the value of α is otherwise interpreted as that which would obtain if the population obeyed mass action. In particular, the quantities in Equation (2) are calculated globally, ignoring the spatial structure that played a role in their generation. The extent to which αb exceeds 1 then measures the effect of spatial structure in conspiring to allow a higher density of cells than would accrue without spatial structure. It indicates, in effect, the added degree of protection experienced by cells in a spatial setting. If, for example, the current value is $\alpha b = 5$ in a spatial simulation, this should be interpreted to mean that if the system suddenly transformed to mass action dynamics, the phage progeny from a burst would infect an average of five uninfected cells. (Of course, this excess of infections would be sustained only briefly.) We have qualitatively confirmed this behavior with spatial simulations that had equilibrated by suddenly (in the middle of the simulation run) increasing phage diffusion and allowing cells to move as an approximation to mass action. Finally, in any trial, the maximum possible value of αb is b, but arbitrarily large values of b can be tested for compatibility with cell maintenance.

The observed αb in spatially structured trials is not a measure of cell density directly. However, in the absence of debris attachment ($k_D = 0$) and superinfection of infected cells ($k_I = 0$), it may be used to calculate the equilibrium cell density expected under mass action. From Equation (2), the cell density satisfying $\alpha b = 1$ is

$$\hat{C} = \frac{k_E E}{k_C(b-1)}. \tag{3}$$

$\hat{C}$ provides a constant baseline against which the observed cell density (C_o) may be compared under the above assumptions. The amplification of cell density due to spatial structure (what we will denote as A_g, for the grid model, in anticipation of defining an A for a second model) is thus the ratio of observed cell density to $\hat{C}$:

$$A_g = \frac{C_o}{\hat{C}}. \tag{4}$$

A_g is dimensionless, thus does not depend on cell density units. For convenience, and to emancipate the results from specific values of grid size, cell densities will be measured as the fraction of patches in the grid occupied by cells.

It is evident from inspection of (4) that A_g must have an upper bound ($A_{ub,g}$) whenever cell density has an upper bound. In our model, the upper bound does not arise from grid size, rather it stems from the maximum ratio of cells to EPS:

$$A_{ub,g} = \frac{1}{\hat{C}} = \frac{k_C(b-1)}{k_E E}, \tag{5}$$

where E is measured as the fraction of the grid occupied by EPS and the 1 in $1/\hat{C}$ is for a grid filled with cells.

The foregoing applies only if the causes of phage death are unchanging. When superinfection occurs or debris traps phages,

$$\hat{C} = \frac{k_I I + k_D D + k_E E}{k_C(b-1)}. \tag{6}$$

As I and D are dynamic variables, their values will not generally be the same at the mass action equilibrium as at equilibrium with spatial structure. The calculation of $\hat{C}$ when superinfection and/or debris are admitted, and thus requires some means of determining those values; it may be possible to put bounds on them, however.

6.3. Simulations

6.3.1. Increased Cell Densities Especially with Large Burst Sizes

Any effect of spatial structure on cell density, even relative density, is likely to depend on details of phage and cell biology. To look for generalities that transcend specifics, simulations were studied for each of a variety of EPS levels, burst sizes, diffusion rates, and cell growth rates (Figure 1). There are in fact general trends, especially that spatial structure often leads to higher cell densities than mass action, but only under some conditions, especially large phage burst sizes.

In each trial, our measure of relative cell density, A_g, as well as αb and cell density were averaged over the last 3000 steps of runs lasting 10,000 steps, so that the system should have been approaching its equilibrium behavior and any fluctuations would be averaged out. These trials disallowed superinfection of infected cells and attachment to debris: as explained above, this allows calculation of the cell density expected under mass action ($\hat{C}$). An otherwise equivalent set of trials was run allowing superinfection; the αb values were largely unaffected by superinfection, nearly always differing in the first or second decimal place.

Figure 1 shows averages of A_g from 15 trials with different random number seeds and three initial conditions (the averages shown exclude extinctions). These A_g averages sometimes exceeded 1 by more than an order of magnitude, but were also less than 1 for some parameter combinations (as Figure 1 rounds to the nearest integer, values between 0.5 and 1 are not evident). Not all parameter combinations led to sustained coexistence of bacteria and phage, and parameter combinations leading to extinctions for all 15 trials are omitted from the figure. The largest effects on A_g were from changes in EPS and burst size, but changes in the other parameters also had detectable effects. Some of the effects are easily appreciated; for example, it is expected and observed that higher diffusion rates will shift A_g toward 1, as the system gets closer to mass action—if cells and phage in fact coexist.

As expected from previous work [16,17], these systems did not always go to a static equilibrium. The trials recorded distributions of αb and A_g values for the last 3000 time steps; the distributions were narrow for many parameter combinations but were large for some others. There was no suggestion that high αb or A_g was due to large (or small) oscillations, a point that will become reinforced when considering spatially clustered EPS (below). For example, for trials in the upper right corner of Figure 1C (the highest A_g averages observed), 80% of the αb values from the run were usually contained in a range spanning 1.0 around the average. In general, there was wider variance in αb with larger bursts and small EPS values. Within the same figure panel (the same cell reproduction and diffusion rates), there was wider variance the closer the burst size and EPS values approached the extinction zone in the upper left quadrant, although trials with burst sizes of 2 and 6 typically did not show a wide variance.

All trials in Figure 1 used the same attachment probabilities, k_C and k_E. To see if the patterns generalize, additional trials considered different combinations of attachment rates for three burst sizes and two EPS values (Table 1); diffusion and cell reproduction rates were those of Figure 1C, and superinfection was again precluded. There is overlap in A_g values between burst sizes of 2 and 10 and between 10 and 60. Within an EPS level, the smaller A_g value is associated with the smaller burst (with one exception). However, there does not appear to be any single variable strongly determining A_g value across all variables. It is also clear that both large and small A_g values are not limited to the attachment rates used in Figure 1.

To address the possibility that the observed A_g values are bounded artificially by the model, Table 1 includes the upper-bound A_g value for each set of parameters, $A_{ub,g}$. In some cases, the observed A_g is indeed near its upper bound, raising the possibility that the observed value would be higher with a model structure allowing a higher limit. However, not all high A_g values appear to be constrained in this way. This argument will be addressed further when the model is modified to cluster EPS.

The table includes a parallel set of trials and corresponding A_g values for mass action in the grid model; the ratio of A_g for spatial structure over that for mass action is explicitly the ratio of average cell densities maintained under the two conditions, an empirical comparison that bypasses any use of $\hat{C}$. The major difference between mass action and spatial structure is extinction of the former. For the mass action trials that avoided extinction, none of the spatially structured counterparts had A_g averages as high as 2.0.

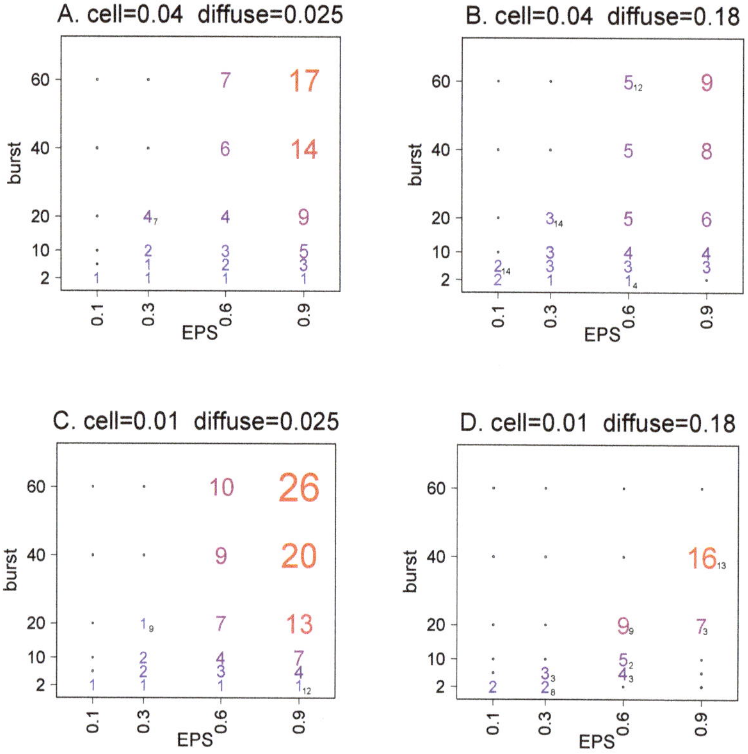

Figure 1. The density of cells maintained in the presence of phage is often increased by spatial structure. Shown in each panel are the A_g values, giving the fold increase in cell density over that with mass action. A_g values are greatly influenced by EPS levels and burst sizes, exceeding 10 only in the upper right quadrant, with large bursts and high EPS densities, and then only for some values of diffusion and cell reproduction rate. Values within each panel give average A_g values from 15 trials each using the same burst and EPS levels, with rate of cell reproduction and phage diffusion rate given at the top of each panel; trials leading to extinction of phage or cells are not included in the averages. EPS was assigned randomly to each patch at the start and remained in the patch for the life of the run; superinfection of infected cells was not allowed ($k_I = 0$), nor was debris attachment ($k_D = 0$). Each trial ran 10,000 time steps, and A was averaged over the last 3000 steps; values are rounded to the nearest integer (values rounded to 1 were often less than 1). A black subscript denotes the number of trials with bacterial and/or phage extinction; a dot indicates that all 15 trials led to extinction. The 'cell=' value given above each panel is the probability that an uninfected cell reproduced at each time step; the 'diffuse=' value is the fraction of phage that left the patch in each time step. In all trials, the adsorption probability to uninfected cells was $k_C = 0.25$, and that to EPS was $k_E = 0.35$.

Table 1. Effect of attachment probabilities on cell density in grid models.

				Spatial		Mass Action		
Burst	**EPS**	k_C	k_E	A_g	*ext*	A_g	*ext*	$A_{ub,g}$
2	0.3	0.05	0.05	1.2	-	1.4	-	3.3
2	0.3	0.05	0.15	0.5	-	1.0	-	1.1
2	0.3	0.05	0.25	0.3	-	-	10	0.7
2	0.3	0.15	0.05	3.1	-	-	10	10.0
2	0.3	0.15	0.15	1.3	-	-	10	3.3
2	0.3	0.15	0.25	0.8	-	1.1	1	2.0
2	0.3	0.25	0.05	5.1	-	-	10	16.7
2	0.3	0.25	0.15	2.2	-	-	10	5.6
2	0.3	0.25	0.25	1.4	-	-	10	3.3
2	0.9	0.05	0.05	—	10	1.0	-	1.1
2	0.9	0.05	0.25	0.2	9	-	10	0.2
2	0.9	0.15	0.05	3.3	8	-	10	3.3
2	0.9	0.15	0.15	1.1	7	1.0	8	1.1
2	0.9	0.15	0.25	0.7	7	-	10	0.7
2	0.9	0.25	0.05	5.5	6	-	10	5.6
2	0.9	0.25	0.15	1.8	7	-	10	1.9
2	0.9	0.25	0.25	1.1	9	-	10	1.1
10	0.3	0.05	0.15	0.8	-	-	10	10.0
10	0.3	0.05	0.25	0.7	-	-	10	6.0
10	0.3	0.15	0.15	1.8	2	-	10	30.0
10	0.3	0.15	0.25	1.3	-	-	10	18.0
10	0.3	0.25	0.15	0.9	4	-	10	50.0
10	0.3	0.25	0.25	2.0	-	-	10	30.0
10	0.9	0.05	0.05	4.4	-	-	10	10.0
10	0.9	0.05	0.15	2.9	-	-	10	3.3
10	0.9	0.05	0.25	1.9	-	1.1	4	2.0
10	0.9	0.15	0.05	9.7	-	-	10	30.0
10	0.9	0.15	0.15	7.7	-	-	10	10.0
10	0.9	0.15	0.25	5.5	-	-	10	6.0
10	0.9	0.25	0.05	14.4	-	-	10	50.0
10	0.9	0.25	0.15	12.2	-	-	10	16.7
10	0.9	0.25	0.25	9.2	-	-	10	10.0
60	0.9	0.05	0.15	8.8	-	-	10	21.9
60	0.9	0.05	0.25	8.2	-	-	10	13.1
60	0.9	0.15	0.15	17.3	-	-	10	65.6
60	0.9	0.15	0.25	18.0	-	-	10	39.3
60	0.9	0.25	0.15	31.4	1	-	10	109.3
60	0.9	0.25	0.25	25.0	-	-	10	65.6

Average amplification of cell density (A_g) due to spatial structure compared to the amplification under mass action across a range of EPS values, burst sizes, and attachment probabilities (k_C, k_E). Columns 5 and 6 are for spatial structure, 7 and 8 for mass action. For each combination, the A_g shown in the row is the mean of 10 runs differing in the random seed and spanning 2 different initial concentrations of phage and bacteria (extinctions were excluded from the averages, and superinfection was not allowed). Both EPS values (0.3, 0.9) were tested at each burst size (2, 10, 60) for each possible combination of k_C and k_E in (0.5, 0.15, 0.25); rows are omitted when all 10 trials resulted in extinction for both mass action and spatial structure (17 cases, including all nine trials with a burst of 60 and EPS value of 0.3); numbers of extinctions are otherwise given when more than 0. A_g modestly exceeds 1.0 due to oscillations in density being asymmetric around 1.0. The mass action assumptions were applied in the grid model, so the model parameters are directly comparable except that cells and phage were randomly assigned to locations each generation.

6.3.2. Understanding the Puzzle of Why Larger Phage Burst Sizes Lead to Higher Cell Densities

The results show clearly that some sets of parameter values lead to large elevations of cell density. The next step is to understand how this elevation happens. In particular, some patterns seem to defy intuition, such as why our relative cell density measures (A_g and αb) increase with b when holding other parameters constant. It is clear that increasing burst size will affect whether cells and phage are both maintained indefinitely, but the fact that αb changes with b indicates that some properties of the infection do not scale proportionally with burst. (αb is more easily addressed in this respect than is A_g.) Changing EPS abundance is also expected to affect coexistence, but the reason for its affect on αb is not clear. Understanding this absence of proportionality is potentially critical to understanding the effect of spatial structure on cell density, and is addressed next.

To understand how spatial structure enables A_g (and thus αb) to exceed 1 and why A_g varies with b, additional statistics were calculated for the parameter combinations used in Figure 1C (Table 2). The statistics included (i) losses of phage to EPS, (ii) the spatial association of cells and phage with EPS (probability that an uninfected cell or free phage was found in a patch with EPS), and (iii) the proportion of infections that happened in patches with EPS. As true of Figure 1, (i) all statistics were averaged over the last 3000 time steps of 10,000 step runs, and (ii) all statistics were averaged over all runs that led to coexistence.

One striking observation is that, holding all other parameters constant, increases in burst size led to directly corresponding increases in phage lost to EPS, while the losses to uninfected cells were only slightly affected. Thus, as burst size increased, the fraction of phage lost to EPS increased disproportionately. Proportionality is expected unless the association of phage or cells with EPS is changing.

Table 2. Spatial grid model outcomes with random placement of EPS, no superinfection or debris.

Burst	EPS	A_g	$A_{ub,g}$	αb	P→E	C:E	P:E	I:E
2	0.1	0.9	7.1	0.9	1.0	0.2	0.01	0.10
2	0.3	1.0	2.4	1.0	1.0	0.5	0.02	0.19
2	0.6	1.1	1.2	1.0	1.0	0.6	0.03	0.25
2	0.9	0.8	0.8	0.9	1.0	0.9	0.04	0.27
6	0.3	1.9	11.9	1.6	5.0	0.5	0.02	0.34
6	0.6	2.9	6.0	2.2	5.0	0.8	0.06	0.51
6	0.9	3.9	4.0	2.6	5.0	0.9	0.09	0.59
10	0.3	1.9	21.4	1.8	9.0	0.4	0.02	0.37
10	0.6	4.3	10.7	3.2	9.0	0.8	0.07	0.58
10	0.9	6.9	7.1	4.3	9.0	0.9	0.12	0.69
20	0.3	0.5	45.2	0.5	18.6	0.4	0.02	0.39
20	0.6	6.8	22.6	5.2	19.0	0.8	0.08	0.65
20	0.9	13.2	15.1	8.2	19.0	0.9	0.17	0.79
40	0.6	9.5	46.4	7.8	39.0	0.7	0.08	0.68
40	0.9	20.4	31.0	13.7	39.0	1.0	0.24	0.87
60	0.6	10.1	70.2	8.8	59.0	0.7	0.08	0.67
60	0.9	26.1	46.8	18.4	59.0	1.0	0.28	0.90

For these numerical trials, parameter values and initial conditions were as in Figure 1C. For each combination of burst size and EPS, the output values shown in the row are the means of 15 runs differing in the random seed and spanning three different initial concentrations of phage and bacteria. All four EPS values (0.1, 0.3, 0.6, 0.9) were tested at each burst size (2, 6, 10, 20, 40, 60); values are not shown when all 15 trials resulted in extinction. The numbers of extinctions for the data shown are given in Figure 1. Burst is phage burst size. EPS is the fraction of grid sites containing EPS, assigned randomly. A_g is the magnitude to which total grid cell density is increased above that expected with mass action. P→E is the average number of phage per burst lost to EPS. C:E is the fraction of uninfected cells found in patches with EPS. P:E is the fraction of free phage found in patches with EPS. I:E is the fraction of infections occurring in patches with EPS.

A second observation is that uninfected cells are somewhat associated with EPS (the association is often only modestly greater than the fraction of patches with EPS), whereas free phage are strongly associated with an absence of EPS. These latter observations suggest that spatial structure favors the retention of cells and phage into separate refuges where they are differentially protected from loss.

There are also apparent trends that, as burst size increases, (i) an increasing proportion of all infections happen on patches with EPS, and (ii) phage are increasingly associated with EPS. As burst size increases, the phage appears to be spreading to less protected areas and incurring greater loss.

6.3.3. Reasons for Higher Cell Densities Become Clearer When EPS Is Clumped: Cells Have More Protection from Spatial Structure

The patterns seen in Tables 1 and 2 are somewhat noisy. Those trials assigned EPS randomly to patches across the grid. Although random assignment may be realistic, it may also complicate understanding. Random assignment gives rise to varied and inconsistent boundaries between EPS-containing and EPS-free regions, possibly complicating inferences about associations of phage and cells with EPS. A clustering of EPS into a single area can overcome those difficulties by ensuring that all trials have the same boundaries around EPS. Trials were conducted so that EPS was laid down contiguously within the grid (adjacent rows were filled until the total EPS allotment was reached). This design resulted in a band of EPS on the grid. One straightforward effect of deterministic clustering is that the size of the boundary between EPS and EPS-free zones is now unaffected by the overall level of EPS. Table 3 provides values from a set of runs corresponding to those in Table 2.

Table 3. Spatial grid model outcomes with deterministically clustered EPS, no superinfection or debris.

Burst	EPS	A_g	$A_{ub,g}$	αb	P→E	C:E	P:E	I:E
2	0.1	0.7	7.1	0.8	1.0	1.0	0.00	0.274
2	0.3	0.7	2.4	0.8	1.0	1.0	0.00	0.273
2	0.6	0.7	1.2	0.8	1.0	1.0	0.00	0.276
2	0.9	0.7	0.8	0.8	1.0	1.0	0.00	0.272
6	0.1	3.3	35.7	2.4	4.9	1.0	0.00	0.594
6	0.3	3.5	11.9	2.5	4.9	1.0	0.00	0.593
6	0.6	3.5	6.0	2.5	4.9	1.0	0.00	0.594
6	0.9	3.5	4.0	2.5	5.0	1.0	0.01	0.600
10	0.1	5.8	64.3	3.9	8.9	1.0	0.00	0.699
10	0.3	6.2	21.4	4.1	8.9	1.0	0.00	0.698
10	0.6	6.3	10.7	4.1	8.9	1.0	0.00	0.697
10	0.9	6.3	7.1	4.1	9.0	1.0	0.01	0.703
20	0.1	11.7	135.7	7.6	19.0	1.0	0.00	0.805
20	0.3	12.9	45.2	8.1	19.0	1.0	0.00	0.805
20	0.6	13.2	22.6	8.2	19.0	1.0	0.00	0.805
20	0.9	13.4	15.1	8.3	19.0	1.0	0.01	0.804
40	0.1	22.4	278.6	14.6	39.2	1.0	0.00	0.889
40	0.3	26.0	92.9	16.0	39.2	1.0	0.00	0.889
40	0.6	26.9	46.4	16.3	39.2	1.0	0.00	0.890
40	0.9	27.2	31.0	16.4	39.0	1.0	0.02	0.887
60	0.1	30.5	421.4	20.5	59.4	1.0	0.00	0.935
60	0.3	38.0	140.5	23.5	59.3	1.0	0.00	0.943
60	0.6	40.1	70.2	24.3	59.3	1.0	0.01	0.942
60	0.9	40.8	46.8	24.5	59.0	1.0	0.04	0.939

For these numerical trials, parameter values were as in Figure 1C, except that EPS was laid down deterministically in a single cluster. For each combination of EPS and burst size, the output values shown in the row are the means of 15 trials differing in the random seed, spanning three different initial abundances of phage and cells. The range of values as a percent of the mean obtained from the 15 trials never exceeded 11%, except for P:E (the range reaching as high as 110% of the mean, which was invariably tiny). No extinctions occurred. Notation is as in Table 2.

Patterns are clearer than with random EPS assignment and support intuition about the effect of spatial structure in enabling high cell densities over those with mass action:

1. A_g (αb) is now moderately constant across different EPS levels within the same burst size. The constancy is stronger at smaller burst sizes. This suggests that the width of the EPS zone itself is unimportant to the properties being measured until bursts get large.
2. The span of A_g (αb) values across the table is higher than with random EPS, not profoundly so, and some A_g (αb) are consistently less than 1, even when $A_{ub,g}$ cannot have imposed the low value. Spatial structure does not invariably increase cell density over mass action.

3. Phage and cells coexist over a wider range of parameter values with clustered EPS than with random EPS. There were no extinctions, in contrast to the many extinctions when EPS was placed randomly.
4. The association of cells with EPS and phage avoidance of EPS is more extreme than with random placement of EPS.
5. There is now a consistent trend that increasing burst size increases the fraction of infections occurring in patches with EPS.
6. Within a burst size, the value A_g is far more stable than is the $A_{ub,g}$, suggesting that the observed A_g is not often constrained by the upper bound.

An intuitive interpretation of these results is that free phage and uninfected cells tend to occupy different patches (phages live in EPS-free patches, cells live in patches with EPS: Figure 2). At low burst sizes, phage are lost to EPS at a high enough rate relative to burst that they virtually only persist in patches without EPS, and they amplify when cells migrate into those patches. This pattern can be argued from the fraction of infections that occur in EPS-free patches. As burst size increases, phages increasingly diffuse into zones with EPS, where they encounter otherwise protected cells. However, these successful infections also result in high rates of phage lost to EPS.

Burst sizes measured from infected cells grown in rich media are often much larger than those evaluated here [1]. However, it should first be appreciated that our simulations of spatial structure are two-dimensional, and a smaller burst size will operate in two dimensions than in three. Since our 2D model characterizes the horizontal spread of phage, it is appropriate to think of only a fraction of the full 3D burst contributing to horizontal spread. Since the volume of a thin slice (say of thickness equal to a tenth of the radius) that intersects the center of a sphere of radius r is less than 10% of the volume of the sphere, a full 3D burst B should correspond to an analogous 2D burst of size $b < B/10$. For example, a burst of 60 in two dimensions corresponds to a burst of over 600 in three dimensions.

Nonetheless, trials with burst sizes of 100 and 300 were evaluated for the same EPS levels and adsorption rates as in Table 3. Analyses of these large bursts were reserved for the clumped EPS model because of the repeatability of outcomes provided by this model. The largest A_g values were observed for the EPS levels of 0.9: 48 for a burst of 100 and 66 for a burst of 300. Thus, increasing burst sizes several-fold led to only modest increases in A_g values. As in Table 3, nearly all phage per burst were lost to EPS. All trials with EPS of 0.1 and a burst of 300 went extinct, revealing that phage can indeed overwhelm cells if the EPS is clustered in small enough patches (no extinctions were observed for the smaller bursts in Table 3). Furthermore, strong oscillations were typical of all trials, again suggesting that, with the larger burst sizes, phages are invading deeper into the EPS-protected refuges. These dynamical effects of large burst sizes on extinction and dynamics would likely disappear with sufficiently large grid sizes (much larger than 10,000 patches) because the zones of EPS protection would be larger and thus require phages to traverse greater distances before reaching the centers of the EPS zones. From the perspective of how spatial structure contributes to an elevated density of cells, larger bursts increase the elevation, but much less than proportionally.

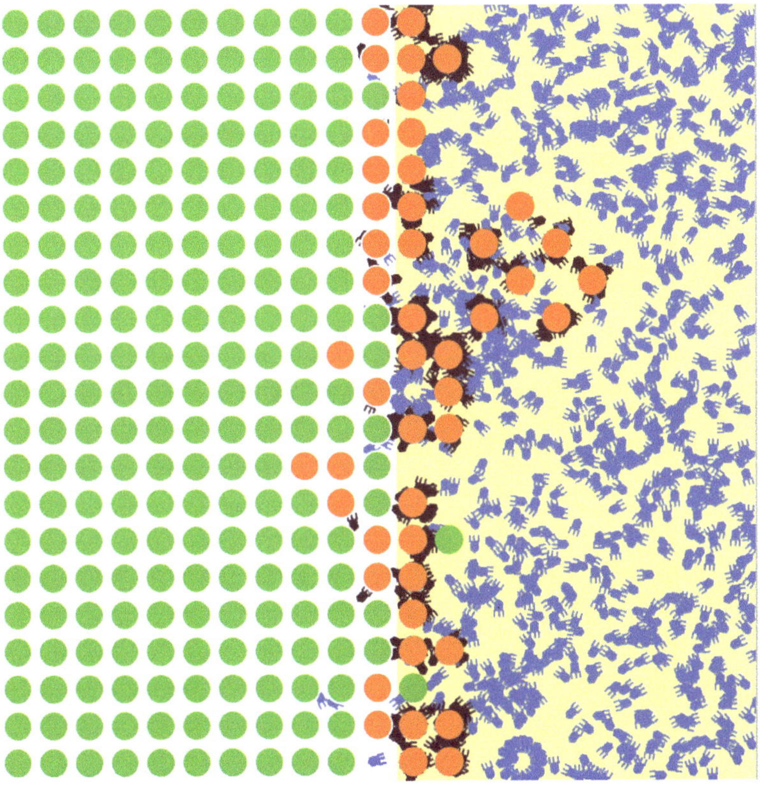

Figure 2. Illustration of self-organization of phages and cells with clumped EPS. White background indicates EPS, yellow is absence of EPS. A green (orange) circle is an uninfected (infected) cell. A blue or black legged icon is a phage (blue is free, black is attached to a cell). Phages are mostly confined to the EPS-free zone and the first row of EPS. Figure was generated from a NetLogo trial with a grid size of 21 × 21, a burst of 20, diffusion step size of 0.45 and attachment probabilities as in Figure 1C. There were 32 phage (partially obscured by cells) in the first three rows of EPS; αb for the entire grid was 7.26.

6.3.4. Average Densities under Differential Equation Mass Action Are Also Sometimes Elevated but Not as Much and for a Different Reason

The analysis so far has compared simulated cell densities under spatial structure to densities expected for mass action equilibrium, except for a few trials in Table 1. It is well known that models of mass action dynamics do not obey equilibrium for wide parameter ranges, instead exhibiting either stable oscillations or accelerating oscillations [29]. It is thus possible that average cell densities under mass action will themselves systematically differ from the expected equilibrium. That is, A-values for mass action may not equal 1, as has been implied above.

Two efforts were undertaken to calculate A-values for mass action: a simulated version of mass action based on adding 'mixing' to the spatial grid model, and a version based on an ODE model. The first mass action model merely modified the simulation code of spatial structure so that phage and cells were each assigned grid positions randomly every time step. However, the comparison of A-values for spatial structure and mass action is most informative when the A-values for spatial structure are well above 1, as those are the only cases in which there appears to be a meaningful effect of spatial structure on cell density. As was shown in Table 1, cell-phage coexistence under mass action was obtained only with parameter values for which the spatial structure A_g values were

1–2. (Increasing the grid size 9-fold did not lead to coexistence for any informative combinations either.) The A-values under mass action were sometimes higher, but the main result is that mass action extinction was always the outcome for parameter combinations leading to even moderate A_g under spatial structure.

The second approach used ordinary differential equation (ODE) models of mass action:

$$\begin{aligned} \dot{C} &= rC(1 - C/K) - \kappa CP, \\ \dot{P} &= b\kappa C_L P_L - \kappa CP - \delta P, \end{aligned} \tag{7}$$

with the "dot" indicating time derivative, parameters in Table 4, and a subscript L indicating the value L time units in the past. The $(1 - C/K)$ term slows bacterial growth as cell density nears K, the carrying capacity.

In contrast to the comparison of mass action and spatially structured trials under the grid model, it is not practical to directly compare cell densities between the grid model and an ODE model because of the much higher cell densities enabled by the ODE model. This realization motivates the use of a parallel A statistic for the ODE model. To wit, equilibrium cell density under this ODE model is

$$\bar{C} = \frac{\delta}{\kappa(b-1)}$$

and hence it is this quantity that observed cell densities are compared to when defining an ODE-based A value:

$$A_{ode} = \frac{C}{\bar{C}}$$

also dimensionless (the subscript *ode* indicating the ODE model). Its upper limit is

$$A_{ub,ode} = K/\bar{C}. \tag{8}$$

Table 4. Model variables and parameters.

Notation	Description	Units
Variables		
C	density of uninfected bacteria	/mL
P	density of phage	/mL
Parameters		
κ	adsorption rate of phage to cells	mL/min
δ	loss rate of phage to EPS	/min
b	burst size of phage	
L	lysis time	min
K	carrying capacity of environment	/mL

Using ODEs presents the additional challenge establishing a correspondence between attachment probabilities in the grid model to attachment rates in the ODE model. To develop such a correspondence, we used the fact that, over a single unit of time (1 min in ODE corresponds to 1 time step in grid-based model), a phage avoids EPS in the grid-based model with probability $1 - k_E E$ and in the ODE model with probability $e^{-\delta}$. Thus, $\delta = -\ln(1 - k_E E)$ is an approximate equivalence. For $k_E = 0.35$ and $E \in 0.1, 0.3, 0.6, 0.9$, this gives a range for δ of (0.04, 0.38), but the range goes down to 0.005 for the lowest k_E and E values used in Table 1. A similar basis was used to obtain equivalence between k_C and κ; in contrast to the equivalence for EPS, however, cell density is not fixed, so it is necessary to choose a density for the equivalence. Here, that density was the maximum for the system (K for the ODE versus 1 in the grid model). For those cell densities and the k_C values used in Table 1, κ was in the range (5×10^{-11}, 3×10^{-10}). For an ODE model scaled per-minute, these are reasonable values [28], although on the low end for some phages [26]. An exact correspondence between mass action and spatial models is not required, of course, because we are interested in whether any realistic mass action process can give high A-values; the conversions derived above merely suggest that the ODE model equivalence lies in established regions of parameter space for phages grown in liquid culture.

ODE numerical trials were run for 50,000 time steps using appropriate parameters (Table 5). Many parameter combinations led to expanding oscillations and premature termination of the run (effective extinction). Coexistence of cells and phage was obtained for many runs as well, typically with stable oscillations. For those, average A_{ode} values ranged from slightly above 1 to 10. The average exceeds 1.0 because of the asymmetry in the range of values: A_{ode} periodically goes up to the limit ($A_{ub,ode}$) but can go no lower than 0 (reflecting an asymmetry in the range of bacterial densities).

The highest A_{ode} averages were associated with the highest oscillations in cell densities (up to 18 orders of magnitude for the trial with an average A_{ode} of 10). No attempt was made to evaluate parameter space comprehensively, as our goal was merely to discover whether sustained oscillations resulted in a deviation of A_{ode} from 1.0. In the absence of oscillations, A_{ode} was 1.0, as expected (one example shown).

Summary of ODE model versus spatial grid model. For the differential equation model, an average A_{ode} above 1 is due entirely to sustained oscillations, whereas for the spatial grid model, an elevated A_g is not from oscillations but is intrinsic to the dynamics. Thus, the mechanism of high A-values are completely different for spatial structure and mass action; in the former, they are intrinsic to the environment and are approximately constant. In the latter, they arise because of oscillations in cell density and the asymmetry of limits on A.

Table 5. A_{ode} values for the ODE mass action model.

A_{ode}	$A_{ub,ode}$	**Burst**	δ	κ	L	r
3.9–4.1	9.0	10	0.1	1×10^{-10}	20	0.03
1.7	4.5	10	0.1	5×10^{-11}	20	0.03
5.8	18.0	10	0.05	1×10^{-10}	20	0.03
1.1	3.0	10	0.3	1×10^{-10}	20	0.03
4.2–4.4	9.0	10	0.1	1×10^{-10}	25	0.03
1.0	1.9	20	0.3	3×10^{-11}	20	0.03
1.2	3.3	20	0.4	7×10^{-11}	20	0.03
5.0	9.5	20	0.2	1×10^{-10}	20	0.03
5.6–5.8	9.8	60	0.3	5×10^{-11}	25	0.04
3.8	7.4	60	0.4	5×10^{-11}	20	0.03
10.1–10.2	39.3	60	0.045	3×10^{-11}	21	0.03

A_{ode} values for a small sample of numerical trials of Equation (7) in which bacteria-phage coexistence was observed for the full 50,000 time units. Parameter combinations leading to extinctions are not shown and often resulted with small changes in a single parameter from a parameter set in which coexistence was otherwise observed. A_{ode} was calculated as the arithmetic mean of cell density divided by $\delta/(\kappa(b-1))$; averages were calculated every 10,000 time units spanning time 10,000 to 50,000, and when the four values differed, the range is given. Parameters used in the trial are defined in Table 4. Carrying capacity K was 10^9 for all trials.

6.3.5. Debris: Adding Greater Reality Does Not Change the Trends

The preceding results from the grid model exclude all mechanisms of phage death except irreversible attachment to EPS, and EPS locations and levels were fixed. Other cell-based mechanisms of phage death are plausible and likely temporary. To consider whether our results continue to hold when other phage death mechanisms are present, we expanded the spatial grid model in include cellular debris—remnants of lysed cells that cause phage to bind irreversibly or inject their genomes non-productively. In mass action (liquid culture), this effect appears to be negligible empirically, as phage concentrations are often stable over months ([1], and personal observations). It is unknown whether debris may constitute a greater element of phage death in biofilms and other structured habitats, but we entertain it as an example of a possibly general phenomenon of local phage death resulting from the lysis of cells. Furthermore, the killing effect of debris may be short-lived, thus difficult to detect empirically, but such effects can be studied in the models.

Following [16,31,32], debris was introduced as infected cells that persisted after death (after lysing). In our trials, they were assigned a fixed lifespan, during which they could act as a phage sink in the same capacity as an infected (but unlysed) cell; α is correspondingly recalculated to include this new loss term, and, because of its inclusion, we can no longer use (3) to calculate an expected cell density under mass action. Our presentation is thus of αb instead of A, but αb is a suitable proxy. Additionally, superinfection of infected cells was allowed in these trials. The main effect of this debris model is that the dead cell is present after burst and thus is an additional source of death in the patch when phage densities are highest. Even limiting debris longevity to a mere two time steps had a huge effect on shifting the source of phage loss from EPS to debris but had only a modest effect on αb (as well as on coexistence) (Table 6, columns were added to indicate phage lost to debris and infected cells). Coexistence of phage and cells was typically not observed when debris was present and EPS was absent (not shown), but this outcome is necessarily sensitive to debris longevity (our trials assumed a moderately short life for debris).

Table 6. Random EPS with debris lasting two steps, superinfection allowed.

EPS	Burst	αb	P→C	P→I	P→E	P→D	C:E	P:E	I:E
0.1	6	2.2	1.08	0.71	1.58	2.58	0.17	0.01	0.097
0.3	6	2.4	1.07	0.60	1.83	2.50	0.49	0.02	0.218
0.6	6	2.8	1.11	0.74	1.69	2.46	0.68	0.04	0.257
0.9	6	2.7	1.14	0.83	1.61	2.46	0.90	0.05	0.255
0.1	10	0.3	1.13	1.22	3.51	4.28	0.14	0.00	0.100
0.3	10	3.1	1.13	0.92	3.85	4.10	0.48	0.02	0.272
0.6	10	3.9	1.17	1.04	3.80	3.99	0.76	0.05	0.356
0.9	10	4.4	1.25	1.35	3.44	3.96	0.90	0.07	0.343
0.3	20	3.9	1.24	1.62	9.06	8.09	0.43	0.02	0.297
0.6	20	6.4	1.26	1.42	9.60	7.72	0.77	0.07	0.451
0.9	20	8.7	1.40	1.86	9.16	7.59	0.91	0.11	0.461
0.3	40	4.2	1.47	2.97	19.71	16.12	0.36	0.02	0.272
0.6	40	10.4	1.38	1.88	21.62	15.13	0.76	0.08	0.501
0.9	40	16.4	1.56	2.16	21.62	14.67	0.93	0.17	0.558
0.6	60	13.5	1.47	2.30	33.67	22.55	0.75	0.08	0.510
0.9	60	22.4:	1.63	2.23	34.45	21.69	0.95	0.20	0.616

αb values and other properties of dynamics when debris is included and superinfection of infected cells is allowed. Dead cells persisted for two time steps after cell lysis and acted as a phage sink during this time (adsorption to debris was the same as to live cells, 0.25). Parameter values were otherwise as in Figure 1C. For each combination of EPS and burst size, the output values shown in the row are the means of 15 trials differing in the random seed and using three different initial densities of cells and phage. All four EPS values were tested at each burst size; values are not shown when all 15 trials resulted in extinction. For those rows shown, 10 extinctions occurred for (EPS = 0.9, burst =6), 13 extinctions for (0.1, 10), and two extinctions each for (0.9, 10) and (0.3, 40). Ranges of the 15 values as a per cent of the mean were mostly less than 20% and never exceeded 42%, except that the range of αb values was almost as large as the mean for (0.3, 40); some of those trials experienced large variation in αb values with occasional low numbers of cells. Notation as in Table 2, with $P \rightarrow I$ indicating the approximate number of phage per burst lost to infected cells and $P \rightarrow D$ indicating the loss to debris. In contrast to Tables 1–3, A_g is not provided because the baseline calculation of equilibrium cell density for mass action includes terms that depend on dynamics.

7. Discussion

Phage and their hosts exist in a predator–prey relationship, the dynamics of which have been modeled for over half a century. These models have assumed population structures of well-mixed environments (mass action), both for mathematical convenience and because laboratory studies of phage have used conditions that represent mass action—flasks in shakers and chemostats. However, it is increasingly evident that bacteria grown in biofilms and other spatial contexts are able to persist at much higher densities than apparent from the models, and it is not clear why. This study used a computational approach to investigate the simple question of whether and how spatially structured cell and phage growth might allow higher equilibrium cell densities than with the well mixed conditions of mass action. This question is motivated by empirical observations suggesting that genetically sensitive cells are often profoundly more protected from phage when grown with structure (e.g., biofilms or aggregates) than when grown in liquid. By uncovering the mechanisms behind these high densities, it may be possible to improve the prospects for phage therapy.

Our main findings are:

1. Spatial structure sometimes, but not always, led to cell densities above those maintained at equilibrium under mass action. However, average cell densities under mass action were also often greater than expected at equilibrium. Any effect of spatial structure in elevating cell densities thus appears to be less than an order of magnitude.
2. The mechanisms of 'elevated' cell densities are different between spatial structure and mass action. The effect of spatial structure appears to stem from phage and cells dynamically sorting to occupy different patches in the environment, with cells in patches that otherwise kill phage, and phage occupying patches that did not kill them but were largely free of cells. The elevation under mass action arises from sustained oscillations, due to a large dynamic range for $A > 1$ but A being bounded to lie above 0.
3. Under spatial structure, increasing burst size was usually observed to increase the relative cell density—to increase the effect of spatial structure in raising cell density—holding other parameters constant. However, a high abundance of environmental protection (EPS) contributed to relative cell density; phage diffusion rates, cell reproduction rates and attachment rates also had influences.
4. The burst size effect was shown to result from a curious effect of the spatial segregation between phages and cells. At higher burst sizes, phages increasingly invaded refuges occupied by cells and suffered proportionally greater losses. Thus, the per capita phage loss to EPS was higher with higher burst sizes, thus accounting for their poorer efficacy in suppressing cell density.

7.1. Back to Nature: Do Our Spatial Models Explain What We Observe?

Our efforts were primarily to look for mechanisms that might promote high cell densities as observed in nature. Having found possible mechanisms, the question then turns whether those mechanisms do indeed operate in nature. The latter question is empirical and is a far greater challenge than merely identifying possible mechanisms. Understanding of the empirical side of phage-bacterial dynamics with spatial structure is rudimentary, and our discussion of it is correspondingly speculative. It is premature to suggest that the mechanisms promoting high cell density in our models are empirically important, but they at least suggest directions of inquiry. Indeed, a recent study accounts for bacterial colony survival amid phage attack merely by considering the rate of colony growth versus the rate of phage penetration; when the colony reaches a certain size before phage encounter, it grows faster than the rate at which phage can penetrate—due in no small part to the large number of phages infecting the same cell in the close confines of the bacterial colony [33]. In their model, therefore, cells persist in spatial structure because phages are slow to invade the structure and because many different phage infect the same cell—an effect we intentionally excluded in most of our trials.

The largest effect of spatial structure on cell density observed in our trials is well short of the apparent effects of spatial structure observed in some empirical systems. Furthermore, mass action models were also observed to maintain average cell densities above the expected equilibrium, albeit that this elevated average arises from oscillations. In some natural systems, cells are maintained at densities several orders of magnitude above those in liquid systems. It could be that the cell density increase under spatial structure observed in a numerical trial is artificially bounded by the construct of the model, hence that a more realistic model would exhibit a far higher equilibrium cell density. While a larger grid size or allowing multiple cells per site could increase the dynamic range of A, we speculate that a fully 3D system would better capture the large cell densities seen in biofilms that are subjected to phage attack. Imagining our clustered EPS zone as a 2D slice of a biofilm, a 3D version would have a two-dimensional (surface) interface between the protected and unprotected regions. This surface would be more permeable to phage incursions, but the potential gain in cell density in the EPS zone when going from a 2D to a 3D model could vastly increase the dynamic range of A.

An alternative interpretation is that empirically high cell densities arise with spatial structure mostly from mechanisms other than those considered here. At one extreme, cells grown with structure may be resistant to infection. This resistance need not even be genetic [23–26]. Resistance could stem from changes in gene expression that arise when cells are attached to surfaces. Such gene expression changes could lower phage receptor densities or could lead to the secretion of protective layers.

Alternatively, cell protection with spatial structure could be an automatic consequence of limited diffusion and not even involve changes in gene expression. Thus, if cells normally secrete diffusible substances that can form gradients or protective boundaries, spatial structure would allow those gradients to form and protect cells from all sides, whereas liquid culture would not. In contrast, our models allowed protection purely from phage death: cells could escape phage merely because phage were killed before they could attack cells. That phage death was spatially structured, allowing cells to associate with refuges within that structure. Spatial structure offers many possible mechanisms of cellular escape from phages, and our models point a direction toward more biologically comprehensive processes. Empirical progress in understanding bacterial escape will obviously be useful in directing further modeling efforts.

7.2. Our Models in Context

Whereas it is straightforward to measure an average cell density with spatial structure any time cells and phage are maintained, it is difficult to use the same approach to determine the cell density that would obtain under the same conditions if phage and cells were fully mixed: the dynamic ranges of cell and phage densities are limited in the simulations, and the oscillations that typically accompany mass action dynamics lead to extinctions in finite populations, even when the average densities are well above the extinction threshold. We thus developed a metric for calculating the equilibrium cell density expected under mass action that could be compared to the cell densities observed in many of the simulations.

For cells to persist amid phages, the cells must either be fully protected from infection (i.e., some form of resistance, genetic or otherwise), or phages must die often enough to keep from overwhelming the cells. We explored the latter process here. Many of our observations as regards dynamics with spatial structure are similar to those of [17], but we took the analysis one step farther by making a comparison of the effect of spatial structure versus mass action on cell density. Another difference is that we did not impose an intrinsic phage death rate, instead allowing phages to die either from sticking to spatially static substances that could in principle be produced by cells (exopolysaccharide, or EPS), or from infection of 'debris,' represented here as short-lived parts of dead cells that persist after lysis (inspired by [31,32]). EPS, which is fixed spatially in our model, and thereby allows cells and phages to differentially organize around them, is similar to the fixed refuge model in [17], the main difference being that we have a specific mechanism for inhibiting phage growth. In our model, superinfection results in phage loss; results in our Figure 1 and Tables 1–3 specifically precluded superinfection, but parallel trials that allowed superinfection yielded similar outcomes. In [16], superinfection is beneficial to phage since it is assumed to inhibit lysis with a resultant increase in burst size; in [17], there is no superinfection.

Our chief interest in this study was to evaluate the effect of spatial structure on long-term or equilibrium cell density, comparing it to the density expected under mass action. For the purpose of evaluating the effect of spatial structure on phage-bacterial coexistence, Heilman et al. [17] provided a direct comparison of coexistence under the the two conditions. However, oscillations in cell and phage densities under mass action often led to extinction in the grid-based simulations of mass action we attempted, except in cases for which spatial structure appeared to have a small or no elevating effect on cell density. To evaluate the effect of mass action on cell density for cases of interest, we used an ordinary differential equation model, with parameters chosen to correspond to those of the spatial structure model.

To evaluate the effect of spatial structure on cell density, compared to mass action, we used the well-known principle that, in populations at reproductive equilibrium, each individual merely replaces itself, on average—each offspring has one successful offspring during its lifetime (in asexual populations). For a phage with burst size b, this means that for each infection of a cell that survives to burst, only one of those b progeny will itself establish a surviving infection. We defined α as the ratio of successful infections divided by all sources of free phage loss, hence this equilibrium condition is $\alpha b = 1$. Under some conditions easily implemented in numerical trials, this equilibrium condition can be used to calculate an equilibrium cell density for mass action. The dimensionless statistic A was then used as the ratio of observed cell density over the equilibrium cell density under mass action—the 'amplification' effect of spatial structure. This statistic could be derived for the grid model (with or without spatial structure) and for the ODE model of mass action, allowing easy comparisons of the effect of different structures.

Across different parameter combinations in the model of spatial structure, grid-based A_g values ranged from slightly less than 1 to nearly 30. Thus, spatial structure sometimes conspired to reduce cell density below that maintained with mass action, but also commonly led to an elevation of cell density—depending on parameter values. However, a similar elevation of average density was also observed under mass action whenever the dynamics exhibited sustained oscillations.

A large effect on A was from burst size (b). It was not immediately clear why increasing burst size should increase the effect of spatial structure on cell density, so various metrics of phage dynamics were analyzed, and a simple explanation was found. The environmental structure allows cells to reside in protected areas (those with EPS) and phages to exist in death-free areas (those without EPS). This is a type of self organization due to the different causes of death for cells (phage kill them) and phage (EPS kills them). When this organization is established, infections result from cells growing into unprotected areas and/or phage diffusing into zones in which they are rapidly killed but where cells reside. The balance between these two processes shifts as burst size is increased—a larger burst means that phages diffuse further into protected-cell zones, but at a cost that more phage progeny are killed. It is also clear that large burst sizes result in the EPS-free zone being essentially devoid of bacteria; this is reflected in a large fraction of infections being limited to the EPS zone. In contrast, with low burst sizes, cells are growing into unprotected zones, where they are killed by phages and where phages do not die (as in Figure 2). In the case of no superinfection or debris attachment, it is also clear that the denominator in A decreases as a function of b.

7.3. Caveats

One potentially important omission from our models is local variation in cell growth rate (as might be mediated by variation in resource concentration). Bacterial growth is known to be important to phage growth (e.g., [1]), with starved cells reducing burst sizes and increasing times to lysis [34]; a change in susceptibility of cell populations at high density requires non-standard models and leads to alternative stable states of the bacterial system even with mass action [35]. Biofilms are thought to be highly structured for resources and consequent cell growth rates [9,36]. To what extent starvation of cells or delayed spread of phages contributes to high cell densities is not addressed by our model but is certainly a worthy avenue of further analysis. Also excluded are temperate phages, whereby infection can lead to a viable cell carrying the phage genome (a lysogen); dynamics of temperate phages with spatial structure presents a fundamentally different set of challenges [37].

The theory advanced here motivates the empirical search for phage death mechanisms, especially those that operate with spatial structure. We yet know little of how rapidly phage are inactivated by exopolysaccharides, outer membrane vesicles, or other materials produced in situ. Such measurements will be difficult when phages are actively growing on live cells in structured environments, but it should be possible to inactivate cells while leaving the structure intact, then measuring the effect on phages.

8. Methods: Simulation Model Basics

Three computer programs were used to model spatial dynamics: a program written in C, a program written in Python, and a program written in NetLogo. Due to its superior runtime and versatility, the C program was used for all results presented; the Python program was written to verify the C program results. The Netlogo program was used early in the study to visualize spatial dynamics and develop intuition about the processes. All three models are broadly similar to those in [16,17].

The C code was also adapted to model mass-action dynamics in a grid model. This version of mass action allows an "apples-to-apples" comparison of mass action and spatial dynamics on the same computational platform—including finite population size and identical parameters. With finite population size, the grid based mass action model is stochastic and thus differs somewhat from an ODE-based mass action model. (The randomness actually disappears in the limit as population size goes to infinity.) Aside from the randomness and heightened probability of extinction due to finite population size in simulations of grid-based mass action, they produce similar behavior to numerical solutions of ODE-based mass action.

C program for spatial grid model. The spatial C program was typically run with a 100 × 100 grid of patches with no boundary effects (migration on a torus). Figure 1 and Tables 1–3 were generated using this program. All phage, infected cells, dead cells, and EPS were assigned to a patch, and all interactions of phage within a patch occurred with other entities in that patch. A patch could harbor at most one cell (infected or uninfected), but in runs allowing debris (dead cells), a dead cell could occur in a patch with an infected or uninfected cell. Independent phage infection probabilities were assigned to the entities of EPS, cells, infected cells, and dead cells, such that a phage could remain uninfected or infect only one of the other entities. Once infected, cells had a finite lifespan (20 steps).

Within a time step, phage migration from a patch was limited to its eight neighbors, with probabilities according to a truncated symmetric, bivariate normal distribution centered on the patch and with a single variance parameter, as follows. Writing

$$f(x,y) = \frac{1}{2\pi\sigma^2} e^{-\frac{(x^2+y^2)}{2\sigma^2}}, \tag{9}$$

if $F(z) = P(Z \leq z)$ denotes the cumulative distribution of the standard (1-dimensional) normal, we have the following probabilities for phage diffusion to the eight patches (of side length 1) in the basic neighborhood:

center patch: A^2 (no diffusion),
each "orthogonal" neighboring patch: AB,
each diagonal patch: B^2,
where $A = 2F(0.5/\sigma) - 1$ and $B = F(1.5/\sigma) - F(0.5/\sigma)$.
These values were normalized by dividing each by $C = A^2 + 4AB + 4B^2$ to give the fractions of phage diffusing and remaining in the central patch.

In our trials, most of the probability was to remain on the central patch (no diffusion), so a phage was unlikely to move to a neighboring patch in a single time step. Phage diffusion was calculated deterministically (assigning appropriate fractions of the phage in a patch to that patch and the eight neighboring patches), but the overall net effect of migration on the patch was converted to an integral value by assigning any decimal fraction to 0 or 1 with a random draw in proportion to its magnitude.

Cell reproduction was permitted in every time step, each cell's reproductive fate chosen randomly according to a fixed probability, and independently of other cells' fates. Cells could reproduce only if one or more of their eight neighboring patches were unoccupied by a live cell (infected or uninfected), and preference was given that a daughter cell move into an orthogonal (off-diagonal) patch. All runs began with cells distributed randomly to 30% of patches and phage distributed randomly to 30% of patches (a patch getting phage received a burst size of phage).

C program for mass action grid model. For mass action, the C program was altered in three ways: (i) after burst and before new infections were allowed, all individuals in the entire population of phage were randomly assigned to patches in the grid; (ii) localized phage diffusion was turned off; and (iii) after cell reproduction, the entire population of infected and uninfected cells was reassigned to new patches, with at most one cell (infected or not) per patch. All other aspects of the mass-action C code are identical to those in the spatial C code, allowing us to assess the effects of spatial structure using the same computational platform. There was no simulation of mass-action dynamics in the case of spatially clumped EPS since only the amount of EPS makes a difference in this case.

Python program for spatial grid model. The second spatial simulation, written in Python, assumed a 20×20 grid of patches without boundary effects. This simulation served as a prototype for the C simulation, and operates similarly with some exceptions. During each time step in the simulation, following a randomized order, each patch executed cell lysis (if applicable), cell reproduction, infection, and phage diffusion. Then, the simulation repeated the same steps in the next randomly-selected patch until all patches were updated for that time step. Contrast this process with the C simulation, where a single event (e.g., lysis) executed across all patches before the next type of event (e.g., reproduction) executed. In the Python simulation, phage and cells only diffused to orthogonal patches. Allowing for diagonal diffusion did not qualitatively impact the simulation results, as long as both phage and cells followed similar diffusion rules. Early simulations in which cells were allowed to diffuse diagonally (but phage were not) decreased the proportion of infections (I:E) that occurred in EPS under deterministic EPS clustering, and also made I:E sensitive to EPS abundance. Such disparity in diffusion capabilities of phage and cells was determined to be unrealistic, so in the C version of the program, both cells and phage were allowed to diffuse both orthogonally and diagonally. In summary, the differences between the Python and C simulations are minor, and both simulations produced comparable output.

NetLogo program for spatial grid model. The third spatial simulation, written in the agent-based platform NetLogo, assumed a 51×51 grid of patches without boundary effects. This discrete-time simulation updates all patches simultaneously according to probabilities that are based on the current configuration. It is similar to the C simulation except for the following: (a) individual phage diffuse randomly and independently by taking steps in random directions with a prescribed step size; (b) nutrient-dependent cell growth and lysis, where an initial allocation of nutrient was provided and then replenished periodically by pulsing in fresh nutrient across the grid (though the simulations used here had nutrient pulsing every time step to match the nutrient-independent dynamics of the other two simulations); (c) the offspring of a reproducing cell is placed at one of the eight neighboring patches as long as there is space available. Reproduction is suppressed whenever all these local patches are at their carrying capacity; and (d) an approximation to mean-field dynamics is simulated by using large phage step size and random placement of cell offspring (but no subsequent cellular diffusion). Trends observed with the NetLogo program were similar to those with the other two programs.

The choices of a 20×20 grid size for the Python simulation, a 51×51 grid size for the NetLogo simulation, and a 100×100 grid size for the C simulation were made because of computational constraints but are arbitrary. An increase in grid size moderately increased αb in some conditions and decreased it in others. However, the magnitude of these changes was small, and the larger grid size simulations showed smaller variances in αb than in smaller grid size simulations. For example, in one set of simulations with the C program (EPS = 0.9, burst = 60, random placement of EPS), αb was 18.19, 17.98, 17.96 at grid sizes of 30×30, 100×100, and 300×300, respectively. Thus, the choice of grid size does not affect the overall trends in αb described here.

Numerical ODE trials were carried out with Mathematica 11.1.0 (Wolfram Research Inc., Champaign, IL, USA) using NDSolve.

9. Conclusions

Phages are predators of bacteria. Their predator-prey dynamics have been studied for decades in the ideal conditions of liquid culture, where a reasonable agreement has been obtained between models and observations. More recent studies of phages and bacteria grown on surfaces and other 'structured' environments suggest that bacterial densities are often much higher than expected from liquid culture results.

Our study focused on the simple question of how spatial structure alone might allow densities of sensitive cells to be maintained at higher levels than in liquid. Our approach relied on computational models in which bacteria could escape phage only by residing adjacent to environmental phage traps, such as exopolysaccharide or cellular debris that irreversibly binds phage. We found that these types of environments could enable an elevation of cell density in which phage and cells self-organized into different regions of the environment: cells persisted in protected areas, phages persisted in areas that lacked phage-killing agents. However, the magnitude to which cell densities were elevated was always less than 2 orders of magnitude, often less than one order—and less than reported in empirical contexts. Other mechanisms are thus needed to account for bacterial survival amid phage attack in structured environments.

Acknowledgments: We thank Benji Oswald for assistance with the IBEST computer cluster. We are pleased to acknowledge the following grant support for this work: to B.R.J. and J.J.B.: National Institutes of Health Grants R01 GM 088344 and GM 122079; to K.C. and C.S.: National Science Foundation UBM (DMS-1029485); to S.MK.: Center for Modeling Complex Interactions at the University of Idaho (NIH grant P20GM104420), and the IBEST Computational Resources Core (NIH grant UL1 TR000423).

Author Contributions: J.J.B. and S.M.K. conceived of the problem, the general approach and were responsible for all analytical work. All authors contributed to one or more simulation codes and carried out trials. The manuscript was written by J.J.B. and S.M.K.

Conflicts of Interest: The authors declare no conflict of interest. The founding sponsors had no role in the design of the study; in the collection, analyses, or interpretation of data; in the writing of the manuscript, and in the decision to publish the results.

Abbreviations

The following abbreviations are used in this manuscript:

ODE	ordinary differential equations
EPS	exopolysaccharide

References

1. Adams, M.H. *Bacteriophages*; Interscience Publishers: New York, NY, USA, 1959.
2. Bohannan, B.J.M.; Lenski, R.E. Linking genetic change to community evolution: Insights from studies of bacteria and bacteriophage. *Ecol. Lett.* **2000**, *3*, 362–377.
3. Weitz, J.S.; Hartman, H.; Levin, S.A. Coevolutionary arms races between bacteria and bacteriophage. *Proc. Natl. Acad. Sci. USA* **2005**, *102*, 9535–9540.
4. Donlan, R.M.; Costerton, J.W. Biofilms: Survival mechanisms of clinically relevant microorganisms. *Clin. Microbiol. Rev.* **2002**, *15*, 167–193.
5. Briandet, R.; Lacroix-Gueu, P.; Renault, M.; Lecart, S.; Meylheuc, T.; Bidnenko, E.; Steenkeste, K.; Bellon-Fontaine, M.N.; Fontaine-Aupart, M.P. Fluorescence correlation spectroscopy to study diffusion and reaction of bacteriophages inside biofilms. *Appl. Environ. Microbiol.* **2008**, *74*, 2135–2143.
6. Alhede, M.; Kragh, K.N.; Qvortrup, K.; Allesen-Holm, M.; van Gennip, M.; Christensen, L.D.; Jensen, P.O.; Nielsen, A.K.; Parsek, M.; Wozniak, D.; et al. Phenotypes of Non-Attached Pseudomonas aeruginosa Aggregates Resemble Surface Attached Biofilm. *PLoS ONE* **2011**, *6*, e27943.
7. Hanlon, G.W.; Denyer, S.P.; Olliff, C.J.; Ibrahim, L.J. Reduction in exopolysaccharide viscosity as an aid to bacteriophage penetration through Pseudomonas aeruginosa biofilms. *Appl. Environ. Microbiol.* **2001**, *67*, 2746–2753.

8. Sutherland, I.W.; Hughes, K.A.; Skillman, L.C.; Tait, K. The interaction of phage and biofilms. *FEMS Microbiol. Lett.* **2004**, *232*, 1–6.
9. Xavier, J.B.; Foster, K.R. Cooperation and conflict in microbial biofilms. *Proc. Natl. Acad. Sci. USA* **2007**, *104*, 876–881.
10. Lu, T.K.; Collins, J.J. Dispersing biofilms with engineered enzymatic bacteriophage. *Proc. Natl. Acad. Sci. USA* **2007**, *104*, 11197–11202.
11. Cornelissen, A.; Ceyssens, P.J.; T'Syen, J.; Van Praet, H.; Noben, J.P.; Shaburova, O.V.; Krylov, V.N.; Volckaert, G.; Lavigne, R. The T7-related Pseudomonas putida phage phi-15 displays virion-associated biofilm degradation properties. *PLoS ONE* **2011**, *6*, e18597.
12. Hosseinidoust, Z.; Tufenkji, N.; van de Ven, T.G.M. Formation of biofilms under phage predation: Considerations concerning a biofilm increase. *Biofouling* **2013**, *29*, 457–468.
13. Soothill, J. Use of bacteriophages in the treatment of *Pseudomonas aeruginosa* infections. *Expert Rev. Anti-Infect. Ther.* **2013**, *11*, 909–915.
14. Darch, S.E.; Kragh, K.N.; Abbott, E.A.; Bjarnsholt, T.; Bull, J.J.; Whiteley, M. Phage Inhibit Pathogen Dissemination by Targeting Bacterial Migrants in a Chronic Infection Model. *mBio* **2017**, *8*, doi:10.1128/mBio.00240-17.
15. Schmerer, M.; Molineux, I.J.; Bull, J.J. Synergy as a rationale for phage therapy using phage cocktails. *PeerJ* **2014**, *2*, e590.
16. Heilmann, S.; Sneppen, K.; Krishna, S. Sustainability of virulence in a phage-bacterial ecosystem. *J. Virol.* **2010**, *84*, 3016–3022.
17. Heilmann, S.; Sneppen, K.; Krishna, S. Coexistence of phage and bacteria on the boundary of self-organized refuges. *Proc. Natl. Acad. Sci. USA* **2012**, *109*, 12828–12833.
18. Taylor, B.P.; Penington, C.J.; Weitz, J.S. Emergence of increased frequency and severity of multiple infections by viruses due to spatial clustering of hosts. *Phys. Biol.* **2016**, *13*, 066014.
19. Robb, S.M.; Woods, D.R.; Robb, F.T. Phage growth characteristics on stationary phase Achromobacter cells. *J. Gen. Virol.* **1978**, *41*, 265–272.
20. Los, M.; Golec, P.; Łoś, J.M.; Weglewska-Jurkiewicz, A.; Czyz, A.; Wegrzyn, A.; Wegrzyn, G.; Neubauer, P. Effective inhibition of lytic development of bacteriophages lambda, P1 and T4 by starvation of their host, Escherichia coli. *BMC Biotechnol.* **2007**, *7*, doi:10.1186/1472-6750-7-13.
21. Manning, A.J.; Kuehn, M.J. Contribution of bacterial outer membrane vesicles to innate bacterial defense. *BMC Microbiol.* **2011**, *11*, 258.
22. Erez, Z.; Steinberger-Levy, I.; Shamir, M.; Doron, S.; Stokar-Avihail, A.; Peleg, Y.; Melamed, S.; Leavitt, A.; Savidor, A.; Albeck, S.; et al. Communication between viruses guides lysis-lysogeny decisions. *Nature* **2017**, *541*, 488–493.
23. Lenski, R.E. Dynamics of interactions between bacteria and virulent bacteriophage. *Adv. Microb.Ecol.* **1988**, *10*, 1–44.
24. Chapman-McQuiston, E.; Wu, X.L. Stochastic receptor expression allows sensitive bacteria to evade phage attack. Part I: Experiments. *Biophys. J.* **2008**, *94*, 4525–4536.
25. Chapman-McQuiston, E.; Wu, X.L. Stochastic receptor expression allows sensitive bacteria to evade phage attack. Part II: Theoretical analyses. *Biophys. J.* **2008**, *94*, 4537–4548.
26. Bull, J.J.; Vegge, C.S.; Schmerer, M.; Chaudhry, W.N.; Levin, B.R. Phenotypic resistance and the dynamics of bacterial escape from phage control. *PLoS ONE* **2014**, *9*, e94690.
27. Campbell, A. Conditions for the existence of bacteriophage. *Evolution* **1961**, *15*, 143–165.
28. Levin, B.R.; Stewart, F.M.; Chao, L. Resource—Limited growth, competition, and predation: A model and experimental studies with bacteria and bacteriophage. *Am. Nat.* **1977**, *977*, 3–24.
29. Weitz, J.S. *Quantitative Viral Ecology: Dynamics of Viruses and Their Microbial Hosts*; Princeton University Press: Oxford, UK, 2015.
30. Charnov, E.L. *Life History Invariants: Some Explorations of Symmetry in Evolutionary Ecology*; Oxford University Press: Oxford, UK, 1993.
31. Aviram, I.; Rabinovitch, A. Dynamical types of bacteria and bacteriophages interaction: Shielding by debris. *J. Theor. Biol.* **2008**, *251*, 121–136.
32. Rabinovitch, A.; Aviram, I.; Zaritsky, A. Bacterial debris-an ecological mechanism for coexistence of bacteria and their viruses. *J. Theor. Biol.* **2003**, *224*, 377–383.

33. Eriksen, R.S.; Svenningsen, S.L.; Sneppen, K.; Mitarai, N. A growing microcolony can survive and support persistent propagation of virulent phages. *Proc. Natl. Acad. Sci. USA* **2018**, *115*, 337–342.
34. Bryan, D.; El-Shibiny, A.; Hobbs, Z.; Porter, J.; Kutter, E.M. Bacteriophage T4 Infection of Stationary Phase *E. coli*: Life after Log from a Phage Perspective. *Front. Microbiol.* **2016**, *7*, 1391.
35. Weitz, J.S.; Dushoff, J. Alternative stable states in host-phage dynamics. *Theor. Ecol.* **2008**, *1*, doi:10.1007/s12080-007-0001-1.
36. Nadell, C.D.; Drescher, K.; Foster, K.R. Spatial structure, cooperation and competition in biofilms. *Nat. Rev. Microbiol.* **2016**, *14*, 589–600.
37. Mitarai, N.; Brown, S.; Sneppen, K. Population dynamics of phage and bacteria in spatially structured habitats using phage λ and *Escherichia coli*. *J. Bacteriol.* **2016**, *198*, 1783–1793.

antibiotics

MDPI

Article

Efficacy of an Optimised Bacteriophage Cocktail to Clear *Clostridium difficile* in a Batch Fermentation Model

Janet Y. Nale [1], Tamsin A. Redgwell [2], Andrew Millard [1] and Martha R. J. Clokie [1,*]

[1] Department of Infection, Immunity and Inflammation, University of Leicester, Leicester LE1 9HN, UK; jn142@le.ac.uk (J.Y.N.); adm39@leicester.ac.uk (A.M.)

[2] School of Life Sciences, University of Warwick, Coventry CV4 7AL, UK; T.Redgwell@warwick.ac.uk

* Correspondence: mrjc1@le.ac.uk; Tel.: +44-116-252-2959

Received: 31 December 2017; Accepted: 6 February 2018; Published: 13 February 2018

Abstract: *Clostridium difficile* infection (CDI) is a major cause of infectious diarrhea. Conventional antibiotics are not universally effective for all ribotypes, and can trigger dysbiosis, resistance and recurrent infection. Thus, novel therapeutics are needed to replace and/or supplement the current antibiotics. Here, we describe the activity of an optimised 4-phage cocktail to clear cultures of a clinical ribotype 014/020 strain in fermentation vessels spiked with combined fecal slurries from four healthy volunteers. After 5 h, we observed ~6-log reductions in *C. difficile* abundance in the prophylaxis regimen and complete *C. difficile* eradication after 24 h following prophylactic or remedial regimens. Viability assays revealed that commensal enterococci, bifidobacteria, lactobacilli, total anaerobes, and enterobacteria were not affected by either regimens, but a ~2-log increase in the enterobacteria, lactobacilli, and total anaerobe abundance was seen in the phage-only-treated vessel compared to other treatments. The impact of the phage treatments on components of the microbiota was further assayed using metagenomic analysis. Together, our data supports the therapeutic application of our optimised phage cocktail to treat CDI. Also, the increase in specific commensals observed in the phage-treated control could prevent further colonisation of *C. difficile*, and thus provide protection from infection being able to establish.

Keywords: *Clostridium difficile*; *Clostridium difficile* infection; bacteriophages; phage therapy; microbiome; in vitro fermentation model

1. Introduction

Antimicrobial resistance is a global health threat to clinical practice and public health [1–4]. It is estimated that the continued rise in multidrug resistance (MDR) will cause 10 million people to die worldwide by 2050 and cost 100 trillion USD [5]. To effectively control bacterial infections, novel effective antimicrobials with target specificity and high efficiency are urgently needed [6–8]. Although bacteriophages or phages (viruses which specifically lyse bacteria) were first isolated over 100 years ago, for a long period they were mainly the focus of fundamental research. However, particularly over the last decade, there has been an increasing interest in the isolation, characterisation and development of phages for therapeutic use in humans, animals, and plants [9–13]. This revived interest is mainly driven by problems associated with ineffective antibiotics. These natural bacterial predators have the potential to provide a safe and suitable supplement, or replacement for antibiotics because of their specificity and amplification at the site of infection [14–17]. Indeed, phage products have been developed for medical use, and some can be found as over-the-counter medicines and are used as decontamination agents in food industries [10,18–20].

Clostridium difficile is a notorious nosocomial bacterium that remains a major cause of infectious diarrhea, with high morbidity and mortality in the elderly and in immunocompromised patients worldwide [21–25]. *C. difficile* surveillance for the United Kingdom showed that there were 19,269 reported cases of *C. difficile* infection (CDI) and 488 (~3%) fatalities in 2015. In the US, ~500,000 CDI cases are reported annually, with approximate 30,000 deaths, 20% recurrent rates, and an estimated treatment cost of ~$10,000 per case [21,22,26]. CDI is becoming increasingly difficult to treat because of the emergence of severe and antibiotic-resistant ribotypes, and very limited treatment options [6,27,28]. Currently, only three antibiotics are available on the market for CDI treatment. Metronidazole is cheap, largely effective, and is recommended for initial use in moderate or non-severe episodes [29–31]. However, there are problems associated with its efficacy to treat some important prevalent and clinically relevant ribotypes, resistance has also been seen towards this antibiotic, and it is associated with health-related complications such as low birth weight [30,32,33]. Vancomycin is the antibiotic of choice for moderate to severe CDI but its use, particularly if long-term, can promote the emergence of vancomycin-resistant enterococci [34,35]. Also reduced susceptibility have been reported in *C. difficile* leading to recurrent infection, thus it is suboptimal [36,37]. Fidaxomicin is a highly specific antibiotic that has been shown to be effective when vancomycin treatment has failed [38] but it is expensive ($3360 compared to $1273 for vancomycin or $21.90 for metronidazole—all per course) and may not be cost-effective for some strain-specific CDIs [39–41]. This complex relationship between *C. difficile* and antibiotics is compounded by the fact that they generally have a detrimental impact on the gut microbiota, which leads to dysbiosis that then enables *C. difficile* to colonise the gut and cause disease. Therefore, there is a clear need to develop additional antimicrobials with increased target specificity in order to efficiently remove this pathogen but leave other components of the gut microbiota intact [6,10].

Previous reports have described the isolation of phages that specifically target *C. difficile* and demonstrated the use of different in vivo and in vitro models to test the specificity and efficacy of the phages to selectively eradicate this bacterium [42–46]. The commonly used in vivo model for CDI and *C. difficile* phage therapy is the hamster model, which is useful as hamsters demonstrate the classical CDI clinical symptoms seen in humans [47,48]. However, the model is difficult to use because of the exquisite sensitivity of hamsters to *C. difficile* toxins, high costs, and inherent technical issues associated with working with these animals [49,50]. Therefore, alternative models such as the wax moth *Galleria mellonella* larva have been developed as suitable replacement models to probe many aspects of CDI phage therapy [46]. Other models that have been used to study *C. difficile* phage therapy are the in vitro gut and batch fermentation models [44,51]. Although these models have been developed to study the gut microbiome and pharmokinetics of antibiotics, very few studies have applied them to study *C. difficile* phage therapy [38,50,52].

The four myoviruses CDHM1, 2, 5, and 6 used in this study were isolated from the environment and were well characterised in our laboratory [45,53]. This optimised phage cocktail was the first phage mix shown to completely clear *C. difficile* in pure cultures and it was also shown to prevent biofilm formation in vitro. In addition, the phages reduced colonisation in vivo in both hamster and wax moth larva CDI models [45,46]. The data obtained from these models provided novel insights into the therapeutic applications of these phages to treat CDI. However, more information is needed in order to determine the specificity of this phage set to *C. difficile*, and to establish their ability to clear the target pathogen in the presence of competitive pressure from other components of the human gut microbiota. Indeed, no previous publications have examined the potential impact of the application of *C. difficile* phages on the human microbiome. Therefore, we developed and present results from an in vitro phage therapy assay using a batch fermentation model. To do this, we obtained human feces from a specific age profile of healthy volunteers (with full ethical consent) in order to examine the impacts of our phage set on a wide range of human microbiota. The work was designed to: (i) determine the efficacy of our optimised phage cocktail to clear a clinically relevant ribotype 014/020 strain in the presence of the gut microbiota, (ii) test the efficiency of the phages using prophylactic or remedial regimens in the targeted eradication of the bacterium in the batch fermentation model, and (iii) ascertain the potential synergistic or antagonistic effect of phage application on culturable and unculturable components of the human gut microbiome.

2. Results

2.1. Individual Donors Have a Unique Microbiome Composition

To determine the specificity and efficacy of the phages to clear *C. difficile* in the presence of competitive pressure from representative human gut microbiota, we spiked five fermentation vessels containing a minimal medium with combined fecal slurries obtained from four healthy volunteers [54–56]. The donors were comprised of individuals from diverse ethnic and age groups (a 70-year-old white British woman, 44-year-old black woman, 17-year-old black girl, and 7-year-old white British boy) to capture a wide range of human gut microbial diversity. Prior to mixing the fecal slurries together, we determined the microbiome composition from the individual donors by resuspending the fecal matter in the minimal medium and enumerating the bacteria present on selective agar media targeting five commonly occurring gut commensals [44,51]. We observed that approximately 10^5–10^6 CFU/mL (colony-forming unit per milliliter) of enterococci counts were detected from all the four donors. Similar counts were observed with the lactobacilli group, except that the abundance of this bacterium was very low, hence it was undetectable in the teenager. Relatively higher counts were observed in the total anaerobes and enterobacteria, which ranged from 10^6 to 10^7 CFU/mL in all the donors. The bifidobacterial counts were quite variable, from very low counts of ~10^3 CFU/mL in the teenager, to 10^5 and 10^6 CFU/mL in the infant and adult, respectively, and 10^7 CFU/mL in the elderly donor lady. Interestingly, but not unexpectedly, we did not recover *C. difficile* from any of the donors. When the fecal matter was mixed together and assayed, it was observed that the total anaerobes and enterobacterial numbers were the highest with ~10^6 CFU/mL, but ~10^5 CFU/mL was being contributed by the enterococci, lactobacilli, and bifidobacteria (Figure 1).

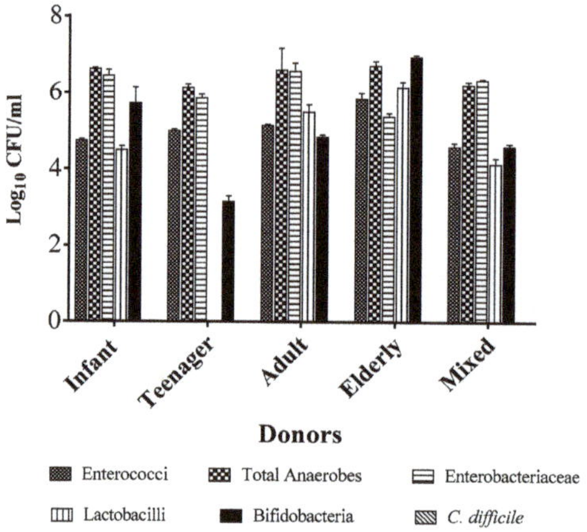

Figure 1. Contributory culturable bacterial counts from each of the individual donors and final cumulative counts of each bacterium added to the fermentation vessels. The bacteria present in the fecal sample of each donor were determined by recovery on selective medium for each bacterial grouping, after which, the samples were mixed together in relatively equal amounts and used to prime the fermentation vessels. The data was analysed using GraphPad Prism 7. Error bars are SEMs of three biological replicates.

2.2. Phages Cleared *C. difficile* in the Batch Fermentation Model

We determined the ability of the phage cocktail to clear CD105LC2 (ribotype 014/020 clinical strain) in the batch fermentation model using two treatments. In the prophylactic regimen, the fecal slurries were exposed to a dose of the phage cocktail followed by a mixture of the phages and bacteria, and subsequently by two doses of the phage cocktail (Table 1). In the first 5 h following bacteria and phage exposure, we observed a ~6-log reduction of CFU/mL of *C. difficile* counts in the prophylactic regimen compared to the bacterial control. When the fermentation vessels were treated remedially, the phage cocktail was added after culturing the bacteria for 5 h. At the 24 h time point (19 h post-phage treatment in the remedial regimen), *C. difficile* was eradicated in both the prophylactic and remedial treatment vessels, and this observation remained consistent from this time until the end of the experiment (72 h). As expected, *C. difficile* was not detected in the untreated (vessel 1) and the phage (vessel 3) controls. However, in the *C. difficile*-only control (vessel 2), we observed that after 5 h of incubation, *C. difficile* numbers began to drop from ~10^7 CFU/mL (at 5 h) to ~10^4 CFU/mL at 36, 48, and 72 h, respectively (Figure 2A).

Table 1. Bacteria and phage treatment regimens for the gut fermentation vessels.

Fermentation Vessels	Treatments	Time to Dose (h)						
		−2	0	5	24	36	48	72
1	Control untreated	-	M	M	M	M	M	-
2	*C. difficile* control	-	B	M	M	M	M	-
3	Control phage	-	P	P	P	P	M	-
4	Prophylaxis	P	P+B	P	P	M	M	-
5	Remedial	-	B	P	P	P	P	-

Five vessels containing combined fecal slurries from four healthy volunteers were treated with 2 mL each of 6–8 × 10^8 CFU/mL of *C. difficile* culture (B), 2–6 × 10^9 PFU/mL (plaque-forming units per milliliter) of phage cocktail (P), and/or minimal medium (M) at the time points shown above [51]. At each time point, 2 mL of samples from the vessels were removed and replaced with an equal volume of the bacteria, phage or medium. The time points were selected based on our prior in vitro data on the phages, which showed that the phages maintained clearance of CD105LC2 cultures at the first 5 and 24 h time points [45]. The additional 36 and 48 h time points were based on previous fermentation studies [51].

2.3. *Impact of Phage Treatmens on the Viability of other Components of the Culturable Gut Microbiota*

After establishing the efficacy of the phage cocktail to clear *C. difficile* in the batch fermentation model, we investigated their impact on five common major bacterial groups in the human gut. We did this by conducting viability assays on selective media for bifidobacteria, enterococci, enterobacteria, lactobacilli, and total anaerobes in the five fermentation vessels at all the time points examined [44,51].

Bifidobacterial numbers were relatively constant in both the treatment regimens and the three controls throughout the experiment. The ~10^5 CFU/mL of bacteria observed at 0 h decreased to ~10^4 CFU/mL in all the treatment regimens as well as in the controls at 5 h. The bacterial numbers remained relatively stable at this level until the end of the experiment (72 h). There was no significant difference in the number of bacteria left in all the treatment vessels at the end of the 72 h time period of the experiment (p = 0.05) (Figure 2B).

The enterococci abundance showed distinct changes depending on the treatment. Relatively equal numbers, ~10^5 CFU/mL of bacteria, were observed at the beginning of the experiment (0 h) in all the treatment vessels, and this number remained consistent until 24 h. After this time, the bacterial numbers dropped to ~10^4 CFU/mL in vessels 1 and 2, which corresponded to the untreated and *C. difficile* controls, respectively, both not treated with the phages. The numbers for this group of bacteria continued to drop in vessels 1 and 2, and after 72 h, only ~10^3 CFU/mL of bacteria were recovered. However, in all the phage-treated vessels comprising the phage control (vessel 3), the prophylactic (vessel 4), and the remedial regimens (vessel 5), the enterococci detected remained relatively stable at ~10^5 CFU/mL from 5 h to 72 h (Figure 2C).

For the enterobacteria, the numbers increased from ~10^6 to 10^8 CFU/mL within the first 5 h of the experiment in all the fermentation vessels. After 24 h, the bacterial numbers remained stable in the phage-only treated control, but lower numbers (10^7 CFU/mL) were observed in the untreated vessel

(vessel 1) and the prophylaxis vessel (vessel 3). The remedial and the bacterial control vessels had even lower numbers (10^6 CFU/mL). After 24 h, higher enterobacterial numbers (~10^8 CFU/mL) were seen in the phage-only treated vessel (vessel 3), which remained stable throughout the experiment. In the other vessels (vessels 1, 2, 4, and 5), however, the bacterial counts remained lower (at ~10^6 CFU/mL) than in the phage-treated vessel at 24 h until the experiment was terminated. At the end of the experiment, ~2-log CFU/mL higher numbers of enterobacteria were observed in the phage-only treated vessel (vessel 3), compared to the other vessels (Figure 2D).

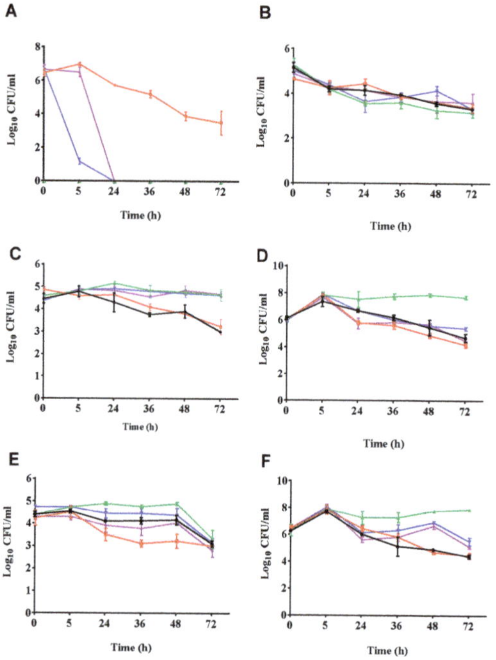

Figure 2. Impact of phage treatment on *C. difficile* and other components of the gut microbiota. The impact of phage treatment was ascertained by recovering the bacteria on selective media for (**A**) *C. difficile*; (**B**) bifidobacteria; (**C**) enterococci; (**D**) enterobacteria; (**E**) lactobacilli; (**F**) total anaerobes. The bacterial counts of the different treatment vessels and time points are presented. Black lines, vessel 1, untreated slurries; red lines, vessel 2, *C. difficile* control; green lines, vessel 3, phage-only-treated control; blue lines, vessel 4, prophylaxis regimen, and purple lines, vessel 5, remedial regimen. The data was analysed using GraphPad Prism 7. Error bars are SEMs of 3 biological replicates.

We also assayed for lactobacilli counts in all the treatment vessels over the time points. We observed that equal numbers (~10^5 CFU/mL) of the bacteria were detected in the beginning of the experiment at 5 h in all the vessels. At the 24 h time point, the phage-only treated and the prophylactic-treated vessels had higher bacterial numbers than the remedial regimen, the *C. difficile*, and the untreated control vessels. In all the vessels, the bacterial numbers remained stable at this level and at the subsequent three time points of 24, 36, and 48 h, but steadily declined to ~10^3 CFU/mL at 72 h (Figure 2E).

The final bacterial group assayed on selective media was the total anaerobes. As observed for the enterococci, the total number of anaerobes increased markedly from ~10^6 CFU/mL at 0 h to 10^8 CFU/mL at 5 h in all the vessels. However, the total number of anaerobes (~10^7 CFU/mL) in the phage-only treated control vessel (vessel 3) was ~1-log CFU/mL higher than in the other four treatment vessels (vessels 1, 2, 4, and 5) at 24 and 36 h. At 48 h however, the total anaerobe count was higher in the phage-only treatment (vessel 3), with ~10^8 CFU/mL recovered, followed by the prophylaxis and remedial regimen vessels with ~10^6 CFU/mL of bacteria detected. The control untreated slurries (vessel 1) and the *C. difficile* control (vessel 2) had relatively lower numbers (~10^5 CFU/mL) at 72 h (Figure 2F).

2.4. Metagenomics Analysis of the Impact of Phage Treatment on the Total Microbiome within the Gut Fermentation Vessels

The viability assays confirmed a complete eradication of *C. difficile* at the 24 h after phage treatment in both the prophylactic and remedial fermentation vessels. The beginning of an effect on five other bacterial groups was also observed. To probe these observations more deeply and to determine the impact on the components of the microbiota, including those that cannot be cultured, the total DNA from the five treatment vessels at the 24 h time point was extracted and used as a template for whole metagenomics analysis. The total reads per vessel at 24 h were mapped to relevant sequences representing all three domains of life (Table 2). The percentage of reads for each domain was normalised against the total number of reads found in each vessel. The highest percentage of reads for bacteria was found in vessel 5, with 99.11%, whereas the lowest percentage of reads was found in vessel 1, with 97.76% reads. We observed that Firmicutes, Bacteroidetes, Proteobacteria, Actinobacteria, Cyanobacteria, Euryarchaeota, Verrucomicrobia, Deinococcus-Thermus, Spirochaetes and Synergistetes abundances were consistently the most abundant among the bacterial phyla examined, irrespective of the treatment vessel (Table S1, Figure 3A–D,Fi). Although the individual groups of bacteria remained consistent in all treatments, their abundances varied considerably in the vessels. We observed that percent reads mapped to Actinobacteria were higher in the none-phage-treated vessels (vessel 1, 26.2% and vessel 2, 26.9%) compared to vessels 3 (22.4%), 4 (23.8%), and 5 (25.3), which corresponded to phage-treated vessels (Table S1, Figure 3A–E). This pattern in all the vessels was also observed for the Bacteroidetes, for which the reads found in non-phage-treated vessels were higher compared to those in phage-treated vessels (Table S1). In contrast, the Deinococcus levels were higher in vessels 3, 4, and 5, which contained the phages, but reduced in non-phage-treated vessels (vessels 1 and 2). The Firmicutes and Verrucomicrobia had reduced abundance in the phage-only-treated slurries compared to the other four vessels (Table S1). Conversely, the Cyanobacteria, Enterobacteriaceae, and Proteobacteria abundance was elevated at 24 h in the phage-only-treated control vessel compared to the other four vessels. The abundance of the Spirochaetes in the prophylaxis treatment regimens was comparable to the level found in the untreated slurries (vessel 1). Consistent with our viability assays, we observed that Bifidobacteriaceae, Enterobacteriaceae, Lactobacillales as well as the Coriobacteriaceae, Bacteroidaceae, Porphyromonadaceae, Rikenellaceae, Eubacteriaceae, Lachnospiraceae, Rhizobiales, Desulfovibrionales and Ruminococcacea abundances were considerably high in vessel 3 from the metagenomics data (Figure 3A–D,Fi).

Table 2. Reads mapped to the three domains of life.

Domain	Clade Reads in Each Vessel (%)				
	V1	V2	V3	V4	V5
Bacteria	89,526 (97.76)	74,116 (99.05)	23,061 (98.89)	96,888 (98.77)	103,406 (99.11)
Archaea	2030 (2.217)	646 (0.8633)	19 (0.08148)	926 (0.944)	764 (0.7323)
Viruses	18 (0.01966)	64 (0.08553)	240 (1.029)	284 (0.2895)	165 (0.1581)
Total	91,574	74,826	23,320	98,098	104,335

Whole genome sequencing was conducted on DNA samples extracted at the 24 h time point. The data was analysed using Pavian.

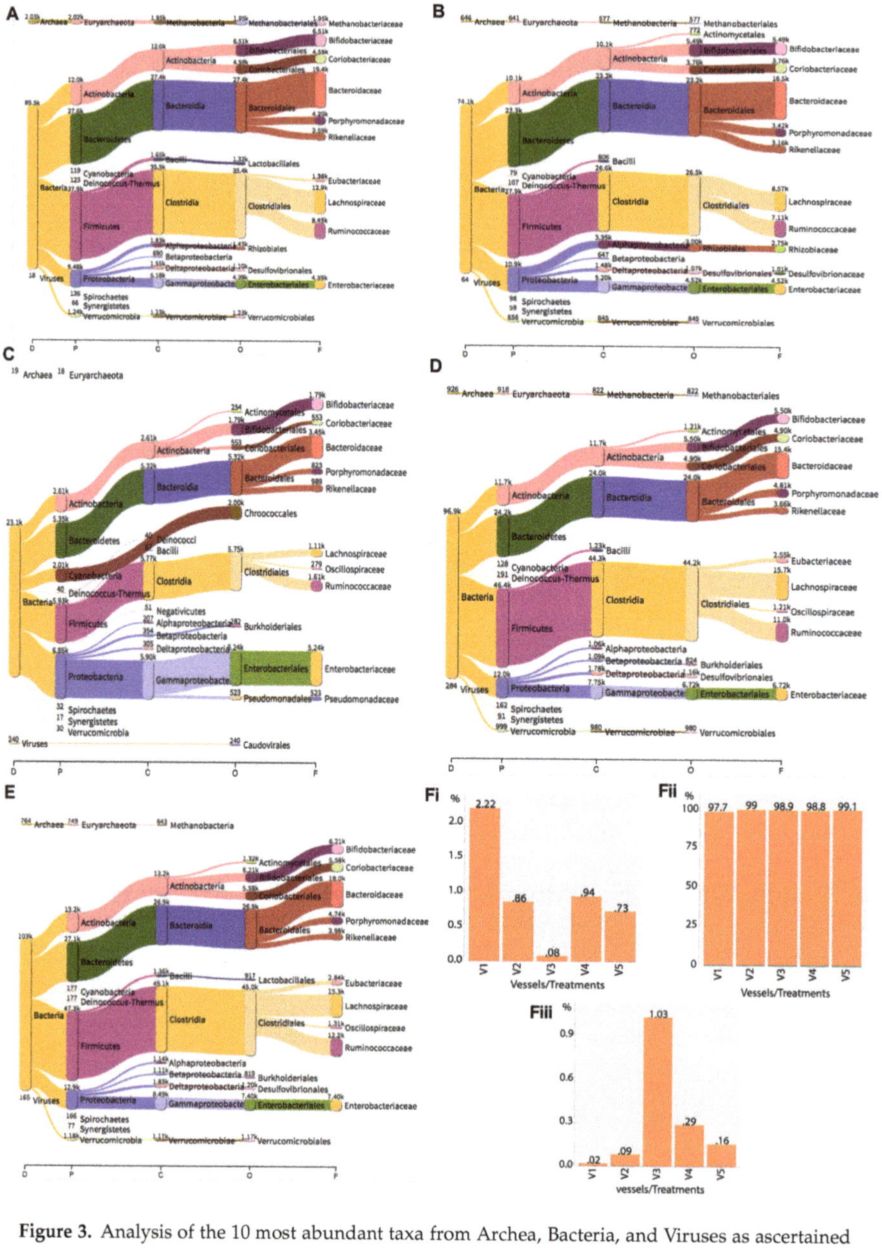

Figure 3. Analysis of the 10 most abundant taxa from Archea, Bacteria, and Viruses as ascertained by the metagenomics data. Total genomic DNA was extracted at 24 h time point from (**A**) vessel 1, untreated slurries; (**B**) vessel 2, *C. difficile* control; (**C**) vessel 3, phage-only-treated control; (**D**) vessel 4, prophylaxis regimen; (**E**) vessel 5, remedial regimen. The samples were prepared using NexteraXT sample preparation kit and sequenced on MiSeq platform using V2 (2 × 250 bp) chemistry. The resulting fastq files were trimmed with Sickle, and the metagenomes were assembled using megahit. An overview of the 10 most abundant taxa: Phyla (P), Classes (C), order (O), and family (F) are shown for each treatment vessel, as visualised using Pavian. The percent reads mapped to Archaea, Bacteria, and Viruses in the vessels at 24 h are shown in (**Fi**), (**Fii**), and (**Fiii**), respectively.

The percentage of reads mapped to Archaea was found to be highest in the untreated vessel (vessel 1), with 2.2% reads mapped to this domain in this vessel at the 24 h time point. The remaining vessels, V2–5, had low percent reads (0.08–0.94) mapped to Archaea compared to the untreated vessel 1. The phylum Euryarchaeota, consisting of the family Methanobacteriaceae was consistently found in all the vessels, although its abundance was considerably lower in the phage-only-treated vessel (vessel 3) (Figure 3Fii).

Consistent with the other two phyla, we found reads corresponding to viruses to be represented in all the treatment vessels. For the percentages of reads which mapped to the viruses, we found the highest level in vessel 3 (phage-only-treated vessel) with 1.029% of the total reads, and lower levels in all other treatment vessels (vessel 1, 2, 4, and 5). Vessel 1 had the lowest percent viral reads (0.0196) compared to vessels 2, 3, 4, and 5, which had 0.08553, 1.029, 0.2895, and 0.1581%, respectively (Figure 3Fiii).

3. Discussion

The need for alternative therapeutics to combat antibiotic resistance is clear [5,57]. The challenges posed by MDR are current and serious and so cannot be ignored. Consequently, significant resources are being channeled towards understanding the root causes of MDR and developing effective approaches to tackle the associated health threats [8,58,59]. There is an urgent clinical unmet need for novel treatments for *C. difficile*, the causal agent of CDI [6,60]. Recent reviews on past, current, and future options for CDI treatment concluded that phage therapy has significant potential as a treatment because of its specificity, amplification at infection sites, and minimal deleterious impact to the gut microbiome [6,16,43]. The development of phage treatments for this pathogen has been hampered by the lack of strictly virulent phages. Furthermore, although past data has demonstrated the efficacy of *C. difficile* phages to clear the bacterium in vitro and in vivo, there is the lack of preclinical data to ascertain the impact of these phages on the human gut microbiome [45,46]. Previously, we developed the first effective *C. difficile* phage cocktail, which consists of four well-characterised, broad host range myoviruses [45,46]. In this study, we showed as a proof of concept that the cocktail can effectively clear a clinically prevalent *C. difficile* ribotype isolate in a batch fermentation model, and that their application promotes the growth of other human gut commensals.

The two previous reports on *C. difficile* phage therapy using in vitro gut models used a batch fermentation model over a 48 h time period [51] or a three-component continuous colon model over a 35 day period [44]. In both assays, the effect of one phage, ΦCD27, to clear NCTC11204, a ribotype 001 strain, was investigated. The two reports are consistent with our observation that the prophylactic regimen is more effective at clearing *C. difficile* than the remedial regimen. Although both previous data sets showed a significant reduction (~6-log CFU/mL) in the prophylaxis treatment, *C. difficile* (~1-log CFU/mL) was still detected in the 48 h report [51], and up to 10^8 CFU/mL of cells or spores were detected in the 35 day report at the end of the assays [44]. Whilst ΦCD27 was shown to be active in the prophylactic regimen, the regrowth observed could be attributed either to resistance-developing, an inherent insensitivity of the bacterium to the phages, or the generation of resistant lysogenic clones, as shown when the bacterium was treated with the phage at a multiplicity of infection (MOI) of 7 [51]. Since all published *C. difficile* phages are temperate and encode an integrase gene, which mediates their integration into host genomes, the development of lysogenic mutants using a single phage for therapy occurs, as shown in previous reports [45,47,51]. In our previous work and in work presented here, we have demonstrated that the impacts of lysogeny and/or phage or antibiotic resistance are mitigated by the application of an optimised diverse phage cocktail [45,46]. *C. difficile* was fully lysed regardless of whether a prophylactic or remedial regimen was applied, and remained undetectable till the end of the experiment (72 h), as shown in this study (Figure 2A) and in our previous in vitro models [45,46]. The phages used here exhibit a complementary effect, whereby resistant or lysogenic clones produced by one phage or antibiotic treatment become susceptible to infection by another phage in the mix, leading to complete eradication of *C. difficile* in vitro and significant reduction of colonisation in vivo [45,46].

The advantages of using a phage cocktail was also demonstrated in our remedial regimen, where we observed a ~7-log CFU/mL reduction in *C. difficile* counts, and the bacterium was completely eliminated from vessel 5 (remedial regimen) within 24 h of the post-phage treatment without any detectable regrowth (Figure 2A). In contrast, there was only a >1-log CFU/mL reduction when the single phage ΦCD27 was applied remedially at an MOI of 10 at this time and at subsequent time points, and no observable impact was reported when the phage was applied at an MOI of 7 [51]. The lower remedial impact of *C. difficile* phages was also observed in other in vitro assays [45–48]. Obviously, these data further support the fact that phage therapy for CDI will require the application of optimised phage combinations, as previously reported for treating other bacterial species [13,61]. Although our phage cocktail was optimised for the ribotype 014/020, work is currently ongoing in our laboratory to determine suitable phage combinations for other prevalent and severe ribotypes.

In our fermentation model, we have for the first time studied the impact of phages on microbiomes derived from fecal samples from four healthy volunteers from different age groups and ethnicities to prime the fermentation vessels. Although we used fecal matter from healthy live volunteers from different age groups, other work used human fecal matter of deceased individuals [62] or healthy elderly individuals [44,51,52,56]. Whilst the elderly group reflects the majority of people commonly predisposed to CDI because of their weak immune system, other age groups are also susceptible to the pathogen, though to a lesser extent [63,64]. In addition, human gut microbiomes have been shown to vary greatly and be shaped by individual lifestyles, age groups, and geographical regions, and have been studied in combined emulsions (Figure 1) [54,55,65].

Although *C. difficile* is often a natural human commensal, we did not recover it from the fecal samples of our donors using our viability assays. *C. difficile* is generally considered to be an opportunistic bacterium and could remain in the gut of a healthy individual without causing disease until there is a disruption of the microbial balance (dysbiosis) through antibiotic use, triggering *C. difficile* to colonise the gut and causing disease [6,66,67]. *C. difficile* could possibly be present in the guts of our donors but in low abundance, hence it could not be detected in the feces [67]. In addition, the decline in *C. difficile* counts, as observed in the bacterial control after 24 h, has been reported previously and could reflect the response of the bacterium outside the natural gut environment or the depletion of nutrients in the medium used [44,51].

Our observations from the viability assays and metagenomic analysis show that both the prophylactic or remedial phage regimens did not have a significant detrimental impact on the five bacterial groups examined and concur with other previous phage therapy assays [44,51]. Similarly, we did not observe a huge difference in the abundance of bacteria in the phage-treated vessels and controls in the metagenomics data, and this clearly supports the advantages of phage therapy over antibiotics [68]. Because of their specificity, phages are able to infect the targeted bacteria, preserving the commensal niche as opposed to chemical antibiotics, which have a broader activity and may induce superinfection by some species such as *C. difficile* [15,69].

Prior to our work, no data had examined the impact of just *C. difficile* phages on other components of the microbiota during therapy. The two previously published fermentation model reports examined the impact of phages on *C. difficile* clearance and the corresponding impact on other commensals but did not examine the effect of phages alone on the gut microbiota [44,51]. Here, we showed that the abundances of certain gut commensals were elevated and restored to the initial levels of the donor samples during phage administration, and this clearly links to the high viral abundance found in vessel 3 (Table 2). This observation strongly suggests that the phages could promote the growth of natural human bacteria, providing health benefits, and thus could protect the gut from *C. difficile* colonisation [69]. The possible roles of phages in the restoration of the gut microbiomes of CDI patients have previously been reported and may provide biological insights into the mechanisms of fecal transplantation. For example, previous data showed that fecal matter containing higher diversity of Caudovirales led to increased richness, diversity, and evenness of these viral particles when transplanted to CDI patients. A concomitant increase in the abundance of other gut commensals (such as Proteobacteria and Actinobacteria) and the resultant resolution of CDI were also observed in the majority of the recipient patients [70]. Similarly, the administration of fecal filtrates from healthy humans via nasojejunal tubes restored the normal stool habits and eliminated CDI symptoms in five symptomatic chronic relapsing CDI patients in another study [71].

4. Materials and Methods

4.1. Bacterial Isolates and Phage Cocktail Used in This Study

In this study, two *C. difficile* isolates were examined. The first, CD105HE1, is an environmental isolate of ribotype 076 and was used as the propagating host for the phages [53]. The second bacterial isolate, CD105LC2, was the test strain for the gut fermentation model and belonged to the clinically prevalent ribotype 014/020 [45]. The bacterial isolates were routinely cultured on brain heart infusion (BHI) agar (Oxoid, Hampshire, UK) supplemented with 7% defibrinated horse blood (Thermo Scientific, Hampshire, UK) for 48 h prior to use, or stored in cryogenic storage tubes (Abtek Biologicals Ltd., Liverpool, UK) at −80 °C. The bacterial culture used for the gut fermentation experiments was produced by inoculating a single colony of the test bacterium in 5 mL of pre-reduced fastidious anaerobic broth (BioConnections, Knypersley, UK) and incubating anaerobically (10% H_2, 5% CO_2 and 85% N_2, Don Whitley Scientific, West Yorkshire, UK) at 37 °C for 18–24 h. A 1:100 dilution of the overnight culture was prepared in 10 mL BHI broth, incubated until OD_{550}~0.25–0.3 (~10^8 CFU/mL) was attained, and used to inoculate the fermentation vessels. All liquid culture media were pre-reduced anaerobically at 37 °C for at least 1 h prior to use.

An optimised phage cocktail containing four *C. difficile* myoviruses, CDHM1, 2, 5, and 6 was used for phage therapy in this study. The individual phages were isolated from the environment and characterised previously in our laboratory [45,46], and propagated individually in liquid cultures of the environmental isolate CD105HE1 to produce 10^{10} PFU/mL of infective phage particles [45,53]. Prior to use, the phages were diluted to 10^9 PFU/mL in BHI and mixed in equal proportions to constitute the cocktail, which was kept at 4 °C for short-term storage or in 25% glycerol for long-term storage at −80 °C.

4.2. Gut Fermentation Model Set-Up

The gut fermentation model examined here was adapted from previous *C. difficile* phage studies [51] with slight modifications. The fermentation vessels were comprised of five 250 mL capacity Duran bottles containing 135 mL of a minimal medium containing 0.2% peptone, yeast extract, $NaHCO_2$, and Tween 80, 0.01% NaCl, 0.004% each of K_2HPO_4 and $KHPO_4$, 0.001% of $MgSO_4{\cdot}7H_2O$, $CaCl{\cdot}2H_2O$, and vitamin K (in 5% aqueous solution), 0.005% each of Cysteine HCl and bile salts, and 0.0002% haemin (dissolved in 400 μL of 1 M NaOH). The pH of the medium was adjusted to and maintained at ~6.8 throughout the experiment, using filter sterilised NaOH and HCl. The medium was pre-reduced anaerobically at 37 °C for 24 h before use.

Freshly voided fecal samples were collected from four donors comprising a heathy White infant British boy (7 years old), a Black African teenage girl (17 years old), a Black African adult lady (44 years old) and a White British Elderly lady (70 years old). All donors were healthy at the time of sample collection and had not had antibiotics for 6 months prior to the time of sampling [56]. The fecal matter was passed at will into sterile plastic bowls before being transferred to Elkay 30 mL polystyrene transport tubes fortified with spoons. The samples were immediately stored under ice, and analysed within 2 h of collection. Under anaerobic conditions, approximately 5 g (approximately one spoonful within the Elkay tube) of each stool sample was diluted in 20 mL of the minimal media and mixed by inversion until all fecal materials were completely suspended to form a fecal emulsion or slurry.

One milliliter of each sample slurry was taken to determine the donors' contributory microbiota. To do this, the culturable bacteria present in each fecal sample were ascertained by a viability assay on selective media for *C. difficile* (Brazier's CCEY medium, BioConnections, Knypersley, UK), bifidobacteria (BSM medium, Sigma, Steinheim, Germany), total anaerobe (Wilkins-Chalgren anaerobic agar, Oxoid, Hampshire UK), lactobaccilli (Rogosa agar, Oxoid, Hampshire, UK), enterobacteria (MacConkey agar, Sigma, Steinheim, Germany), and Gram-positive cocci (Slanetz-Bartley agar, Oxoid, Hampshire, UK), prepared according to the manufactures' recommendations. Afterwards, equal volumes of the fecal slurries were thoroughly mixed together, and 15 mL of the combined slurries was added to each of the fermentation vessels and further pre-reduced (anaerobically at 37 °C) for 2 h (−2 h, Table 1). The fermentation vessels were continuously agitated throughout the duration of the experiment using a sterile magnetic stirrer set, at 100 rpm.

4.3. Bacteria and Phage Treatment of the Fermenation Vessels

The five fermentation vessels were treated with 2 mL each of 6–8 × 10^8 CFU/mL culture of the test strain CD105LC2 (B), 2–6 × 10^9 PFU/mL of phage cocktail (P), and/or the minimal medium (M) at the time points shown in Table 1. At each time point, a 2 mL sample was removed from each of the treatment vessels and replaced with an equal volume of either the phage, the bacteria, or the minimum media, as appropriate (Table 1). The 2 mL samples were used for bacterial enumeration and DNA extraction for that time point. Vessel 1 (control untreated) contained the minimal medium and the fecal slurry only, and, at time points 0, 5, 24, 36, and 48 h, 2 mL of the pre-reduced minimal medium was added. In vessel 2 (*C. difficile* control), the bacterial culture was added at time point 0 h, and subsequently at the 5, 24, and 36 h time points, and at 48 h 2 mL of the minimal medium was added. For vessel 3 (control phage), the phages were added at 0, 5, 24, and 36 h, and at 48 h 2 mL of the medium was added. In the prophylactic regimen (vessel 4), the phages were added during the pre-reduction time (−2 h, on the basis of our prior data on phage pre-treatment time during the prophylaxis regimen [46]) and followed by the mixture of the phage and bacterial inocula, added at the 0 h time point. Afterwards, the phages were added at 5 and 24 h, and the medium at 36 and 48 h in vessel 4. For the remedial regimen (vessel 5), the bacterial inoculum was added at the 0 h time point, followed by the phage cocktail at the 5, 24, 36, and 48 h time points. The experiment was terminated at 72 h, and at this time the bacterial numbers were enumerated as described above.

4.4. DNA Extraction, Metagenomics Sequencing, and Analysis

The DNA was extracted from 1 mL of the 2 mL aliquot taken from the treated vessels using FastDNA spin kit for feces (MP Biomedicals, Santa Ana, CA, USA). The DNA quality was ascertained using Nanodrop One (Thermo Fisher Scientific, Madison, WI, USA) and Qubit ds DNA HS Assay kit with Qubit 3 fluorometer (Invitrogen, Carlsbad, CA, USA). Total genomic DNA was prepared using the NexteraXT DNA sample preparation kit (Illumina, San Diego, CA, USA). Sequencing was performed on the MiSeq platform using V2 (2 × 250 bp) chemistry. The resulting fastq files were trimmed with Sickle using default parameters, and metagenomes were assembled using megahit. An overview of sample diversity for each metagenome was obtained using Kraken [72] and visualised using Pavian [73]. The reads were mapped against the genomes of interest using BWA-MEM [74]).

5. Conclusions

Taken together, our data showed that the phage cocktail examined specifically cleared *C. difficile* in both prophylactic and remedial regimens despite the competitive pressure imposed by the diverse human microbiota. Furthermore, a therapy using the optimised phages altered the human commensal bacteria such that specific bacterial groups associated with a healthy gut microbiota dominated. In summary, the use of phages removed the target pathogen and favorably modified the model gut microbiome. Both of these outcomes would be beneficial if the phages were to be used therapeutically, so our data supports their further development as therapeutic agents.

Supplementary Materials: The following are available online at http://www.mdpi.com/2079-6382/7/1/13/s1, Table S1: Percent bacterial families analysed from the metagenomics data using Pavian.

Acknowledgments: The project was funded by AmpliPhi Biosciences in collaboration with the University of Leicester to develop the phages for therapeutic purpose for CDI. The funders had no role in study design, data collection, and analysis or preparation of the manuscript.

Author Contributions: J.Y.N., M.R.J.C., and A.M. conceived and designed the experiments; J.Y.N. performed the gut fermentation assays; A.M. and T.A.R. conducted the metagenomics sequencing; J.Y.N., T.A.R., and A.M. analysed the data; J.Y.N. wrote the paper; J.Y.N., T.A.R., A.M., and M.R.J.C. edited the paper and approved the final version to be submitted.

Conflicts of Interest: Although this work was funded by AmpliPhi Biosciences and under the terms of a license agreement to the University of Leicester headed by MRJC, the license agreement has since been terminated, and thus there is no conflict of interest in terms of license revenue shares. Similarly, all the other authors declare that the research was conducted in the absence of any commercial or financial relationships that could be construed as a potential conflict of interest.

Patents: The phages described are part of a Leicester patent, pending. European Patent Application No 13759275.4 and US Patent Application No. 14/423284.

Ethics: All ethical measures relating to collecting, handling of fecal material, and disposing of waste were observed (University of Leicester Ethics Reference: 10564-jn142-infectionimmunityinflamm).

References

1. De Kraker, M.E.A.; Stewardson, A.J.; Harbarth, S. Will 10 million people die a year due to antimicrobial resistance by 2050? *PLoS Med.* **2016**, *13*, e1002184. [CrossRef] [PubMed]
2. De Kraker, M.E.A.; Davey, P.G.; Grundmann, H. Mortality and hospital stay associated with resistant *Staphylococcus aureus* and *Escherichia coli* bacteremia: Estimating the burden of antibiotic resistance in Europe. *PLoS Med.* **2011**, *8*, e1001104. [CrossRef] [PubMed]
3. Stewardson, A.J.; Allignol, A.; Beyersmann, J.; Graves, N.; Schumacher, M.; Meyer, R.; Tacconelli, E.; De Angelis, G.; Farina, C.; Pezzoli, F.; et al. The health and economic burden of bloodstream infections caused by antimicrobial-susceptible and non-susceptible Enterobacteriaceae and *Staphylococcus aureus* in European hospitals, 2010 and 2011: A multicentre retrospective cohort study. *Euro Surveill.* **2016**, *21*. [CrossRef] [PubMed]
4. Naylor, N.R.; Pouwels, K.B.; Hope, R.; Green, N.; Henderson, K.L.; Knight, G.M.; Atun, R.; Robotham, J.V.; Deeny, S. A national estimate of the health and cost burden of *Escherichia coli* bacteraemia in the hospital setting: The importance of antibiotic resistance. *bioRxiv* **2017**. [CrossRef]
5. O'Neil, J. Antimicrobial Resistance: Tackling a Crisis for the Health and Wealth of Nations. The Review on Antimicrobial Resistance 2014. Available online: https://amr-review.org/sites/default/files/AMR%20Review%20Paper%20-%20Tackling%20a%20crisis%20for%20the%20health%20and%20wealth%20of%20nations_1.pdf (accessed on 31 December 2017).
6. Zucca, M.; Scutera, S.; Savoia, D. Novel avenues for *Clostridium difficile* infection drug discovery. *Expert Opin. Drug Discov.* **2013**, *8*, 459–477. [CrossRef] [PubMed]
7. Wise, R.; Blaser, M.J.; Carrs, O.; Cassell, G.; Fishman, N.; Guidos, R.; Levy, S.; Powers, J.; Norrby, R.; Tillotson, G.; et al. The urgent need for new antibacterial agents. *J. Antimicrob. Chemother.* **2011**, *66*, 1939–1940. [CrossRef] [PubMed]
8. Rai, J.; Randhawa, G.K.; Kaur, M. Recent advances in antibacterial drugs. *Int. J. Appl. Med. Res.* **2013**, *3*, 3–10.
9. Kutter, E.M.; Kuhl, S.J.; Abedon, S.T. Re-establishing a place for phage therapy in Western medicine. *Future Microbiol.* **2015**, *10*, 685–688. [CrossRef] [PubMed]
10. Abedon, S.T.; Kuhl, S.J.; Blasdel, B.G.; Kutter, E.M. Phage treatment of human infections. *Bacteriophage* **2011**, *1*, 66–85. [CrossRef] [PubMed]
11. Buttimer, C.; McAuliffe, O.; Ross, R.P.; Hill, C.; O'Mahony, J.; Coffey, A. Bacteriophages and bacterial plant diseases. *Front. Microbiol.* **2017**, *8*, 34. [CrossRef] [PubMed]
12. Cisek, A.A.; Dabrowska, I.; Gregorczyk, K.P.; Wyzewski, Z. Phage therapy in bacterial infections treatment: One hundred years after the discovery of bacteriophages. *Curr. Microbiol.* **2016**, *74*, 277–283. [CrossRef] [PubMed]

13. Regeimbal, J.M.; Jacobs, A.C.; Corey, B.W.; Henry, M.S.; Thompson, M.G.; Pavlicek, R.L.; Quinones, J.; Hannah, R.M.; Ghebremedhin, M.; Crane, N.J.; et al. Personalized therapeutic cocktail of wild environmental phages rescues mice from *Acinetobacter baumannii* wound infections. *Antimicrob. Agents Chemother.* **2016**, *60*, 5806–5816. [CrossRef] [PubMed]
14. Sulakvelidze, A.; Alavidze, Z.; Morris, J.G.J. Bacteriophage therapy. *Antimicrob. Agents Chemother.* **2001**, *45*, 649–659. [CrossRef] [PubMed]
15. Loc-Carrillo, C.; Abedon, S.T. Pros and cons of phage therapy. *Bacteriophage* **2011**, *1*, 111–114. [CrossRef] [PubMed]
16. Wittebole, X.; De Roock, S.; Opal, S.M. A historical overview of bacteriophage therapy as an alternative to antibiotics for the treatment of bacterial pathogens. *Virulence* **2014**, *5*, 226–235. [CrossRef] [PubMed]
17. Speck, P.; Smithyman, A. Safety and efficacy of phage therapy via the intravenous route. *FEMS Microbiol. Lett.* **2015**, *363*. [CrossRef] [PubMed]
18. Brüssow, H. What is needed for phage therapy to become a reality in Western medicine? *Virology* **2012**, *434*, 138–142. [CrossRef] [PubMed]
19. Slopek, S.; Weber-Dabrowska, B.; Dabrowski, M.; Kucharewicz-Krukowska, A. Results of bacteriophage treatment of suppurative bacterial infections in the years 1981–1986. *Arch. Immunol. Ther. Exp.* **1987**, *35*, 569–583.
20. EBI Food Safety. FDA and USDA Extend GRAS Approval for LISTEX for All Food Products. 2007. Available online: http://www.ebifoodsafety.com/en/news-2007.aspx (accessed on 31 December 2017).
21. Kwon, J.H.; Olsen, M.A.; Dubberke, E.R. The morbidity, mortality, and costs associated with *Clostridium difficile* infection. *Infect. Dis. Clin. N. Am.* **2015**, *29*, 123–134. [CrossRef] [PubMed]
22. Olsen, M.A.; Young-Xu, Y.; Stwalley, D.; Kelly, C.P.; Gerding, D.N.; Saeed, M.J.; Mahé, C.; Dubberke, E.R. The burden of *Clostridium difficile* infection: Estimates of the incidence of cdi from U.S. Administrative databases. *BMC Infect. Dis.* **2016**, *16*. [CrossRef] [PubMed]
23. Dubberke, E.R.; Olsen, M.A. Burden of *Clostridium difficile* on the healthcare system. *Clin. Infect. Dis.* **2012**, *55*, S88–S92. [CrossRef] [PubMed]
24. Bauer, M.P.; Notermans, D.W.; van Benthem, B.H.B.; Brazier, J.S.; Wilcox, M.H.; Rupnik, M.; Monnet, D.L.; van Dissel, J.T. *Clostridium difficile* infection in Europe: A hospital-based survey. *Lancet* **2011**, *377*, 63–73. [CrossRef]
25. Boyle, N.M.; Magaret, A.; Stednick, Z.; Morrison, A.; Butler-Wu, S.; Zerr, D.; Rogers, K.; Podczervinski, S.; Cheng, A.; Wald, A.; et al. Evaluating risk factors for *Clostridium difficile* infection in adult and pediatric hematopoietic cell transplant recipients. *Antimicrob. Resist. Infect. Control* **2015**, *4*. [CrossRef] [PubMed]
26. Lessa, F.C.; Mu, Y.; Bamberg, W.M.; Beldavs, Z.G.; Dumyati, G.K.; Dunn, J.R.; Farley, M.M.; Holzbauer, S.M.; Meek, J.I.; Phipps, E.C.; et al. Burden of *Clostridium difficile* infection in the United States. *N. Engl. J. Med.* **2015**, *372*, 825–834. [CrossRef] [PubMed]
27. Lessa, F.C.; Gould, C.V.; McDonald, L.C. Current status of *Clostridium difficile* infection epidemiology. *Clin. Infect. Dis.* **2012**, *55*, S65–S70. [CrossRef] [PubMed]
28. Kociolek, L.K.; Gerding, D.N. Breakthroughs in the treatment and prevention of *Clostridium difficile* infection. *Nat. Rev. Gastroenterol. Hepatol.* **2016**, *13*, 150–160. [CrossRef] [PubMed]
29. DuPont, H.L. Diagnosis and management of *Clostridium difficile* infection. *Clin. Gastroenterol. Hepatol.* **2013**, *11*, 1216–1223. [CrossRef] [PubMed]
30. Shah, D.; Dang, M.D.; Hasbun, R.; Koo, H.L.; Jiang, Z.D.; DuPont, H.L.; Garey, K.W. *Clostridium difficile infection*: Update on emerging antibiotic treatment options and antibiotic resistance. *Expert Rev. Anti-Infect. Ther.* **2010**, *8*, 555–564. [CrossRef] [PubMed]
31. Debast, S.B.; Bauer, M.P.; Kuijper, E.J. European society of clinical microbiology and infectious diseases: Update of the treatment guidance document for *Clostridium difficile* infection. *Clin. Microbiol. Infect.* **2014**, *20*, 1–26. [CrossRef] [PubMed]
32. Moura, I.; Spigaglia, P.; Barbanti, F.; Mastrantonio, P. Analysis of metronidazole susceptibility in different *Clostridium difficile* PCR ribotypes. *J. Antimicrob. Chemother.* **2013**, *68*, 362–365. [CrossRef] [PubMed]
33. Koss, C.A.; Baras, D.C.; Lane, S.D.; Aubry, R.; Marcus, M.; Markowitz, L.E.; Koumans, E.H. Investigation of metronidazole use during pregnancy and adverse birth outcomes. *Antimicrob. Agents Chemother.* **2012**, *56*, 4800–4805. [CrossRef] [PubMed]

34. Nelson, R.L.; Suda, K.J.; Evans, C.T. Antibiotic treatment for *Clostridium difficile*-associated diarrhoea in adults. *Cochrane Database Syst. Rev.* **2017**, 3. [CrossRef] [PubMed]
35. Poduval, R.D.; Kamath, R.P.; Corpuz, M.; Norkus, E.P.; Pitchumoni, C.S. *Clostridium difficile* and vancomycin-resistant enterococcus: The new nosocomial alliance. *Am. J. Gastroenterol.* **2000**, *95*, 3513–3515. [CrossRef] [PubMed]
36. Bauer, M.P.; Kuijper, E.J.; van Dissel, J.T. European Society of Clinical Microbiology and Infectious Diseases (ESCMID): Treatment guidance document for *Clostridium difficile* infection (CDI). *Clin. Microbiol. Infect.* **2009**, *15*, 1067–1079. [CrossRef] [PubMed]
37. Baines, S.D.; Wilcox, M.H. Antimicrobial resistance and reduced susceptibility in *Clostridium difficile*: Potential consequences for induction, treatment, and recurrence of *C. difficile* infection. *Antibiotics* **2015**, *4*, 267–298. [CrossRef] [PubMed]
38. Chilton, C.H.; Crowther, G.S.; Freeman, J.; Todhunter, S.L.; Nicholson, S.; Longshaw, C.M.; Wilcox, M.H. Successful treatment of simulated *Clostridium difficile* infection in a human gut model by fidaxomicin first line and after vancomycin or metronidazole failure. *J. Antimicrob. Chemother.* **2014**, *69*, 451–462. [CrossRef] [PubMed]
39. Bartsch, S.M.; Umscheid, C.A.; Fishman, N.; Lee, B.Y. Is fidaxomicin worth the cost? An economic analysis. *Clin. Infect. Dis.* **2013**, *57*, 555–561. [CrossRef] [PubMed]
40. Stranges, P.M.; Hutton, D.W.; Collins, C.D. Cost-effectiveness analysis evaluating fidaxomicin versus oral vancomycin for the treatment of *Clostridium difficile* infection in the United States. *Value Health* **2013**, *16*, 297–304. [CrossRef] [PubMed]
41. Cruz, M.P. Fidaxomicin (Dificid), a novel oral macrocyclic antibacterial agent for the treatment of *Clostridium difficile*–associated diarrhea in adults. *Pharmacol. Ther.* **2012**, *37*, 278–281.
42. Hargreaves, K.R.; Clokie, M.R.J. *Clostridium difficile* phages: Still difficult? *Front. Microbiol.* **2014**, *5*. [CrossRef] [PubMed]
43. Sangster, W.; Hegarty, J.P.; Stewart, D.B. Phage therapy for *Clostridium difficile* infection: An alternative to antibiotics? *Semin. Colon Rectal Surg.* **2014**, *25*, 167–170. [CrossRef]
44. Meader, E.; Mayer, M.J.; Steverding, D.; Carding, S.R.; Narbad, A. Evaluation of bacteriophage therapy to control *Clostridium difficile* and toxin production in an in vitro human colon model system. *Anaerobe* **2013**, *22*, 25–30. [CrossRef] [PubMed]
45. Nale, J.Y.; Spencer, J.; Hargreaves, K.R.; Buckley, A.M.; Trzepiński, P.; Douce, G.R.; Clokie, M.R.J. Bacteriophage combinations significantly reduce *Clostridium difficile* growth in vitro and proliferation in vivo. *Antimicrob. Agents Chemother.* **2016**, *60*, 968–981. [CrossRef] [PubMed]
46. Nale, J.Y.; Chutia, M.; Carr, P.; Hickenbotham, P.; Clokie, M.R.J. 'Get in early'; biofilm and wax moth (*Galleria mellonella*) models reveal new insights into the therapeutic potential of *Clostridium difficile* bacteriophages. *Front. Microbiol.* **2016**, *7*. [CrossRef] [PubMed]
47. Ramesh, V.; Fralick, J.A.; Rolfe, R.D. Prevention of *Clostridium difficile*-induced ileocecitis with bacteriophage. *Anaerobe* **1999**, *5*, 69–78. [CrossRef]
48. Govind, R.; Fralick, J.A.; Rolfe, R.D. In vivo lysogenization of a *Clostridium difficile* bacteriophage ϕCD119. *Anaerobe* **2011**, *17*, 125–129.
49. Price, A.B.; Larson, H.E.; Crow, J. Morphology of experimental antibiotic-associated enterocolitis in the hamster: A model for human pseudomembranous colitis and antibiotic-associated diarrhoea. *Gut* **1979**, *20*, 467–475. [CrossRef] [PubMed]
50. Best, E.L.; Freeman, J.; Wilcox, M.H. Models for the study of *Clostridium difficile* infection. *Gut Microbes* **2012**, *3*, 145–167. [CrossRef] [PubMed]
51. Meader, E.; Mayer, M.J.; Gasson, M.J.; Steverding, D.; Carding, S.R.; Narbad, A. Bacteriophage treatment significantly reduces viable *Clostridium difficile* and prevents toxin production in an in vitro model system. *Anaerobe* **2010**, *16*, 549–554. [CrossRef] [PubMed]
52. Chilton, C.H.; Crowther, G.S.; Todhunter, S.L.; Nicholson, S.; Freeman, J.; Chesnel, L.; Wilcox, M.H. Efficacy of surotomycin in an in vitro gut model of *Clostridium difficile* infection. *J. Antimicrob. Chemother.* **2014**, *69*, 2426–2433. [CrossRef] [PubMed]
53. Hargreaves, K.R.; Kropinski, A.M.; Clokie, M.R.J. What does the talking?: Quorum sensing signalling genes discovered in a bacteriophage genome. *PLoS ONE* **2014**, *9*, e85131. [CrossRef] [PubMed]

54. Lloyd-Price, J.; Abu-Ali, G.; Huttenhower, C. The healthy human microbiome. *Genome Med.* **2016**, *8*. [CrossRef] [PubMed]
55. Browne, H.P.; Forster, S.C.; Anonye, B.O.; Kumar, N.; Neville, B.A.; Stares, M.D.; Goulding, D.; Lawley, T.D. Culturing of 'unculturable' human microbiota reveals novel taxa and extensive sporulation. *Nature* **2016**, *533*, 543–546. [CrossRef] [PubMed]
56. Baines, S.D.; Chilton, C.H.; Crowther, G.S.; Todhunter, S.L.; Freeman, J.; Wilcox, M.H. Evaluation of antimicrobial activity of ceftaroline against *Clostridium difficile* and propensity to induce *C. difficile* infection in an in vitro human gut model. *J. Antimicrob. Chemother.* **2013**, *68*, 1842–1849. [CrossRef] [PubMed]
57. Gould, I.M. The epidemiology of antibiotic resistance. *Int. J. Antimicrob. Agents* **2008**, *32*, S2–S9. [CrossRef] [PubMed]
58. Roberts, M.C.; McFarland, L.V.; Mullany, P.; Mulligan, M.E. Characterization of the genetic basis of antibiotic resistance in *Clostridium difficile*. *J. Antimicrob. Chemother.* **1994**, *33*, 419–429. [CrossRef] [PubMed]
59. Abhilash, M.; Vidya, A.G.; Jagadevi, T. Bacteriophage therapy: A war against antibiotic resistant bacteria. *Internet J. Altern. Med.* **2009**, *7*, 1.
60. Bauer, M.P.; van Dissel, J.T. Alternative strategies for *Clostridium difficile* infection. *Int. J. Antimicrob. Agents* **2009**, *33*, S51–S56. [CrossRef]
61. Wall, S.K.; Zhang, J.; Rostagno, M.H.; Ebner, P.D. Phage therapy to reduce preprocessing salmonella infections in market-weight swine. *Appl. Environ. Microbiol.* **2010**, *76*, 48–53. [CrossRef] [PubMed]
62. Macfarlane, G.T.; Macfarlane, S.; Gibson, G.R. Validation of a three-stage compound continuous culture system for investigating the effect of retention time on the ecology and metabolism of bacteria in the human colon. *Microb. Ecol.* **1998**, *35*, 180–187. [CrossRef] [PubMed]
63. Taslim, H. *Clostridium difficile* infection in the elderly. *Acta Medica Indonesiana* **2009**, *41*, 148–151. [PubMed]
64. McGowan, K.L.; Kader, H.A. *Clostridium difficile* infection in children. *Clin. Microbiol. Newsl.* **1999**, *21*, 49–53. [CrossRef]
65. Lagier, J.C.; Million, M.; Hugon, P.; Armougom, F.; Raoult, D. Human gut microbiota: Repertoire and variations. *Front. Cell. Infect. Microbiol.* **2012**, *2*. [CrossRef] [PubMed]
66. Antharam, V.C.; Li, E.C.; Ishmael, A.; Sharma, A.; Mai, V.; Rand, K.H. Intestinal dysbiosis and depletion of butyrogenic bacteria in *Clostridium difficile* infection and nosocomial diarrhea. *J. Clin. Microbiol.* **2013**, *51*, 2884–2892. [CrossRef] [PubMed]
67. Miyajima, F.; Roberts, P.; Swale, A.; Price, V.; Jones, M.; Horan, M.; Beeching, N.; Brazier, J.; Parry, C.; Pendleton, N.; et al. Characterisation and carriage ratio of *Clostridium difficile* strains isolated from a community-dwelling elderly population in the United Kingdom. *PLoS ONE* **2011**, *6*, e22804. [CrossRef] [PubMed]
68. Chan, B.K.; Abedon, S.T.; Loc-Carrillo, C. Phage cocktails and the future of phage therapy. *Future Microbiol.* **2013**, *8*, 769–783. [CrossRef] [PubMed]
69. Britton, R.A.; Young, V.B. Interaction between the intestinal microbiota and host in *Clostridium difficile* colonization resistance. *Trends Microbiol.* **2012**, *20*, 313–319. [CrossRef] [PubMed]
70. Zuo, T.; Wong, S.H.; Lam, K.; Lui, R.; Cheung, K.; Tang, W.; Ching, J.Y.L.; Chan, P.K.S.; Chan, M.C.W.; Wu, J.C.Y.; et al. Bacteriophage transfer during faecal microbiota transplantation in *Clostridium difficile* infection is associated with treatment outcome. *Gut* **2017**. [CrossRef] [PubMed]
71. Ott, S.J.; Waetzig, G.H.; Rehman, A.; Moltzau-Anderson, J.; Bharti, R.; Grasis, J.A.; Cassidy, L.; Tholey, A.; Fickenscher, H.; Seegert, D.; et al. Efficacy of sterile fecal filtrate transfer for treating patients with *Clostridium difficile* infection. *Gastroenterology* **2017**, *152*, 799–811. [CrossRef] [PubMed]
72. Wood, D.E.; Salzberg, S.L. Kraken: Ultrafast metagenomic sequence classification using exact alignments. *Genome Biol.* **2014**, *15*. [CrossRef] [PubMed]
73. Breitwieser, F.P.; Salzberg, S.L. Pavian: Interactive analysis of metagenomics data for microbiomics and pathogen identification. *bioRxiv* **2016**. [CrossRef]
74. Heng, L. Aligning Sequence Reads, Clone Sequences and Assembly Contigs with Bwa-Mem. Available online: https://arxiv.org/abs/1303.3997 (accessed on 31 December 2017).

antibiotics

MDPI

Article

Contributions of Net Charge on the PlyC Endolysin CHAP Domain

Xiaoran Shang [1] and Daniel C. Nelson [1,2,*]

1 Institute for Bioscience and Biotechnology Research, Rockville, MD 20850, USA; sxr520@umd.edu
2 Department of Veterinary Medicine, University of Maryland, College Park, MD 20742, USA
* Correspondence: nelsond@umd.edu; Tel.: +1-240-314-6249

Received: 29 April 2019; Accepted: 25 May 2019; Published: 28 May 2019

Abstract: Bacteriophage endolysins, enzymes that degrade the bacterial peptidoglycan (PG), have gained an increasing interest as alternative antimicrobial agents, due to their ability to kill antibiotic resistant pathogens efficiently when applied externally as purified proteins. Typical endolysins derived from bacteriophage that infect Gram-positive hosts consist of an N-terminal enzymatically-active domain (EAD) that cleaves covalent bonds in the PG, and a C-terminal cell-binding domain (CBD) that recognizes specific ligands on the surface of the PG. Although CBDs are usually essential for the EADs to access the PG substrate, some EADs possess activity in the absence of CBDs, and a few even display better activity profiles or an extended host spectrum than the full-length endolysin. A current hypothesis suggests a net positive charge on the EAD enables it to reach the negatively charged bacterial surface via ionic interactions in the absence of a CBD. Here, we used the PlyC CHAP domain as a model EAD to further test the hypothesis. We mutated negatively charged surface amino acids of the CHAP domain that are not involved in structured regions to neutral or positively charged amino acids in order to increase the net charge from -3 to a range from +1 to +7. The seven mutant candidates were successfully expressed and purified as soluble proteins. Contrary to the current hypothesis, none of the mutants were more active than wild-type CHAP. Analysis of electrostatic surface potential implies that the surface charge distribution may affect the activity of a positively charged EAD. Thus, we suggest that while charge should continue to be considered for future engineering efforts, it should not be the sole focus of such engineering efforts.

Keywords: endolysin; PlyC CHAP; protein net charge; CBD-independent; FoldX

1. Introduction

Bacteriophage (phage) endolysins are peptidoglycan (PG) hydrolases produced by phage at the end of a lytic cycle [1]. In the presence of holins, pore-forming proteins, endolysins can pass the cytoplasmic membrane to degrade the PG layer of the cell wall, resulting in the lysis of the bacteria and release of new progeny virions [2]. These enzymes are also capable of destroying the Gram-positive bacterial PG from outside the cell as recombinant proteins [3]. Due to the physical barrier of the outer membrane, exogenously added endolysins usually cannot access the PG of Gram-negative bacteria. However, engineered endolysins with cationic or membrane-disrupting peptides have been reported to successfully kill Gram-negative bacteria "from without" [4]. Consequently, endolysins are novel antimicrobial agents and can be used to treat both Gram-positive and Gram-negative antibiotic-resistant bacterial infections because their mode of action is not inhibited by traditional resistance mechanisms [3].

Endolysins derived from phage that infect Gram-positive hosts have very similar modular structures with one or more N-terminal enzymatically-active domains (EADs) and a C-terminal cell wall binding domain (CBD) [5]. The EADs that cleave covalent bonds in the PG are conserved into five mechanistic classes: muramidases, glucosaminidases, N-acetylmuramyl-L-alanine amidases, endopeptidase, and lytic transglycosylases. In contrast, the CBDs possess no enzymatic activity but rather function to bind to specific ligands on the cell wall, which are usually secondary wall carbohydrates or teichoic acid moieties. Thus, the endolysin host range is often dictated by the specificity of the CBD, which is either broad-spectrum, targeting molecules harbored by a bacterial genus or multiple genera, or narrow-spectrum, targeting molecules shared by a single species or serovar [6,7]. The CBDs have been shown to be essential for function of EADs in a number of modular endolysins, including PlyGRCS [8], PlySs2 [9], PlyB [10], Cpl-1 [11], and PlyB30 [12].

Whereas many EADs require the presence of the CBD for binding and subsequent activity, some EADs can bind the bacterial surface independently of the CBD, and a few even have increased enzymatic activity compared to the full-length endolysin. One example is the staphylococcal phage endolysin, LysK. The LysK EAD, a cysteine-histidine amidohydrolase/endopeptidase (CHAP), alone displays higher lytic activity against staphylococci than the full-length LysK [13]. Similarly, when the Group B streptococcal phage endolysin, PlyGBS, was truncated to remove the CBD from the EAD, a ~20 fold increase in specific activity was noted compared to PlyGBS [14]. Moreover, without the constraining binding properties of the CBD, some EADs from modular endolysins show an extended host range compared to their parental full-length endolysins. Examples include the EAD of the *Bacillus anthracis* phage endolysin, PlyL [15] and the EAD of the *Clostridium difficile* phage endolysin, CD27L [16].

The reason(s) as to why some EADs can target and lyse the PG in the absence of a CBD, whereas the presence of a CBD is an absolute requirement for activity in other EADs, is unknown. However, a thought-provoking study by Low et al. [15] suggested that a net positive charge of an EAD enables it to function independently of its CBD, presumably through ionic interactions with the bacterial surface, which characteristically has a net negative charge due to surface carbohydrates. This conceptual understanding was then applied by the authors to endolysin bioengineering studies. For example, the EAD of a *Bacillus subtilis* prophage endolysin, XlyA, had a net charge (Z) of -3 at neutral pH and displayed no lytic activity against *B. subtilis* cells in the absence of its CBD. Site-directed mutagenesis of five non-cationic residues to lysine (K) produced a shift in net charge from Z = −3 to Z = +3, and the mutated XlyA EAD alone was able to lyse *B. subtilis* cells at a rate nearly identical to that of full-length XlyA. In a separate study, the addition of a positively-charged peptide enhanced the lytic activity of the λSa2lys endolysin [17], suggesting the positive charges may increase the avidity of the enzyme for the bacterial surface.

In the present work, we sought to validate Low's hypothesis. The model EAD for this study is the CHAP domain from the PlyC endolysin [18]. This EAD possesses potent catalytic activity and is amenable to engineering, as it has been subjected to mutational analysis to improve thermostability [19] and has been used as the EAD in chimeragenesis projects incorporating different CBDs (i.e., ClyR [20] and ClyJ [21]). The homolog of the PlyC CHAP domain via a structural DALI search is the LysK CHAP domain, which is known to harbor improved activity compared to full-length LysK. However, the PlyC CHAP catalytic domain, in contrast, loses most (~99%) lytic activity in the absence of PlyCB (i.e., the CBD of PlyC) [18]. The net charge of the LysK CHAP is Z = +1, whereas the net charge of the PlyC CHAP is Z = −3. Therefore, the PlyC CHAP is a good candidate to test Low's hypothesis, proposing that a conversion of net charge on an EAD will enable it to display lytic activity in the absence of a CBD.

2. Results

2.1. Library of PlyC CHAP Mutants

The PlyC CHAP domain (i.e., the C-terminal EAD of PlyCA comprising amino acid 309-465) was isolated from the PlyC holoenzyme crystal structure and edited in PyMOL (atomic coordinates were only available for amino acid 310-464). Five surface and unstructured residues, Asp-311, Asp-355, Asp-363, Asp-429, and Asp-450, were selected for mutagenesis (Figure 1). Through different combinations of point mutations incorporating either a neutral charge (i.e., alanine) or a positive charge (i.e., lysine) in place of each aspartic acid residue, a library of 192 mutant candidates harboring a net charge between +1 and +7 was computationally generated.

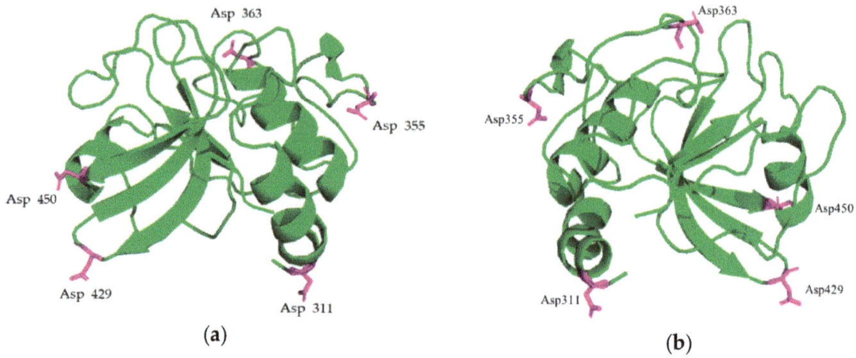

Figure 1. Mutation sites of PlyC CHAP. (**a**) 3.3 Å resolution of the PlyC CHAP crystal structure. The magenta-colored amino acids represent the potential mutation sites. (**b**) 180° horizontal rotation of (a). The mutation sites are solvent exposed, not structured in α-helix or β-sheets, and form no interactions with other residues.

2.2. Prediction of the Properly Folded PlyC CHAP Mutants via $\Delta\Delta G_{FoldX}$

FoldX is a computational biology tool developed for rapid evaluation of the effect of mutations on stability, folding, and protein dynamics [22]. FoldX was used to narrow down the 192 mutants via a change in free energy of the mutant relative to the wild-type (WT) protein ($\Delta\Delta G_{FoldX} = \Delta G_{mut} - \Delta G_{WT}$). A negative $\Delta\Delta G_{FoldX}$ ($\Delta\Delta G_{FoldX} < 0$) suggests that the mutation is more stable than the WT protein and should fold properly. However, 79% of the mutants were predicted to have a positive $\Delta\Delta G_{FoldX}$, meaning these mutations had destabilizing effects (Figure 2). At each charge category (Z = +1 to Z = +7), only the mutants possessing the largest predicted negative $\Delta\Delta G_{FoldX}$ were chosen to be made (Table 1). Notably, the selected +6 charged and +7 charged CHAP mutants contained either neutral or positive $\Delta\Delta G_{Foldx}$, probably due to the high number of required mutations (Table 1).

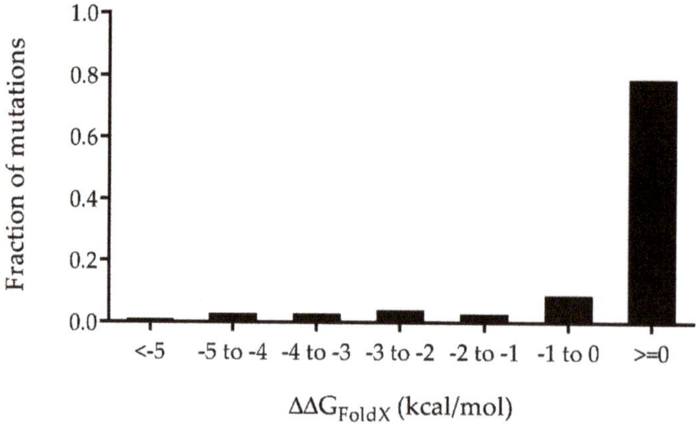

Figure 2. Distribution of the predicted change in folding free energy ($\Delta\Delta G_{foldX}$) for all 192 possible CHAP mutants calculated with FoldX3.0. Mutations with $\Delta\Delta G_{foldX} < 0$ are expected to retain the same folding characteristics as WT CHAP.

Table 1. PlyCA CHAP Selected Mutations.

Mutant Name	pI (Isoelectric Point)	Net Charge (Z) at pH 7.4	Point Mutations	$\Delta\Delta G_{FoldX}$ (kcal/mol) = ΔG_{Mut}-ΔG_{WT}
CHAP WT	6.11	−3	N/A	0
CHAP +1	7.89	+1	D311K:D355K	−5.32
CHAP +2	8.29	+2	D311K:D355K:D429A	−4.61
CHAP +3	8.59	+3	D311K:D355K:D363K	−4.32
CHAP +4	8.88	+4	D311K:D355K:D363K:D429A	−4.13
CHAP +5	9.11	+5	D311K:D355K:D363K:D429K	−2.49
CHAP +6	9.30	+6	D311K:D355K:D363K:D429A:D450K	0.01
CHAP +7	9.43	+7	D311K:D355K:D363K:D429K:D450K	1.11

2.3. Protein Solubility and Purity

All chosen mutants were expressed and purified by nickel affinity chromatography. The 6x His-tag at the N-terminus of each protein, which might affect the net surface charge in solution, was cleaved at an engineered thrombin cleavage site before further purification. The SDS-PAGE gel after His-tag removal suggested that PlyC CHAP mutants were pure and as soluble as WT CHAP (Figure 3). PlyC CHAP+7 displayed as a double band in the SDS-PAGE may indicate degradation of the protein since the positive $\Delta\Delta G_{FoldX}$ (Table 1) suggests this mutant is slightly unstable.

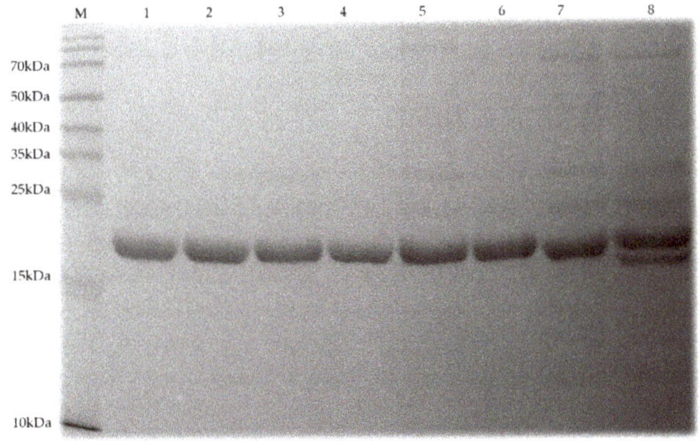

Figure 3. SDS-PAGE analysis of WT PlyC CHAP and its mutants. The solubility and purity of each enzyme after the His-tag cleavage were assessed via a 7.5% SDS-PAGE gel. The lanes correlate to: (M) BioRad protein markers; (**1**) PlyC CHAP WT; (**2**) PlyC CHAP +1; (**3**) PlyC CHAP +2; (**4**) PlyC CHAP +3; (**5**) PlyC CHAP +4; (**6**) PlyC CHAP +5; (**7**) PlyC CHAP +6, and; (**8**) PlyC CHAP +7.

2.4. In Vitro PlyC CHAP Mutants' Activity

PlyC is one of the most potent endolysins studied to date [23] and the PlyC CHAP domain is known to require its CBD for full activity. However, despite the Z = −1 charge, the PlyC CHAP domain does retain a very small (<1% of PlyC), but measurable and reproducible lytic activity against sensitive streptococcal species [18]. A turbidity reduction assay was used to benchmark the lytic activity of PlyC CHAP mutants to WT PlyC CHAP. However, none of the CHAP mutants displayed increased lytic activity compared with WT using *Streptococcus pyogenes* D471 as host over a broad concentration range (Figure 4). Nonetheless, the data revealed several interesting aspects. First, the CHAP mutants with net +1 and +2 surface charges (i.e., CHAP+1 and CHAP+2) showed the same lytic activity as WT CHAP, which suggests that the positive charge alone does not affect lytic activity. Second, in the low concentration ranges (< 16 μg/mL), the CHAP mutants with +3 to +7 surface charges (CHAP+3 to CHAP+7) were virtually devoid of lytic activity, but as the concentration increased, they had similar lytic activity to WT CHAP as well as CHAP+1 and CHAP+2. The activity noted for CHAP WT compared to the full PlyC holoenzyme is consistent with previous data [18], and presumably represents activity resulting from random collisions of CHAP with the cell wall in the absence of the PlyC CBD.

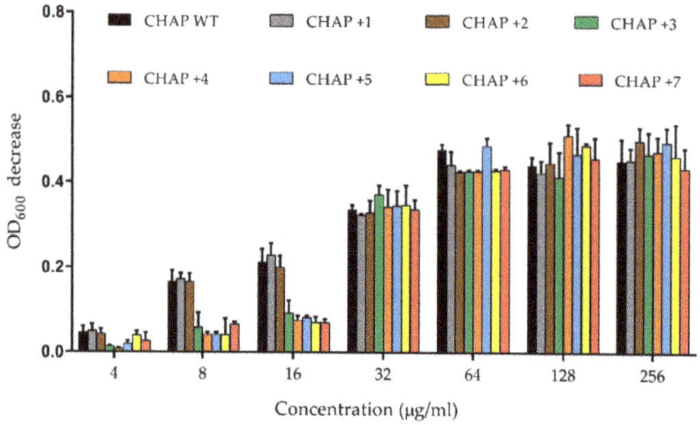

Figure 4. In vitro lytic activity against *S. pyogenes* D471. The different concentrations of enzymes were added to the overnight bacterial cultures. The OD_{600} was recorded every 15 seconds for 1 hour. The OD_{600} decrease, which is the net change in OD_{600} between the PBS treated control and enzyme treatment for 1 hour, was represented as the enzyme activity. The experiment was conducted in triplicate, and the error bars represent the standard deviation.

2.5. Analysis of PlyC CHAP Electrostatic Surface Potential

The surface charge distributions were then examined through CCP4MG software [24]. The active-site residues (C333 and H420) of the PlyC CHAP are located in a neutral groove, which remains unchanged in CHAP mutants (Figure 5A, neutral groove on far right column). Although an overall increased surface charge indicates an increased positive electrostatic potential in the CHAP mutants, the regions accumulating the positive surface potential are evenly distributed over the entire CHAP surface (Figure 5A). Low et al. [15] did not imply any relationship between the relative position of the active-site and the distribution of positive charge in their work. However, when we examined the surface potential of their enzyme, XlyA, and its mutant, XlyA+5K, we did notice an accumulation of positive surface potential near the negative charged active-site (Figure 5B, negative groove on far right column). Thus, a simple conversion of the surface charge on an EAD may not be adequate by itself to create CBD-independent lytic activity.

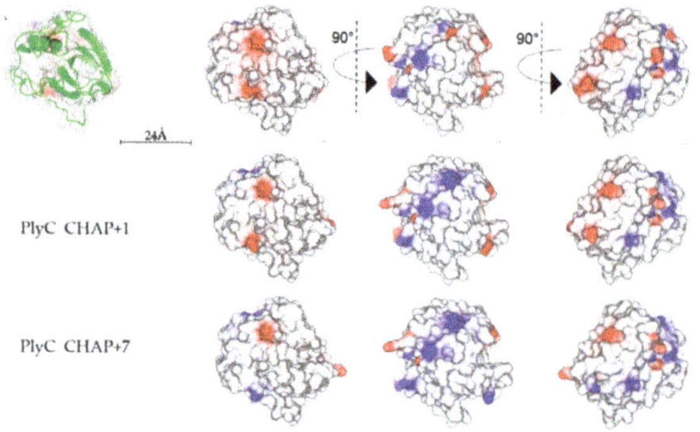

(**a**)

Figure 5. *Cont.*

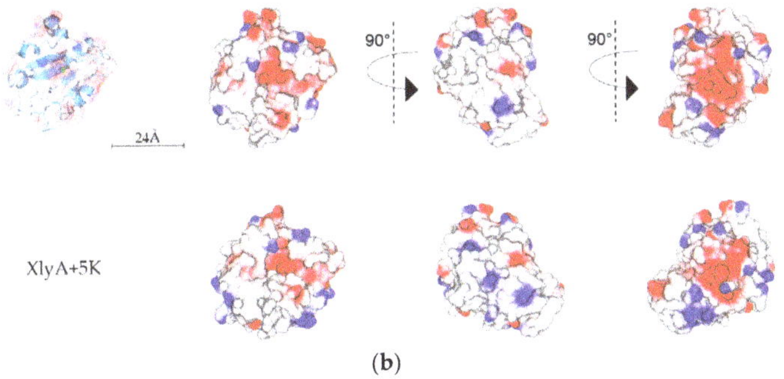

Figure 5. CCP4MG generated electrostatic surface potential maps of PlyC CHAP, XlyA, and their mutants. Surfaces are color-coded according to electrostatic potential (calculated by the Poisson-Boltzmann solver within CCP4MG). The color of the surface represents the electrostatic potential on the protein surface, going from blue (potential of +10kT/e) to red (potential of -10kT/e). (**a**) Electrostatic surface potential of PlyC CHAP WT, CHAP +1, and CHAP +7 in different orientations. The active-site of PlyC CHAP is in a neutral groove observed in the far right column. (**b**) Electrostatic surface potential of XlyA and XlyA+5K in different orientations. The active-site of XlyA is in a negative groove observed in the far right column.

3. Discussion

Bacteriophage have optimized endolysins for lytic activity at high concentrations from within the bacterial intracellular environment through coevolution with bacterial hosts to ensure phage release and survival. In contrast, these enzymes have not evolved to be used exogenously to produce lysis "from without" and, therefore, may not be fully optimized for this application. As such, there exists an engineering potential for these enzymes to be modified to increase their activity, alter host range, or overcome complex extracellular environments [25]. As a growing amount of research focuses on the modular design and crystal structures of endolysins, structure-based rational engineering approaches, such as chimeragenesis and structure-guided mutagenesis, can be used to produce engineered endolysins with desirable antimicrobial properties [26].

Chimeragenesis is a method used to exchange functional modules (i.e., EADs and CBDs) for better activities and/or an altered or expanded host range. This engineering approach has been exploited by nature itself through horizontal gene transfer, such as the pneumococcal endolysin Pal whose EAD is most similar to the endolysin of *Lactococcus lactis* phage BK5-T, whereas its CBD is homologous to other pneumococcal endolysins [27]. A well-studied chimeric endolysin that extends the host range of the parental enzymes is ClyR [20,28]. This enzyme is a fusion of a PlyC CHAP EAD and the PlySs2 CBD. While each parental enzyme has a narrow host range to select streptococcal species, the ClyR chimera possesses a broad activity against almost all streptococcal species as well as staphylococcal and enterococcal species. Other engineered chimeras, including ClyS [29], Cpl-711 [30], Csl2 [31], and PL3 [32] result in endolysins that displayed enhanced antimicrobial activity over either parental enzyme. Some chimeric endolysin, such as λSA2-E-Lyso-SH3b and λSA2-E-LysK-SH3b include multiple EADs from different endolysins and result in both extended host range and increased antimicrobial activity [33,34].

Structure-guided mutagenesis is expected to expand bioengineering efforts toward modulating endolysin properties as more high resolution X-ray crystallography and NMR structures are solved. The atomic coordinates can be used for computational modeling to predict mutations that can affect enzymatic activity, substrate binding affinity, and interactions with the solvent interface or intra- or inter-domain interactions that may affect stability of the enzyme. Notably, several algorithms, such as Rosetta [35] and FoldX [22,36], can be used to rapidly assess point mutations *in silico* and calculate the effect these mutations have on protein stability through the approximation of Gibbs free energy (ΔG).

The top resulting mutants can then be made and characterized experimentally. Such an approach was recently done to introduce a thermostabilizing mutation, T406R, to the EAD of PlyC. This point mutant introduced a critical hydrogen bond between two domains and resulted in a 16 fold increase of half-life at 45 °C [19].

Low et al. [15] used similar site-directed mutagenesis techniques to introduce five lysine residues to non-structured, non-catalytic regions of the XlyA endolysin EAD, effectively changing the charge from Z = -3 to Z = +3. Whereas the XlyA EAD had no lytic activity against *B. subtilis* in the absence of its CBD, the mutant EAD, XlyA+5K, possessed WT XlyA activity independent of the CBD. The authors postulated that the net positive charge allowed for interaction with the "continuum of negative charge" on the secondary cell wall polymers on the bacterial surface, thus obviating the need for a CBD. To contrast this gain-of-function experiment, the authors also tested a loss-of-function engineering approach using the PlyBa04 EAD, which carries a net charge of Z = +1. Unlike the native XlyA EAD, the PlyBa04 EAD is able to lyse its target, *B. cereus*, as efficiently as the full-length endolysin. However, after site-directed mutagenesis to change the net charge from Z = +1 to Z = −3, the resulting PlyBa04 EAD mutant showed negligible lytic activity towards *B. cereus*. The authors concluded that engineering a reversal of net charge on an EAD could be used to either create or eliminate CBD dependence, which could be used in future bioengineering studies to fine-tune endolysin activity. Two years later, a separate study by Diez-Martinez et al. [37] with the Cpl-7 pneumococcal endolysin supported Low's charge hypothesis. These authors introduced charge substitutions to 15 amino acids of the Cpl-7 CBD resulting in a change from Z = −14.93 to Z = +3.0 at neutral pH. The mutant, named Cpl-7S, displayed significantly increased lytic activity against several streptococcal strains compared to WT Cpl-7.

The ultimate goal of our study was to create an engineered endolysin that is simple (i.e., one catalytic domain) and works on a very broad host range (i.e., does not require a CBD, meaning the EAD alone defines host range). Toward this end, we sought to engineer the PlyC CHAP domain to be such an enzyme using engineering principles guided by the findings of Low et al. [15]. Toward this end, we successfully made a range of positively charged CHAP mutants. The crystal structure of PlyC provided a model for selecting the potential point mutations. The computational tool, FoldX, helped narrow down the candidates from a total of 192 to 40. The ones that were most stable among the 40 candidates were used for cloning, protein expression, and purification. The achievement of this experimental design suggests that computational tools, like FoldX, can be used in the upstream evaluation providing a rationale for the selected mutations.

Our results indicated that none of the positively charged CHAP mutants displayed higher lytic activity than WT CHAP. Thus, at least for the PlyC CHAP, the hypothesis developed by Low et al. [15] is not supported. Nonetheless, while our approach was similar to Low et al. [15] in that mutations were rationally selected based on being solvent/surface accessible, did not participate in hydrogen bonding or salt bridges, and were not part of ordered structures, our mutations were globally distributed over the surface of the PlyC CHAP whereas Low's mutations were more tightly grouped around the XlyA active-site. Additionally, we note that the PlyC CHAP active-site is located along a neutrally charged groove and the XlyA active-site is within an anionic pocket. It is presently not known if these differences account for the opposing results of our work with those of Low et al. [15]. We suggest that while the charge hypothesis should continue to be acknowledged for future engineering efforts, it should not be the sole focus and other characteristics (i.e., surface charge distribution, nature of active-site electrostatics, etc.) should also be taken into consideration.

4. Materials and Methods

4.1. Bacterial Strains and Culture Conditions

Streptococcus pyogenes D471 was cultured from a −80 °C frozen stock and grown in Todd Hewitt broth supplemented with 1% yeast extract (THY) without shaking at 37 °C. *E. coli* strains DH5α and

BL21 (DE3) were grown in Luria-Bertani (LB) broth. When needed, kanamycin (50 μg/mL) was added to the media. All bacterial cultures were grown at 37 °C in a shaking incubator unless otherwise stated.

4.2. In Silico Modeling of PlyC CHAP Mutants

The crystal structure coordinates were obtained from the Protein Data Bank for PlyC (4F88), XlyA (3RDR), and XlyA+5K (3HMB). The strategy used to change the net charge (Z) of the CHAP domain was to substitute negatively charged amino acids, aspartic acid (D) and glutamic acid (E), that were surface exposed and not involved in structured regions (i.e., α-helix or β-sheet) to neutral (alanine (A)) or positively charged (lysine (K)) amino acids to increase the Z score from −3 to +1 through +7 at pH 7.4. The net charges were calculated from the online Protein Calculator Version 3.4 [38]. PyMOL (The PyMOL Molecular Graphics System, Version 1.7.4 Schrödinger, LLC) was used to identify surface amino acids and a library of 192 CHAP mutant candidates was established following the outlined strategy. Before further validation, the CHAP mutant candidates had their side-chain orientation optimized using the FoldX 3.0 Repair PBD command [36]. The resulting coordinates were then processed by FoldX 3.0 for calculating the free energy change of the mutants ($\Delta\Delta G_{FoldX}$). The desirable mutants possessed $\Delta\Delta G_{FoldX} < 0$ kcal/mol ($\Delta\Delta G_{FoldX} = \Delta G_{mut} - \Delta G_{WT}$) and the mutants with the largest negative $\Delta\Delta G_{FoldX}$ were then picked for experimental study. The electrostatic surface potential was imaged using CCP4MG [24].

4.3. Cloning and Site-Directed Mutagenesis

The primers used in this study are listed in Table 2. The WT gene of PlyCA CHAP domain was amplified from pBAD24::*plyC* [39] and cloned via NdeI and BamHI sites into pET28a, as the template for the mutagenesis. The Change-IT^{TM} Multiple Mutation Site-Directed Mutagenesis Kit from Affymetrix was used to generate all mutants. Each mutation was designed to be in the middle of a 30 nucleotide phosphorylated forward primer and the mutagenesis followed instructions provided by the manufacturer of the kit. The resulting mutants were confirmed by sequencing (Macrogen, Rockville, MD, USA) before being transformed into *E. coli* BL21 (DE3) for protein expression.

Table 2. Primer information.

Plasmid	Template	Primer	Sequence
CHAP D311K	pET28a::*chap*	XS3	5′-ATGGGGTCTAAAAGAGTTGCAGCAAAC-3′
CHAP D355K	pET28a::*chap*	XS4	5′-TCATACTCAACAGGTAAACCAATGCTACCGTTA-3′
CHAP D363K	pET28a::*chap*	XS5	5′-CTACCGTTAATTGGTAAAGGTATGAACGCTCAT-3′
CHAP D429K	pET28a::*chap*	XS6	5′-ATTGAAAGCTGGTCAAAAACTACCGTTACAGTC-3′
CHAP D429A	pET28a::*chap*	XS8	5′-ATTGAAAGCTGGTCAGCGACTACCGTTACAGTC-3′
CHAP D450K	pET28a::*chap*	XS7	5′-ATACGCAGCACCTATAAACTTAACACATTCCTA-3′

4.4. Protein Expression and Purification

The overnight cultures of *E. coli* BL21 (DE3) harboring the WT PlyCA CHAP domain or mutants were sub-cultured 1:100 into 1.5 L LB supplemented with kanamycin in a 4 L baffled Erlenmeyer flask at 37 °C in a shaking incubator. The culture was induced at mid-log phase (about 4 h) with 1 mM isopropyl β-D-1-thiogalactopyranoside (IPTG) and incubated at 18 °C overnight. The next morning, *E. coli* BL21 (DE3) bacterial cells were harvested at 5,000 rpm, resuspended in PBS, pH 7.4, sonicated, and clarified via centrifugation at 12,000 rpm at 4 °C. The soluble portion of the cell lysate was applied to a Ni-NTA resin column (Thermo Fisher Scientific, Waltham, MA, USA). The 6X His-tagged protein was washed and eluted using a gradient of imidazole from 20 mM to 500 mM in PBS buffer, pH 7.4. The protein purity was assessed on a 7.5% SDS-PAGE gel before dialysis to remove the imidazole. The 6x His-tag was removed using the Thrombin Cleavage Capture Kit (EMD Millipore, Burlington, MA, USA) according to the protocol provided by the manufacturer.

4.5. *In Vitro* PlyC CHAP Activity

The activities of the PlyC CHAP domain and its mutants were evaluated via a spectrophotometric-based turbidity reduction assay as described previously [6]. An overnight culture of *S. pyogenes* D471 was harvested at 4500 rpm for 10 min, washed twice and resuspended in PBS, pH 7.4 buffer to reach an $OD_{600} = 2.0$. In a flat-bottom 96-well plate (Thermo Fisher Scientific), bacterial cells were mixed 1:1 with equimolar amounts of the PlyC CHAP domain or its mutants and the OD_{600} was monitored every 15 sec for 1 hour at 37 °C using a SpectraMax 190 spectrophotometer (Molecular Devices, San Jose, CA, USA). The OD_{600} decrease was calculated after 1 hour treatment by calculating the difference in OD_{600} readings between the PBS treated bacteria and the enzyme treated bacteria. Each assay was conducted in triplicate.

5. Conclusions

Bacteriophage-encoded endolysins offer an emerging alternative to conventional antibiotics. These enzymes function to degrade the bacterial peptidoglycan, producing fatal osmotic lysis in susceptible Gram-positive bacteria. A cell wall binding domain (CBD) is often, but not always, required for activity. In some endolysins, the catalytic domain alone is sufficient for lysis. In these examples, it has been noted that the catalytic domains have a net positive charge, suggesting they can interact with the negatively charged bacterial surface in the absence of a CBD. Protein engineering studies support this hypothesis by showing that a change in net charge of a catalytic domain, from negative to positive, correlated with CBD independence. However, similar experiments on the PlyC endolysin catalytic domain reported here do not support the hypothesis. While not discounting the charge hypothesis, other factors, such as distribution of surface charges, may also affect the CBD dependence of an endolysin and should be considered for future protein engineering efforts.

Author Contributions: X.S. and D.C.N. conceived and designed the experiments; X.S. performed the experiments and analyzed the data; the manuscript was written by X.S. and D.C.N. The overall editing was performed by D.C.N.

Funding: This research was funded by the National Institute of Food and Agriculture grant number 2015-06925.

Conflicts of Interest: The authors declare no conflict of interest. The funders had no role in the design of the study; in the collection, analyses, or interpretation of data; in the writing of the manuscript or in the decision to publish the results.

References

1. Fischetti, V.A. Exploiting what phage have evolved to control gram-positive pathogens. *Bacteriophage* **2011**, *1*, 188–194. [CrossRef] [PubMed]
2. Young, R. Bacteriophage lysis: mechanism and regulation. *Microbiol. Rev.* **1992**, *56*, 430–481.
3. Fischetti, V.A.; Nelson, D.; Schuch, R. Reinventing phage therapy: are the parts greater than the sum? *Nat. Biotechnol.* **2006**, *24*, 1508–1511. [CrossRef] [PubMed]
4. Briers, Y.; Walmagh, M.; Grymonprez, B.; Biebl, M.; Pirnay, J.P.; Defraine, V.; Michiels, J.; Cenens, W.; Aertsen, A.; Miller, S.; et al. Art-175 is a highly efficient antibacterial against multidrug-resistant strains and persisters of Pseudomonas aeruginosa. *Antimicrob. Agents Chemother.* **2014**, *58*, 3774–3784. [CrossRef] [PubMed]
5. Oliveira, H.; Melo, L.D.; Santos, S.B.; Nobrega, F.L.; Ferreira, E.C.; Cerca, N.; Azeredo, J.; Kluskens, L.D. Molecular aspects and comparative genomics of bacteriophage endolysins. *J. Virol.* **2013**, *87*, 4558–4570. [CrossRef]
6. Nelson, D.C.; Schmelcher, M.; Rodriguez-Rubio, L.; Klumpp, J.; Pritchard, D.G.; Dong, S.; Donovan, D.M. Endolysins as antimicrobials. *Adv. Virus Res.* **2012**, *83*, 299–365. [CrossRef] [PubMed]
7. Broendum, S.S.; Buckle, A.M.; McGowan, S. Catalytic diversity and cell wall binding repeats in the phage-encoded endolysins. *Mol. Microbiol.* **2018**, *110*, 879–896. [CrossRef]

8. Linden, S.B.; Zhang, H.; Heselpoth, R.D.; Shen, Y.; Schmelcher, M.; Eichenseher, F.; Nelson, D.C. Biochemical and biophysical characterization of PlyGRCS, a bacteriophage endolysin active against methicillin-resistant Staphylococcus aureus. *Appl. Microbiol. Biotechnol.* **2015**, *99*, 741–752. [CrossRef]
9. Huang, Y.; Yang, H.; Yu, J.; Wei, H. Molecular dissection of phage lysin PlySs2: integrity of the catalytic and cell wall binding domains is essential for its broad lytic activity. *Virol. Sin.* **2015**, *30*, 45–51. [CrossRef]
10. Porter, C.J.; Schuch, R.; Pelzek, A.J.; Buckle, A.M.; McGowan, S.; Wilce, M.C.; Rossjohn, J.; Russell, R.; Nelson, D.; Fischetti, V.A.; et al. The 1.6 A crystal structure of the catalytic domain of PlyB, a bacteriophage lysin active against Bacillus anthracis. *J. Mol. Biol.* **2007**, *366*, 540–550. [CrossRef]
11. Sanz, J.M.; Diaz, E.; Garcia, J.L. Studies on the structure and function of the N-terminal domain of the pneumococcal murein hydrolases. *Mol. Microbiol.* **1992**, *6*, 921–931. [CrossRef] [PubMed]
12. Donovan, D.M.; Lardeo, M.; Foster-Frey, J. Lysis of staphylococcal mastitis pathogens by bacteriophage phi11 endolysin. *FEMS Microbiol. Lett.* **2006**, *265*, 133–139. [CrossRef] [PubMed]
13. Horgan, M.; O'Flynn, G.; Garry, J.; Cooney, J.; Coffey, A.; Fitzgerald, G.F.; Ross, R.P.; McAuliffe, O. Phage lysin LysK can be truncated to its CHAP domain and retain lytic activity against live antibiotic-resistant staphylococci. *Appl. Environ. Microbiol.* **2009**, *75*, 872–874. [CrossRef] [PubMed]
14. Cheng, Q.; Fischetti, V.A. Mutagenesis of a bacteriophage lytic enzyme PlyGBS significantly increases its antibacterial activity against group B streptococci. *Appl. Microbiol. Biotechnol.* **2007**, *74*, 1284–1291. [CrossRef]
15. Low, L.Y.; Yang, C.; Perego, M.; Osterman, A.; Liddington, R. Role of net charge on catalytic domain and influence of cell wall binding domain on bactericidal activity, specificity, and host range of phage lysins. *J. Biol. Chem.* **2011**, *286*, 34391–34403. [CrossRef] [PubMed]
16. Mayer, M.J.; Garefalaki, V.; Spoerl, R.; Narbad, A.; Meijers, R. Structure-based modification of a Clostridium difficile-targeting endolysin affects activity and host range. *J. Bacteriol.* **2011**, *193*, 5477–5486. [CrossRef]
17. Rodriguez-Rubio, L.; Chang, W.L.; Gutierrez, D.; Lavigne, R.; Martinez, B.; Rodriguez, A.; Govers, S.K.; Aertsen, A.; Hirl, C.; Biebl, M.; et al. 'Artilysation' of endolysin lambdaSa2lys strongly improves its enzymatic and antibacterial activity against streptococci. *Sci. Rep.* **2016**, *6*, 35382. [CrossRef]
18. McGowan, S.; Buckle, A.M.; Mitchell, M.S.; Hoopes, J.T.; Gallagher, D.T.; Heselpoth, R.D.; Shen, Y.; Reboul, C.F.; Law, R.H.; Fischetti, V.A.; et al. X-ray crystal structure of the streptococcal specific phage lysin PlyC. *Proc. Natl. Acad. Sci. USA* **2012**, *109*, 12752–12757. [CrossRef]
19. Heselpoth, R.D.; Yin, Y.; Moult, J.; Nelson, D.C. Increasing the stability of the bacteriophage endolysin PlyC using rationale-based FoldX computational modeling. *Protein Eng. Des. Sel.* **2015**, *28*, 85–92. [CrossRef]
20. Yang, H.; Linden, S.B.; Wang, J.; Yu, J.; Nelson, D.C.; Wei, H. A chimeolysin with extended-spectrum streptococcal host range found by an induced lysis-based rapid screening method. *Sci. Rep.* **2015**, *5*, 17257. [CrossRef]
21. Yang, H.; Gong, Y.; Zhang, H.; Etobayeva, I.; Miernikiewicz, P.; Luo, D.; Li, X.; Zhang, X.; Dabrowska, K.; Nelson, D.C.; et al. ClyJ, a novel pneumococcal chimeric lysin with a CHAP catalytic domain. *Antimicrob. Agents Chemother.* **2019**. [CrossRef] [PubMed]
22. Schymkowitz, J.; Borg, J.; Stricher, F.; Nys, R.; Rousseau, F.; Serrano, L. The FoldX web server: an online force field. *Nucleic Acids Res.* **2005**, *33*, W382–W388. [CrossRef] [PubMed]
23. Nelson, D.; Loomis, L.; Fischetti, V.A. Prevention and elimination of upper respiratory colonization of mice by group A streptococci by using a bacteriophage lytic enzyme. *Proc. Natl. Acad. Sci. USA* **2001**, *98*, 4107–4112. [CrossRef] [PubMed]
24. McNicholas, S.; Potterton, E.; Wilson, K.S.; Noble, M.E. Presenting your structures: the CCP4mg molecular-graphics software. *Acta Crystallogr. D Biol. Crystallogr.* **2011**, *67*, 386–394. [CrossRef] [PubMed]
25. Schmelcher, M.; Donovan, D.M.; Loessner, M.J. Bacteriophage endolysins as novel antimicrobials. *Future Microbiol.* **2012**, *7*, 1147–1171. [CrossRef] [PubMed]
26. Heselpoth, R.D.; Swift, S.M.; Linden, S.B.; Mitchell, M.S.; Nelson, D.C. Enzybiotics: Endolysins and Bacteriocins. In *Bacteriophages—Biology, Technology, Therapy*; Harper, D., Abedon, S.T., Burrowes, B., McConvill, M., Eds.; Springer International Publishing: Dordrecht, The Netherlands, 2018.
27. Sheehan, M.M.; Garcia, J.L.; Lopez, R.; Garcia, P. The lytic enzyme of the pneumococcal phage Dp-1: a chimeric lysin of intergeneric origin. *Mol. Microbiol.* **1997**, *25*, 717–725. [CrossRef]
28. Yang, H.; Bi, Y.; Shang, X.; Wang, M.; Linden, S.B.; Li, Y.; Li, Y.; Nelson, D.C.; Wei, H. Antibiofilm activities of a novel chimeolysin against Streptococcus mutans under physiological and cariogenic conditions. *Antimicrob. Agents Chemother.* **2016**, *60*, 7436–7443. [CrossRef]

29. Daniel, A.; Euler, C.; Collin, M.; Chahales, P.; Gorelick, K.J.; Fischetti, V.A. Synergism between a novel chimeric lysin and oxacillin protects against infection by methicillin-resistant Staphylococcus aureus. *Antimicrob. Agents Chemother.* **2010**, *54*, 1603–1612. [CrossRef]
30. Diez-Martinez, R.; De Paz, H.D.; Garcia-Fernandez, E.; Bustamante, N.; Euler, C.W.; Fischetti, V.A.; Menendez, M.; Garcia, P. A novel chimeric phage lysin with high in vitro and in vivo bactericidal activity against Streptococcus pneumoniae. *J. Antimicrob. Chemother.* **2015**, *70*, 1763–1773. [CrossRef]
31. Vazquez, R.; Domenech, M.; Iglesias-Bexiga, M.; Menendez, M.; Garcia, P. Csl2, a novel chimeric bacteriophage lysin to fight infections caused by Streptococcus suis, an emerging zoonotic pathogen. *Sci. Rep.* **2017**, *7*, 16506. [CrossRef]
32. Blazquez, B.; Fresco-Taboada, A.; Iglesias-Bexiga, M.; Menendez, M.; Garcia, P. PL3 amidase, a tailor-made lysin constructed by domain shuffling with potent killing activity against pneumococci and related species. *Front. Microbiol.* **2016**, *7*, 1156. [CrossRef] [PubMed]
33. Schmelcher, M.; Powell, A.M.; Becker, S.C.; Camp, M.J.; Donovan, D.M. Chimeric phage lysins act synergistically with lysostaphin to kill mastitis-causing Staphylococcus aureus in murine mammary glands. *Appl. Environ. Microbiol.* **2012**, *78*, 2297–2305. [CrossRef] [PubMed]
34. Becker, S.C.; Foster-Frey, J.; Stodola, A.J.; Anacker, D.; Donovan, D.M. Differentially conserved staphylococcal SH3b_5 cell wall binding domains confer increased staphylolytic and streptolytic activity to a streptococcal prophage endolysin domain. *Gene* **2009**, *443*, 32–41. [CrossRef] [PubMed]
35. Leaver-Fay, A.; Tyka, M.; Lewis, S.M.; Lange, O.F.; Thompson, J.; Jacak, R.; Kaufman, K.; Renfrew, P.D.; Smith, C.A.; Sheffler, W.; et al. ROSETTA3: an object-oriented software suite for the simulation and design of macromolecules. *Methods Enzymol.* **2011**, *487*, 545–574. [CrossRef] [PubMed]
36. Guerois, R.; Nielsen, J.E.; Serrano, L. Predicting changes in the stability of proteins and protein complexes: a study of more than 1000 mutations. *J. Mol. Biol.* **2002**, *320*, 369–387. [CrossRef]
37. Diez-Martinez, R.; de Paz, H.; Bustamante, N.; Garcia, E.; Menendez, M.; Garcia, P. Improving the lethal effect of cpl-7, a pneumococcal phage lysozyme with broad bactericidal activity, by inverting the net charge of its cell wall-binding module. *Antimicrob. Agents Chemother.* **2013**, *57*, 5355–5365. [CrossRef]
38. PROTEIN CALCULATOR v3.4. Available online: http://protcalc.sourceforge.net (accessed on 4 January 2019).
39. Nelson, D.; Schuch, R.; Chahales, P.; Zhu, S.; Fischetti, V.A. PlyC: a multimeric bacteriophage lysin. *Proc. Natl. Acad Sci. USA* **2006**, *103*, 10765–10770. [CrossRef] [PubMed]

antibiotics

MDPI

Article

Fighting Fire with Fire: Phage Potential for the Treatment of *E. coli* O157 Infection

Cristina Howard-Varona [1,†], Dean R. Vik [1,†], Natalie E. Solonenko [1], Yueh-Fen Li [1], M. Consuelo Gazitua [1], Lauren Chittick [1], Jennifer K. Samiec [1], Aubrey E. Jensen [1], Paige Anderson [1], Adrian Howard-Varona [1], Anika A. Kinkhabwala [2], Stephen T. Abedon [1,*] and Matthew B. Sullivan [1,3,*]

1 Department of Microbiology, The Ohio State University, Columbus, OH 43210, USA; howard-varona.2@osu.edu (C.H.-V.); vik.1@buckeyemail.osu.edu (D.R.V.); solonenko.2@osu.edu (N.E.S.); li.918@osu.edu (Y.-F.L.); consuelogazitua@gmail.com (M.C.G.); chittick.3@osu.edu (L.C.); Jennifer.Samiec@osumc.edu (J.K.S.); aubrey.jensen9@gmail.com (A.E.J.); anderson.2805@buckeyemail.osu.edu (P.A.); ahowardv11@gmail.com (A.H.-V.)
2 EpiBiome, Inc., 29528 Union City blvd, Union City, CA 94587, USA; anikaak@gmail.com
3 Department of Civil, Environmental and Geodetic Engineering, The Ohio State University, Columbus, OH 43210, USA
* Correspondence: abedon.1@osu.edu (S.T.A.); sullivan.948@osu.edu (M.B.S.)
† The author contributed equally to this work.

Received: 23 October 2018; Accepted: 14 November 2018; Published: 16 November 2018

Abstract: Hemolytic–uremic syndrome is a life-threating disease most often associated with Shiga toxin-producing microorganisms like *Escherichia coli* (STEC), including *E. coli* O157:H7. Shiga toxin is encoded by resident prophages present within this bacterium, and both its production and release depend on the induction of Shiga toxin-encoding prophages. Consequently, treatment of STEC infections tend to be largely supportive rather than antibacterial, in part due to concerns about exacerbating such prophage induction. Here we explore STEC O157:H7 prophage induction in vitro as it pertains to phage therapy—the application of bacteriophages as antibacterial agents to treat bacterial infections—to curtail prophage induction events, while also reducing STEC O157:H7 presence. We observed that cultures treated with strictly lytic phages, despite being lysed, produce substantially fewer Shiga toxin-encoding temperate-phage virions than untreated STEC controls. We therefore suggest that phage therapy could have utility as a prophylactic treatment of individuals suspected of having been recently exposed to STEC, especially if prophage induction and by extension Shiga toxin production is not exacerbated.

Keywords: Antibiotic-resistant bacteria; bacteriophage therapy; phage therapy; lysogenic conversion; prophage induction; read recruitment; shiga toxin

1. Introduction

Prophages are bacteriophage (phage) genomes that replicate alongside their bacterial host's genome until induced to produce viral particles. This carriage state, termed a lysogenic cycle, is characteristic of temperate phages (as opposed to strictly lytic, or virulent, phages), and the prophage-carrying bacterial host is termed a lysogen. Recent reviews provide information on the diverse and impactful biology and distribution of temperate phages, along with methods for temperate phage detection [1–3]. One impact of temperate phage biology is lysogenic conversion: the modification of a host phenotype by prophage genes, including genes encoding bacterial virulence factors [4–6].

Notable among prophage-encoded virulence factors are exotoxins, such as those associated with the O157:H7 serotype of Shiga-toxigenic *Escherichia coli* (STEC) [7]. STEC O157:H7 is a polylysogenic human pathogen, often derived from ruminant gastrointestinal tracks and known for its capacity to encode two Shiga toxins, dubbed Stx1 and Stx2 [8,9]. These are generally encoded by the Shiga-toxigenic prophages 933V and 933W, respectively [10–12]. Of these, only the lamboid 933W prophage appears capable of inducing, and does so spontaneously [11,13–17]. This induction and the associated lytic cycle are a prerequisite for Shiga toxin production and release [18–20]. Shiga toxin release during STEC O157:H7 infection can lead to hemorrhagic colitis and hemolytic–uremic syndrome (HUS), which damages kidney nephrons of the STEC-infected human patients [20–22], but causes little to no pathogenesis in ruminants [23].

Certain antibiotics that induce the STEC SOS response also can induce Shiga-toxigenic prophages, resulting in new intracellular Shiga toxin production and subsequent phage lysis-associated toxin release [4,11,16,20,24–26]. Thus, prophage induction, in addition to bacterial lysis, drives increases of Shiga toxin within STEC-infected individuals, and prophage-inducing antibiotics therefore are not recommended for STEC treatment. Consequently, STEC killing via other non-prophage inducing methods—even lytic mechanisms, such as through infection by strictly lytic phages—should serve as viable STEC treatment. Treatment using non-Shiga-toxigenic phages (phage therapy) should not in itself give rise to an increased degree of patient exposure to Shiga toxin than would occur without such non-inductive lysis. Furthermore, lysogen killing by means that do not induce prophages should curtail future induction events, which presumably will result in less overall Shiga toxin production.

Based on the above assumptions, we reasoned that lysis of STEC O157:H7 by strictly lytic phages might eliminate STEC O157:H7 without further contributing to Shiga toxin production. If true, then such lytic phages might be employed as a means of anti-STEC treatment, and by extension as anti-Shiga-toxigenic phage agents—in effect an anti-temperate phage form of phage therapy.

Here we test this hypothesis through in vitro experiments designed to explore the use of strictly lytic phages, unrelated to Shiga toxin-encoding prophages, as anti-STEC bactericidal agents, in order to assess the potential impact of phage therapy on the production of Shiga-toxigenic 933W phages by *E. coli* O157:H7.

2. Results

2.1. Detecting Spontaneous Prophage Induction

From the American Type Culture Collection (ATCC—identifier ATCC43895) we acquired the STEC serotype O157:H7 whose genome sequence is published under strain EDL933 [11,12]. In order to have an up-to-date genome sequence (herein termed STEC), we re-sequenced our working strain and identified prophage regions with the online tool PHASTER [27] (Supplementary Materials). This confirmed the working strain as largely identical to the published EDL933 at 100% average nucleotide identity (ANI) with only a ~1% difference in genome length (see Supplementary Materials, Table S1). Predicted prophage content between STEC and EDL933 was also largely congruent, with the small variation observed likely due to differences in sequencing and assembly methodology (Supplementary Materials, Figure S1).

With a fully-sequenced working strain, we then assessed spontaneous prophage induction in STEC as follows. STEC cultures were grown in triplicate for 5 h, treated with chloroform for 2.5 h to lyse the cells and release encapsidated phage DNA, and 0.2 μm-filtered to remove cells and large cellular debris. Samples were then treated with DNase to minimize free DNA and enrich for encapsidated DNA. The DNA was then extracted and sequenced, and the resulting reads were mapped to the STEC genome, including prophage regions. Given that most free bacterial DNA was removed with DNase, elevated read recruitment across the entirety of any prophage region would indicate induction and subsequent encapsidation of the prophage region(s). This read recruitment methodology is especially useful for identifying which prophages are induced within polylysogens, as previously shown [28–30].

Mean read recruitment coverage values were calculated per host or prophage region and normalized by the sequencing depth and the sequence length of either the 933W genomic region (59,338 bp) or the

STEC genome without the 933W prophage (5,499,692 bp). This revealed that prophage 933W, which encodes the Stx2 genes and is responsible for much of STEC's pathogenesis [20,31], had substantially higher mean coverage (4675×) than either the rest of the host genome (0.09×) or other prophage regions (0.12×), and that this elevated read coverage encompassed nearly all (95%) of the 933W genome (Figure 1, Table 1, and Supplementary Materials). We interpret this as evidence for spontaneous induction and encapsidation of 933W in this STEC strain, a finding consistent with prior work that describes prophage 933W as a highly spontaneously inducible prophage [11,13–17].

With this qualitative screening identifying only the 933W prophage as having been induced, we sought to quantify 933W phage production as a product of spontaneous induction via a quantitative PCR (qPCR) approach targeting the Shiga toxin gene *stx2a* encoded by 933W. To this end, we grew and sampled STEC as done for the whole-genome induction screen above, and found that the prophage 933W-encoded *stx2a* copy number increased eight-fold from the start to the end of the aforementioned 7.5 h experiment (~10^5 to 8×10^5 per µL of filtrate) (Figure 2). This corroborates the sequence-based indication of prophage 933W spontaneous induction and implies ongoing induction over the course of culture incubation, since encapsidated DNA was present in somewhat smaller amounts at the start of the incubation. Prophage 933W induction, therefore, should be quantitatively reducible by preventing ongoing lysogen growth, such as may be accomplished in the course of phage therapy.

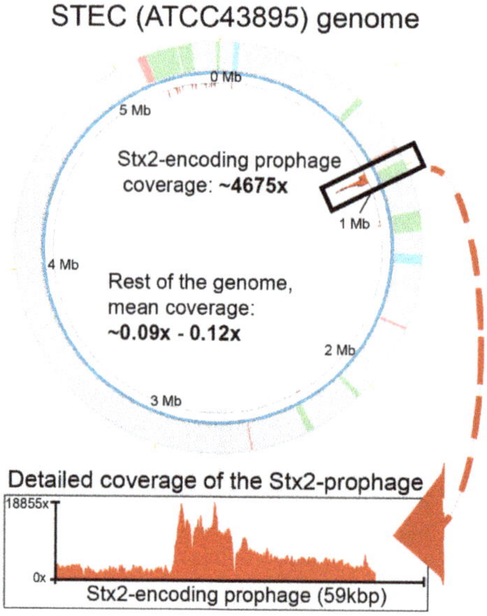

Figure 1. Phage 933W is the only prophage that is spontaneously induced. Shown here is the read mapping from sequenced *Escherichia coli* (STEC) cultures in biological triplicates. The circular plot represents the host genome, with the PHASTER-predicted prophages in colors (pink, red, or blue) in the outer circle, as well as the reads mapped to the entire genome. Prophage 933W is covered ~4675 times on average throughout its entire length, whereas the rest of the non-prophage and prophage genomic regions are covered, on average, 0.09 and 0.12 times, respectively. The prophage 933W region and read-mapping to such a region is amplified below the circular plot to show that the entire prophage length is covered by reads, and their proportion. Detailed information of the reads can be found in Table 1 and in the Supplementary Materials, Dataset.

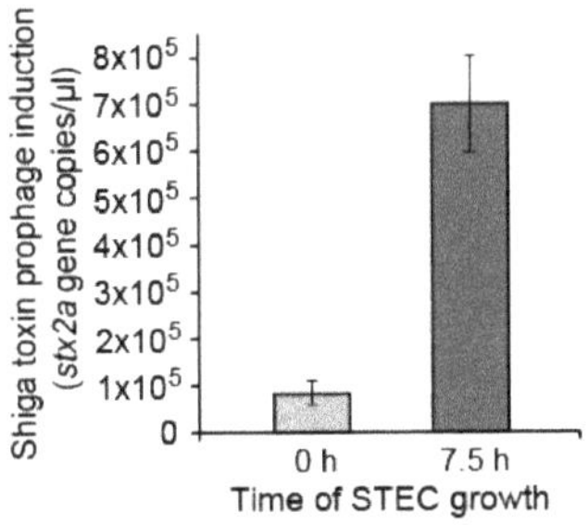

Figure 2. Quantification of prophage 933W induction in uninfected STEC cultures via quantitative PCR (qPCR). Primers are used against the *stx2* subunit *a* gene at 0 and 7.5 h of STEC growth. The former represents a transfer of cells from an overnight growth into fresh media, and the latter represents when cell growth is stopped and the DNA harvesting procedure begins (see Methods). The average of three biological replicates and their error is plotted on the graph. The difference between the two time points is significantly different (*t*-test, $p < 0.05$). Data from this experiment can be found in the Supplementary Materials (see Dataset).

Table 1. Coverage of uninfected ATCC43895 (STEC)'s prophage and non-prophage regions. Represented is the raw coverage, the normalized coverage (to sequencing depth and region length), and the final transformed coverage (multiplied by 10^{11} for better reading) of each of the 17 prophages and the non-prophage regions of STEC, in biological triplicates.

Lysate	Prophage or Not?	Genomic Entity	Raw Coverage	Coverage Normalized by Sequencing Depth and Entity Length	Final Adjusted Coverage (Raised to 10^{11})
No phage control ATCC43895 (STEC)-Replicate 1	Prophages	#1-58370-85143	0.069	5.69×10^{-14}	0.01
		#2-648527-680910	0.067	5.50×10^{-14}	0.01
		#3 and 4-911029-938407 bp	0.835	6.89×10^{-13}	0.07
		#5 (Stx2)-973564-1032902 bp	585.776	4.41×10^{-8}	4407.33
		#6 and 7-1202175-1293616 bp	2.606	2.15×10^{-12}	0.22
		#8-1390536-1436457 bp	0.018	1.47×10^{-14}	0
		#9-1708731-1719671 bp	0.002	1.57×10^{-15}	0
		#10-2054278-2078426 bp	0.167	1.38×10^{-13}	0.01
		#11 (Stx1)-2302225-2335340 bp	8.057	6.65×10^{-12}	0.66
		#12-2579647-2589259 bp	0	0	0
		#13, 14 and 15-5103469-5282316 bp	1.229	1.01×10^{-12}	0.1
		#16-5286283-5348617 bp	0.274	2.26×10^{-13}	0.02
		#17-5449904-5468395 bp	0.24	1.98×10^{-13}	0.02
	Non-prophage	Host genome, non-prophage	0.852	7.04×10^{-13}	0.07
No phage control ATCC43895 (STEC)-Replicate 2	Prophages	#1-58370-85143	0.15	8.91×10^{-14}	0.01
		#2-648527-680910	0.514	3.05×10^{-13}	0.03
		#3 and 4-911029-938407 bp	1.883	1.12×10^{-12}	0.11
		#5 (Stx2)-973564-1032902 bp	881.779	4.77×10^{-8}	4774.17
		#6 and 7-1202175-1293616 bp	4.605	2.74×10^{-12}	0.27
		#8-1390536-1436457 bp	0.118	7.00×10^{-14}	0.01
		#9-1708731-1719671 bp	0.22	1.30×10^{-13}	0.01
		#10-2054278-2078426 bp	0.372	2.21×10^{-13}	0.02
		#11 (Stx1)-2302225-2335340 bp	14.645	8.70×10^{-12}	0.87
		#12-2579647-2589259 bp	0.097	5.73×10^{-14}	0.01
		#13, 14 and 15-5103469-5282316 bp	2.283	1.36×10^{-12}	0.14
		#16-5286283-5348617 bp	0.437	2.59×10^{-13}	0.03
		#17-5449904-5468395 bp	0.562	3.34×10^{-13}	0.03
	Non-prophage	Host genome, non-prophage	1.743	1.04×10^{-12}	0.1

Table 1. *Cont.*

Lysate	Prophage or Not?	Genomic Entity	Raw Coverage	Coverage Normalized by Sequencing Depth and Entity Length	Final Adjusted Coverage (Raised to 10^{11})
No phage control ATCC43895 (STEC)-Replicate 3	Prophages	#1-58370-85143	0.226	1.48×10^{-13}	0.01
		#2-648527-680910	0.299	1.96×10^{-13}	0.02
		#3 and 4-911029-938407 bp	1.379	9.03×10^{-13}	0.09
		#5 (Stx2)-973564-1032902 bp	811.189	4.84×10^{-8}	4844.55
		#6 and 7-1202175-1293616 bp	4.466	2.93×10^{-12}	0.29
		#8-1390536-1436457 bp	0.089	5.82×10^{-14}	0.01
		#9-1708731-1719671 bp	0.117	7.66×10^{-14}	0.01
		#10-2054278-2078426 bp	0.399	2.61×10^{-13}	0.03
		#11 (Stx1)-2302225-2335340 bp	13.974	9.15×10^{-12}	0.92
		#12-2579647-2589259 bp	0.277	1.82×10^{-13}	0.02
		#13, 14 and 15-5103469-5282316 bp	2.031	1.33×10^{-12}	0.13
		#16-5286283-5348617 bp	0.345	2.26×10^{-13}	0.02
		#17-5449904-5468395 bp	0.541	3.54×10^{-13}	0.04
	Non-prophage	Host genome, non-prophage	1.681	1.10×10^{-12}	0.11

2.2. *Fighting Prophage Induction with Phage Treatment*

Given that 933W induction is known to be associated with Shiga toxin production in *E. coli* O157:H7 [20], we next considered whether treatment using exogenously supplied, strictly lytic phages could reduce lysogen numbers without exacerbating prophage induction. We used the T4-like phages p000v and p000y that we previously isolated and sequenced [32], and which we here characterized for their infection of STEC (Supplementary Materials: Figures S2 and S3, Dataset). We then grew and sampled STEC as described above, except we also added either of these exogenous phages to the STEC culture at ratios of roughly 4–6 phages per target bacterium (multiplicity of infection (MOI): ~4–6), where initial infective titers were ~6.4×10^8 and ~4.6×10^8 plaque-forming units per ml for phages p000v and p000y, respectively. Indeed, by the end of the experiment, addition of these phages had decreased the levels of 933W prophage induction, as quantified by qPCR. Namely, while the qPCR-measured ratio of *stx2a* copies per μL between 7.5 and 0 h was ~8 without phage (Figure 2), with the addition of phages p000v and p000y it decreased to ~0.3× and ~0.4×, respectively (Figure 3, Table 2). Thus, these results show that exogenous, strictly lytic phages reduce *stx2a* copies (a proxy for 933W prophage induction) and suggest that Shiga toxin production would also be reduced, due to both no further stimulation of prophage induction upon lytic phage infection, on the one hand, and reduction in the number of lysogens present on the other.

Table 2. Summary of the prophage induction quantification obtained by qPCR in uninfected and infected STEC cultures, as presented in Figures 2 and 3.

***Stx2a* Copies Per μL During STEC Growth with and without Phage**				
Phage	**0 h**	**7.5 h**	**Ratio**	**MOI**
None	8.39×10^4	6.97×10^5	8.31	NA
p000v	6.93×10^3	2.09×10^3	0.30	6.43
p000y	8.00×10^3	2.88×10^3	0.36	4.61

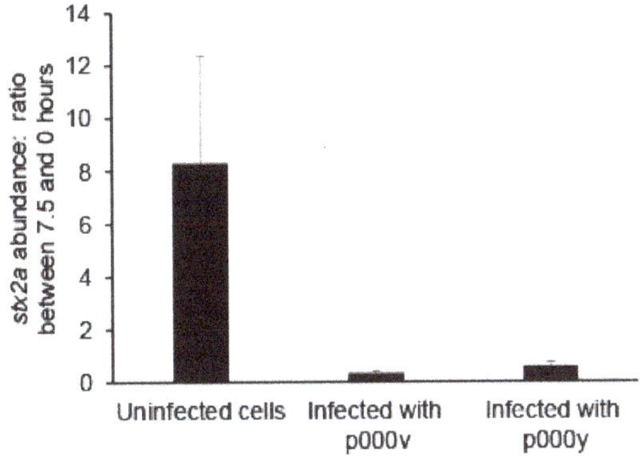

Figure 3. Quantification of prophage 933W induction in phage-infected STEC cultures via qPCR. The Shiga toxin (*stx2* subunit *a*) gene abundance in prophage 933W is measured at 0 and 7.5 h post-phage addition to STEC cell cultures at multiplicities of infection (MOIs) of ~4.6 (for phage p000y) and ~6.4 (for phage p000v). Represented is the ratio of such *stx2a* abundance between 7.5 and 0 h of STEC growth, using the average of the biological replicates and their error, either in the absence (left most bar in the graph) or presence (the other two bars) of phages. The differences between in the absence of phages (uninfected cells) and in the presence of phages (infected with p000v or p000y) are statistically significant ($p < 0.05$).

3. Discussion

The primary question regarding the potential for using phage therapy to treat pathogenic lysogens is whether such treatment might exacerbate patient exposure to toxins produced upon prophage induction. For Shiga-toxigenic *E. coli* O157:H7 in particular, Shiga toxin production and release is associated with prophage induction, mostly prophage 933W [14,19,33–35]. Consequentially, treatment options for STEC infections are largely supportive rather than antimicrobial for tackling Shiga toxin production and patient exposure [36,37].

There are three related routes by which Shiga toxin exposure could occur (Figure 4). First, the standard route (point 1a, Figure 4) is through prophage induction, resulting in Shiga toxin (Stx) gene expression followed by Shiga toxin release via phage-induced bacterial lysis [20,38]. Thus, it is crucial to avoid treatments that can lead to additional prophage induction, which can result from certain antibiotic uses [20,39].

A second route of Shiga toxin release (point 2a, Figure 4) may occur via artificial lysis of induced lysogens by exogenous phages, if such lysogens are capable of becoming infected and sustaining a second bacteriolytic phage infection. This could accelerate cell lysis and thus toxin release if the exogenous phage has a faster replication cycle or is otherwise competitively superior to the prophage. Alternatively, co-infection by an exogenous phage and induced prophage may confound either of the phages' replication cycle, thereby delaying the time to cell lysis. Both instances could reasonably attenuate toxin production overall due to the reduction of either the duration or the efficiency of prophage expression, thus reducing toxin translation.

A third route (points 3a and 3b, Figure 4) may occur when the induced and then released Shiga-toxigenic temperate phage lytically infects other *E. coli* not already lysogenized by Shiga-toxigenic phages, which would consequently enable these non-STEC bacteria to express Shiga toxin [40–42]. This latter route may not be easily blocked if sufficient numbers of these alternative hosts are present and support substantial Shiga-toxigenic phage population growth (point 3b, Figure 4). Based on our results, Shiga-toxin amplification from such infections may, however, be curtailed by intervening with phage treatment prior to lysogen induction and resulting Shiga-toxigenic phage production.

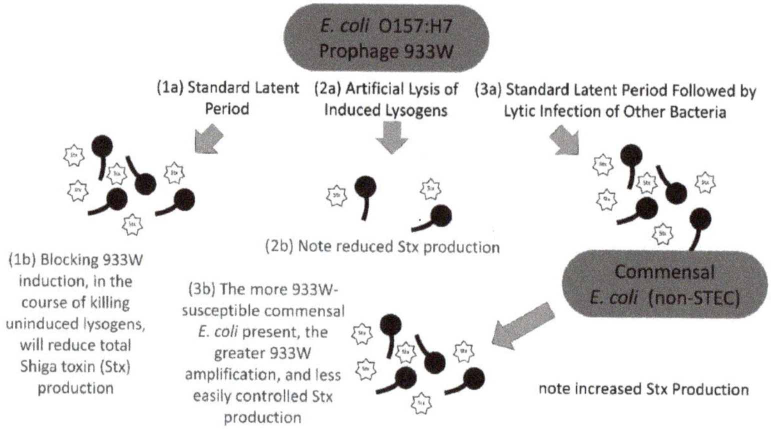

Figure 4. Different routes towards Shiga toxin (Stx) release: (1) 933W prophage induction followed by normal lytic cycles; (2) artificial lysis, for example by exogenous phage, of induced lysogens, resulting in truncated lytic cycles; and (3) subsequent lytic infection of non-Shiga toxigenic *E. coli* strains giving rise to more lytic cycles. Greater Stx production (stars in the figure) can occur given artificial induction of *E. coli* O157:H7 lysogens, but this both is not explicitly illustrated in the figure and is distinct from artificial lysis of already induced lysogens (2). Given the linkage between 933W induction and Shiga toxin production, the killing of *E. coli* O157:H7 lysogens without inducing the 933W prophage should result in reductions in future 933W induction events (Figure 3 and Table 2) along with subsequent reductions in Shiga toxin production.

Here we have confirmed that an exogenously supplied obligately lytic phage "treatment" can interfere with the spontaneous production of Shiga-toxigenic prophages encoded by an *E. coli* O157:H7 strain. The mechanism of reduction in prophage induction presumably is due to the killing of prophage-containing lysogens, apparently prior to natural or artificially triggered induction. It remains unconfirmed, however, how such phage treatment will impact Shiga toxin production or release. It is likely, though, as Shiga-toxogenic prophage induction is tightly coupled to Shiga toxin production [20,38], that phage treatment of *E. coli* O157:H7-exposed patients at the very least should mitigate Shiga toxin production by killing prophage-carrying Shiga-toxigenic lysogens.

To most effectively treat such Shiga-toxigenic pathogens, future research will need to explore several areas. First, it is not known to what extent, or with what variability, different treatment phages can impact the lytic cycles of already-induced lysogens (Figure 4, 2a). Additionally, recent research with environmental phage–host systems depicts the importance of also considering the host's response, given that they are often the ones driving the infection outcomes instead of the phages [43–45]. Second, it needs to be determined whether rapid treatment-phage-mediated *E. coli* O157:H7 killing is achievable in situ. It is likely, however, that achieving relatively high in-situ phage titers, e.g., 10^8 per ml or higher [46], would be required to attain such rapid treatment-phage impact, while substantial reductions in overall Stx production will require early initiation of treatment, such as in response to suspected rather than confirmed pathogen exposure (i.e., so-called "inundative" and prophylactic phage treatment, respectively). Third, while phage therapy is generally considered as a safe treatment, given the relative lack of toxicities and side effects, especially during oral delivery [47], further verification is needed prior to generalizing clinical implementation. Generally, these issues point to a broader "pharmacologically aware" approach to the development of any phage-based *E. coli* O157:H7 infection treatment, involving iteration between continued in vitro and in vivo as well as in silico studies. In this vein, the observations reported here are consistent with *E. coli* O157:H7 phage treatment likely not giving rise to negative outcomes, as can stem from the exacerbation of 933W prophage induction.

4. Materials and Methods

Raw data is provided in the Dataset, and additional methods can be found in the Supplementary Materials.

4.1. Bacterial Strain and Phages Used in This Study

The Shiga-toxigenic *E. coli* serotype O157:H7 (STEC) used in this study was obtained from the American Type Culture Collection (ATCC) under identifier 43895, which is published as EDL933 under GenBank accession numbers CP008957 and CP008958 [11,12]. The T4-like Myoviridae phages p000v and p000y are described elsewhere [32], and can be found in the Cyverse data repository [48] under DOI 10.7946/P2HP89 (https://www.doi.org/10.7946/P2HP89), and in GenBank under accession numbers MK047717 and MK047718, respectively.

4.2. Cell Growth

Bacteria were streaked onto TSA (Tryptic Soy Agar, 40 g/L, Ward's Cat. 38-1010) plates from glycerol stocks, grown overnight at room temperature (RT), and held at 4 °C. Single colonies were then inoculated in TSB (Tryptic Soy Broth, 30 g/L, Ward's Cat. 38-1012) and grown shaking at ~200 rpm at 37 °C overnight.

4.3. Lysates for Phage Amplification

An overnight bacterial culture was diluted 1:50 in fresh TSB and grown shaking at ~200 rpm at 37 °C until the optical density (OD) reached ~0.3 (2.94×10^8 CFUs/mL). Phages were added to 10–50 mL of the host culture at a low MOI (10^{-6}–0.1), and incubated shaking at ~200 rpm at 37 °C for ~5 h. Chloroform was added to the infection at 1% (*v/v*) and incubated shaking at ~70 rpm at RT for 2 h. The chloroform was allowed to settle for 30 min, and the aqueous phase was 0.2 µm-filtered to remove any remaining cells. Some lysates were also concentrated via polyethylene glycol (PEG)-precipitation. For those, both NaCl (6.5 g) and PEG-8000 (10 g) were added per 100 mL lysate. This was incubated overnight at 4 °C, then centrifuged at 10,000 *g* in a Beckman J2-MC centrifuge (Beckman Coulter, Brea, CA, USA) for 10 min. The supernatant was removed, and the pellet resuspended in phage buffer (4 g NaCl, 0.1 g gelatin, 10 mL 1 M Tris Base (pH 7.6), and 1 mL 1 M $MgSO_4$ per L).

4.4. DNA Extraction

The STEC strain obtained from ATCC (ATCC43895) was sequenced. For that, its genomic DNA was extracted using the ZymoBIOMICS DNA mini kit (Zymo Research, Irvine, CA, USA) following the manufacturer's protocol. Similarly, sequenced phage infections and phage-free cultures from which prophage induction was assessed were also sequenced. The DNA of these samples was extracted using the Phage DNA Isolation Kit (Norgen Biotek Corp., Thorold, ON, Canada) following the manufacturer's protocol. Any remaining host DNA was degraded by adding 10 µL (20 U) of DNase I from the RNase-free DNase I kit (Norgen Biotek Corp., Thorold, ON, Canada) prior to proteinase K treatment. Bacterial host DNA concentrations were quantified using the Qubit 3.0 Fluorometer and the Qubit dsDNA High Sensitivity Kit (Thermo Fisher Scientific, Waltham, MA, USA).

4.5. Library Preparation and Illumina Miseq Sequencing of Phage and Host Genomes

Extracts from the previous step were prepared for sequencing on the Illumina MiSeq platform (Illumina, San Diego, CA, USA) using the Nextera XT Library Preparation Kit (Illumina, San Diego, CA, USA) according to the manufacturer's protocol (Part # 15031942, revision D). The magnetic bead normalization step was replaced with a manual normalization step, based on library concentration and average size as measured by the Qubit 3.0 Fluorometer and Qubit dsDNA High Sensitivity Kit (Thermo Fisher Scientific, Waltham, MA, USA) and the Fragment Analyzer (AATI, Ankeny, IA,

USA), respectively. Paired-end sequencing was performed using the MiSeq Reagent v3 (600 cycle) kit (Illumina, San Diego, CA, USA).

4.6. Whole-Genome Sequencing and Read-Mapping to Assess Prophage Induction

Two sample types were obtained for sequencing whole bacterial and phage genomes, mapping reads to such genomes, and assessing prophage induction: phage-free and phage-infected bacterial cultures. The phage-infected samples were lysates grown as described in the "Lysates for Phage Amplification" section. The phage-free samples were mock lysates prepared as the infection samples, but without phages and in 30 mL containing 3.5 mL of phage buffer. After DNA extraction and library preparation procedures as described, samples were sequenced via the MiSeq technology described.

4.7. Read Mapping and Visualization of Prophage Induction

Reads from each of the lysates were mapped to the STEC and respective phage genome using the Burrows–Wheeler Aligner (BWA) [49] version 0.7.13, with default parameters. The resulting SAM files were converted to BAM files using samtools [50] version 1.3.1. Coverage across either phage or host genome was calculated using the Bayesian Analysis of Macroevolutionary Mixtures (BaMM) [51], software version 1.4.1, with the parse tool and the "tpmean" setting. Coverage values were then normalized by the number of reads that mapped to the virulent phages or STEC (i.e., the sequencing depth), as inferred by the samtools version 1.3.1 flagstat tool, and by the genome length of either prophage 933W, the virulent phages, or STEC without the prophage 933W. All depth- and length-normalized coverage values were then multiplied by 10^{11} to derive more comprehensible whole-genome coverage values. Coverage values per base were visualized by creating bedgraph files, using the bedtools [52] version 2.27.1 package and the genomecov -bg option. These bedgraph files were then uploaded to the Integrative Genomics Viewer (IGV) version 2.4.6 package [53] and Circos [54] version 0.69.

4.8. qPCR of Phage Lysates' DNA

The OD of an overnight bacterial culture was read to determine the volume containing 10^{10} cells, which was then added to 100 mL of TSB. This was grown shaking at ~200 rpm at 37 °C; the OD was read after ~30 min and then every 10 min until the reading was 0.25–0.3. Phages were added to 2–5 mL of the host culture at MOIs lower than 0.1 or close to 6. A 0.3–1 mL sample was taken immediately and 0.2 µm-filtered to remove any cells, then stored at 4 °C. The infected culture was then incubated shaking at ~200 rpm at 37 °C for 5 h. Chloroform was added to the infection at 1% (*v/v*) and incubated shaking at ~70 rpm at RT for 2.5 h. The chloroform was allowed to settle for 20 min, and the aqueous phase was 0.2 µm-filtered to remove cells. Another 0.3–1 mL sample was taken and stored at 4 °C. DNA was extracted from the two filtered samples. First, the viral DNA was inactivated using DNase in a ratio of 1 µL of DNase to 9 µL of sample. Ethylenediamine tetraacetic acid (EDTA) and Ethylene glycol tetraacetic acid (EGTA) were added at 100 mM to inactivate the DNase. After the DNA was inactivated, extraction was continued using a Wizard DNA Clean-up Kit (Cat. #A7181, Promega Corporation, Madison, WI, USA). Then, 1 mL of DNA clean-up resin was added to each sample, and they were mixed by inversion. The samples were put into a syringe and pushed through a Wizzard minicolumn (Cat. #A7211, Promega Corporation, Madison, WI, USA), followed by the addition of 2 mL 80% isopropanol pushed through the column, 1 mL at a time. Samples were centrifuged for 2 min at 10,000 *g* to remove any excess isopropanol. Each sample was eluted in 50 µL of Tris EDTA (TE) that had been warmed to 80 °C. At the addition of TE, each sample was briefly vortexed and then centrifuged at 10,000 *g* for 30 s to elute the DNA. The samples were then analyzed for their prophage content via qPCR.

To run the qPCR, 2 µL of DNA extracted from a phage lysate was used as the template in the 15 µL qPCR reaction that contained 1× Perfecta SYBR Green FastMix (Quanta Biosciences, Gaithersburg, MD, USA) and 300 nM of each of the forward and reverse primers targeting prophage 933W (gene *stx2a* forward primer: 5′-ATGTGGCCGGGTTCGTTAAT-3′; reverse primer: 5′-TGCTGTCCGTTGTCATGGAA-3′). The qPCR reaction was carried out with the following thermocycler conditions: initial denaturation and enzyme activation at 95 °C for 5 min, 40 cycles of denaturation at 95 °C for 30 s, annealing at 57 °C for 30 s, and extension at 72 °C for 30 s, followed by one cycle of 95 °C for 15 s, 57 °C for 15 s, and 95 °C for 15 s for the dissociation curve. Fluorescence signal was collected at the end of the extension step and of the ramping period of dissociation curve. Serial dilution of the genomic DNA of strain STEC was used to generate the standard curve ($R^2 > 0.99$).

5. Conclusions

With the rise in antibiotic resistance in bacterial pathogens, phage therapy presents a promising alternative for treatment. Importantly, though, many of pathogens contain prophages that not only are commonly the source of pathogenesis [3,6], but have also been shown to impact cell-virus communication systems [55]. Thus, as phage therapy advances, it will be important to investigate the impacts of exogenous phages on prophage induction. Here we have provided a first step towards such investigation with whole-genome sequencing and PCR-based quantification approaches that enable a genome-wide view and quantification of what prophages are induced under both exogenous phage-free and phage-rich environments. Future work should investigate the levels of the Shiga toxin under such scenarios, as well as the impacts of phage treatment plus stressors that can induce lysogens, such as antibiotics. Additionally, research from environmental phage-host systems provides invaluable insight into 'phage-host biology' in nature by showing that even in prophage-free bacteria exogenous phages are not always efficient at infecting [44,45]; important when considering phage candidates for therapy. Altogether, advancing knowledge of phage-prophage-host interactions should provide a baseline for engineering phages [56–58] as well as inform what phages to choose to make phage cocktails that can eradicate bacterial pathogens.

Supplementary Materials: The following are available online at http://www.mdpi.com/2079-6382/7/4/101/s1, Figure S1: Prophages in STEC (ATCC43895) and EDL933, and comparison between them; Figure S2: Liquid-based infection of STEC by phages p000v and p000y; Figure S3: Adsorption of p000v and p000y to ATCC43895 (STEC); Table S1: Comparison between ATCC43895 (STEC) and EDL933.

Author Contributions: C.H.-V., N.E.S., L.C., M.C.G., Y.-F.L., J.K.S., A.E.J., P.A., and A.H.-V. contributed to designing and executing experiments. C.H.-V. and D.R.V. contributed to most data analysis and figure/table generation, and with S.T.A. and M.B.S. to the writing of the manuscript. A.A.K. contributed to methods writing and sequencing efforts, as well to providing partial funding, along with Ohio State University.

Funding: Research reported in this publication was partly supported by the Infectious Diseases Institute Seed Grant program through Ohio State University (OSU) to M.B.S., an National Institutes of Health (NIH)/National Institute of Allergy and Infectious Diseases (NIAID) award # 1-T32-AI-112542 to C.H.-V. through the Infectious Disease Institute at OSU, as well as the Bill and Melinda Gates Foundation Grand Challenges Award OPP1139917 to Epibiome.

Acknowledgments: We thank Michelle Davison and Rebecca Lu at Epibiome for DNA extraction and sequencing, respectively, as well as Zachary Hobbs and Ryan Honaker from Epibiome for providing the phages. We also thank the following undergraduates for their help with initial phage–host optimizations: Alice L. Herneisen, Catherine E. Johnson, John M. Thomas, and Storm A. Mohn. We also thank OSU PhD rotation students Siavash Azari, Sravya Kovvali, and Yiwei Liu.

Conflicts of Interest: Epibiome is a phage therapy start-up company, but all data were independently analyzed at OSU and none of the authors from OSU have financial ties with Epibiome.

References

1. Penades, J.R.; Chen, J.; Quiles-Puchalt, N.; Carpena, N.; Novick, R.P. Bacteriophage-mediated spread of bacterial virulence genes. *Curr. Opin. Microbiol.* **2015**, *23*, 171–178. [CrossRef]
2. Feiner, R.; Argov, T.; Rabinovich, L.; Sigal, N.; Borovok, I.; Herskovits, A.A. A new perspective on lysogeny: Prophages as active regulatory switches of bacteria. *Nat. Rev. Microbiol.* **2015**, *13*, 641–650. [CrossRef]
3. Howard-Varona, C.; Hargreaves, K.R.; Abedon, S.T.; Sullivan, M.B. Lysogeny in nature: Mechanisms, impact and ecology of temperate phages. *ISME J.* **2017**, *11*, 1511–1520. [CrossRef]

4. Fortier, L.-C.; Sekulovic, O. Importance of prophages to evolution and virulence of bacterial pathogens. *Virulence* **2013**, *4*, 354–365. [CrossRef]
5. Davies, E.V.; Winstanley, C.; Fothergill, J.L.; James, C.E. The role of temperate bacteriophages in bacterial infection. *FEMS Microbiol. Lett.* **2016**, *363*, fnw015. [CrossRef]
6. Touchon, M.; Bernheim, A.; Rocha, E.P.C. Genetic and life-history traits associated with the distribution of prophages in bacteria. *ISME J.* **2016**, *10*, 2744–2754. [CrossRef]
7. Kaper, J.B.; O'Brien, A.D. Overview and Historical Perspectives. *Microbiol. Spectr.* **2014**, *2*. [CrossRef]
8. Brabban, A.D.; Hite, E.; Callaway, T.R. Evolution of foodborne pathogens via temperate bacteriophage-mediated gene transfer. *Foodborne Pathog. Dis.* **2005**, *2*, 287–303. [CrossRef]
9. Melton-Celsa, A.R. Shiga Toxin (Stx) Classification, Structure, and Function. *Microbiol. Spectr.* **2014**, *2*. [CrossRef]
10. Plunkett, G.; Rose, D.J.; Durfee, T.J.; Blattner, F.R. Sequence of Shiga Toxin 2 Phage 933W from *Escherichia coli* O157:H7: Shiga Toxin as a Phage Late-Gene Product. *J. Bacteriol.* **1999**, *181*, 1767–1778.
11. Perna, N.T.; Plunkett, G.; Burland, V.; Mau, B.; Glasner, J.D.; Rose, D.J.; Mayhew, G.F.; Evans, P.S.; Gregor, J.; Kirkpatrick, H.A.; et al. Genome sequence of enterohaemorrhagic *Escherichia coli* O157:H7. *Nature* **2001**, *409*, 529–533. [CrossRef]
12. Latif, H.; Li, H.J.; Charusanti, P.; Palsson, B.O.; Aziz, R.K. A gapless, unambiguous genome sequence of the enterohemorrhagic *Escherichia coli* O157:H7 strain EDL933. *Genome Announc.* **2014**, *2*. [CrossRef]
13. Colon, M.P.; Chakraborty, D.; Pevzner, Y.; Koudelka, G.B. Mechanisms that determine the differential stability of Stx(+) and Stx(−) lysogens. *Toxins (Basel)* **2016**, *8*, 96. [CrossRef]
14. Iversen, H.; L'Abee-Lund, T.M.; Aspholm, M.; Arnesen, L.P.S.; Lindback, T. Commensal *E. coli* Stx2 lysogens produce high levels of phages after spontaneous prophage induction. *Front. Cell. Infect. Microbiol.* **2015**, *5*, 5. [CrossRef]
15. Livny, J.; Friedman, D.I. Characterizing spontaneous induction of Stx encoding phages using a selectable reporter system. *Mol. Microbiol.* **2004**, *51*, 1691–1704. [CrossRef]
16. Herold, S.; Siebert, J.; Huber, A.; Schmidt, H. Global expression of prophage genes in *Escherichia coli* O157:H7 strain EDL933 in response to norfloxacin. *Antimicrob. Agents Chemother.* **2005**, *49*, 931–944. [CrossRef]
17. Asadulghani, M.; Ogura, Y.; Ooka, T.; Itoh, T.; Sawaguchi, A.; Iguchi, A.; Nakayama, K.; Hayashi, T. The defective prophage pool of *Escherichia coli* O157: Prophage–prophage interactions potentiate horizontal transfer of virulence determinants. *PLoS Pathog.* **2009**, *5*, e1000408. [CrossRef]
18. Herold, S.; Karch, H.; Schmidt, H. Shiga toxin-encoding bacteriophages–genomes in motion. *Int. J. Med. Microbiol.* **2004**, *294*, 115–121. [CrossRef]
19. Łoś, J.M.; Łoś, M.; Węgrzyn, G.; Węgrzyn, A. Differential efficiency of induction of various lambdoid prophages responsible for production of Shiga toxins in response to different induction agents. *Microb. Pathog.* **2009**, *47*, 289–298. [CrossRef]
20. Los, J.M.; Los, M.; Wegrzyn, G. Bacteriophages carrying Shiga toxin genes: Genomic variations, detection and potential treatment of pathogenic bacteria. *Future Microbiol.* **2011**, *6*, 909–924. [CrossRef]
21. Pacheco, A.R.; Sperandio, V. Shiga toxin in enterohemorrhagic *E. coli*: Regulation and novel anti-virulence strategies. *Front. Cell. Infect. Microbiol.* **2012**, *2*, 81. [CrossRef] [PubMed]
22. Obrig, T.G.; Karpman, D. Shiga toxin pathogenesis: Kidney complications and renal failure. *Curr. Top. Microbiol. Immunol.* **2012**, *357*, 105–136. [CrossRef] [PubMed]
23. Pruimboom-Brees, I.M.; Morgan, T.W.; Ackermann, M.R.; Nystrom, E.D.; Samuel, J.E.; Cornick, N.A.; Moon, H.W. Cattle lack vascular receptors for *Escherichia coli* O157:H7 Shiga toxins. *Proc. Natl. Acad. Sci. USA* **2000**, *97*, 10325–10329. [CrossRef] [PubMed]
24. Kruger, A.; Lucchesi, P.M.A. Shiga toxins and stx phages: Highly diverse entities. *Microbiology* **2015**, *161*, 451–462. [CrossRef] [PubMed]
25. Wagner, P.L.; Waldor, M.K. Bacteriophage control of bacterial virulence. *Infect. Immun.* **2002**, *70*, 3985–3993. [CrossRef] [PubMed]
26. Freedman, S.B.; Xie, J.; Neufeld, M.S.; Hamilton, W.L.; Hartling, L.; Tarr, P.I.; Nettel-Aguirre, A.; Chuck, A.; Lee, B.; Johnson, D.; et al. Shiga Toxin-Producing *Escherichia coli* Infection, Antibiotics, and Risk of Developing Hemolytic Uremic Syndrome: A Meta-analysis. *Clin. Infect. Dis.* **2016**, *62*, 1251–1258. [CrossRef] [PubMed]
27. Arndt, D.; Grant, J.R.; Marcu, A.; Sajed, T.; Pon, A.; Liang, Y.; Wishart, D.S. PHASTER: A better, faster version of the PHAST phage search tool. *Nucleic Acids Res.* **2016**, *44*, W16–W21. [CrossRef] [PubMed]

28. Hertel, R.; Rodríguez, D.P.; Hollensteiner, J.; Dietrich, S.; Leimbach, A.; Hoppert, M.; Liesegang, H.; Volland, S. Genome-based identification of active prophage regions by next generation sequencing in bacillus licheniformis DSM13. *PLoS ONE* **2015**, *10*, e0120759. [CrossRef] [PubMed]
29. Utter, B.; Deutsch, D.R.; Schuch, R.; Winer, B.Y.; Verratti, K.; Bishop-Lilly, K.; Sozhamannan, S.; Fischetti, V.A. Beyond the chromosome: The prevalence of unique extra-chromosomal bacteriophages with integrated virulence genes in pathogenic Staphylococcus aureus. *PLoS ONE* **2014**, *9*, e100502. [CrossRef] [PubMed]
30. Chen, F.; Wang, K.; Stewart, J.; Belas, R. Induction of multiple prophages from a marine bacterium: A genomic approach. *Appl. Environ. Microbiol.* **2006**, *72*, 4995–5001. [CrossRef] [PubMed]
31. Ogura, Y.; Mondal, S.I.; Islam, M.R.; Mako, T.; Arisawa, K.; Katsura, K.; Ooka, T.; Gotoh, Y.; Murase, K.; Ohnishi, M.; et al. The Shiga toxin 2 production level in enterohemorrhagic *Escherichia coli* O157:H7 is correlated with the subtypes of toxin-encoding phage. *Sci. Rep.* **2015**, *5*, 16663. [CrossRef] [PubMed]
32. Howard-Varona, C.; Vik, D.R.; Solonenko, N.E.; Gazitua, M.C.; Hobbs, Z.; Honaker, R.W.; Kinkhabwala, A.A.; Sullivan, M.B. Whole-genome sequence of phages p000v and p000y infecting the bacterial pathogen Shigatoxigenic *Escherichia coli*. *Press. Genome Announc.* **2018**.
33. Zhang, X.; McDaniel, A.D.; Wolf, L.E.; Keusch, G.T.; Waldor, M.K.; Acheson, D.W. Quinolone antibiotics induce Shiga toxin-encoding bacteriophages, toxin production, and death in mice. *J. Infect. Dis.* **2000**, *181*, 664–670. [CrossRef] [PubMed]
34. Los, J.M.; Los, M.; Wegrzyn, A.; Wegrzyn, G. Altruism of Shiga toxin-producing *Escherichia coli*: Recent hypothesis versus experimental results. *Front. Cell. Infect. Microbiol.* **2012**, *2*, 166. [CrossRef] [PubMed]
35. Tyler, J.S.; Beeri, K.; Reynolds, J.L.; Alteri, C.J.; Skinner, K.G.; Friedman, J.H.; Eaton, K.A.; Friedman, D.I. Prophage induction is enhanced and required for renal disease and lethality in an EHEC mouse model. *PLoS Pathog.* **2013**, *9*, e1003236. [CrossRef] [PubMed]
36. Tarr, P.I. *Escherichia coli* O157:H7: Clinical, diagnostic, and epidemiological aspects of human infection. *Clin. Infect. Dis.* **1995**, *20*, 1–10. [CrossRef] [PubMed]
37. Thorpe, C.M. Shiga toxin-producing *Escherichia coli* infection. *Clin. Infect. Dis.* **2004**, *38*, 1298–1303. [CrossRef] [PubMed]
38. Wagner, P.L.; Neely, M.N.; Zhang, X.; Acheson, D.W.; Waldor, M.K.; Friedman, D.I. Role for a phage promoter in Shiga toxin 2 expression from a pathogenic *Escherichia coli* strain. *J. Bacteriol.* **2001**, *183*, 2081–2085. [CrossRef] [PubMed]
39. Krysiak-Baltyn, K.; Martin, G.J.O.; Stickland, A.D.; Scales, P.J.; Gras, S.L. Computational models of populations of bacteria and lytic phage. *Crit. Rev. Microbiol.* **2016**, *42*, 942–968. [CrossRef] [PubMed]
40. Gamage, S.D.; Patton, A.K.; Strasser, J.E.; Chalk, C.L.; Weiss, A.A. Commensal bacteria influence *Escherichia coli* O157:H7 persistence and Shiga toxin production in the mouse intestine. *Infect. Immun.* **2006**, *74*, 1977–1983. [CrossRef] [PubMed]
41. Acheson, D.W.; Reidl, J.; Zhang, X.; Keusch, G.T.; Mekalanos, J.J.; Waldor, M.K. In vivo transduction with shiga toxin 1-encoding phage. *Infect. Immun.* **1998**, *66*, 4496–4498. [PubMed]
42. Schmidt, H.; Bielaszewska, M.; Karch, H. Transduction of enteric *Escherichia coli* isolates with a derivative of Shiga toxin 2-encoding bacteriophage phi3538 isolated from *Escherichia coli* O157:H7. *Appl. Environ. Microbiol.* **1999**, *65*, 3855–3861. [PubMed]
43. Doron, S.; Fedida, A.; Hernandez-Prieto, M.A.; Sabehi, G.; Karunker, I.; Stazic, D.; Feingersch, R.; Steglich, C.; Futschik, M.; Lindell, D.; et al. Transcriptome dynamics of a broad host-range cyanophage and its hosts. *ISME J.* **2016**, *10*, 1437–1455. [CrossRef] [PubMed]
44. Howard-Varona, C.; Roux, S.; Dore, H.; Solonenko, N.E.; Holmfeldt, K.; Markillie, L.M.; Orr, G.; Sullivan, M.B. Regulation of infection efficiency in a globally abundant marine Bacteriodetes virus. *ISME J.* **2017**, *11*, 284–295. [CrossRef] [PubMed]
45. Howard-Varona, C.; Hargreaves, K.R.; Solonenko, N.E.; Markillie, L.M.; White, R.A.; Brewer, H.M.; Ansong, C.; Orr, G.; Adkins, J.N.; Sullivan, M.B. Multiple mechanisms drive phage infection efficiency in nearly identical hosts. *ISME J.* **2018**, *12*, 1605–1618. [CrossRef] [PubMed]
46. Abedon, S.T. Phage therapy: Eco-physiological pharmacology. *Scientifica (Cairo)* **2014**, *2014*, 581639. [CrossRef] [PubMed]
47. Sarker, S.A.; Berger, B.; Deng, Y.; Kieser, S.; Foata, F.; Moine, D.; Descombes, P.; Sultana, S.; Huq, S.; Bardhan, P.K.; et al. Oral application of *Escherichia coli* bacteriophage: Safety tests in healthy and diarrheal children from Bangladesh. *Environ. Microbiol.* **2017**, *19*, 237–250. [CrossRef] [PubMed]

48. Merchant, N.; Lyons, E.; Goff, S.; Vaughn, M.; Ware, D.; Micklos, D.; Antin, P. The iPlant collaborative: Cyberinfrastructure for enabling data to discovery for the life sciences. *PLoS Biol.* **2016**, *14*, e1002342. [CrossRef] [PubMed]
49. Li, H.; Durbin, R. Fast and accurate short read alignment with Burrows-Wheeler transform. *Bioinformatics* **2009**, *25*, 1754–1760. [CrossRef] [PubMed]
50. Li, H.; Handsaker, B.; Wysoker, A.; Fennell, T.; Ruan, J.; Homer, N. The sequence alignment/map format and SAMtools. *Bioinformatics* **2009**, *25*, 2078–2079. [CrossRef] [PubMed]
51. Rabosky, D.L.; Grundler, M.; Anderson, C.; Title, P.; Shi, J.J.; Brown, J.W.; Huang, H.; Larson, J.G. BAMMtools: An R package for the analysis of evolutionary dynamics on phylogenetic trees. *Methods Ecol. Evol.* **2014**, *5*, 701–707. [CrossRef]
52. Quinlan, A.R.; Hall, I.M. BEDTools: A flexible suite of utilities for comparing genomic features. *Bioinformatics* **2010**, *26*, 841–842. [CrossRef] [PubMed]
53. Thorvaldsdottir, H.; Robinson, J.T.; Mesirov, J.P. Integrative genomics viewer (IGV): High-performance genomics data visualization and exploration. *Brief. Bioinform.* **2013**, *14*, 178–192. [CrossRef] [PubMed]
54. Krzywinski, M.I.; Schein, J.E.; Birol, I.; Connors, J.; Gascoyne, R.; Horsman, D.; Jones, S.J.; Marra, M.A. Circos: An information aesthetic for comparative genomics. *Genome Res.* **2009**. [CrossRef] [PubMed]
55. Erez, Z.; Steinberger-Levy, I.; Shamir, M.; Doron, S.; Stokar-Avihail, A.; Peleg, Y.; Melamed, S.; Leavitt, A.; Savidor, A.; Albeck, S.; et al. Communication between viruses guides lysis–lysogeny decisions. *Nature* **2017**, *541*, 488–493. [CrossRef] [PubMed]
56. Yoichi, M.; Abe, M.; Miyanaga, K.; Unno, H.; Tanji, Y. Alteration of tail fiber protein gp38 enables T2 phage to infect *Escherichia coli* O157:H7. *J. Biotechnol.* **2005**, *115*, 101–107. [CrossRef] [PubMed]
57. Yosef, I.; Goren, M.G.; Globus, R.; Molshanski-Mor, S.; Qimron, U. Extending the host range of bacteriophage particles for DNA transduction. *Mol. Cell* **2017**, *66*, 721.e3–728.e3. [CrossRef] [PubMed]
58. Roach, D.R.; Debarbieux, L. Phage therapy: Awakening a sleeping giant. *Emerg. Top. Life Sci.* **2017**, *1*, 93–103. [CrossRef]

antibiotics

MDPI

Case Report

Resolving Digital Staphylococcal Osteomyelitis Using Bacteriophage—A Case Report

Randolph Fish [1], Elizabeth Kutter [2,*], Daniel Bryan [2], Gordon Wheat [3] and Sarah Kuhl [4]

1 PhageBiotics Research Foundation and Grays Harbor Community Hospital, Aberdeen, WA 98520, USA; rcfish4@gmail.com
2 PhageBiotics Research Foundation, The Evergreen State College, 2700 Evergreen Parkway NW, Olympia, WA 98505, USA; dwbryan@gmail.com
3 PhageBiotics Research Foundation, Saint Peter Hospital Family Medicine Residency, Olympia, WA 98505, USA; gwheat12@gmail.com
4 VA Northern California, Muir Road, Martinez, CA 94553, USA; sarahkuhl52@gmail.com
* Correspondence: kutterb@evergreen.edu; Tel.: +1-360-867-6523

Received: 28 August 2018; Accepted: 19 September 2018; Published: 2 October 2018

Abstract: Infections involving diabetic foot ulcers (DFU) are a major public health problem and have a substantial negative impact on patient outcomes. Osteomyelitis in an ulcerated foot substantially increases the difficulty of successful treatment. While literature suggests that osteomyelitis in selected patients can sometimes be treated conservatively, with no, or minimal removal of bone, we do not yet have clear treatment guidelines and the standard treatment failure fallback remains amputation. The authors report on the successful treatment, with a long term follow up, of a 63 YO diabetic female with distal phalangeal osteomyelitis using bacteriophage, a form of treatment offering the potential for improved outcomes in this era of escalating antibiotic resistance and the increasingly recognized harms associated with antibiotic therapy.

Keywords: Bacteriophages; diabetic foot ulcer; osteomyelitis; phage therapy; *Staphylococcus aureus*

1. Introduction

Diabetic foot infections are a major public-health problem in the US, with high morbidity and long-term costs. We frequently see an impaired wound healing creating disability and amputations, especially in cases complicated by osteomyelitis; this is partly due to antibiotic resistance. The most common infective organism is *Staphylococcus aureus* [1]. Antibiotic treatment of the infection often has poor results and no clear treatment guidelines exist [2]. In one-quarter to one-third of diabetic osteomyelitis cases, the infection is localized in the bones of the toes [3,4].

When treatment for osteomyelitis fails, amputation is the usual option. Prolonged antibiotic-only treatment protocols are sometimes used to try and spare the limb, but are seldom successful [3,4]. Other proposed technologies for osteomyelitis treatment involve topical or local injection of biodegradable materials such as antibiotic-releasing bioactive bone filler materials, antibiotic impregnated collagen sponges or implanted poly-methyl-methacrylate beads [5–10]. Dillon injected antibiotics into the tissue surrounding the bone and used a pneumatic boot to reduce edema and increased the local circulation in ischemic feet and legs, reporting a 91.1% ulcer healing success rate and avoiding amputations in one study [11]. Additionally, under evaluation is the protein synthesis inhibitor fusidic acid, used in conjunction with other antibiotics [12]. However, no treatment has been widely adopted that can reliably obviate the need for amputation. This article demonstrates the successful use of *Staphylococcus aureus*-specific bacteriophage injected into a distal toe phalanx and surrounding tissue to resolve the soft tissue infection and osteomyelitis within the phalanx, leading to the complete resolution of the infected ulcer and the osteomyelitis. The patient was seen three years later for a separate problem and we were able to x-ray the original affected toe again, demonstrating that the osteomyelitis was indeed resolved. We have previously reported on a series of six other cases of diabetic toe infections treated with phage after appropriate antibiotics and debridement had failed. (See Discussion).

2. Materials and Methods

The phage used is a commercial preparation of staphylococcal phage Sb-1. It was isolated in 1977 for detailed characterization and clinical application from a long-used wound therapy cocktail at the Eliava Institute, located in Tbilisi, Georgia. It is a *Staphylococcus aureus*-specific relative of the phages approved by the FDA for dealing with *Listeria monocytogenes* in ready-to-eat foods. Over a decade ago, Sb-1 was completely sequenced and extensively studied under a special US Department of Health and Human Services Biotechnology Engagement Program (DHHS-BTEP) grant, so we can rule out this phage directly encoding either pathogenicity islands or known transduction mechanisms [13,14]. The phage used here is grown in minimal medium, column purified and sealed in sterile vials in 10-mL aliquots. It was brought into this country as part of a research agreement between the Eliava Institute Phage Production Center and the PhageBiotics Research Foundation, Olympia, Washington, for the purpose of this type of compassionate-use case study.

3. Clinical Treatment

The patient was a 63-year-old Caucasian female who presented three months after she developed an ulcer of the distal right second toe. Previous treatment under her primary care provider involved occasional use of topically-applied over-the-counter antibiotic ointment with normal hygiene. Her medical history was positive for type 2 diabetes mellitus, with associated diabetic neuropathy, COPD, hypertension, hyperlipidemia, anxiety disorder, and depression. Physical examination revealed a blood pressure of 160/74 mmHg, pulse of 65 bpm, respirations 20/min and a temperature of 97.4 °F. Pulses measured +1/5 at the dorsalis pedis artery (very weak but palpable) and +3/5 (normal) at the posterior tibial artery. Her legs demonstrated two mm pitting edema. She was morbidly obese, with a BMI of 41.48 and average blood glucose level of 223 mg/dl based on the HgA1c level. Orthopedically, the patient had a contraction deformity of all lesser toes, without bunion deformity. X-rays of the foot revealed osteolysis of the distal phalanx of the right second toe, read as positive for osteomyelitis.

On presentation, she demonstrated an edematous, erythematous, contracted right second mallet toe with ulceration at the distal tip measuring $0.6 \times 0.9 \times 0.3$ cm, probing to the bone (Figure 1). The ulcer bed was primarily fibrin and adipose tissue, with a spotting of red granular tissue. The margins were undermined and the ulcer was without odor. A tissue culture revealed methicillin-sensitive *Staphylococcus aureus*, sensitive to antibiotics except for penicillin.

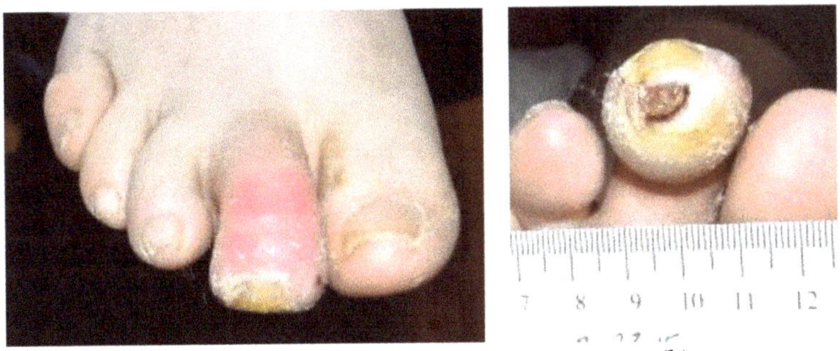

(a) (b)

Figure 1. Images of the patient's ulcer on presentation. (**a**) Top view of foot. (**b**) View of ulcer at the distal tip.

Treatment options presented to the patient included excision of the distal phalanx along with follow-up antibiotics for at least ten days, or a standard six-week intravenous antibiotic course, together with care of the toe ulcer. She refused both recommendations. After discussion of remaining possibilities, she accepted the use of bacteriophage, as she found it the least objectionable. The treatment consisted of injecting bacteriophage into the soft tissue surrounding the distal phalanx, using a standard injection pattern surrounding the base of the distal phalanx, and inserting directly into the bone as much as possible. The toe was also offloaded with a toe crest and the ulcer treated with standard good wound care. Antibiotics were not used initially.

A week into this treatment, the erythema seemed to be increasing, suggesting the infection was worsening. At this time she also started on levofloxacin 500 mg, to which the bacteria were sensitive. At day seven, we noted no changes in the amount or intensity of the erythema or reduction of edema, suggesting that the antibiotic was not helping, so it was discontinued. No further antibiotics were given.

4. Results

The injections began on 23 March 2015 and consisted of 0.7 cc of the highly purified Eliava Institute commercial staphylococcal bacteriophage once weekly for seven weeks, for a total of 4.9 cc. The ulcer healed and the injections were discontinued on 11 May 2015. Serial x-rays demonstrated the re-ossification of the distal phalanx over time (Figure 2). The erythema and edema slowly decreased over the period of treatment, and continued to decrease after the injection treatment was discontinued. The patient was officially discharged on 24 June 2015, but kindly returned for x-rays to follow the progression of the bone healing.

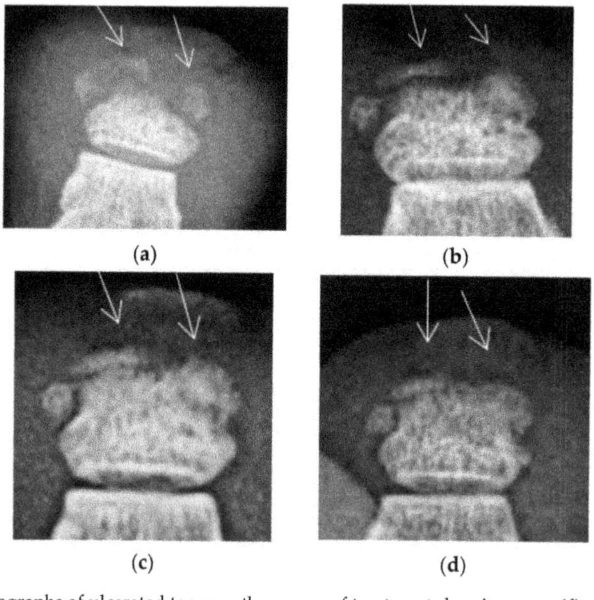

Figure 2. Radiographs of ulcerated toe over the course of treatment showing re-ossification of the toe. (**a**) Radiograph taken on 20 April 2015. (**b**) Radiograph taken on 24 July 2015. (**c**) Radiograph taken on 24 June 2015. (**d**) Radiograph taken on 22 September 2015. Note areas of re-ossification from photo (**a**) to photos (**b**,**c**) and (**d**) (arrows).

Long Term Results

Ulcer recurrence in the diabetic population is high, with literature estimating a 40% recurrence rate within one year of the original ulcer healing, and a rate of 60% within three years. Indeed, the best predictor of a recurrent ulcer in a diabetic patient is the occurrence of the first ulcer [15]. In the case of the above patient, she returned to the wound clinic twice in the two years following the resolution of the original ulcer with a single ulcer on a toe of each foot, although on a different toe on the right foot. In April of 2018, she returned a third time with an ulcer of the same toe as the original (Figure 3a), in the same location, although smaller and more superficial. The digital contraction was never resolved and the mallet toe continued weight-bearing beyond physiological limits, causing ulcer recurrence without any sign of recurrent osteomyelitis. This new ulcer was evaluated with x-ray looking for recurrent osteomyelitis. The radiologist read the film as free of osteomyelitis, (Figure 3b) and no further staph phage was used. The ulcer was resolved quickly by using a simple toe crest to stop the pressure on the tip of the toe (Figure 3c). The odd "notch" in the distal phalanx does suggest some original necrosis of the bone and is seen in the earlier films, but we note the re-ossification of the balance of the phalanx when compared to the photos taken three years earlier.

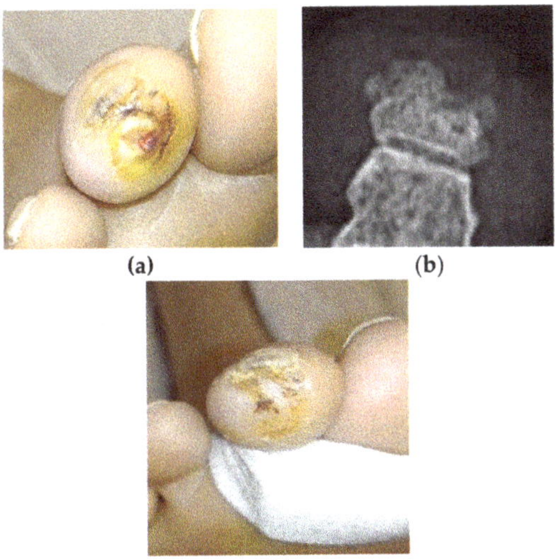

(a) (b)

(c)

Figure 3. Progression of healing of the ulcer. (**a**) Image of ulcer at the distal tip taken on 30 April 2018. (**b**) Radiograph taken on 30 April 2018. (**c**) Image of closed ulcer at the distal tip taken on 8 May 2018.

5. Discussion

Here we demonstrate the feasibility of adding bacteriophage to the standard care of diabetic foot ulcers with osteomyelitis, with the successful outcome suggesting the potential for long-term resolution of such infections with phage therapy. We have previously reported the successful treatment of a series of six cases of diabetic toe infection with clinical osteomyelitis [16]. In every case the soft tissue infection and osteomyelitis cleared rapidly. The wounds healed without recurrence indicating successful treatment using this single staph phage, with no further antibiotic therapy. A seventh successful case involved a seriously gangrenous toe salvaged with careful debridement plus the use of this phage to prevent infection.

This case focused on compassionate use of bacteriophage in a patient with Staphylococcal osteomyelitis who refused amputation and/or long-term antibiotics. Only after discussion of the usual treatment options and after the patient refused conventional treatment, bacteriophage treatment was offered as an alternative. After informed consent and following the principals of the Declaration of Helsinki, the treatment was carried out at her next seven weekly visits. Due to the fact that her treatments were carried out in a private physician's office, no IRB approval was available or necessary. It appears highly unlikely that the successful treatment of this osteomyelitis-complicated ulcer occurred as a result of the one week of levofloxacin early in the treatment, as much longer courses are generally needed for resolution of osteomyelitis if it resolves at all, especially when there is no clinical response to the antibiotic in the first week. The long term follow up also demonstrates that the osteomyelitis was truly resolved, as there was no evidence of disease when she returned three years later, even in the face of a second ulcer in the same location, and with resolution of the ulcer for the previous three years.

There are important microbiological and other properties of this genus of lytic staphylococcal phages that make them the ideal choice for studying phage treatment of chronic wounds. They have an unusually broad spectrum for various strains of *S. aureus*, and there is very little indication of resistance to these phages even after many years of clinical use across the former Soviet Union. Topical phages can penetrate deeply into infected areas even in the presence of poor circulation, as well as through the thick biofilms common in such chronic wounds. Well-studied phages that have repeatedly been shown to be obligatorily lytic through both classical microbiological techniques and genetic sequencing, such as the staphylococcal phage used in this study, do not pose a significant threat of directly transferring genetic factors that engender virulence or resistance.

The reintroduction of phage therapy into Western medicine has long been hampered by the difficulty in getting funding for clinical trials due to uncertainties in the regulatory frameworks for such live virus products and some challenging intellectual property issues. The patentability of many obvious phage applications is compromised by the decades-long clinical use of these lytic phages and the very common availability of the phage in the natural world. At least initially, public-private-academic-governmental partnerships appear to be badly needed in order to adequately evaluate the potential for phage therapy. We are working to improve the understanding of the microbiology of diabetic foot infections and the effects of the staph phage on subsequent wound healing by studying the microbial communities with metagenomics data, as discussed by Spichler et al. [17].

Despite innovative thinking and intensive research, comprehensive answers to the accelerating problem of antibiotic resistance still elude us. Further investigations, including controlled clinical trials, are urgently needed to assess the broader potential usefulness of phage as a complementary, narrow-spectrum topical antibacterial, either as an alternative to antibiotics or together with a shorter, more targeted antibiotic treatment. Thus the addition of phage therapy could help prevent antibiotic resistance, treat resistant infections and reduce antibiotic microbiome injury. Wound infections, especially diabetic foot infections with osteomyelitis, offer an excellent opportunity to demonstrate the potential of phage therapy to heal wounds more quickly and effectively while reducing the harms associated with antibiotics.

Author Contributions: Randolph Fish was the treating physician, and was assisted in the planning, writing and editing of this paper by the other listed authors

Funding: This research received no external funding; the phage preparation was simply added to the standard treatment protocol.

Acknowledgments: We are very grateful to Eliava Biopreparations, Inc. and to Mzia Kutateladze for the highly purified Staph bacteriophage Sb1 used in this treatment.

Conflicts of Interest: The authors declare no conflict of interest.

References

1. Oates, A.; Bowling, F.L.; Boulton, A.J.; McBain, A.J. A molecular and culture-based assessment of the microbial diversity of the diabetic chronic foot wounds and contralateral skin sites. *J. Clin. Microbiol.* **2012**, *50*, 2263–2271. [CrossRef] [PubMed]
2. Spellberg, B.; Lipsky, B. Systemic Antibiotic Therapy for Chronic Osteomyelitis in Adults. *Clin. Infect. Dis.* **2012**, *54*, 393–407. [CrossRef] [PubMed]

3. Yadlapalli, N.G.; Vaishnav, A.; Sheehan, P. Conservative Management of Diabetic Foot Ulcers Complicated by Osteomyelitis. *Wounds* **2002**, *14*, 31–35.
4. Levin, M.D. Management of the Diabetic Foot: Preventing Amputation. *South. Med. J.* **2002**, *95*, 10–20. [CrossRef] [PubMed]
5. Lázaro-Martínez, J.; Aragón-Sánchez, J.; García-Morales, E. Antibiotics Versus Conservative Surgery for Treating Diabetic Foot Osteomyelitis: A Randomized Comparative Trial. *Diabetes Care* **2014**, *37*, 789–795. [CrossRef] [PubMed]
6. Brin, Y.; Golenser, J.; Mizrahi, B.; Maoz, G.; Domb, A.J.; Peddada, S.; Tuvia, S.; Nyska, A.; Nyska, M. Treatment of Osteomyelitis in Rats by Injection of Degradable Polymer Releasing Gentamicin. *J. Control. Release* **2008**, *131*, 121–127. [CrossRef] [PubMed]
7. Shirtliff, M.; Calhoun, J.; Mader, J. Experimental Osteomyelitis Treatment with Antibiotic-Impregnated Hydroxyapatite. *Clin. Orthop. Relat. Res.* **2002**, *401*, 239–247. [CrossRef]
8. Beardmore, A.; Brooks, D.; Wenke, J.; Thomas, D. Effectiveness of Local Antibiotic Delivery with an Osteoinductive and Osteoconductive Bone-Graft Substitute. *J. Bone Jt. Surg.* **2005**, *87*, 107–112. [CrossRef] [PubMed]
9. Roukis, T.S. Lesser toe salvage with external fixation and autogenous bone grafting: A case series. *Foot Ankle Spec.* **2010**, *3*, 108–111. [CrossRef] [PubMed]
10. Lipsky, B.; Kuss, M.; Edmonds, M.; Reyzelman, A.; Segal, F. Topical Application of a Gentamicin-Collagen Sponge Combined with Systemic Antibiotic Therapy for the Treatment Diabetic Foot Infections of Moderate Severity. *J. Am. Podiatr. Med. Assoc.* **2012**, *102*, 223–232. [CrossRef] [PubMed]
11. Dillon, R. Successful Treatment of Osteomyelitis and Soft Tissue Infections by Local Antibiotic Injections and End-diastolic Pneumatic Compression Boot. *Ann. Surg.* **1986**, *204*, 643–649. [CrossRef] [PubMed]
12. Atkins, B.; Gottlieb, T. Fusidic acid in bone and joint infections. *Int. J. Antimicrob. Agents* **1999**, *12* (Suppl. 2), S79–S93. [CrossRef]
13. Kvachadze, L.; Balarjishvili, N.; Meskhi, T.; Tevdoradze, E.; Skhirtladze, N.; Pataridze, T.; Adamia, R.; Topuria, T.; Kutter, E.; Rohde, C.; et al. Evaluation of lytic activity of staphylococcal bacteriophage Sb-1 against freshly isolated clinical pathogens. *Microb. Biotechnol.* **2011**, *4*, 643–650. [CrossRef] [PubMed]
14. Lobocka, M.; Hejnowicz, M.; Dabrowski, K.; Gozdek, A.; Kosakowski, J.; Witkowska, M.; Ulatowska, M.I.; Weber-Dąbrowska, B.; Kwiatek, M.; Parasion, S.; et al. Genomics of staphylocollal Twort-like phages-potential therapeutics of the post-antibiotic era. *Adv. Virus Res.* **2012**, *83*, 143–216. [CrossRef] [PubMed]
15. Armstrong, D.; Boulton, A.; Bus, A. Diabetic Foot Ulcers and Their Recurrence. *NEJM* **2017**, *376*, 2367–2375. [CrossRef] [PubMed]
16. Fish, R.; Kutter, E.; Wheat, G.; Blasdel, B.; Kutateladze, M.; Kuhl, S. Bacteriophage Treatment of Intransient Diabetic Toe Ulcers-A Case Series. *J. Wound Care* **2016**, *25*, S27–S33. [CrossRef] [PubMed]
17. Spichler, A.; Hurwitz, B.L.; Armstrong, D.G.; Lipsky, B.A. Microbiology of diabetic foot infection: From Louis Pasteur to 'crime scene investigation'. *BMC Med.* **2015**, *13*, 2. [CrossRef] [PubMed]

antibiotics

MDPI

Article

Exploring the Effect of Phage Therapy in Preventing *Vibrio anguillarum* Infections in Cod and Turbot Larvae

Nanna Rørbo [1], Anita Rønneseth [2], Panos G. Kalatzis [1,3], Bastian Barker Rasmussen [4], Kirsten Engell-Sørensen [5], Hans Petter Kleppen [6], Heidrun Inger Wergeland [2], Lone Gram [4] and Mathias Middelboe [1,*]

1 Marine Biological Section, University of Copenhagen, 3000 Helsingør, Denmark; nanna_ir@hotmail.com (N.R.); panos.kalatzis@bio.ku.dk (P.G.K.)
2 Department of Biology, University of Bergen, 5020 Bergen, Norway; anita.ronneseth@uib.no (A.R.); heidrun.wergeland@uib.no (H.I.W.)
3 Institute of Marine Biology, Biotechnology and Aquaculture, Hellenic Centre for Marine Research, 71003 Heraklion, Greece
4 Department of Biotechnology and Biomedicine, Technical University of Denmark, 2800 Kongens Lyngby, Denmark; bbara@bio.dtu.dk (B.B.R.); gram@bio.dtu.dk (L.G.)
5 Fishlab, 8270 Højbjerg, Denmark; kes@fishlab.dk
6 ACD Pharmaceuticals AS, 8376 Leknes, Norway; hans.kleppen@acdpharma.com
* Correspondence: mmiddelboe@bio.ku.dk; Tel.: +45-3532-1991

Received: 30 December 2017; Accepted: 10 May 2018; Published: 16 May 2018

Abstract: The aquaculture industry is suffering from losses associated with bacterial infections by opportunistic pathogens. *Vibrio anguillarum* is one of the most important pathogens, causing vibriosis in fish and shellfish cultures leading to high mortalities and economic losses. Bacterial resistance to antibiotics and inefficient vaccination at the larval stage of fish emphasizes the need for novel approaches, and phage therapy for controlling *Vibrio* pathogens has gained interest in the past few years. In this study, we examined the potential of the broad-host-range phage KVP40 to control four different *V. anguillarum* strains in Atlantic cod (*Gadus morhua* L.) and turbot (*Scophthalmus maximus* L.) larvae. We examined larval mortality and abundance of bacteria and phages. Phage KVP40 was able to reduce and/or delay the mortality of the cod and turbot larvae challenged with *V. anguillarum*. However, growth of other pathogenic bacteria naturally occurring on the fish eggs prior to our experiment caused mortality of the larvae in the unchallenged control groups. Interestingly, the broad-spectrum phage KVP40 was able to reduce mortality in these groups, compared to the nonchallenge control groups not treated with phage KVP40, demonstrating that the phage could also reduce mortality imposed by the background population of pathogens. Overall, phage-mediated reduction in mortality of cod and turbot larvae in experimental challenge assays with *V. anguillarum* pathogens suggested that application of broad-host-range phages can reduce *Vibrio*-induced mortality in turbot and cod larvae, emphasizing that phage therapy is a promising alternative to traditional treatment of vibriosis in marine aquaculture.

Keywords: *Vibrio anguillarum*; phage therapy; aquaculture; fish larvae; challenge trials

1. Introduction

Vibrionaceae is a genetic and metabolic diverse family of heterotrophic bacteria which are widespread in aquatic environments around the world [1]. Several vibrios are able to infect a wide range of aquatic animals and constitute therefore a large problem in aquaculture [2]. One of the most important is *Vibrio anguillarum*, which causes the disease vibriosis and is responsible for large-scale losses in the aquaculture industry [3,4]. Chemotherapy against vibriosis is associated with a major concern due to the risk of antibiotic-resistance developing in the pathogenic bacteria [5]. Vaccines against vibrio have been successful in preventing disease [6,7], however, they are often not useful at the larval stage, as the immune system is not fully developed. Therefore, alternative methods for the control and treatment of *V. anguillarum* infections in fish larvae and fry are needed. The use of bacteriophages (phages) has been explored in several studies as a treatment of pathogens in aquaculture [4,8–13]. Pereira et al. [4] and Mateus et al. [11] did in vitro assays with phages infecting different bacteria responsible for the diseases vibriosis and furunculosis and showed that both single-phage suspensions and phage cocktails could inactivate the bacteria [4,11]. However, often regrowth of phage tolerant bacteria was observed within 24 h after phage treatment [11,13]. Phage addition to shrimp larvae infected with *V. harveyi* caused a reduction in the pathogen load and significantly increased shrimp survival compared to untreated controls groups as well as parallel treatments with antibiotics [8,9]. Another study on zebrafish larvae infected with *V. anguillarum* [12] also found significantly enhanced larvae survival after phage addition. Successful phage treatment in Atlantic salmon (*Salmo salar* L.) infected with *V. anguillarum* strain PF4 was found for phage CHOED, resulting in complete elimination of pathogen-induced mortality when phages were added at a high multiplicity of infection [10]. Together, the previous experimental approaches demonstrate that phage therapy can be a feasible alternative method to control specific *Vibrio* pathogens in aquaculture. However, the use of phages is complicated by the fact that multiple strains of the *Vibrio* pathogens with different phage susceptibility patterns may coexist in aquaculture environments [14]. The implications of strain diversity for the efficiency of phage control may be overcome either by combining several phages which target a broad range of pathogenic hosts, or to use a broad-host-range phage which can infect multiple strains within a given species or even multiple species [15]. The phage KVP40 represents a broad-host-range phage which infects at least eight species of *Vibrio* sp. (*V. parahaemolyticus*, *V. alginolyticus*, *V. natriegens*, *V. cholerae*, *V. mimicus*, *V. anguillarum*, *V. splendidus*, and *V. fluvialis*) and one *Photobacterium* sp. (*P. leignathi*) [16]. All of these species contain a 26-kDa outer membrane protein named OmpK, which is a receptor for phage KVP40 [17].

The application of phages for controlling pathogens may be hampered by the development of phage resistance in the bacteria [18], and several mechanisms have been described in *V. anguillarum* which can eliminate or reduce bacterial sensitivity to phages and thus limit the efficiency and duration of phage control [19].

The aim of this study was to examine the effect of phage KVP40 on the survival of turbot and cod larvae challenged with four different *V. anguillarum* strains. Larval mortality and abundance of bacteria and phages were quantified to determine the potential of using phage KVP40 to control *V. anguillarum* infections during the early larval stage. In general, phage KVP40 was able to reduce or delay the mortality of both turbot and cod larvae in all the challenge trials and reduce larval mortality imposed by the background population of pathogens.

The results demonstrated that phage KVP40 reduced the mortality imposed by the added pathogens as well as other *Vibrio* pathogens already present in the environment during the initial 1–4 days of the experiment, emphasizing the potential of using phages to reduce turbot and cod mortality at the larval stage.

2. Results

2.1. Phage Effect on Turbot Mortality in Vibrio Challenge Trials

2.1.1. Turbot Challenge Trial 1

In general, larval mortality was high in all treatments, including the nonchallenged controls where a maximum mortality of 86% (i.e., 103 dead larvae out of 120) was found (Figure 1), indicating that the eggs were associated with unknown bacterial pathogens prior to the challenge trial. Challenging the turbot eggs with *V. anguillarum* resulted in higher mortalities for all four strains (Figure 1), emphasizing that the added *V. anguillarum* pathogens increased larval mortality. Strain PF430-3 was the most virulent of the four strains, with 100% larval mortality after 3 days, whereas strains PF7, 90-11-286, and 4299 caused 97%–100% mortality after 4 and 5 days of challenge. Subsequent quantification of the abundance of colony forming bacteria in the water used for transportation of the fish eggs confirmed the presence of a microbial community associated with the eggs (see Section 2.5).

Despite the presence of other pathogen communities associated with the eggs/larvae, addition of phage KVP40 had a significant positive effect on larval survival in all the challenge treatments during all or part of the trials. When challenged with strain PF430-3, the maximum relative reduction in mortality was 29% ($p < 0.05$) one day after phage addition (Figure 1a; Table 1). The delay in mortality only lasted for 3 days, and the mortality reached almost 100% mortality at day 5 (Figure 1a). When challenged with strain PF7 or strain 90-11-286 (Figure 1b,c), the maximum phage-induced reduction in mortality was 47% obtained 1 and 2 days (Table 1), respectively, after addition of KVP40 and a significant effect of the phage on mortality was observed for 3–4 days ($p < 0.05$). The effect of phage addition was largest in the treatment group with strain 4299, where the larval mortality remained below 66% throughout the 8-day trial, corresponding to an average of 36% reduction in larval mortality compared to larvae challenged with *V. anguillarum* ($p < 0.05$) (Figure 1d).

Interestingly, larval mortality in the KVP40 controls (addition of phage but not *V. anguillarum*) showed the lowest larval mortality, reaching 65% at day 4 and remaining at that level (Figure 1). This significant reduction in mortality compared to the nonchallenged control (i.e., 86% mortality in larvae not exposed to *V. anguillarum* or phage) suggested that phage KVP40 was able to control part of the unknown pathogen community, thereby increasing the larval survival. This was later confirmed by analysis of phage susceptibility of bacteria initially associated with the eggs (see Section 2.5 below).

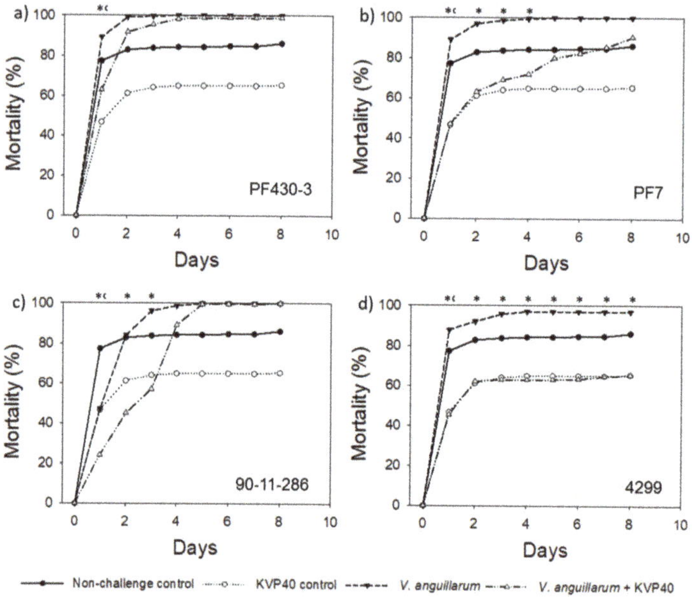

Figure 1. Cumulative percent mortality over time in turbot challenge trial 1: (**a**) strain PF430-3; (**b**) strain PF7; (**c**) strain 90-11-286; (**d**) strain 4299. Significant difference in mortality between cultures "*V. anguillarum*" and "*V. anguillarum* + KVP40" for individual time points is indicated by *. Significant difference in mortality between cultures "Nonchallenge control" and "KVP40 control" is indicated by ^{c}.

Table 1. Overview of the percent reduction in mortality caused by phage KVP40 addition in the four experiments. The maximum relative reduction and reduction at the end of the experiment (final) is shown.

	Relative Reduction * in Larval Mortality in the Presence of Phages (%)							
	Turbot Challenge Trial				Cod Challenge Trial			
V. anguillarum Strains	1		2		1		2	
	Max.	Final	Max.	Final	Max.	Final	Max.	Final
PF430-3	29	N/S [1]	60	N/S [1]	79	N/S [1]	86	N/ [1]
PF7	47	N/S [1]	53	N/S [1]	75	43	59	32
90-11-286	47	N/S [1]	92	N/S [1]	−119	N/S [1]	49	N/S [1]
4299	48	33	45	N/S [1]	N/D [2]	N/D [2]	82	72

* The relative reduction in mortality is calculated as difference in mortality between *V. anguillarum* and *V. anguillarum* + phage treatment, divided by the mortality in the *V. anguillarum* treatment. [1] N/S: not significant, [2] N/D: not determined.

2.1.2. Turbot Challenge Trial 2

The relatively high fraction of low-quality eggs and high mortality in the control group led us to repeat the challenge experiments in an attempt to optimize the egg quality and in order to verify the indications of positive effects of phages for larval mortality in replicate experiments.

Also in the second challenge trial with turbot larvae, a high mortality (71%) was observed in the nonchallenged control groups after 5 days (Figure 2), indicating pathogenic effects of the bacterial background community in the turbot eggs. In contrast to turbot challenge trial 1, the mortality caused by the background bacteria was not observed immediately, and mortality in the control groups gradually increased during the first 4 days, indicating growth of the pathogenic bacteria. Addition of *V. anguillarum* strains increased larval mortality in all four treatments, resulting in mortalities between 72% and 98% after 4–5 days of incubation. As in turbot challenge trial 1, addition of phage KVP40 had significant positive effects on the larval survival. However, in this case, the phage addition delayed the mortality by 2–4 days relative to the treatment with *V. anguillarum* alone.

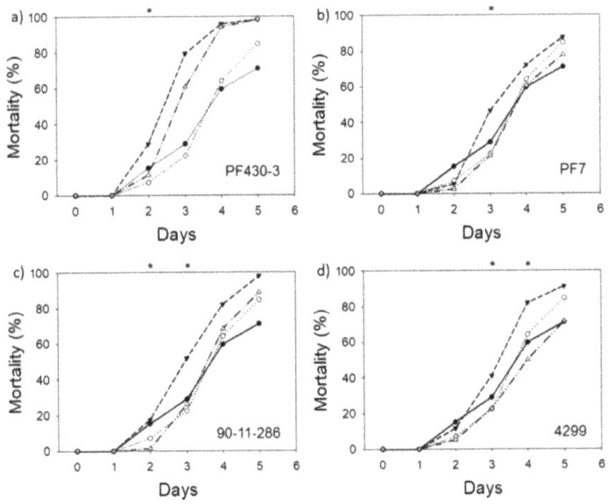

Figure 2. Cumulative percent mortality over time in turbot challenge trial 2: (**a**) strain PF430-3; (**b**) strain PF7; (**c**) strain 90-11-286; (**d**) strain 4299. Significant difference in mortality between cultures "*V. anguillarum*" and "*V. anguillarum* + KVP40" for individual time points is indicated by *.

When challenged with strain PF430-3, the addition of phages reduced mortality from 29% to 11% 2 days after phage addition (Figure 2a), corresponding to a maximum phage-mediated reduction in mortality of 60% ($p < 0.05$, Table 1). The delay in mortality lasted until day 4, where mortality approached 100% mortality as in the treatment without phage (Figure 2a). Phage addition to the larvae challenged with strain PF7 and strain 90-11-286 resulted in a significant 3-day delay in mortality with a maximum reduction in mortality of 53% and 92%, respectively, after 2–3 days relative to the larvae challenged with *V. anguillarum* alone ($p < 0.05$ (Figure 2b,c; Table 1). As in the turbot challenge trial 1, the larvae challenged with strain 4299 were best protected by phage addition, with a maximum relative reduction in mortality of 45% ($p < 0.05$) obtained 3 days after phage addition (Table 1), and a continued reduction in larval mortality of 22% relative to the larvae challenged with bacteria alone throughout the experiment (Figure 2d).

2.2. *Abundance of Bacteria and Phages in Turbot Challenge Trial 2*

In all the treatments in turbot challenge trial 2, the total count of colony forming bacteria (CFU) increased exponentially over time for the first 2–4 days (Figure 3).

The number of infective KVP40 phages increased about 100-fold reaching $1–5 \times 10^{10}$ PFU mL^{-1} in all the treatment groups where KVP40 was added, with no significant differences between cultures with and without the addition of *Vibrio* pathogens. This indicated that the background bacteria supported phage proliferation and that addition of *V. anguillarum* only had a minor effect on phage production.

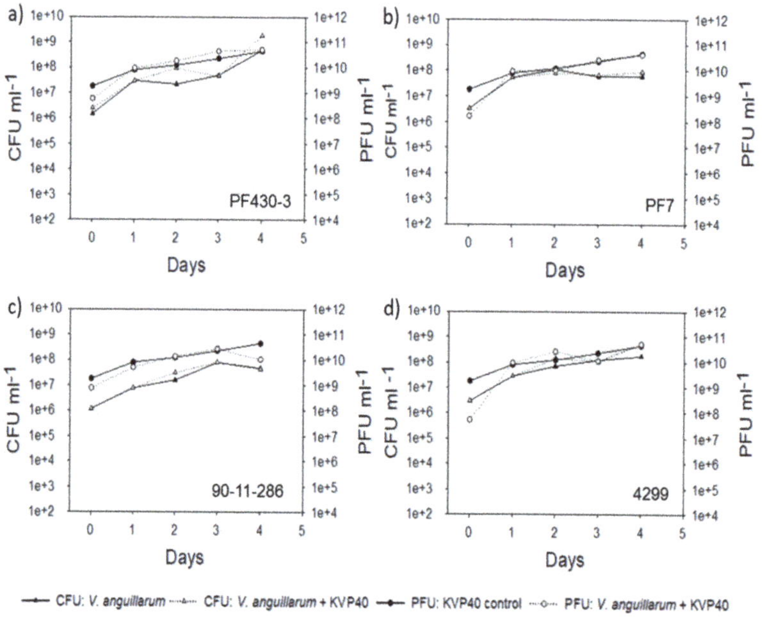

Figure 3. Bacterial abundance (CFU mL^{-1}) and phage abundance (PFU mL^{-1}) in turbot challenge trial 2: (**a**) strain PF430-3; (**b**) strain PF7; (**c**) strain 90-11-286; (**d**) strain 4299.

2.3. *Phage Effect on Cod Mortality in Vibrio Challenge Trials*

2.3.1. Cod Challenge Trial 1

The cod larvae mortality in the nonchallenged controls remained low throughout the trial (<10%) (Figure 4), and the addition of *Vibrio anguillarum* strains increased mortality significantly (Figure 4).

Strain PF430-3 and strain 90-11-286 increased mortality to 82% and 78%, respectively, after 11 days (Figure 4a,c), whereas the mortality was 41% in the treatment with strain PF7 (Figure 4b).

The addition of phage KVP40 had significant positive effects on larval survival in the larvae exposed to strain PF430-3 and strain PF7. For strain PF430-3, the mortality was reduced from 24% to 5% in the phage added cultures after 5 days, corresponding to maximal relative reduction in mortality by phage KVP40 of 79% compared to the larvae only challenged with *V. anguillarum* ($p < 0.05$; Table 1). The significant phage-induced reduction in mortality lasted to day 8 (Figure 4a). Phage KVP40 addition to strain PF7 reduced relative larval mortality by 75% compared to the larvae only challenged with *V. anguillarum* ($p < 0.05$) after 8 days (Table 1), and the significant phage-mediated reduction in mortality remained throughout the 11-day trial (Figure 4b). Surprisingly, the addition of phage KVP40 increased larval mortality significantly in the cultures challenged with strain 90-11-286 with a maximum increase in mortality of 119 ($p < 0.05$) reached at day 6 (Figure 4c; Table 1). The negative effect of phage addition was significant from day 5 to day 10, with the mortality reaching 100% in the phage treated cultures at day 11.

Despite the low mortality in the nonchallenged control treatment, the reduced larval mortality in the phage KVP40 controls (addition of phage but not *V. anguillarum*) (<7%) compared with the nonchallenged control group without phages again indicated a positive effect of the phages in reducing the original pathogenic bacterial load in the trials.

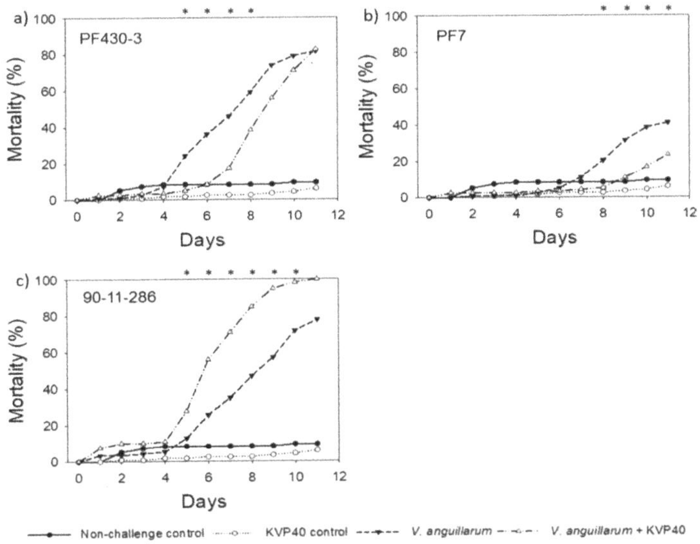

Figure 4. Cumulative percent mortality over time in cod challenge trial 1: (**a**) strain PF430-3; (**b**) strain PF7; (**c**) strain 90-11-286. Significant difference in mortality between cultures "*V. anguillarum*" and "*V. anguillarum* + KVP40" for individual time points is indicated by *.

2.3.2. Cod Challenge Trial 2

As for the turbot experiments, the challenge trials with cod were repeated to examine the reproducibility of the first results using a new batch of eggs. The second challenge trial with cod larvae confirmed the high virulence of strains PF430-3 and 90-11-286 obtained in cod challenge trial 1, whereas strain PF7 caused less mortality in cod challenge trial 2. Strain 4299 was not very virulent to the cod larvae (Figure 5). A gradual increase in mortality was observed in larvae challenged with strains PF430-3, PF7, and 90-11-286, which reached mortalities of 74% to 91% after 11 days post challenge (Figure 5a–c). Challenge with strain 4299 did not increase mortality compared to the nonchallenged control level, suggesting that this strain had very low virulence to cod (Figure 5d). The nonchallenged control showed an increase in mortality from 5% to 15% between days 2 and 3, followed by a more gradual increase to 35% mortality at day 11 (Figure 5).

Addition of phage KVP40 had a significant positive effect on cod larvae survival in all the treatments (Table 1). In the larvae challenged with strain PF430-3, phage addition kept larval mortality below 27% for 6 days, with a maximum reduction in mortality of 86% ($p < 0.05$) obtained 4 days after phage addition (Table 1). The reduced mortality lasted from day 2 to day 9, and after day 10 the mortality reached almost the same level as in the cultures without phages (Figure 5a). When challenged with strain PF7, the maximal effect of phage addition was a reduction in mortality of 59% ($p < 0.05$) obtained 6 days after phage addition (Table 1). The delay in mortality lasted throughout the trial, with the difference being significant from day 5 and onwards (Figure 5b). In the treatments challenged with strain 90-11-286, the maximal reduction in mortality was 49% ($p < 0.05$) obtained 6 days after phage addition (Table 1). The mortality then increased but remained below the nonphage treated group throughout the experiment (Figure 5c). Phage KVP40 very efficiently reduced mortality of larvae challenged with strain 4299, with a maximum reduction of 82% ($p < 0.05$) after 5 days (Table 1), and a significant reduction in mortality (mortality always < 12%) throughout the trial (Figure 5d).

The relatively high initial mortality in the nonchallenged control from day 1 to day 3 compared with corresponding nonchallenge control group in cod challenge trial 1, and compared with the lower and more gradual increase in mortality in the group challenged with strain 4299, suggested the presence of a high fraction of low-quality eggs in this specific control group. As in the previous trials, the phage-added controls showed a lower mortality than in the nonchallenged controls, again suggesting a positive effect of phage KVP40 in controlling other pathogens growing up during the trials (Figure 5).

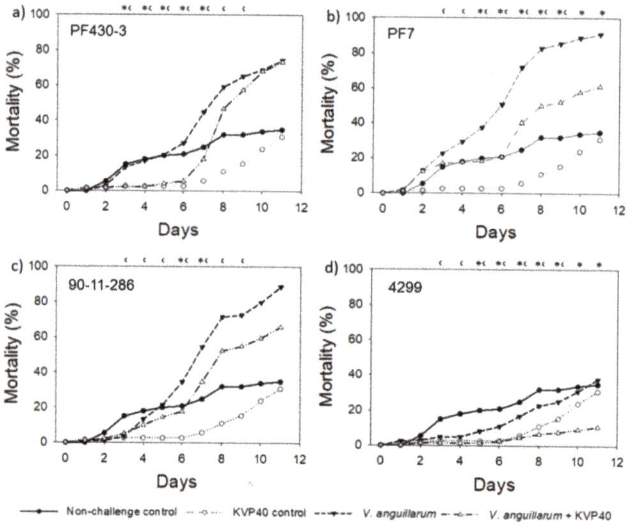

Figure 5. Cumulative percent mortality over time in cod challenge trial 2: (**a**) strain PF430-3; (**b**) strain PF7; (**c**) strain 90-11-286; (**d**) strain 4299. Significant difference in mortality between cultures "*V. anguillarum*" and "*V. anguillarum* + KVP40" for individual time points is indicated by *. Significant difference in mortality between cultures "Nonchallenge control" and "KVP40 control" is indicated by ^{c}.

2.4. Abundance of Bacteria and Phages in Cod Challenge Trials

2.4.1. Cod Challenge Trial 1

The total abundance of colony forming microorganisms increased approximately 10-fold in all *Vibrio* challenged larval groups from approx. 10^5 to 10^6 CFU mL^{-1} (Figure 6). Addition of phages only reduced the bacterial load in the strain PF7 challenged larval group and only during the first 2 days (Figure 6b). In contrast to this, total CFU counts increased after addition of phage KVP40 in larval groups challenged with strain PF430-3 and strain 90-11-286. Especially in the challenge with strain 90-11-286, a > 10-fold increase in colony forming bacteria was observed (Figure 6c) in accordance with the increased larval mortality in this treatment (Figure 4c). The phage abundance was approximately 10^7 PFU mL^{-1} in all phage-added treatments and remained stable during the 4 days when PFU was measured.

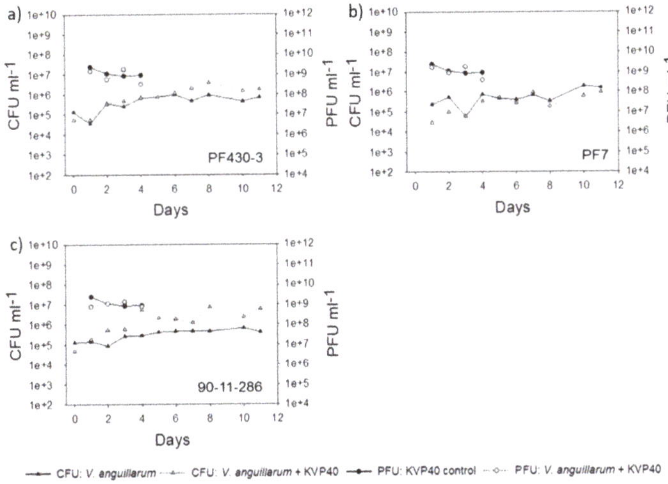

Figure 6. Bacterial abundance (CFU mL^{-1}) and phage abundance (PFU mL^{-1}) in cod challenge trial 1: (**a**) strain PF430-3; (**b**) strain PF7; (**c**) strain 90-11-286.

2.4.2. Cod Challenge Trial 2

The *V. anguillarum* load was approximately 10-fold higher in the second than in the first cod challenge trial and the CFU counts were approximately 10^6 CFU mL^{-1} in the *Vibrio* challenged groups (Figure 7).

In all the groups, addition of phage KVP40 reduced the bacterial counts significantly from day 0. In the groups challenged with strain PF430-3 and strain PF7, a significant phage-mediated reduction (approximately 1 log reduction) in the *V. anguillarum* pathogens was maintained for the first 8–9 days, followed by an increase in total CFU which then reached values close to the bacteria-alone group at day 11 (Figure 7a,b). For the group challenged with strain 90-11-286, phage reduction of the *Vibrio* pathogen was rather short. After 3 days, the bacterial abundance had reached the same level as in the bacteria-only group (Figure 7c). In the group challenged with strain 4299, the addition of phage KVP40 caused a 100-fold reduction in total CFU counts, indicating a strong phage control of the pathogen. However, after day 8, total bacterial cell counts increased 100-fold and reached numbers similar to the group without phage (Figure 7d). Phages were added at an initial concentration of 1.75×10^9 PFU mL^{-1} and the abundance of phage remained stable throughout the trial, both in the absence and presence of the *Vibrio* hosts (Figure 7).

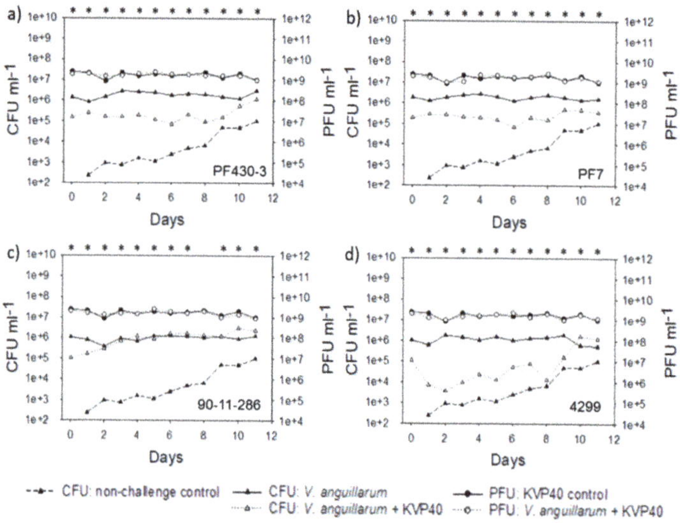

Figure 7. Bacterial abundance (CFU mL^{-1}) and phage abundance (PFU mL^{-1}) in cod challenge trial 2: (**a**) strain PF430-3; (**b**) strain PF7; (**c**) strain 90-11-286; (**d**) strain 4299. Significant difference in CFU between cultures "CFU: *V. anguillarum*" and "CFU: *V. anguillarum* + KVP40" for individual time points is indicated by *.

2.5. Abundance and Phage KVP40 Susceptibility of Bacterial Background Communities Associated with the Turbot Eggs

During the second turbot trial, the abundance of colony-forming bacteria in water used for transportation of the fish eggs was determined to shed light on the observed positive effect of phage KVP40 on unchallenged control groups. Different general and *Vibrio*-promoting growth media were used. In all the experiments, there was a high load of bacteria associated with the eggs, and a general increase in their abundance over time was found (Table 2). The high abundance of colonies growing on TCBS plates (up to >10^8 CFU mL^{-1}) indicated that a large fraction of these background communities were presumptive *Vibrio* or *Vibrio*-related species.

Table 2. Abundance of the bacterial background community (CFU mL^{-1}) associated with the fish eggs, in turbot challenge trial 2, and cultured on different media. Day 0: water the eggs were transported in for 24 h; Day 11: water in the wells of the live nonchallenged larvae.

Growth Substrate	Day 0 (CFU mL^{-1})	Day 11 (CFU mL^{-1})
LB media	2×10^7	9.39×10^6
TCBS media	2×10^6	1.5×10^8
Marine agar	N/D [1]	2.89×10^8

[1] N/D: not determined.

The susceptibility to phage KVP40 was tested in 40 isolates obtained from the water containing the turbot eggs during transportation used for challenge trial 1 by quantification of the growth reduction relative to a control culture without phage KVP40 (Figure 8). The results showed that 35 out of 40 isolated showed a growth reduction, indicating that the majority of the colony-forming cells originating from the water used for transporting the eggs were susceptible to phage KVP40.

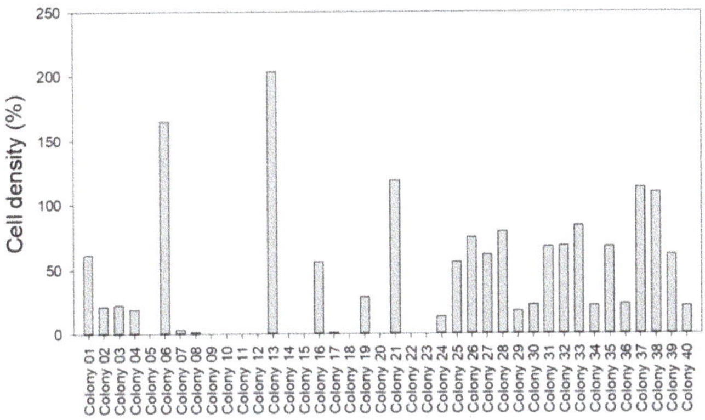

Figure 8. Quantification of phage KVP40-induced inhibition/promotion of cell growth in cultures of bacteria isolated from water used for transport of eggs used in turbot challenge trial 1. Phage-induced growth inhibition/promotion was determined as the percent cell density in cultures added phage KVP40 relative to control cultures without phage KVP40 (100%) after 3 h incubation.

3. Discussion

In general, the addition of phage KVP40 reduced or delayed the mortality of turbot and cod larvae challenged with *V. anguillarum*, with the largest effect observed for strain 4299, where the relative turbot and cod mortality was reduced by 22–33% and 72%, respectively, by the end of the experiment. In most of the challenges, the positive effect of phage KVP40 addition on larval survival was maintained throughout the incubation period. However, incubation with strain PF430-3 showed a temporary effect of phage addition on mortality and larval mortality reached the same level as in the bacterial challenges (without phage) after 4–10 days. Since the phage was maintained in high concentrations throughout the experiment, it is likely that strain PF430-3 was protected against infection, which supports previous observations that strain PF430-3 can reduce its susceptibility to phage KVP40 by forming aggregates or biofilm, creating spatial refuges [20].

In addition to the specific *V. anguillarum* pathogens, other pathogens already associated with the fish eggs prior to the experiments were present in the experiments. This allowed an assessment of the effects of phages on both the mortality caused by the *V. anguillarum* strains and the mortality imposed by the natural background pathogen communities. The decrease in mortality recorded for all the phage controls (without *V. anguillarum*) compared to the nonchallenge controls (without phage and *V. anguillarum*) demonstrated a strong effect of phage KVP40 on the initial bacterial pathogen communities associated with the eggs. This was supported by the observation that >85% of the isolated colonies originating from the background bacterial community were susceptible to phage KVP40.

Despite the large fraction of phage susceptible strains, the bacterial abundance increased in all the incubations over time, and only in cod challenge trial 2 did addition of phage KVP40 reduce the bacterial abundance for multiple days. This suggested that during the experiment, pathogens that were not infected by KVP40 (i.e., non-*Vibrio* pathogens and possibly phage-resistant *V. anguillarum* strains) replaced the phage susceptible strains, and thus were the main cause of mortality in the experiments. This was supported by the increased effects of phages on mortality in cod challenge trial 2, where the eggs were pretreated with 25% glutaraldehyde. These results emphasized that the growth of other pathogens than *V. anguillarum* was the main cause of mortality in the experiments that were not pretreated with glutaraldehyde, and that phage KVP40 was able to significantly reduce mortality imposed by the added *V. anguillarum* strains.

Consequently, even though the presence of a bacterial background pathogen community masked the effect of phage KVP40 on the added *V. anguillarum* strains, it at the same time provided a more realistic demonstration of how the addition of phage KVP40 will affect an infected aquaculture system. These results emphasized the potential of phage KVP40 to control not only the added host strains but also a broader range of pathogens present in the rearing facilities. Similar results were obtained for the two broad-host-range KVP40-like phages φSt2 and φGrn1 infecting the fish pathogens *V. alginolyticus* [21]. These phages were able to reduce the natural *Vibrio* load present in *Artemia* live feed cultures used in fish hatcheries. The current study is, however, the first demonstration of a positive effect of phage application on larval survival by reducing the natural microbiota, rather than exclusively focusing on the effects of one added pathogen. While the composition of the background microbiota was not analyzed in the current study, previous studies have found that bacterial communities associated with cod and turbot eggs in rearing units were dominated by *Pseudomonas*, *Alteromonas*, *Aeromonas*, and *Flavobacterium* [22], but also *Vibrio* has been shown to be prevalent in these environments [23]. In our study, the high fraction of bacteria growing on *Vibrio*-selective TCBS medium combined with the high susceptibility to phage KVP40 suggested that the background bacterial community was dominated by *Vibrio* or *Vibrio*-related species, as the phage KVP40 has been shown to infect at least eight *Vibrio* species and one *Photobacterium* [16]. This was also supported by preliminary analysis of the microbiome associated with the turbot eggs used in challenge trial 2, which showed dominance of *Vibrio* species (Dittmann, unpublished results). The differences in mortality in the control treatments (nonchallenged control and KVP40 control) between different experiments may therefore reflect differences in the composition of background bacterial community, representing differences in virulence and KVP40 susceptibility. Further, higher incubation temperature of the turbot than cod eggs may also have increased bacteria-induced mortality in the turbot experiments. In one of the treatments (cod challenge trial 1 with strain 90-11-286), addition of phage KVP40 increased larval mortality (Figure 4c). Specific secondary metabolites or toxins released during cell lysis may potentially inhibit larval growth [24]. However, since this was not observed in any of the other treatments, it is not likely that the viral lysates affected the cod larvae. Alternatively, the viral lysates may have stimulated growth of other specific pathogens already present in the experiment, as also indicated by the enhanced bacterial growth in the phage added culture (Figure 6c). Previous studies have shown that lysogenization of *V. harveyi* with phage VHS1 increased the virulence of the bacterium against black tiger shrimp (*Penaus monodon*) by the phage encoded toxin associated with hemocyte agglutination ([25]). There has not been any indication of lysogenization of *Vibrio* pathogens with phage KVP40, and the production of a KVP40-encoded toxin is therefore not a likely explanation for the observed increase in larval mortality in this experiment.

Our results support previous attempts to control pathogens in aquaculture by use of phages. A challenge trial in Atlantic salmon using *V. anguillarum* strain PF4, a close relative to strain PF430-3 used in the current study [13], showed 100% survival using the phage CHOED, independent of the original multiplicity of infection (MOI) [10]. The efficiency of this phage on fish survival compared to the current study most likely relates to the fact that larger fish are more robust against infections by co-occurring pathogens than larvae. A delay in mortality after phage addition was also observed by Imbeault et al. [26] and Verner-Jeffreys et al. [27] in brook trout and Atlantic salmon, respectively, infected with *A. salmonicida* using different phages. While Imbeault et al. [26] were able to delay the onset of disease and reduce the mortality to 10%, Verner-Jeffreys et al. [27] also demonstrated a delay in the mortality, but only observed a temporary effect of the phages in survival.

Previous in vivo challenge studies with a positive outcome of phage therapy were conducted on >5 day old larvae [12] or fish averaging 15–25 grams [10], while our study was conducted on eggs which hatched during the course of the challenge trials. Eggs and newly hatched larvae are more sensitive to the infection by pathogenic *V. anguillarum* and other pathogens than late stages due to the inefficient protection provided by the intestinal microflora associated with their gut mucosa, which constitutes a primary barrier [28]. Despite the general frailty of newly hatched larvae, we demonstrated a significant phage-mediated reduction in mortality of cod and turbot larvae in experimental challenge trials with *V. anguillarum* pathogens in combination with the natural pathogenic bacteria associated with the incubated fish eggs. These results emphasize that phage therapy is a promising approach to reduce pathogen load and mortality in marine larviculture.

4. Materials and Methods

4.1. Bacterial Strains and Growth Conditions

The four *V. anguillarum* strains—PF4303-3, PF7, 90-11-286, and 4299—used in this study were isolated in Chile, Denmark, and Norway [10,13,29,30]. The bacteria were stored at −80 °C in Luria-Bertani (LB) medium with 15% glycerol. Before each assay, the strains were inoculated on LB plates and grown overnight at 24 °C. Then, one colony was transferred to 4 mL LB medium and grown overnight at 24 °C with agitation (200 rpm).

4.2. Phage Infectivity and Production

The broad-host-range phage KVP40 [16], which previously has been shown to infect the *V. anguillarum* strains PF430-3, 90-11-286, and 4299 [13], was tested on *V. anguillarum* strain PF7 using the double-layer agar assay [14] with minor modifications. The double-layer agar assay in brief: 100 µL phage lysate was mixed with 300 µL bacterial cells and incubated for 30 min at 24 °C. The mixture was added to 4 mL of 45 °C top agar (LB with 0.4% agar) and poured onto a LB 1.5% agar plate, which was placed for incubation at 24 °C overnight. The next day, the presence of phages in the form of clear plaques in the top agar was detected. KVP40 was produced and purified by ACD Pharmaceuticals AS (Leknes, Norway).

4.3. Eggs and Larvae

Eggs from turbot and cod were used in the challenge trials. The eggs for turbot challenge trial 1 were obtained from Stolt Sea Farm (Galicia, Spain), with 48 h of transport before conducting the challenge trial at the University of Bergen (Bergen, Norway). The eggs for turbot challenge trial 2 were obtained from France Turbot, hatchery L'Epine (Noirmoutier Island, France), with 24 h of transport before conducting the challenge trial at the Technical University of Denmark (Lyngby, Denmark). The eggs for cod challenge trial 1 and cod challenge trial 2 were obtained from the Institute of Marine Research, Austevoll Research Station (Storebø, Norway), with 1 hour of transport before conducting the challenge trial at the University of Bergen (Bergen, Norway). The eggs in cod challenge trial 2 were disinfected with 25% glutaraldehyde at the Institute of Marine Research, Austevoll Research Station before being transported to the University of Bergen for the challenge trial.

4.4. Phage Therapy Assays

Challenge trials with turbot and cod larvae were established as outlined in Table 3. For each of the *V. anguillarum* strains tested, eggs were distributed in 10 24-well dishes with 2 mL sterile filtered (0.2 µm) and autoclaved, oxygenated 80% sea water and 1 egg well^{-1}. In group 1 (*V. anguillarum* only), five 24-well plates were inoculated with 100 µL *V. anguillarum* culture in each well. Prior to addition, the bacterial culture had been grown overnight, washed twice in sterile sea water (ssw), and resuspended in ssw to a final concentration of 0.5–1 × 10^6 CFU mL^{-1}. In group 2 (*V. anguillarum* + phage KVP40), five 24-well plates were inoculated with *V. anguillarum* as above and 50 µL of phage KVP40 was added to each well to a final concentration of 0.5–8 × 10^8 PFU mL^{-1}, resulting in a multiplicity of infection (MOI) of ~5–100. The five 24-well plates in group 3 (nonchallenged control) were only inoculated with 100 µL autoclaved, oxygenated 80% ssw, whereas in group 4 (phage KVP40 control), each well also contained 50 µL of phage KVP40. Plates were then incubated in an air-conditioned room of 15.5 °C and 5.5 °C for turbot and cod, respectively, which are optimal conditions for larval development in the two species. The eggs in groups 1, 2, and 4 had bacteria and/or phages added to them immediately after their distribution in the wells (=day 0 of the experiment). Due to large variation in the viability of the eggs used for the experiment, the challenge trials were done twice for both fish species in an attempt to confirm the results at different egg qualities. The challenge trials lasted for 8 days for turbot challenge trial 1, 5 days for turbot challenge trial 2, and for

11 days for cod. The mortality was monitored daily. The quality of the eggs varied considerably depending on transportation time and handling, resulting in differences in egg mortality prior to hatching. The initial egg mortality was calculated for each 24-well plate and then averaged for all 50 24-well plates used in the individual experiments. Of the 1200 eggs used in each experiment, the average fraction of eggs that died prior to hatching amounted to 0% and 30.3% in turbot challenge trials 1 and 2, respectively, and 4.9% and 23.2% in cod challenge trials 1 and 2, respectively. These eggs were excluded from the analysis. The effect of phage addition on larval mortality was calculated as a relative reduction [31], corresponding to the reduction in mortality in treatments to which both phage KVP40 and *V. anguillarum* were added relative to the mortality in treatments with *V. anguillarum* alone (i.e., the difference in mortality between the two treatments in percentage of the mortality in the incubations without phage.

Table 3. Experimental design and addition *V. anguillarum* and phage KVP40.

Group	Treatment	*V. anguillarum* (CFU mL^{-1})	Phage KVP40 (PFU mL^{-1})	Replicate Wells
1	*V. anguillarum* only	$0.5–1 \times 10^6$	-	5×24 wells $\times$ 4 strains
2	*V. anguillarum* + phage KVP40	$0.5–1 \times 10^6$	$0.5–12 \times 10^8$	5×24 wells $\times$ 4 strains
3	Nonchallenge control	-	-	5×24 wells
4	Phage KVP40 control	-	$0.5–12 \times 10^8$	5×24 wells

The concentration of bacteria and phages was monitored daily except in turbot challenge trial 1, where neither was monitored. In turbot challenge trial 2, the concentrations were only monitored for half of the experiment, while the phage concentration was only monitored for 3 days in cod challenge trial 1. To determine the bacterial concentration, dilutions were inoculated on LB agar plates (in cod challenge trial 2, the dilutions were inoculated on marine agar plates and on selective thiosulfate-citrate-bile salts-sucrose (TCBS) plates), which incubated overnight at 24 °C. To determine the phage concentration, the double-layer agar assay was used as described earlier. The culture medium was LB, the host strain was *V. anguillarum* strain PF430-3 Δ*vanT* [19], and the plates were incubated overnight at 24 °C.

4.5. Bacterial Background Community and Susceptibility Assays

In order to characterize the bacterial background, different media were used in the challenge trials. The water used for the transport of the eggs in turbot challenge trial 1 was spread on TCBS plates at day 4. A total of 40 colonies were picked and transferred to LB medium and grown overnight at 24 °C with agitation (200 rpm). The bacteria had their optical density at 600 nm (OD_{600}), measured using Novaspec Plus Visible Spectrophotometer after 1 hour in the presence and in the absence of KVP40. The sterile 80% sea water with the live nonchallenged control larvae in turbot challenge trial 2 were inoculated on LB, TCBS, and marine agar plates at day 11. The plates incubated overnight at 24 °C before determining the bacterial concentration. Throughout cod challenge trial 2, the bacterial concentration was determined on both marine agar and TCBS plates.

4.6. Statistical Analysis

Differences between challenged larvae with and without phage therapy and between the controls (nonchallenge control and KVP40 control) for each time point were analyzed by chi-squared tests using the software R (R foundation for statistical computing). A value of $p < 0.05$ were considered statistically significant.

5. Conclusions

The significant positive effect of phage KVP40 on larval survival during hatching and initial growth observed in the current experiment demonstrates the potential in using phages to reduce pathogen load in cod and turbot hatcheries and may also be a strategy to improve egg quality and survival during transport from egg producers to hatcheries. It is obvious, however, that the effect of the phage addition on mortality is temporary, and we suggest that a more efficient and long-term control of the pathogens may be obtained using a cocktail of different phages that target a broader range of pathogens.

Author Contributions: N.R., A.R. and M.M. designed the experiments; N.R. and A.R. performed turbot challenge trial 1 and cod challenge trial 1, P.G.K. and B.B.R. performed turbot challenge trial 2, N.R., A.R., P.G.K. and B.B.R. performed cod challenge trial 2; N.R. and M.M. analyzed the data; K.E.-S., H.P.K., H.I.W., L.G. and M.M. contributed reagents/materials/analysis tools; N.R. and M.M. wrote the paper with contributions from all authors.

Acknowledgments: The study was supported by the Danish Council for Strategic Research (ProAqua project 12-132390) and the Danish Research Council for Independent Research (Project # DFF-7014-00080).

Conflicts of Interest: The authors declare no conflict of interest.

References

1. Thompson, F.L.; Iida, T.; Swings, J. Biodiversity of Vibrios. *Microbiol. Mol. Biol. Rev.* **2004**, *68*, 403–431. [CrossRef] [PubMed]
2. Actis, L.A.; Tolmasky, M.E.; Crosa, J.H. Vibriosis. In *Fish Diseases and Disorders*; Woo, P.T.K., Bruno, D.W., Eds.; CAB International: Oxfordshire, UK, 2011; pp. 570–605.
3. Frans, I.; Michiels, C.W.; Bossier, P.; Willems, K.A.; Lievens, B.; Rediers, H. Vibrio anguillarum as a fish pathogen: Virulence factors, diagnosis and prevention. *J. Fish Dis.* **2011**, *34*, 643–661. [CrossRef] [PubMed]
4. Pereira, C.; Silva, Y.J.; Santos, A.L.; Cunha, A.; Gomes, N.C.M.; Almeida, A. Bacteriophages with potential for inactivation of fish pathogenic bacteria: Survival, host specificity and effect on bacterial community structure. *Mar. Drugs* **2011**, *9*, 2236–2255. [CrossRef] [PubMed]
5. Karunasagar, I.; Pai, R.; Malathi, G.R.; Karunasagar, I. Mass mortality of Penaeus monodon larvae due to antibiotic-resistant Vibrio harveyi infection. *Aquaculture* **1994**, *128*, 203–209. [CrossRef]
6. Bricknell, I.R.; Bowden, T.J.; Verner-Jeffreys, D.W.; Bruno, D.W.; Shields, R.J.; Ellis, A.A.E. Susceptibility of juvenile and sub-adult Atlantic halibut (*Hippoglossus hippoglossus* L.) to infection by Vibrio anguillarum and efficacy of protection induced by vaccination. *Fish Shellfish Immunol.* **2000**, *10*, 319–327. [CrossRef] [PubMed]
7. Mikkelsen, H.; Lund, V.; Larsen, R.; Seppola, M. Vibriosis vaccines based on various sero-subgroups of Vibrio anguillarum O2 induce specific protection in Atlantic cod (*Gadus morhua* L.) juveniles. *Fish Shellfish Immunol.* **2011**. [CrossRef] [PubMed]
8. Vinod, M.G.; Shivu, M.M.; Umesha, K.R.; Rajeeva, B.C.; Krohne, G.; Karunasagar, I.; Karunasagar, I. Isolation of Vibrio harveyi bacteriophage with a potential for biocontrol of luminous vibriosis in hatchery environments. *Aquaculture* **2006**. [CrossRef]
9. Karunasagar, I.; Shivu, M.M.; Girisha, S.K.; Krohne, G.; Karunasagar, I. Biocontrol of pathogens in shrimp hatcheries using bacteriophages. *Aquaculture* **2007**, *268*, 288–292. [CrossRef]
10. Higuera, G.; Bastías, R.; Tsertsvadze, G.; Romero, J.; Espejo, R.T. Recently discovered Vibrio anguillarum phages can protect against experimentally induced vibriosis in Atlantic salmon, Salmo salar. *Aquaculture* **2013**, *392*, 128–133. [CrossRef]
11. Mateus, L.; Costa, L.; Silva, Y.J.; Pereira, C.; Cunha, A.; Almeida, A. Efficiency of phage cocktails in the inactivation of Vibrio in aquaculture. *Aquaculture* **2014**, *424*, 167–173. [CrossRef]
12. Silva, Y.J.; Costa, L.; Pereira, C.; Mateus, C.; Cunha, Â.; Calado, R.; Gomes, N.C.M.; Pardo, M.A.; Hernandez, I.; Almeida, A. Phage therapy as an approach to prevent Vibrio anguillarum infections in fish larvae production. *PLoS ONE* **2014**. [CrossRef] [PubMed]
13. Tan, D.; Gram, L.; Middelboe, M. Vibriophages and their interactions with the fish pathogen vibrio anguillarum. *Appl. Environ. Microbiol.* **2014**, *80*, 3128–3140. [CrossRef] [PubMed]
14. Stenholm, A.R.; Dalsgaard, I.; Middelboe, M. Isolation and characterization of bacteriophages infecting the fish pathogen *Flavobacterium psychrophilum*. *Appl. Environ. Microbiol.* **2008**, *74*, 4070–4078. [CrossRef] [PubMed]
15. Letchumanan, V.; Chan, K.G.; Pusparajah, P.; Saokaew, S.; Duangjai, A.; Goh, B.H.; Ab Mutalib, N.S.; Lee, L.H. Insights into bacteriophage application in controlling vibrio species. *Front. Microbiol.* **2016**, *7*. [CrossRef] [PubMed]

16. Matsuzaki, S.; Tanaka, S.; Koga, T.; Kawata, T. A broad-host-range vibriophage, KVP40, isolated from sea water. *Microbiol. Immunol.* **1992**. [CrossRef]
17. Inoue, T.; Matsuzaki, S.; Tanaka, S. A 26-kDa outer membrane protein, OmpK, common to Vibrio species is the receptor for a broad-host-range vibriophage, KVP40. *FEMS Microbiol. Lett.* **1995**, *125*, 101–105. [CrossRef] [PubMed]
18. Labrie, S.J.; Samson, J.E.; Moineau, S. Bacteriophage resistance mechanisms. *Nat. Rev. Microbiol.* **2010**, *8*, 317–327. [CrossRef] [PubMed]
19. Tan, D.; Svenningsen, S.L.; Middelboe, M. Quorum sensing determines the choice of antiphage defense strategy in *Vibrio anguillarum*. *MBio* **2015**, *6*, e00627. [CrossRef] [PubMed]
20. Tan, D.; Dahl, A.; Middelboe, M. Vibriophages differentially influence biofilm formation by *Vibrio anguillarum* strains. *Appl. Environ. Microbiol.* **2015**, *81*, 4489–4497. [CrossRef] [PubMed]
21. Kalatzis, P.G.; Bastías, R.; Kokkari, C.; Katharios, P. Isolation and characterization of two lytic bacteriophages, φst2 and φgrn1; Phage therapy application for biological control of vibrio alginolyticus in aquaculture live feeds. *PLoS ONE* **2016**, *11*. [CrossRef] [PubMed]
22. Hansen, G.H.; Olafsen, J.A. Bacterial colonization of cod (*Gadus morhua* L.) and halibut (*Hippoglossus hippoglossus*) eggs in marine aquaculture. *Appl. Environ. Microbiol.* **1989**, *55*, 1435–1446. [PubMed]
23. Austin, B. Taxonomy of bacteria isolated from a coastal marine fish-rearing unit. *J. Appl. Bacteriol.* **1982**, *53*, 253–268. [CrossRef]
24. Goodridge, L.D. Designing phage therapeutics. *Curr. Pharm. Biotechnol.* **2010**, *11*, 15–27. [CrossRef] [PubMed]
25. Khemayan, K.; Prachumwat, A.; Sonthayanon, B.; Intaraprasong, A.; Sriurairatana, S.; Flegel, T.W. Complete genome sequence of virulence-enhancing siphophage VHS1 from Vibrio harveyi. *Appl. Environ. Microbiol.* **2012**, *78*, 2790–2796. [CrossRef] [PubMed]
26. Imbeault, S.; Parent, S.; Lagacé, M.; Carl, F.; Blais, J. Using bacteriophages to prevent furunculosis caused by Aeromonas salmonicida in farmed brook trout. *J. Aquat. Anim. Health* **2006**, *18*, 203–214. [CrossRef]
27. Verner-Jeffreys, D.W.; Algoet, M.; Pond, M.J.; Virdee, H.K.; Bagwell, N.J.; Roberts, E.G. Furunculosis in Atlantic salmon (*Salmo salar* L.) is not readily controllable by bacteriophage therapy. *Aquaculture* **2007**, *270*, 475–484. [CrossRef]
28. Hansen, G.H.; Olafsen, J.A. Bacterial interactions in early life stages of marine cold water fish. *Microb. Ecol.* **1999**, *38*, 1–26. [CrossRef] [PubMed]
29. Skov, M.N.; Pedersen, K.; Larsen, J.L. Comparison of pulsed-field gel electrophoresis, ribotyping, and plasmid profiling for typing of *Vibrio anguillarum* serovar O1. *Appl. Environ. Microbiol.* **1995**, *61*, 1540–1545. [PubMed]
30. Mikkelsen, H.; Schrøder, M.B.; Lund, V. Vibriosis and atypical furunculosis vaccines; efficacy, specificity and side effects in *Atlantic cod, Gadus morhua* L. *Aquaculture* **2004**. [CrossRef]
31. Ranganathan, P.; Pramesh, C.; Aggarwal, R. Common pitfalls in statistical analysis: Absolute risk reduction, relative risk reduction, and number needed to treat. *Perspect. Clin. Res.* **2016**, *7*. [CrossRef] [PubMed]

antibiotics

MDPI

Article

Protective Effects of Bacteriophages against *Aeromonas hydrophila* Causing Motile Aeromonas Septicemia (MAS) in Striped Catfish

Tuan Son Le [1,2], Thi Hien Nguyen [3], Hong Phuong Vo [3], Van Cuong Doan [3], Hong Loc Nguyen [3], Minh Trung Tran [3], Trong Tuan Tran [3], Paul C. Southgate [4] and D. İpek Kurtböke [1,*]

1 GeneCology Research Centre, Faculty of Science, Health, Education and Engineering, University of the Sunshine Coast, 90 Sippy Downs Drive, Sippy Downs, QLD 4556, Australia; tuan.son.le@research.usc.edu.au

2 Research Institute for Marine Fisheries, 224 Le Lai, Ngo Quyen, Hai Phong 180000, Vietnam

3 Research Institute for Aquaculture No. 2, 116 Nguyen Dinh Chieu, District 1, Ho Chi Minh 700000, Vietnam; nguyenhien05@gmail.com (T.H.N.); vohongphuong@gmail.com (H.P.V.); vancuongdisaqua@gmail.com (V.C.D.); hongloc@gmail.com (H.L.N.); trung16893@yahoo.com.vn (M.T.T.); tuantran_695@yahoo.com.vn (T.T.T.)

4 Australian Centre for Pacific Islands Research and Faculty of Science, Health, Education and Engineering, University of the Sunshine Coast, Maroochydore, QLD 4556, Australia; psouthgate@usc.edu.au

* Correspondence: ikurtbok@usc.edu.au; Tel: +61-07-5430-2819

Received: 28 December 2017; Accepted: 23 February 2018; Published: 25 February 2018

Abstract: To determine the effectivity of bacteriophages in controlling the mass mortality of striped catfish (*Pangasianodon hypophthalmus*) due to infections caused by *Aeromonas* spp. in Vietnamese fish farms, bacteriophages against pathogenic *Aeromonas hydrophila* were isolated. *A. hydrophila*-phage 2 and *A. hydrophila*-phage 5 were successfully isolated from water samples from the Saigon River of Ho Chi Minh City, Vietnam. These phages, belonging to the *Myoviridae* family, were found to have broad activity spectra, even against the tested multiple-antibiotic-resistant *Aeromonas* isolates. The latent periods and burst size of phage 2 were 10 min and 213 PFU per infected host cell, respectively. The bacteriophages proved to be effective in inhibiting the growth of the *Aeromonas* spp. under laboratory conditions. Phage treatments applied to the pathogenic strains during infestation of catfish resulted in a significant improvement in the survival rates of the tested fishes, with up to 100% survival with MOI 100, compared to 18.3% survival observed in control experiments. These findings illustrate the potential for using phages as an effective bio-treatment method to control Motile Aeromonas Septicemia (MAS) in fish farms. This study provides further evidence towards the use of bacteriophages to effectively control disease in aquaculture operations.

Keywords: *Aeromonas hydrophila*; Motile Aeromonas Septicemia; MAS; multiple-antibiotic-resistance; bacteriophage; biological control; striped catfish (*Pangasianodon hypophthalmus*)

1. Introduction

Striped catfish (*Pangasianodon hypophthalmus*) is one of the most important farmed fish species, especially in Vietnam, Thailand, Cambodia, Laos and, more recently, the Philippines and Indonesia [1]. Vietnam supplied 90% of catfish production with a value of US$1.1 to 1.7 billion in 2015. Motile Aeromonas Septicemia (MAS), also called haemorrhage disease or red spot disease, causes great losses for farmers (up to 80% mortality) and presents in fish with clinical signs of haemorrhages on the head, mouth, and at the base of fins, a red, swollen vent, and the presence of pink to yellow ascitic fluid [2]. *Aeromonas hydrophila*, *Aeromonas caviae*, and *Aeromonas sobria* species were often isolated from diseased catfish, and new species such as *Aeromonas dhakensis* and *Aeromonas veronii* were also reported by using molecular methods based on the sequencing of the *rpo*D gene [3].

Multiple antibiotic resistance (MAR) of *A. hydrophila* strains has been reported in different countries. Vivekanandhan et al. [4] tested 319 strains of *A. hydrophila* isolated from fish and prawns in South India and indicated that all of them were resistant to methicillin, rifampicin, bacitracin, and novobiocin (99%). Moreover, 21 *Aeromonas* spp. isolated from carp showed resistance to ampicillin and penicillin [5]. Recently, Thi et al. [6] tested antibiotic resistance of 30 strains of *A. hydrophila* isolated from diseased striped catfish in the Mekong Delta from January 2013 to March 2014. The study found that *A. hydrophila* isolates were highly resistant to tetracycline and florfenicol and were completely resistant to trimethoprim, sulfamethoxazole, ampicillin, amoxicillin, and cefalexine.

ALPHA JECT ® Panga 2 vaccine, protecting against *Edwardsiella ictaluri* and *A. hydrophila*, has been approved for market in Vietnam since the early 2017 (https://www.pharmaq.no/updates/pharmaq-fish-va/). However, the cost-effectiveness of vaccine use in catfish production is another obstacle in intensive catfish production. Moreover, the development of a commercial vaccine against *A. hydrophila* has been slow because *A. hydrophia* is biochemically and serologically heterogeneous [7]. Therefore, there is a need for effective, environmentally safe control measures for managing MAS in catfish.

One approach has been the use of bacteriophages (phages) to control pathogenic bacteria in aquaculture operations. Recently, studies related to the use of phages specific to *A. hydrophila* in aquaculture have gained attention. Hsu et al. [8] isolated two *A. hydrophila* phages and three *Edwardsiella tarda* phages to treat disease in eels (*Anguilla japonica*) in vitro. The phages reduced bacterial density by about 1000 times after 2 h when the MOI was 11.5 at 25 °C in the fluid environment. El-Araby et al. [9] demonstrated the effectiveness of bacteriophage ZH1 and ZH2 treatment against *A. hydrophila* in Tilapia, improving the survival rates by up to 82%.

However, so far, treatments using bacteriophages against pathogens causing MAS in catfish have not been studied extensively. The objective of this study was, therefore, to isolate bacteriophages infective in pathogenic *A. hydrophila* with a long-term objective to eradicate this disease-causing pathogen in aquaculture operations.

2. Results

2.1. Prophage Induction

No reduction in the optical density of bacterial suspension treated with Mitomycin C (Table S1, Supplementary Materials) and no clear zones from the spot technique were observed. Therefore, it was concluded that there was no prophage in *A. hydrophila* N17.

2.2. Antibiotic Susceptibility

All isolates were completely (100%) resistant to oxytetracycline, ampicillin, gentamycin and amoxicillin/clavulanic acid, enrofloxacin, and bactrim. Nearly all isolates (83.3%) were resistant to kanamycin and 33.3% were resistant to tetracycline, doxycycline, and ciprofloxacin (Table 1).

Table 1. Antibiogram profile of the *Aeromonas hydrophila* strains tested.

Antibiotics	Number of Resistant Isolates (n = 6)
Tetracycline	2
Oxytetracycline	6
Gentamycin	6
Kanamycin	5
Bactrim (SMX/TMP)	6
Doxycycline	2
Enrofloxacin	6
Amoxicillin/clavulanic acid	6
Ampicillin	6
Ciprofloxacin	2

2.3. Isolation and Characterization of Bacteriophages

The A. hydrophila-phage 2 (or Φ2) and A. hydrophila-phage 5 (or Φ5) were successfully isolated against the propagation hosts used (Figure 1 and Table 2).

Φ2 had an isometric head of 129 nm in diameter with a tail sheath 173 nm long and 15 nm wide. Φ5 was composed of: (i) an isometric head of 120 nm in diameter, (ii) a tail sheath of 198 nm in length and 15 nm in width. All of the phages had contractile tails (Figure 1 and Table 2).

Table 2. Characteristics of bacteriophages against *A. hydrophila* strains.

Φ	Concentration PFU/mL	Head (nm)		Neck (nm)		Tail Sheath (nm)		Genus
		L	W	L	W	L	W	
2	10^9		129	10	15	173	15	*Spounalikevirus*
5	10^{10}		120	15	15	198	15	*Spounalikevirus*

W: width; L: length.

Both phages produced clear plaques with diameters of 0.1 mm (Figure 1).

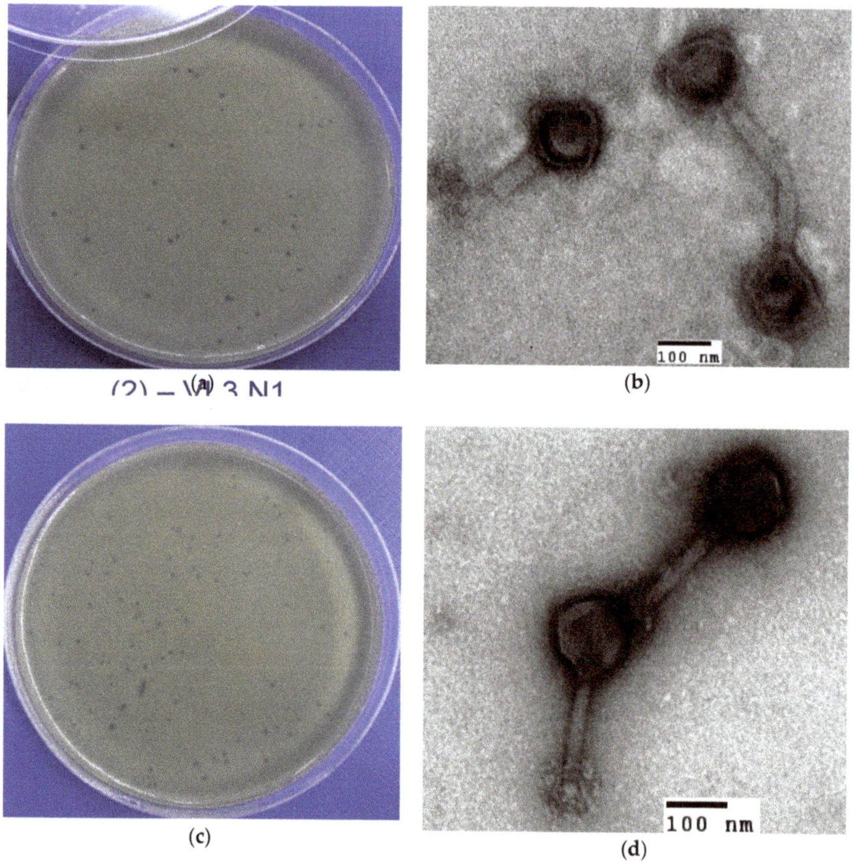

Figure 1. Plaque formation and microphotograph of *A. hydrophila* phages. (**a**,**b**) Φ2 and (**c**,**d**) phage Φ5.

The genome size of the phage isolates was above 20 kb. The genomic material of the isolated phages was not digested by Mung bean nuclease and RNase A. Since Mung bean nuclease specifically cuts single-stranded nucleic acids of both DNA and RNA, it was concluded that the genomic DNA of both phages was double-stranded. RNA nucleic acids are degraded by RNase A, therefore, the nucleic acids of Φ2 and Φ5 were determined as double-stranded DNA (dsDNA) (Figure 2). The phages Φ2 and Φ5 belong to the Myoviridae family.

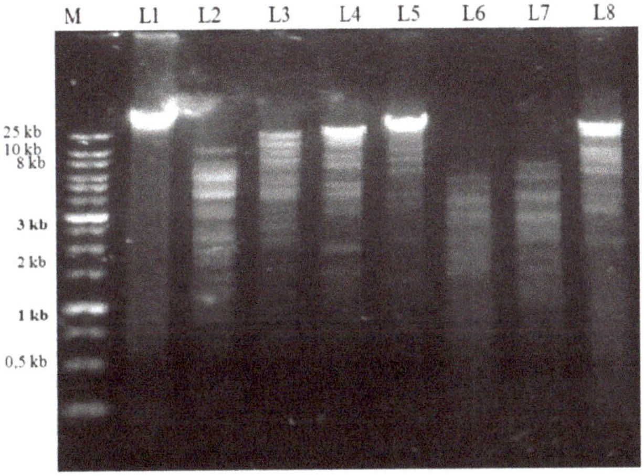

Figure 2. Restriction enzyme-digested fragments of the genomic DNA of *A. hydrophila*-phage 2. Footnote: Lane M: 1kb Plus Opti-DNA Marker (ABM, Canada); Lane L1: genomic DNA of Φ2; Lanes L2–L8: genomic DNA of Φ2 digested with EcoRV; EcoRI; NcoI; SalI; MspI; XmnI; KpnI, restriction enzymes respectively.

2.4. Host Range

Phage 2 and phage 5 were found to inhibit the growth of all *A. hydrophila* strains tested. None of the other 27 species was found to be susceptible to these phages (Table S2, Supplementary Materials).

2.5. Adsorption Rate of Phages and One-Step Growth Curve

The number of free phages in suspension decreased over time, as illustrated in the adsorption curve (Figure 3a). At 40 min, the percentage of Φ2-infected bacteria was over 90%.

The one-step growth experiment (Figure 3b) results revealed that the latent period and burst size of Φ2 were 10 and 213 PFU per infected host cell, respectively.

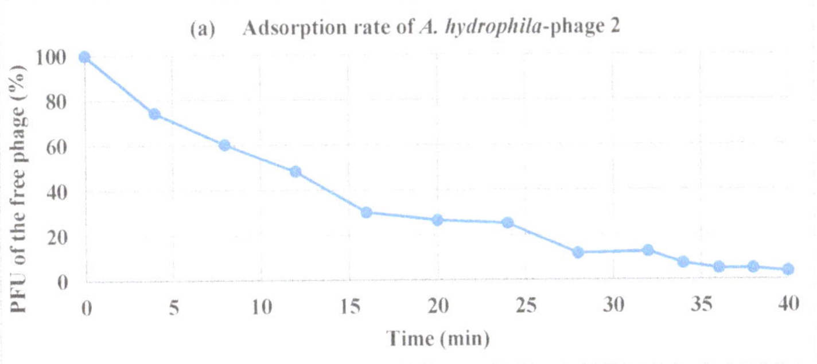

Figure 3. *Cont.*

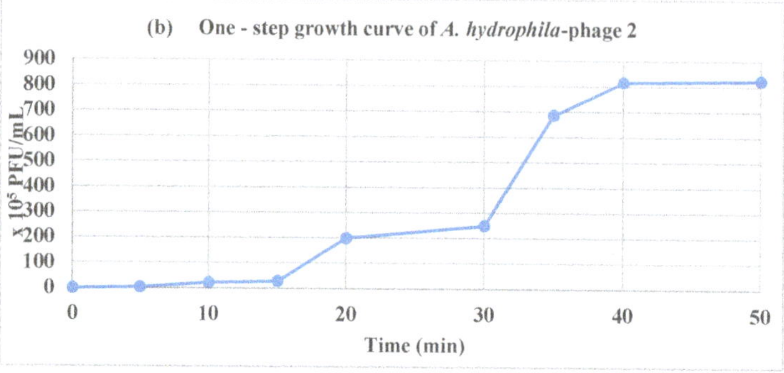

Figure 3. (**a**) Adsorption rate and (**b**) one-step growth curves of Φ2.

2.6. Inactivation of Aeromonas Species in Vitro

The bacterial concentration (OD_{550nm} values) of the uninfected control (only *A. hydrophila* N17) increased continuously during 18 h of incubation. In contrast, during the infection with Φ2 at MOI 1, MOI 0.1, and MOI 0.01 bacterial growth began to be inhibited at 1, 2, and 2.5 h, respectively, and the inhibition was maintained up to 8 h (Figure 4a). Then, the bacterial concentration increased as a consequence of the development of phage-resistant *A. hydrophila* cells.

The lowest OD_{550nm} value was 0.177 ± 0.023 after 4 h of incubation of Φ5 at MOI 0.1. There was a significant decline in the bacterial concentration (MOI 0.01, 0.1, and 1) in the first 3 h, followed by low level stabilization in the next 1, 2, and 4 h for MOI 1, 0.1, and 0.01, respectively (Figure 4b). Then, the bacterial concentration underwent a turnaround because of the development of phage-resistant *A. hydrophila* cells.

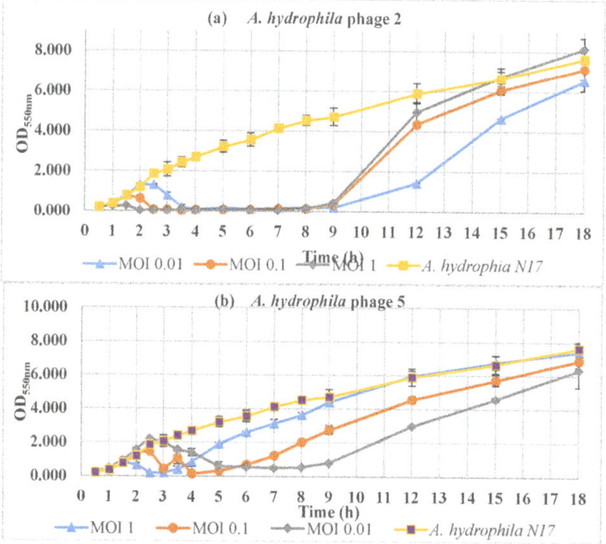

Figure 4. Inactivation of *A. hydrophila* N17 by the phages (**a**) Φ2 and (**b**) Φ5 at different MOI (0.01, 0.1 and 1).

2.7. Phage Treatment of Infected Fish

The negative control 1 (fishes with no injection) and negative control 2 (fishes injected with the growth medium filtered to remove bacterial cells) showed no mortality of catfish (Figure 5), indicating that the uninfected, control medium did not have any detrimental effect on fish health.

Catfish in the positive control groups (infected with *A. hydrophila* N17) that were not treated with bacteriophages started to die at a constant rate starting from post-infection day two, with a cumulative mortality rate of 81.67 ± 2.36% (Figure 5).

In contrast, the fish treated with the phages showed lower mortality rates at each different MOI ($p < 0.01$). While no mortality was observed in the groups treated with MOI 100, the cumulative mortalities in the other groups were 45% (MOI 1) and 68.33 ± 2.36% (MOI 0.01) at the end of the eight-day experiment (Figure 5).

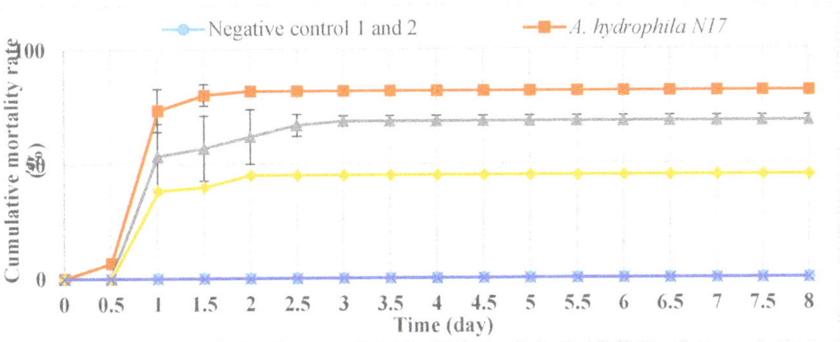

Figure 5. Cumulative mortality rates (%) of striped catfishes obtained in challenging experiments using *A. hydrophia* N17 and the phage cocktail at the different MOIs (0.01, 0.1, and 1). The ratio of Φ2 to Φ5 in a phage cocktail was 1:1.

3. Discussion

The findings of this study demonstrate that the examined *Aeromonas* spp. were resistant to multiple antibiotics and were thus able to cause high mortality rates in catfish in Vietnam, in spite of the use of various antibiotic treatments. In the bacteriophage treatments, however, Φ2 and Φ5 were able to lyse all tested *A. hydrophila* strains, displaying strong inhibition also of the virulent *A. hydrophila* strains carrying many virulence genes. Therefore, Φ2 and Φ5 are promising candidates for the application of a phage therapy to control *Aeromonas* infection in catfish.

Phage Φ2 and Φ5 were found to belong to the *Myoviridae* family, and our findings are in line with those of Ackermann [10] who indicated that 33 of a total of 43 *Aeromonas* phages he investigated were tailed and belonged to the *Myoviridae* family. Recently, other *Aeromonas* phage studies against different *Aeromonas* species by Haq et al. [11], Jun et al. [12], and Kim et al. [13] also reported that all phages they identified belonged to the *Myoviridae* family. Therefore, *Myoviridae* family members are most likely to be abundant in natural environments.

There was a correlation between the diameter of the plaques observed and the latent period and burst size for the *A. hydrophila* phage. The Φ2 had a short latent period (10 min), and these findings are in line with another study conducted by Anand et al. [14] who found that *Aeromonas* phage BPA 6 had a latent period of 10 min and a burst size of 244 PFU/cell.

The different MOI of Φ2 and Φ5 caused different bacterial growth patterns. The higher the MOI value, the sooner phage-resistant bacterial cells appeared. A similar result was noted by Kim et al. [13] for the phage PAS 1 against an *Aeromonas salmonicida* strain, indicating that bacterial resistance appeared after 3, 6, and 24 h at MOIs 10, 1, and 0.1, respectively.

Several *Aeromonas* phages, such as Aeh1, Aeh2, AH1 have also been reported [12,15,16]. However, there have been few reports demonstrating the successful use of phages for the treatment of *Aeromonas* infections in catfish. The treatment of catfish by an intraperitoneal (IP) injection illustrated significant protective effects, which increased the relative percentages of the survival rates observed for fish compared to the controls when the MOI increased. Our study revealed that in the MOI-100 experiment the relative percentage survival was 100%. The study of Jun et al. [12] showed that the relative percentage survival of fish treated with *A. hydrophila* phages pAh6-C and pAh1-C was 16.67 $\pm$ 3.82% and 43.33 $\pm$ 2.89%, respectively, when the fish were injected with the bacterium (2.6 $\times$ 10^7 CFU/fish). However, the labour-intensive and time-consuming mode of delivery of bacteriophages can constitute a disadvantage for the treatment of fish by IP injection in catfish farms. Therefore, further studies should be conducted into whether phage treatments are effective when an on-farm oral method of administration is evaluated. With the use of bioreactors, large volumes of bacteriophages can be produced for bacteriophage incorporation into fish feed. Moreover, the survival of phages and their persistent survival on or in fish, as well as in phage-coated feed preparations should be studied under different environmental factors (e.g., temperature, salt concentration) to determine whether phages are able to persist and effectively reduce *Aeromonas* spp. levels in fish farms. In conclusion, this study demonstrates that phage treatment of *Aeromonas* spp. might be an effective tool to improve the survival of farmed catfish affected by MAS.

4. Materials and Methods

4.1. *Aeromonas* Species

Bacterial isolates stored at the Research Institute for Aquaculture No. 2 (Ho Chi Minh City, Vietnam) and the ATCC type strains of the pathogens are listed in Table S2. Isolates were previously obtained from diseased catfish in farms in the south of Vietnam (Table S2).

4.2. *Prophage Induction*

In order to choose an *Aeromonas* species as a propagation host for phage isolation, *A. hydrophila* N17 was subjected to a prophage induction test. The *Aeromonas* species was cultured in 10 mL fresh Luria-Bertani (or LB) broth (Sigma-Aldrich, St. Louis, MO, USA) and incubated at 30 °C on an orbital shaker operating at 150 rpm until reaching an OD_{550nm} of 0.2. Mitomycin C (Sigma-Aldrich) was added to a final concentration of 1 µg/mL and 5 µg/mL, and again the bacterial suspension was incubated at 30 °C on an orbital shaker operating at 150 rpm. The cell density of the bacteria (OD_{550nm}) was monitored every 1 h for a 6 h period. At the end of the incubation, the bacterial suspension was centrifuged at 10,000 g for 15 min and filtered through a nitrocellulose filter (0.45 µm, Merck Millipore, Burlington, MA, USA) before spotting the filtrate onto an agar plate seeded with the host bacterium to confirm the presence of viable phage particles. A significant decrease in the cell density (OD_{550nm}) suggested that prophages were released [17,18].

4.3. *Antibiotic Susceptibility*

Antibiotic susceptibility tests of six *A. hydrophila* strains [3] were conducted against 10 different antimicrobial susceptibility discs (OXOID, Hampshire, UK) by the method recommended by the Clinical and Laboratory Standards Institute [19]. The antimicrobial agents tested included tetracycline (30 µg), doxycycline (30 µg), oxytetracycline (30 µg), bactrim (SMX/TMP) (23.75/1.25 µg), gentamycin (40 µg), kanamycin (30 µg), ciprofloxacin (10 µg), enrofloxacin (10 µg), ampicillin (33 µg), and amoxicillin-clavulanic acid (20/10 µg).

The antimicrobial susceptibility of *Aeromonas* species is usually recorded using Enterobacteriaceae breakpoints [20]. Susceptible (S), intermediate resistance (I), and resistant (R) were evaluated according the criteria given in the Performance Standards for Antimicrobial Susceptibility Testing M100-S21

(2017, Table 2A-1, pages 33–39) [19]. Multi-antibiotic resistance (MAR) was recorded when the bacteria resisted to three or more antibiotics [21].

4.4. Isolation and Characterization of Bacteriophages

Phages were isolated from water samples from the Saigon River in the south of Vietnam against *A. hydrophila* N17 and they were purified following the methods described by Jun et al. [12].

Phage titres were determined using both surface spread [22,23] and double-layer [24] agar plaque assay techniques where agar plates were previously seeded with the *Aeromonas* sp. ($\times 10^6$ CFU/mL).

For transmission electron microscopy (TEM): A 200 mesh copper grid was immersed in 40 µL of phage solution for five min before fixing the phage with glutaraldehyde solution (1%) for five min. Then, the phage samples were negatively stained with 5% (*w/v*) uranyl acetate and observed by TEM (JEOL JEM-1010) operating at a voltage of 80 kV at the Vietnam National Institute of Hygiene and Epidemiology. The phage morphology was determined using the criteria of the International Committee on Taxonomy of Viruses (ICTV) (http://www.ictvonline.org/) and Ackermann et al. [25].

Phage genomic DNA extraction and restriction analyses: Phage genomic DNA was extracted using the Phage DNA Isolation Kit (Norgen Biotek Corp, Thorold, Canada). The nature of the nucleic acids was determined by digestion with Mung bean nuclease and RNase A (ThermoFisher Scientific, Waltham, MA, USA) as per the manufacturer's protocols. The genomic DNA phages were digested using the restriction enzymes: EcoRV, EcoRI, NcoI, SalI, MspI, XmnI, and KpnI, as per the manufacturer's instruction (ThermoFisher Scientific). The DNA fragments were then electrophoresed at 120 V for 40 min.

4.5. Host Range

The method was adapted from Le et al. [23] and Goodridge et al. [26] with some modifications described below. The *Aeromonas* spp. (Table S2) were incubated overnight. Then, a 100 µL aliquot of each *Aeromonas* spp. culture (optical density of 0.5 at 550 nm) was spread on brain heart infusion agar (BHIA) (OXOID, UK) and dried for 20 min in a biological safety cabinet Class II. The host range of the phage was determined by pipetting 10 µL of phage preparation (~10^8 PFU/mL) on lawn cultures of the strains. The plates were observed for the appearance of clear zones after incubation at 30 °C after 18 h.

4.6. Adsorption Rate of Phages

Phage adsorption was studied using the method described previously [27]. A phage solution was added to 100 mL of log-phase growing *Aeromonas hydrophila* N17 culture ($\times 10^7$ CFU/mL) in LB broth to get a final MOI of 0.1. The mixture was incubated at 30 °C. An aliquot of 1 mL was collected from the sample every two min over a period of 60 min. The sample was then centrifuged at 4000 g for 15 min, and then the supernatant was diluted with SM buffer + 1% chloroform (http://cshprotocols.cshlp.org/content/2006/1/pdb.rec8111.full?text_only=true). Then, the titers of unabsorbed free phages in the supernatant were determined by the double-layer agar technique, and the results were recorded as percentages of the initial phage counts. The percentages of free phages and the adsorption rates were calculated following the formula of Haq et al. [11].

4.7. One-Step Growth Curve

The phage and bacteria were prepared in the same way as in the adsorption method described above. At 40 min, when the adsorption rate was maximal, the mixture was further incubated at 30 °C with 150 rpm. Samples were collected every 5 min for 120 min and phage titers were determined by the double-layer agar technique. Then, the latent period and burst size were calculated [28].

4.8. Inactivation of *Aeromonas hydrophila* N17 in Vitro

The method used in this study was adapted from Jun et al. [12] and Le et al. [23] with some modifications described below. *A. hydrophila* N17 was streaked onto sheep blood agar (OXOID, UK), incubated at 30 °C overnight, and harvested on LB to have a final concentration of 10^7 CFU/mL. A 10 mL suspension of the *Aeromonas* sp. in LB (around 10^7 CFU/mL) was then mixed with same volume of a phage preparation (concentration of 10^5 to 10^7 PFU/mL) to reach multiplicity of infection (MOI) of 0.01, 0.1, 1 (http://www.bio-protocol.org/e1295). A 20 mL sample of *Aeromonas* sp. in LB ($\sim\times 10^7$ CFU/mL) was used as a control. The mixture was incubated at 30 °C and 150 rpm. Samples were taken every 30 min for 8 h to determine the exact time of the appearance of phage-resistant bacteria, and every 60 min for the next 6 h to determine the increase in the concentration of phage-resistant bacteria. Then, samples were withdrawn every 3 h to the end of the experiment. The concentration of the *Aeromonas* sp. was measured by optical density determination at 550 nm using a spectrophotometer (Thermo Scientific Genesys 20, Waltham, MA, USA).

4.9. Phage Treatment of Infected Fish

A total of 360 healthy catfish (*Pangasianodon hypophthalmus*) (30 g/fish) were divided into 12 groups in 50 L plastic tanks at 30 ± 1 °C. All treatment fishes were infected intraperitoneally with *A. hydrophila* N17 (final concentration: 3.2×10^6 CFU/fish) and were then immediately injected with a phage cocktail (MOI 0.01, 1 and 100). A positive control was composed of fishes injected with *A. hydrophila* N17 only. Negative controls 1 and 2 were fishes with no injection and fishes injected with fluid separated from the broth containing bacteria and medium, respectively. The mixed phage preparation consisted of Φ2 and Φ5.

The mortality rates of the fishes were recorded every 12 h for eight days, and the kidneys of both the dead and surviving fishes were subjected to a bacterial isolation study [3]. Bacteria isolation was carried out from all dead fishes, indicating that the deaths were caused by *A. hydrophila* [3]. All treatments were performed in duplicates.

The animal experiment was conducted according to the animal ethical guidelines of the Vietnamese government (project supported by Vietnam Ministry of Agriculture and Rural Development, 2016–2018, number: 04/TCTS-KHCN-HTQT-DT 2016).

4.10. Statistical analysis

IBM SPSS Statistics 20 software was used to analyze the data. Single factor ANOVA was applied to test for differences in the fish numbers in the *Aeromonas*-infected fishes receiving or not the phage therapy ($p < 0.05$). Standard deviations were calculated in all experiments.

5. Conclusions

The phages Φ2 and Φ5, belonging to the *Myoviridae* family, were successfully isolated and displayed inhibition of the growth of the *A. hydrophila* strains tested. The results obtained from the use of a phage cocktail indicate that phages can be used successfully for the treatment of *Aeromonas* infections in catfish via intraperitoneal injection. Phages may therefore be considered as potential biocontrol agents to combat *Aeromonas* infections in fish farms.

Supplementary Materials: The following are available online at www.mdpi.com/2079-6382/7/1/16/s1, Table S1: The OD_{550nm} value of *A. hydrophila* N17 with the effect of Mitomycin C over the incubation time at 30 °C; Table S2: Host range of the phages against *Aeromonas* species.

Acknowledgments: Authors would like to acknowledge the financial support from the Vietnam Ministry of Agriculture and Rural Development (2016–2018, number: 04/TCTS-KHCN-HTQT-DT). Son Le Tuan gratefully acknowledges MOET-VIED/USC PhD scholarship.

Author Contributions: Tuan Son Le analysed the data and drafted the manuscript. Thi Hien Nguyen, Hong Phuong Vo, Trong Tuan Tran, Hong Loc Nguyen, Van Cuong Doan, Minh Trung Tran and Tuan Son Le conducted the experiments under the guidance of Ipek Kurtböke, Thi Hien Nguyen, and Hong Phuong Vo, D. Ipek Kurtböke and Paul C. Southgate oversaw the preparation of the manuscript.

Conflicts of Interest: The authors declare no conflict of interest.

References

1. Nguyen, N. Improving Sustainability of Striped Catfish (*Pangasianodon hypophthalmus*) Farming in the Mekong Delta, Vietnam through Recirculation Technology. Ph.D. Thesis, Wageningen University, Wageningen, The Netherlands, 2016.
2. Dung, T.; Ngoc, N.; Thinh, N.; Thy, D.; Tuan, N.; Shinn, A.; Crumlish, M. Common diseases of pangasius catfish farmed in Viet Nam. *GAA* **2008**, *11*, 77–78.
3. Hien, N.T.; Lan, M.T.; Anh, P.V.N.; Phuong, V.H.; Loc, N.H.; Trong, C.Q.; Trung, C.T.; Phuoc, L.H. Report "Genetics of *Aeromonas hydrophila* on Catfish". *Research Institute for Aquaculture No. 2.* 2014. Available online: http://www.sinhhoctomvang.vn/ban-tin/chi-tiet/Phat-hien-gen-gay-doc-cua-vi-khuan-Aeromonas-hydrophila-gay-benh-xuat-huyet-tren-ca-tra-106/ (accessed on 7 August 2017). (in Vietnamese).
4. Vivekanandhan, G.; Savithamani, K.; Hatha, A.; Lakshmanaperumalsamy, P. Antibiotic resistance of *aeromonas hydrophila* isolated from marketed fish and prawn of south India. *Int. J. Food Microbiol.* **2002**, *76*, 165–168. [CrossRef]
5. Guz, L.; Kozinska, A. Antibiotic susceptibility of *Aeromonas hydrophila* and *A. sobria* isolated from farmed carp (*cyprinus carpio l.*). *Bull. Vet. Inst. Pulawy* **2004**, *48*, 391–395.
6. Thi, Q.V.C.; Dung, T.T.; Hiep, D.P.H. The current status antimicrobial resistance in *Edwardsiella ictaluri* and *Aeromonas hydrophila* cause disease on the striped catfish farmed in the mekong delta. *Cantho Univ. J. Sci.* **2014**, *2*, 7–14. (In vietnamese)
7. Pridgeon, J.W.; Klesius, P.H. Major bacterial diseases in aquaculture and their vaccine development. *CAB Rev.* **2012**, *7*, 1–16. [CrossRef]
8. Hsu, C.-H.; Lo, C.-Y.; Liu, J.-K.; Lin, C.-S. Control of the eel (*anguilla japonica*) pathogens, *Aeromonas hydrophila* and *Edwardsiella tarda*, by bacteriophages. *J. Fish. Soc. Taiwan* **2000**, *27*, 21–31.
9. El-Araby, D.; El-Didamony, G.; Megahed, M. New approach to use phage therapy against *Aeromonas hydrophila* induced Motile Aeromonas Septicemia in nile tilapia. *J. Mar. Sci. Res. Dev.* **2016**, *6*. [CrossRef]
10. Ackermann, H.-W. 5500 phages examined in the electron microscope. *Arch. Virol.* **2007**, *152*, 227–243. [CrossRef] [PubMed]
11. Haq, I.U.; Chaudhry, W.N.; Andleeb, S.; Qadri, I. Isolation and partial characterization of a virulent bacteriophage ihq1 specific for *Aeromonas punctata* from stream water. *Microb. Ecol.* **2012**, *63*, 954–963. [CrossRef] [PubMed]
12. Jun, J.W.; Kim, J.H.; Shin, S.P.; Han, J.E.; Chai, J.Y.; Park, S.C. Protective effects of the *Aeromonas* phages pah1-c and pah6-c against mass mortality of the cyprinid loach (*misgurnus anguillicaudatus*) caused by *Aeromonas hydrophila. Aquaculture* **2013**, *416*, 289–295. [CrossRef]
13. Kim, J.; Son, J.; Choi, Y.; Choresca, C.; Shin, S.; Han, J.; Jun, J.; Kang, D.; Oh, C.; Heo, S. Isolation and characterization of a lytic *Myoviridae* bacteriophage pas-1 with broad infectivity in *Aeromonas salmonicida*. *Curr. Microbiol.* **2012**, *64*, 418–426. [CrossRef] [PubMed]
14. Anand, T.; Vaid, R.K.; Bera, B.C.; Singh, J.; Barua, S.; Virmani, N.; Yadav, N.K.; Nagar, D.; Singh, R.K.; Tripathi, B. Isolation of a lytic bacteriophage against virulent *Aeromonas hydrophila* from an organized equine farm. *J. Basic Microb.* **2016**, *56*, 432–437. [CrossRef] [PubMed]
15. Chow, M.S.; Rouf, M. Isolation and partial characterization of two *Aeromonas hydrophila* bacteriophages. *Appl. Environ. Microbiol.* **1983**, *45*, 1670–1676. [PubMed]
16. Wu, J.-L.; Lin, H.-M.; Jan, L.; Hsu, Y.-L.; Chang, L.-H. Biological control of fish bacterial pathogen, *Aeromonas hydrophila*, by bacteriophage ah 1. *Fish Pathol.* **1981**, *15*, 271–276. [CrossRef]
17. Fortier, L.-C.; Moineau, S. Morphological and genetic diversity of temperate phages in clostridium difficile. *Appl. Environ. Microbiol.* **2007**, *73*, 7358–7366. [CrossRef] [PubMed]
18. Walakira, J.; Carrias, A.; Hossain, M.; Jones, E.; Terhune, J.; Liles, M. Identification and characterization of bacteriophages specific to the catfish pathogen, *Edwardsiella ictaluri. J. Appl. Microbiol.* **2008**, *105*, 2133–2142. [CrossRef] [PubMed]

19. CLSI. Performance standards for antimicrobial susceptibility testing 27th ed. CLSI supplement m100. *Clinical and Laboratory Standards Institute.* 2017. Available online: http://www.facm.ucl.ac.be/intranet/CLSI/CLSI-2017-M100-S27.pdf (accessed on 10 November 2017).
20. Lamy, B.; Laurent, F.; Kodjo, A.; Roger, F.; Jumas-Bilak, E.; Marchandin, H.; Group, C.S. Which antibiotics and breakpoints should be used for *Aeromonas* susceptibility testing? Considerations from a comparison of agar dilution and disk diffusion methods using Enterobacteriaceae breakpoints. *Eur. J. Clin. Microbiol. Infect. Dis.* **2012**, *31*, 2369–2377. [CrossRef] [PubMed]
21. Daka, D.; Yihdego, D. Antibiotic-resistance *Staphylococcus aureus* isolated from cow's milk in the hawassa area, south Ethiopia. *Ann. Clin. Microbiol. Antimicrob.* **2012**, *11*. [CrossRef] [PubMed]
22. Cerveny, K.E.; DePaola, A.; Duckworth, D.H.; Gulig, P.A. Phage therapy of local and systemic disease caused by *Vibrio vulnificus* in iron-dextran-treated mice. *Infect. Immun.* **2002**, *70*, 6251–6262. [CrossRef] [PubMed]
23. Le, T.S.; Southgate, P.C.; O'Connor, W.; Poole, S.; Kurtböke, D.I. Bacteriophages as biological control agents of enteric bacteria contaminating edible oysters. *Curr. Microbiol.* **2017**. [CrossRef] [PubMed]
24. Paterson, W.; Douglas, R.; Grinyer, I.; McDermott, L. Isolation and preliminary characterization of some *Aeromonas salmonicida* bacteriophages. *J. Fish. Board Canada* **1969**, *26*, 629–632. [CrossRef]
25. Ackermann, H.-W.; Dauguet, C.; Paterson, W.; Popoff, M.; Rouf, M.; Vieu, J.-F. *Aeromonas* bacteriophages: Reexamination and classification. *Ann. Inst. Pasteur Virol.* **1985**, *136*, 175–199. [CrossRef]
26. Goodridge, L.; Gallaccio, A.; Griffiths, M.W. Morphological, host range, and genetic characterization of two coliphages. *Appl. Environ. Microbiol.* **2003**, *69*, 5364–5371. [CrossRef] [PubMed]
27. Phumkhachorn, P.; Rattanachaikunsopon, P. Isolation and partial characterization of a bacteriophage infecting the shrimp pathogen *Vibrio harveyi*. *Afr. J. Microbiol. Res.* **2010**, *4*, 1794–1800.
28. Hyman, P.; Abedon, S.T. Practical methods for determining phage growth parameters. *Methods Mol. Boil.* **2009**, *501*, 175–202.

antibiotics

MDPI

Article

Protein Expression Modifications in Phage-Resistant Mutants of *Aeromonas salmonicida* after AS-A Phage Treatment

Catarina Moreirinha [1,†], Nádia Osório [2,†], Carla Pereira [1,†], Sara Simões [2], Ivonne Delgadillo [3] and Adelaide Almeida [1,*]

1 Departament of Biology & CESAM, Campus Universitário de Santiago, Universidade de Aveiro, 3810-193 Aveiro, Portugal; anacatarinafernandes@gmail.com (C.M.); csgp@ua.pt (C.P.)

2 Escola Superior de Tecnologia da Saúde, Rua 5 de Outubro, SM Bispo. Instituto Politécnico de Coimbra, Apartado 7006, 3046-854 Coimbra, Portugal; nadia.osorio@estescoimbra.pt (N.O.); sarajmsimoes@hotmail.com (S.S.)

3 Departament of Chemistry, QOPNA, University of Aveiro, Campus Universitário de Santiago, 3810-193 Aveiro, Portugal; ivonne@ua.pt

* Correspondence: aalmeida@ua.pt; Tel.: +351-234-370-784

† These authors contributed equally to this work.

Received: 31 January 2018; Accepted: 6 March 2018; Published: 8 March 2018

Abstract: The occurrence of infections by pathogenic bacteria is one of the main sources of financial loss for the aquaculture industry. This problem often cannot be solved with antibiotic treatment or vaccination. Phage therapy seems to be an alternative environmentally-friendly strategy to control infections. Recognizing the cellular modifications that bacteriophage therapy may cause to the host is essential in order to confirm microbial inactivation, while understanding the mechanisms that drive the development of phage-resistant strains. The aim of this work was to detect cellular modifications that occur after phage AS-A treatment in *A. salmonicida*, an important fish pathogen. Phage-resistant and susceptible cells were subjected to five successive streak-plating steps and analysed with infrared spectroscopy, a fast and powerful tool for cell study. The spectral differences of both populations were investigated and compared with a phage sensitivity profile, obtained through the spot test and efficiency of plating. Changes in protein associated peaks were found, and these results were corroborated by 1-D electrophoresis of intracellular proteins analysis and by phage sensitivity profiles. Phage AS-A treatment before the first streaking-plate step clearly affected the intracellular proteins expression levels of phage-resistant clones, altering the expression of distinct proteins during the subsequent five successive streak-plating steps, making these clones recover and be phenotypically more similar to the sensitive cells.

Keywords: phage therapy; *Aeromonas salmonicida*; furunculosis; phage-resistant mutants; proteins; infrared spectroscopy

---

1. Introduction

Aquaculture produces around 30% of the seafood for human consumption, being an increasingly important food fish source worldwide [1]. Generally, fish aquaculture is subjected to greater stress than wild conspecifics, which affects their natural immune system and often favours bacterial infection, especially during early life stages. This happens because of the high organic content and low concentration of dissolved oxygen often recorded in culture water, as well as the proximity of cultured individuals. Thus, opportunistic infections can easily emerge, causing significant economic losses to producers [2].

The Food and Agriculture Organization (FAO) and most aquaculture organizations recommend a decrease, or even the avoidance, of antibiotics in aquaculture, though they are still often used by the industry worldwide [1]. This can lead to the development of resistant bacteria and dispersal of antibiotic resistance in the environment, indirectly affecting bacterial species that are not associated with disease (non-target), allowing resistant strains to enter the human food chain [3,4].

Although vaccination is considered the best approach for the prevention of fish infections, it is practically impossible to employ during fish early life stages, due to their small size and low capacity to develop immunity [5,6]. Consequently, the development and application of innovative treatment technologies are demanded by the fish farming industry in order to increase the efficacy of aquaculture production, by lowering production costs and fish mortality, with reduced environmental impacts.

Aeromonas salmonicida, the causative agent of furunculosis, is a significant fish pathogen in aquaculture. This disease causes high mortality and morbidity in a broad variety of fish, with important economic losses in aquaculture worldwide [7]. The chronic skin ulcers in weakened old fish make them unsuitable for human consumption [8]. The acute form is more common in juveniles and, usually, leads to septicaemia, being fatal in two to three days [9,10].

Phage therapy is an alternative approach to treat fish bacterial infections, being based on the use of bacteriophages (viruses that infect bacteria) to inactivate pathogenic bacteria. Compared to conventional methods such as antibiotics and vaccination, it presents several advantages: (a) phages are target specific; (b) serious or irreversible side effects of phage addition are not known; (c) phage therapy is an environmentally friendly strategy; (d) phages are resistant to various environmental conditions; (e) phage therapy is a flexible, fast and inexpensive technology [11,12]. Consequently, phage therapy appears to be a promising and environmentally friendly methodology to control bacterial infection. However, there are some studies reporting the development of phage-resistance by some bacteria [13–16]. This resistance may be due to the modification or loss of the bacterial cell surface receptors, blocking of the receptors by the bacterial extracellular matrix, production of modified restriction endonucleases that degrade the phage DNA, and inhibition of phage DNA penetration [17]. Additional causes for the development of resistance again bacteriophages are genetic mutations affecting phage receptors, restriction modification or abortive infection associated with the presence of clustered regularly interspaced short palindromic repeats (CRISPRs) in the bacterial genome [17,18]. Apart from genetic, resistance may also be phenotypic, which has been mostly disregarded in the literature [12,19,20]. It has been previously hypothesized that some of the reasons for phenotypic resistance may be: (i) induced, the products of phage-lysed bacteria result in a change in uninfected bacterial gene expression, thus reducing adsorption; (ii) intrinsic, reduced adsorption is due to a physiological or gene expression state that happens prior to the phage introduction; and (ii) dynamic, degradation or blocking of bacterial receptors by phage proteins released during cell lysis [19]. As very little is known about the effects of the phage infection in the bacterial cells, it is important to understand the inactivation mechanisms and the modifications that are induced by bacteriophages in the host cell, in order to obtain knowledge and a solution to the problem of phage-resistant bacteria.

Infrared spectroscopy (IR) has been a valuable method for detection and differentiation of microbial cells. It has also been successfully used to detect modification in proteins and lipids extracted from bacteria after exposure to a stress [21], and to study DNA structure [22]. Another advantage is the possibility of studying the whole cell, without the need to extract cellular components [23,24]. This methodology has already been used to discriminate phage-resistant from phage-susceptible bacteria [15]. The infrared absorbance spectrum represents a "fingerprint" that is characteristic of a chemical or biological substance. The main reasons for the wide acceptance of this method are the speed with which samples can be characterized with almost no handling, the flexibility of the equipment, the minimum sample amount required and the low cost of the analysis [25]. The analytical information from the spectra can be interpreted using a multivariate analysis that relates the spectra obtained with the properties of the object of study, thus facilitating data interpretation [26].

The main objective of this study was to understand the cellular modifications that occur in host targets after phage therapy, using the causative agent of furunculosis, *A. salmonicida*, and its specific phage AS-A as a model.

2. Results

2.1. Detection of Host Sensitivity to Phages after Phage Contact

Firstly, five phage resistant colonies that grew inside a clear spot-test were selected for use in the subsequent steps. These colonies were smaller than the sensitive ones and took three times longer to appear on the petri plates. These colonies were subjected to five successive streak-plating steps. It was observed that the spot-tests were negative (Figure 1A) until the fourth streak-plating step, when the spot tests became positive (Figure 1B). However, efficiency of plating (EOP) results indicated that even after the fourth streak-plating step, phages neither form lysis plaques nor adsorb and replicate in the presence of the phage-resistant clones.

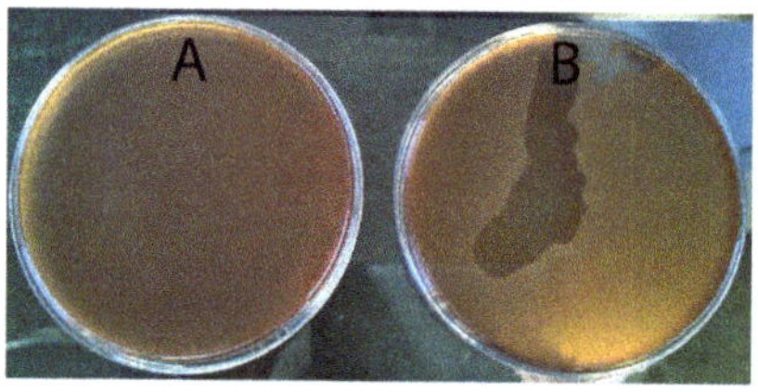

Figure 1. Spot test results using a phage-resistant mutant of phage AS-A and phage AS-A after first (**A**) and fifth streak-plating steps (**B**).

2.2. Infrared Spectroscopy of Whole Cells

The phage resistant clones from the five streaking-steps were analysed by IR spectroscopy in order to understand if there were any detectable differences in cellular components between these clones.

Principal component analysis (Figure 2) of the whole bacterial cells shows two distinct groups. It is visible a good discrimination between control phage-sensitive colonies and resistant colonies after the fourth and fifth streak-plating steps (negative PC1) and colonies from earlier streaking steps, corresponding to days 1, 2 and 3 (positive PC1).

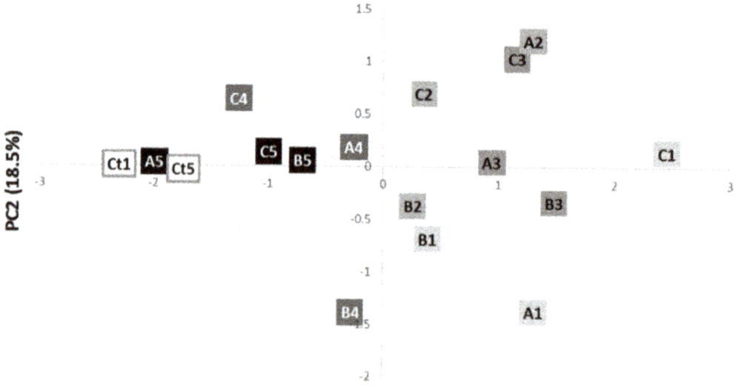

Figure 2. Scores scatter plot of the IR spectra of phage-resistant colonies A, B and C, along the 5 streak plating steps, and control phage sensitive colonies after 1 (Ct1) and 5 (Ct5) streaking steps. The letters correspond to the different colonies (A is colony A; B is colony B; C is colony C) and the numbers to the streaking-plate days (1 is day 1; 2 is day 2; 3 is day 3; 4 is day 4; 5 is day 5).

Analysing the loadings plot profile (Figure 3), there are various peaks that are contributing to the distribution of the samples according to the principal component analysis (PCA). The samples that are located in negative PC1, that is, the later streaking days and the controls that are sensitive to phage are characterized by peaks at 1510 cm^{-1}, 1440 cm^{-1}, 1380 cm^{-1}, 1150 cm^{-1}, 1070 cm^{-1}, 1025 cm^{-1} and 980 cm^{-1}. The samples corresponding to the early streaking steps (1, 2 and 3), located in positive PC1, are characterized by peaks at 1695 cm^{-1}, 1650 cm^{-1}, 1590 cm^{-1}, 1570 cm^{-1}, 1560 cm^{-1}, 1250 cm^{-1} and 1175 cm^{-1}. Table 1 summarizes the infrared spectra peak assignments. It was found that the proteins were the most affected cellular component between phage-sensitive bacteria and phage-resistant bacteria. Taking into account these results, we decided to verify if there was also differential expression of the intracellular proteins in these cases. Phage-resistant clones of day 1, i.e., after one streak-plating step, and phage-resistant clones of day 5, i.e., after five streak-plating steps were chosen to perform protein analysis.

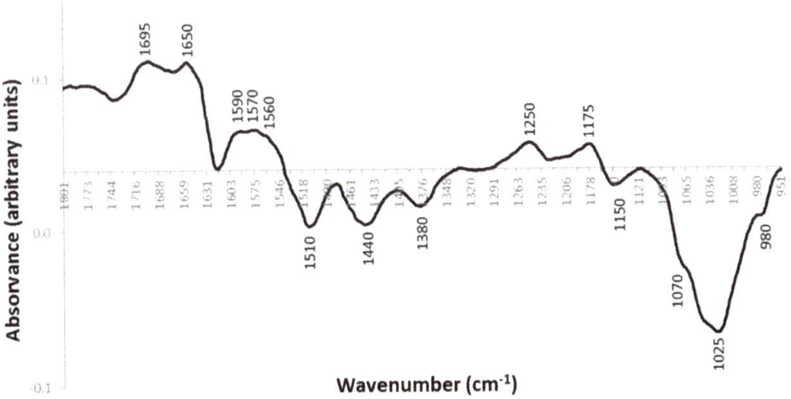

Figure 3. Loadings plot profile of PC1 corresponding to the IR spectra of the phage-resistant colonies A, B and C, along the 5 streak-plating steps, and control phage sensitive colonies.

Table 1. Peaks/regions assignments (wavenumber) from principal component analysis (PCA) loadings plot profile of spectra from colonies of *A. salmonicida* sensitive and resistant to phage AS-A.

PC1 − (cm^{-1})	PC1 + (cm^{-1})	Assignment	Reference
	1695	Amide I—proteins (β-sheet)	[27]
	1650	Amide I—proteins (α-helix)	[23,27]
	1590, 1570, 1560	Amide II—proteins	[27]
1510		Amide II—proteins	[27]
1440		CH_3 bending—proteins (methyl groups)	[28]
1380		COO^-—acids and methyl groups from proteins/CO bonds or deformation of C-H or N-H bonds of proteins	[28,29]
	1250	Amide III—proteins/PO_2^-—phospholipids	[30,31]
	1175	C-O—proteins and glycomaterials	[32,33]
1150		C-O carbohydrates	[33]
1070		PO_2^-—nucleotides	[34]
1025		Carbohydrates	[35]
980		OCH_3—polysaccharides	[36]

2.3. *Differential Expression of the Proteins of the Phage-Resistant Clones (First Streak-Plating and Fifth Streak-Plating)*

In order to try to understand why phage-sensitive bacteria (control) and clones after five streak-plating steps were different from clones after one streak-plating step, 1D SDS-PAGE gels were performed, comparing control and first streak-plating clones (Figure 4A), and comparing control and fifth streak-plating clones (Figure 5A).

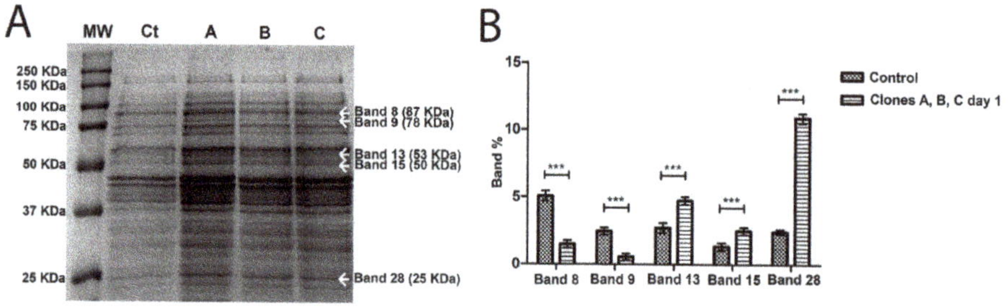

Figure 4. (**A**) SDS PAGE gel of the intracellular proteins of *A. salmonicida* on first streak-plating. MW, molecular weight marker; Ct, control phage-sensitive *A. salmonicida;* A is Colony A of the phage-resistant *A. salmonicida* mutant; B is Colony B of the phage-resistant *A. salmonicida* mutant; C is Colony C of the phage-resistant *A. salmonicida* mutant. The marked bands are the ones that showed differential expression between control and clones A, B and C. Band weight is expressed in kilodalton (KDa). (**B**) Differential expression of the bands, in percentage, comparing Control (phage-sensitive *A. salmonicida*) with clones A, B and C (phage-resistant *A. salmonicida*) after 1 streak-plating steps. *** $p < 0.001$.

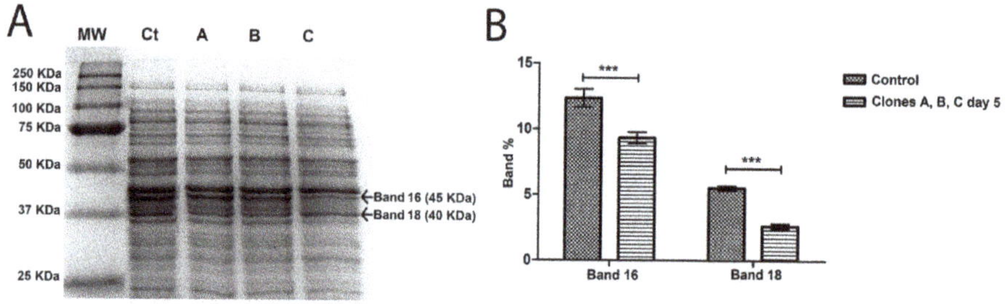

Figure 5. (**A**) SDS PAGE gel of the intracellular proteins of *A. salmonicida* on fifth streak plating. MW, molecular weight marker; Ct, control phage-sensitive *A. salmonicida;* A is Colony A of the phage-resistant *A. salmonicida* mutant; B is Colony B of the phage-resistant *A. salmonicida* mutant; C is Colony C of the phage-resistant *A. salmonicida* mutant. The marked bands are the ones that showed differential expression between control and clones A, B and C. Band weight is expressed in kilodalton (KDa). (**B**) Differential expression of the bands, in percentage, comparing Control (phage-sensitive *A. salmonicida*) with clones A, B and C (phage-resistant *A. salmonicida*) after 5 streak-plating steps. *** $p < 0.001$.

In total, 39 bands were detected and compared, in the control and *A. salmonicida* clones, both in first streak-plating and fifth streak-plating clones. When compared to the control, the bands that were significantly differentially expressed on first streak-plating clones were bands 8, 9, 13, 15 and 28 (Figure 4B). The expression patterns of the bands 8 and 9 in *A. salmonicida* first streak-plating clones tend to be less when compared to the control. However, the 13, 15 and 28 bands tend to have an increased expression compared to the control (Figure 4B). On fifth streak-plating clones, the differentially expressed bands were band 16 and 18 (Figure 5B). All of the bands with differential expression decreased between the control and fifth streak-plating clones.

By using the homology of the molecular weight of the bands with differential expression, consulting the databases referred to in the Section 4, presumptive identification of the proteins was made (Table 2).

Table 2. Presumptive band identification of the 1-D electrophoresis gel of intracellular proteins, associated proteins and their molecular functions.

Band	MW (KDa)	Protein/Gene	Molecular Function
Band 8	87	Phage transcriptional protein (ASA_3866)	Interacts selectively and non-covalently with the DNA with a specific nucleotide composition or with a specific sequence motif or type of DNA.
Band 9	78	Phage shock protein B (pspB, ASA_2424)	Response of the bacteria to a variety of stimuli, including phage infection. It is involved in bacterial protection mechanisms.
Band 13	53	Sec-independent protein translocase proteinTatA (tatA, ASA_3970)	Biological process: controlled liberation of proteins from a cell.
Band 15	50	ASA_P5G151	Unknown function.
Band 16	45	Transposase (VO70_17345, VO70_21745)	Facilitates the transference of genetic material between organisms.
Band 18	40	Toxin-antitoxin system, toxin component (VO68_18510, VO70_09250)	Plasmid maintenance, stress regulation and adaptation, growth control and programmed cellular death.
Band 28	25	Q70WF0, Q70WF0_AERSA	Unknown function.

3. Discussion

The emergence of phage-resistant mutants during phage infection has been reported in many studies [12,37–41], but the mechanisms of bacteria resistance to phages are not yet completely understood. A previous study by our group [41] showed that the agent of furunculosis can be efficiently inactivated by the phage AS-A (reduction of 4 Log CFU·mL^{-1} after 8 h of treatment). However, some bacteria survived the infection by the phage due to the development of phage-resistance [41]. Nevertheless, the frequency of resistance, with a value of 2.24×10^{-4} Log CFU·mL^{-1}, was limited as already reported in previous studies [14,42,43].

So, in our previous study [41] we verified that although a specific phage against the agent of furunculosis can efficiently control the bacterial growth, some phage-resistant bacteria emerge after treatment. In the present study, we observed that the resistant colonies after the fourth and fifth streak-plating steps are clearly distinct from those of the earlier streaking steps (steps 1, 2 and 3). A significant modification in the expression of intracellular proteins was observed when compared with the phage-sensitive bacteria. Moreover, these modifications affect distinct proteins after the first and the fifth streak-plating steps, allowing "lysis from without" (positive spot test) after the forth streak-plating step, contrary to that observed for bacteria from the first, second and third streak-plating steps.

It has been stated in the literature that resistance to phages can be overcome by the phage itself because it evolves along with the host [44]. Moreover, it has also been asserted that resistance to phages entails great costs to the bacteria [45]. In fact, as observed for other phages, colonies of AS-A phage-resistant mutants were smaller than colonies formed by the non-phage added control [14]. These results suggest that the remaining bacterial mutants (forming small size colonies) maintained their viability in the presence of phages but their phenotypes were affected. The decrease in the bacterial size after phage exposure could be a fitness cost, which might contribute to their elimination from the environment faster than their wild-type parents.

In this study, as already observed for other phages [15,16], it was detected that phage-resistant bacteria also mutate after successive streak-plating steps. Although the spot tests showed negative results until the fourth streak-plating step, at the fourth and fifth steps, the spot test was positive, as also observed in other studies [15,16]. These results were confirmed by infrared spectroscopy data of the whole cells. Infrared spectroscopy results show that the spectra obtained from the fourth and fifth streak-plating colonies are similar to ones from phage-sensitive control colonies, suggesting that these colonies are more similar to control phage-sensitive bacteria than the colonies from streak-plating steps 1, 2 and 3. It seems that the resistant bacteria somehow "recovered", being more similar to control bacterial populations, which are sensitive to the phage infection. The infrared peaks that contributed to these results were found to be especially associated with proteins. Taking this into account, we focused further studies on protein analysis with 1D SDS PAGE gels.

Regarding the presumptively identified proteins with differential expression on first streak-plating phage-resistant clones, a decrease in band 8 is noticeable when compared to the control, being the band associated with a phage transcriptional protein with regulation function in the transcription of phage genes [46]. This may be a response by the bacteria to the viral infection, preventing the transcription of the viral genome. Similarly, the expression of the protein corresponding to band 9 in first streak-plating clones decreased when compared to the control. This protein, phage-shock B protein, is involved in a regulation system that responds to aggression, habitually to phage secretins, promoting the defensive response of the bacteria [47]. This protein has been previously detected in the response of other bacteria, however, this response mechanism is not yet completely understood [47,48]. In our case, this protein is less expressed in the phage-resistant clones, which seems contradictory. Nevertheless, it was stated that bacteria synthesise phage shock proteins after being infected with phage, that, in the case of the resistant clones could not happen [49]. Contrarily, the protein associated to band 13, TatA, increased in *A. salmonicida* first streak-plating clones. This protein belongs to the Tat system (twin-arginine translocation) which is responsible for the transport of various substances at the membrane level, against the concentration gradient of the cytoplasm to the extracellular space, namely proteins, being associated with the bacterial pathogenicity in the secretion of virulence factors [50,51]. This increase suggests that these first streak-plating clones could be more virulent than control bacteria. However, some studies have shown that phage-resistant clones are less pathogenic than phage sensitive bacteria [20,52]. This suggests that the increase in the expression of this protein could be associated with other mechanisms not related with pathogenicity.

Regarding the proteins with differential expression on phage-resistant clones in the fifth streak-plating, that have a positive spot-test, band 16 suggests the expression of a transposase that is decreased in these clones when compared to control phage-sensitive bacteria. These type of enzymes facilitates the transference of the genetic material between organisms [53]. The bacteria may have decreased the expression of this protein as a defence mechanism in order to prevent the phage replication. Band 18, corresponds to a toxin-antitoxin system, which is implied in the maintenance of plasmids, stress regulation and adaptation, as well as in growth control and programmed cellular death [54,55]. This system requires the dual activity of a toxin and an antagonistic antitoxin [56]. A decrease in this band in the clones of the fifth streak-plating was found when compared to the control. As this protein decreased in this study, this suggests that in the fifth streak-plating clones, the stress caused by the phage decreased. In fact, the efficiency of plating (EOP) results indicate that the fifth streak-plating clones do not replicate the phage. Other authors [57] have obtained similar results, designating this situation by "lysis from without". The spot test lysis when the phage is not replicated by the host (EOP is zero) has been described as a plausible mechanism which happens when an overload of phage simultaneously infects a bacterium leading to lysis, either from the action of phage lysins or from rapid depletion of the cell resources [58]. As in the spot test the same volume of phage suspension was used and lysis was only observed for the clones of the fourth and fifth streak-plating, so, the hypothesis of rapid depletion of the cells resources does not seems plausible. As stated before, the lysis can be due to the presence of phage lysins. However, it is difficult to understand why the spot test was only positive for the clones of the fourth and fifth streak-plating and not for the clones of the first, second and third streak-plating. However, modifications in the bacterial proteins along the successive streak-plating could allow the clones to recover the sensitivity to the phage lysins. This is in agreement with the infrared spectroscopy results which showed that the fourth and fifth streak-plating clones were similar to the phage-sensitive bacteria (control), but clearly different from those of the first, second and third streak-plating steps. In order to test this hypothesis, further studies are needed.

It would be interesting, for example, to try to correlate IR spectra with the regaining of sensitivity to phage lysins to extract more information from the spectra.

We noticed that the different analysed clones present significant modifications in intracellular proteins related to phage infection, both in the first and fifth streak-plating steps. However, there are more proteins that are differentially expressed in clones of the first streak-plating than in clones of the fifth streak-plating, which is in accordance with infrared spectroscopy results. The fact that the phage-sensitive control bacteria have infrared spectra that are more similar to the fourth and fifth streak-plating clones may be because the cellular envelope, used by the phages to infect the bacteria, became more similar in these cases. This may be related to the fact that the spot test turns positive again for the fourth and fifth streak-plating clones, which might be due to phenotypical similarities in the cell envelope. In our study, phenotypic resistance may have been acquired by phage-resistant cells, showing less pronounced cell modifications than genetic resistance, which would be more definitive. In order to better understand this, more experiments should be done, such as serial dilution spot-tests and EOP tests with varying multiplicity of infection (MOI). In order to confirm these results, the presumptively identified proteins and the non-identified proteins that show differential expression between the clones should be confirmed/identified by methods such as mass spectrometry. In future experiments, it would also be interesting to include the whole cell proteins, which would provide more information. Moreover, since there are some indications of which proteins seem to alter their expression, molecular assays using specific primers for these proteins would be a reliable method to use in order to explore and elucidate the whole process of the clone expression pattern.

4. Materials and Methods

4.1. Bacteria and Phage

The bacteria *A. salmonicida* CECT 894 was used in this study. Fresh plate bacterial cultures were maintained in solid Tryptic Soy Agar medium (TSA; Liofilchem, Roseto degli Abruzzi, Italy) at 4 °C. Before each assay, one isolated colony was aseptically transferred to 10 mL of Tryptic Soy Broth medium (TSB; Liofilchem, Roseto degli Abruzzi, Italy) and was grown overnight at 25 °C. An aliquot of this culture (100 μL) was aseptically transferred to 10 mL of fresh TSB medium (Liofilchem, Roseto degli Abruzzi, Italy) and grown overnight at 25 °C to reach an optical density (O.D. 600) of 0.8, corresponding to about 10^9 cells·mL^{-1}.

Phage AS-A was isolated from sewage water from a lift station of the sewage network of Aveiro, Portugal (station EEIS9 of SIMRIA Multi Sanitation System of Ria de Aveiro) using *A. salmonicida* as host, according to [41]. The phage stocks were stored at 4 °C and 1% chloroform (final volume) (Scharlau, Sentmenat, Spain) was added. The phage suspension titre was determined by the double-layer agar method using TSA (Liofilchem, Roseto degli Abruzzi, Italy) as the culture medium [59]. The plates were incubated at 25 °C for 12 h and the number of lysis plaques was counted. The results were expressed as plaque forming units per millilitre (PFU·mL^{-1}).

4.2. Isolation of *A. salmonicida* Phage-Resistant Mutants

Only bacterial colonies that were resistant to the phage were used (bacteria that developed inside phage plates). For this, bacteria *A. salmonicida* and phage AS-A were plated by the double layer agar method and the plates were incubated for 24 h at 25 °C. After that, several colonies that grew inside the phage plates, thus, resistant to phage infection, were visible. Three individualized colonies (A, B and C) were chosen and used in the subsequent assays.

4.3. Detection of Bacteria Sensitivity to the Phage after One Cycle of Phage Contact

The phage resistant colonies obtained in Section 4.2 were used. The colonies were inoculated in TSB medium for 24 h at 25 °C. After that, the culture was used to perform a spot test and was also plated in TSA medium. This procedure was done 4 more times, making a total of 5 streak plating steps. This procedure was done for the 3 selected colonies.

4.4. *Efficiency of Plating (EOP)*

The efficiency of plating was determined for bacteria that shown positive spot tests (clear lysis area), i.e., for the bacteria from the fourth and fifth streak-plating steps, according to Pereira et al. [15] using the double-agar method [59]. The EOP was calculated (average PFU on target bacteria/average PFU on host bacteria), three independent assays were performed.

4.5. *Phage Adsorption*

The determination of phage adsorption was performed according to Pereira et al. [15]. Briefly, ten microliters of phage suspension of about 10^6 PFU/mL were added to 10 mL of *A. salmonicida* culture of about 10^9 CFU/mL (corresponding to an optical density (600 nm) of 0.8) [60] and incubated at 25 °C. Aliquots of this culture were collected after 0, 5, 10, 15, 20, 25, 30, 40, 50, 60 and 70 min of incubation and chloroform was added to a final concentration of 1%. The mixture was centrifuged at 12,000× *g* for 5 min, after that the supernatants were filtered using 0.2 µL membranes (Millipore, Bedford, VA, USA). The filtrates containing unadsorbed phages were then diluted and titrated. The plates were then incubated at 25 °C and observed after 8 h for plaque formation. The values were calculated as the decrease of phage titre in supernatant (percentage) compared with time zero. Three independent assays were performed.

4.6. *Infrared Spectroscopy*

In order to access the spectral differences of sensitive *A. salmonicida* colonies and phage resistant mutant colonies, mid-infrared spectroscopy was used, as it was previously described [15,24]. They were used for the *A. salmonicida* phage resistant colonies A, B and C (from Section 4.3).

To analyse the whole cells, colonies A, B and C were analysed during the 5 days of streaking (Section 4.3), as well as control sensitive colonies Ct1 and Ct5 (after the 1 and 5 streak plating steps). The colonies were collected with a loop and placed in the crystal of a horizontal single reflection ATR accessory. The colonies were gently air dried and the spectra were acquired.

Spectra were done in a MIR (Bruker ALPHA FTIR spectrometer, Germany) with a resolution of 4 cm^{-1} and 32 scans, in the infrared region (4000 to 600 cm^{-1}). At least 5 replicate spectra were performed for each colony. Mid-infrared spectra were obtained in OPUS format (OPUS 6.5, Bruker, Germany) and transferred via JCAMP.DX format for use in a house-developed data analysis software (CATS build 97). The spectra were SNV (standard normal deviate) corrected previous to multivariate analysis. Principal component analysis (PCA) was done in order to find the major sources of variability in the spectra and to detect groups.

4.7. *Extraction and Quantification of Intracellular Proteins from Phage-Sensitive and Phage-Resistant Bacteria*

The proteins extracts were obtained from the growth until the late exponential phase of the strains (OD 0.9 at 550 nm) in Luria Bertani Broth (Merck, Darmstadt, Germany). The cells were separated from the supernatant by centrifugation at 8000× *g* for 10 min at 4 °C. The protein extractions were made in three independent experiments per each strain and the protein quantification was performed in triplicate.

The cell pellets were washed three times in 10 mM phosphate buffered saline pH 7.4. After that they were resuspended in 1 mL of lysis and protein solubilisation buffer solution (7 M urea, 2 M thiourea, 4% cholamidopropyl dimethylammonio-1-propanesulfonate (CHAPS), 30 mM Tris base, pH 8.5). Crude cell-free extracts were obtained by sonication in ice to minimize protein damage, during a 2 min period, using a 30% duty cycle, 2 s pulses with intervening periods of 3 s. The intracellular protein solution was incubated with 1 mg·mL^{-1} of Dnase I (GE Healthcare, Uppsala, Sweden) and 10 mM of protease inhibitor mix (GE Healthcare, Uppsala, Sweden) for 1 h at 15 °C. The final solution was collected by centrifugation at 20,000× *g* for 40 min at 4 °C and then, the protein concentration was measured using the 2-D Quant Kit (GE Healthcare, Uppsala, Sweden), following the manufacturer's instructions. The procedure was performed in triplicate.

4.8. Protein Separation by 1-D Electrophoresis

Proteins were separated by 12.5% SDS-PAGE [61], in a Mini-PROTEAN 3 Cell (Bio-Rad, Hercules, CA, USA), for 50 min at 150 V. 5 µg/mL of each protein sample were used in this assay. Proteins were visualized by colloidal Coomassie Brilliant BlueG-250 (CBB) staining [62]. Gel images were acquired using the Gel DocTM XR+ (Bio-Rad, Hercules, CA, USA). The comparative analysis of the acquired images was performed in Image Lab v3.0 software (Biorad, Hercules, CA, USA) and based on the optical density measurement of each band. To minimize possible differences in the quantity of the proteins loaded, the results were normalized and expressed as a band percentage, resulting from the value of the optical density of a given band in the total of the bands per lane × 100. The comparison of the differential expression of the intracellular proteins of the different tested *A. salmonicida* clones in the different analysis times was made through a two-way ANOVA, using GraphPad Prism software v7 (USA). The differences were considered statistically significant when $p < 0.05$.

4.9. Presumptive Identification of the Proteins in Differentially Expressed Bands

The molecular weight of the bands that were differentially expressed between control and *A. salmonicida* clones on day 1 and between control and day 5, using the databases UniProtKB (www.uniprot.org) and NCBI (www.ncbi.nlm.nih.gov/pubmed) allowed us to presumptively identify the proteins and their respective function, based on the deposited genome of *Aeromonas salmonicida* A449.

5. Conclusions

A single cycle of phage treatment causes a significant modification in the expression of intracellular proteins of phage-resistant bacterial clones relative to the phage sensitive bacteria, but after successive streaking-plate steps these clones recover and are phenotypically more similar to the sensitive cells. Taking this information into account, this study paves the way for future experiments in order to better understand the bacterial resistance mechanisms to phages.

Acknowledgments: This work was supported by FEDER through COMPETE—Programa Operacional Factores de Competitividade, and by National funding through Fundação para a Ciência e Tecnologia (FCT), within the research projects FCOMP-01-0124-FEDER-013934 and ENV/ES/001048. Financial support was provided to Catarina Moreirinha in the form of a Postdoctoral grant (ENV/ES/001048), and Carla Pereira in the form of a PhD grant (SFRH/BD/76414/2011).

Author Contributions: Catarina Moreirinha, Nádia Osório, Carla Pereira and Sara Simões performed the experiments. Catarina Moreirinha wrote the paper and Nádia Osório and Carla Pereira also contributed to the writing. Ivonne Delgadillo and Adelaide Almeida supervised the work, revised the paper and contributed with reagents and analysis tools.

Conflicts of Interest: The authors declare no conflict of interest.

References

1. FAO. *The State of World Fisheries and Aquaculture*; FAO: Rome, Italy, 2014.
2. Iwama, G.K.; Pickering, A.D.; Sumpter, J.P.; Schreck, C.B. *Fish Stress and Health in Aquaculture*; Cambridge University Press: Cambridge, UK, 2011; Volume 62, p. 279.
3. Furushita, M.; Shiba, T.; Maeda, T.; Yahata, M.; Kaneoka, A.; Takahashi, Y.; Torii, K.; Hasegawa, T.; Ohta, M. Similarity of Tetracycline Resistance Genes Isolated from Fish Farm Bacteria to Those from Clinical Isolates. *Appl. Environ. Microbiol.* **2003**, *69*, 5336–5342. [CrossRef] [PubMed]
4. World Health Organization (WHO). *Use of Antimicrobials Outside Medicine and Resultant Antimicrobial Resistance in Humans*; WHO: Geneva, Switzerland, 2002; Volume 268, p. 2.
5. Duckworth, D.H.; Gulig, P.A. Bacteriophages. *BioDrugs* **2002**, *16*, 57–62. [CrossRef] [PubMed]

6. Vadstein, O. The use of immunostimulation in marine larviculture: Possibilities and challenges. *Aquaculture* **1997**, *155*, 401–417. [CrossRef]
7. Wiklund, T.; Dalsgaard, I. Occurrence and significance of atypical Aeromonas salmonicida in non-salmonid and salmonid fish species: A review. *Dis. Aquat. Organ.* **1998**, *32*, 49–69. [CrossRef] [PubMed]
8. Uhland, F.C.; Martineau, D.; Mikaelian, I.; Canada, S.-L.V. *Maladies des Poissons D'eau Douce du Québec: Guide de Diagnostic*; University Press of Montreal: Montreal, QC, Canada, 2000; ISBN 2760617785.
9. Boyd, J.; Williams, J.; Curtis, B.; Kozera, C.; Singh, R.; Reith, M. Three small, cryptic plasmids from Aeromonas salmonicida subsp. *salmonicida A449*. *Plasmid* **2003**, *50*, 131–144. [CrossRef]
10. Burr, S.E.; Pugovkin, D.; Wahli, T.; Segner, H.; Frey, J. Attenuated virulence of an Aeromonas salmonicida subsp. *salmonicida type III secretion mutant in a rainbow trout model. Microbiology* **2005**, *151*, 2111–2118.
11. Almeida, A.; Cunha, A.; Gomes, N.C.M.; Alves, E.; Costa, L.; Faustino, M.A.F. Phage therapy and photodynamic therapy: Low environmental impact approaches to inactivate microorganisms in fish farming plants. *Mar. Drugs* **2009**, *7*, 268–313. [CrossRef] [PubMed]
12. Vieira, A.; Silva, Y.J.; Cunha, A.; Gomes, N.C.M.; Ackermann, H.-W.W.; Almeida, A.; Cunha, A.; Gomes, N.C.M.; Ackermann, H.-W.W.; Almeida, A. Phage therapy to control multidrug-resistant Pseudomonas aeruginosa skin infections: In vitro and ex vivo experiments. *Eur. J. Clin. Microbiol. Infect. Dis.* **2012**, *31*, 3241–3249. [CrossRef] [PubMed]
13. Gill, J.; Hyman, P. Phage Choice, Isolation, and Preparation for Phage Therapy. *Curr. Pharm. Biotechnol.* **2010**, *11*, 2–14. [CrossRef] [PubMed]
14. Silva, Y.J.; Costa, L.; Pereira, C.; Mateus, C.; Cunha, A.; Calado, R.; Gomes, N.C.M.; Pardo, M.A.; Hernandez, I.; Almeida, A. Phage Therapy as an Approach to Prevent Vibrio anguillarum Infections in Fish Larvae Production. *PLoS ONE* **2014**, *9*, e114197. [CrossRef] [PubMed]
15. Pereira, C.; Moreirinha, C.; Lewickab, M.; Almeida, P.; Clemente, C.; Delgadillo, I.; Romalde, J.L.; Nunes, M.L.; Lewicka, M.; Almeida, P.; et al. Bacteriophages with potential to inactivate Salmonella Typhimurium: Use of single phage suspensions and phage cocktails. *Virus Res.* **2016**, *220*, 179–192. [CrossRef] [PubMed]
16. Pereira, C.; Moreirinha, C.; Lewicka, M.; Almeida, P.; Clemente, C.; Romalde, J.L.; Nunes, M.; Almeida, A. Characterization and in vitro evaluation of new bacteriophages for the biocontrol of Escherichia coli. *Virus Res.* **2017**, *227*, 171–182. [CrossRef] [PubMed]
17. Labrie, S.J.; Samson, J.E.; Moineau, S. Bacteriophage resistance mechanisms. *Nat. Rev. Microbiol.* **2010**, *8*, 317–327. [CrossRef] [PubMed]
18. Heller, K.J. Molecular interaction between bacteriophage and the gram-negative cell envelope. *Arch. Microbiol.* **1992**, *158*, 235–248. [CrossRef] [PubMed]
19. Bull, J.J.; Vegge, C.S.; Schmerer, M.; Chaudhry, W.N.; Levin, B.R. Phenotypic Resistance and the Dynamics of Bacterial Escape from Phage Control. *PLoS ONE* **2014**, *9*, e94690. [CrossRef] [PubMed]
20. Laanto, E.; Bamford, J.J.K.H.; Laakso, J.; Sundberg, L.L.-R. Phage-driven loss of virulence in a fish pathogenic bacterium. *PLoS ONE* **2012**, *7*, e53157. [CrossRef] [PubMed]
21. Santos, A.L.; Moreirinha, C.; Lopes, D.; Esteves, A.C.; Henriques, I.; Almeida, A.; Domingues, M.R.M.; Delgadillo, I.; Correia, A.; Cunha, A.; et al. Effects of UV Radiation on the Lipids and Proteins of Bacteria Studied by Mid-Infrared Spectroscopy. *Environ. Sci. Technol.* **2013**, *47*, 6306–6315. [CrossRef] [PubMed]
22. Taillandier, E.; Liquier, J. *DNA Structures Part A: Synthesis and Physical Analysis of DNA*; Methods in Enzymology; Elsevier: Amsterdam, The Netherlands, 1992; Volume 211, ISBN 9780121821128.
23. Helm, D.; Naumann, D. Identification of some bacterial cell components by FT-IR spectroscopy. *FEMS Microbiol. Lett.* **1995**, *126*, 75–79. [CrossRef]
24. Moreirinha, C.; Nunes, A.; Barros, A.A.; Almeida, A.; Delgadillo, I. Evaluation of the potential of Mid-infrared spectroscopy to assess the microbiological quality of ham. *J. Food Saf.* **2015**, *35*, 270–275. [CrossRef]
25. Blanco, M.; Villarroya, I. NIR spectroscopy: A rapid-response analytical tool. *TrAC Trends Anal. Chem.* **2002**, *21*, 240–250. [CrossRef]
26. Brereton, R. *Chemometrics: Data Analysis for the Laboratoty and Chemical Plant*; Wiley: London, UK, 2003.
27. Barth, A. Infrared spectroscopy of proteins. *Biochim. Biophys. Acta* **2007**, *1767*, 1073–1101. [CrossRef] [PubMed]
28. Alves, E.; Moreirinha, C.; Faustino, M.A.; Cunha, Â.; Delgadillo, I.; Neves, M.G.; Almeida, A. Overall biochemical changes in bacteria photosensitized with cationic porphyrins monitored by infrared spectroscopy. *Future Med. Chem.* **2016**, *8*, 613–628. [CrossRef] [PubMed]

29. Pudziuvyte, B.; Bakiene, E.; Bonnett, R.; Shatunov, P.A.; Magaraggia, M.; Jori, G. Alterations of Escherichia coli envelope as a consequence of photosensitization with tetrakis(*N*-ethylpyridinium-4-yl)porphyrin tetratosylate. *Photochem. Photobiol. Sci.* **2011**, *10*, 1046. [CrossRef] [PubMed]
30. Naumann, D. Infrared and NIR Raman Spectroscopy in Medical Microbiology. *Proc. SPIE* **1998**, *3257*. [CrossRef]
31. Dovbeshko, G.I.; Gridina, N.Y.; Kruglova, E.B.; Pashchuk, O.P. FTIR spectroscopy studies of nucleic acid damage. In *Talanta*; Elsevier: Amsterdam, The Netherlands, 2000; Volume 53, pp. 233–246.
32. Gasper, R.; Dewelle, J.; Kiss, R.; Mijatovic, T.; Goormaghtigh, E. IR spectroscopy as a new tool for evidencing antitumor drug signatures. *Biochim. Biophys. Acta Biomembr.* **2009**, *1788*, 1263–1270. [CrossRef] [PubMed]
33. Smith, B.C. *Infrared Spectral Interpretation: A Systematic Approach*; CRC Press: Boca Raton, FL, USA, 1999; ISBN 9780849324635.
34. Huffman, S.W.; Lukasiewicz, K.; Geldart, S.; Elliott, S.; Sperry, J.F.; Brown, C.W. Analysis of Microbial Components Using LC-IR. *Anal. Chem.* **2003**, *75*, 4606–4611. [CrossRef] [PubMed]
35. Salzer, R.; Siesler, H.W. *Infrared and Raman Spectroscopic Imaging*; Wiley-VCH: Weinheim, Germany, 2014; ISBN 3527336524.
36. Stuart, B. *Infrared Spectroscopy: Fundamentals and Applications*, 2nd ed.; John Wiley & Sons, Ltd.: Chichester, UK, 2004.
37. Kudva, I.T.; Jelacic, S.; Tarr, P.I.; Hovde, C.J.; Youderian, P. Biocontrol of Escherichia coli O157 with Biocontrol of Escherichia coli O157 with O157-Specific Bacteriophages. *Appl. Environ. Microbiol.* **1999**, *65*, 3767–3773. [PubMed]
38. Tomat, D.; Mercanti, D.; Balague, C.; Quiberoni, A. Phage biocontrol of enteropathogenic and shiga toxin-producing escherichia coli during milk fermentation. *Lett. Appl. Microbiol.* **2013**, *57*, 3–10. [CrossRef] [PubMed]
39. Park, S.; Nakai, T. Bacteriophage control of Pseudomonas plecoglossicida infection in ayu. *Dis. Aquat. Organ.* **2003**, *53*, 33–39. [CrossRef] [PubMed]
40. O'Flynn, G.; Coffey, A.; Fitzgerald, G.; Ross, R. The newly isolated lytic bacteriophages st104a and st104b are highly virulent against Salmonella enterica. *J. Appl. Microbiol.* **2006**, *101*, 251–259. [CrossRef] [PubMed]
41. Silva, Y.J.; Moreirinha, C.; Pereira, C.; Costa, L.; Rocha, R.J.M.; Cunha, Â.; Gomes, N.C.M.; Calado, R.; Almeida, A. Biological control of Aeromonas salmonicida infection in juvenile Senegalese sole (Solea senegalensis) with Phage AS-A. *Aquaculture* **2016**, *450*, 225–233. [CrossRef]
42. Filippov, A.A.; Sergueev, K.V.; He, Y.; Huang, X.-Z.; Gnade, B.T.; Mueller, A.J.; Fernandez-Prada, C.M.; Nikolich, M.P. Bacteriophage-resistant mutants in Yersinia pestis: Identification of phage receptors and attenuation for mice. *PLoS ONE* **2011**, *6*, e25486. [CrossRef] [PubMed]
43. Levin, B.R.; Bull, J.J. Population and evolutionary dynamics of phage therapy. *Nat. Rev. Microbiol.* **2004**, *2*, 166–173. [CrossRef] [PubMed]
44. Koskella, B.; Brockhurst, M.A. Bacteria-phage coevolution as a driver of ecological and evolutionary processes in microbial communities. *FEMS Microbiol. Rev.* **2014**, *38*, 916–931. [CrossRef] [PubMed]
45. Scanlan, P.D.; Buckling, A.; Hall, A.R. Experimental evolution and bacterial resistance: (Co)evolutionary costs and trade-offs as opportunities in phage therapy research. *Bacteriophage* **2015**, *5*, e1050153. [CrossRef] [PubMed]
46. Joly, N.; Schumacher, J.; Buck, M. Heterogeneous Nucleotide Occupancy Stimulates Functionality of Phage Shock Protein F, an AAA+ Transcriptional Activator. *J. Biol. Chem.* **2006**, *281*, 34997–35007. [CrossRef] [PubMed]
47. Lloyd, L.J.; Jones, S.E.; Jovanovic, G.; Gyaneshwar, P.; Rolfe, M.D.; Thompson, A.; Hinton, J.C.; Buck, M. Identification of a New Member of the Phage Shock Protein Response in Escherichia coli, the Phage Shock Protein G (PspG). *J. Biol. Chem.* **2004**, *279*, 55707–55714. [CrossRef] [PubMed]
48. Darwin, A.J. The phage-shock-protein response. *Mol. Microbiol.* **2005**, *57*, 621–628. [CrossRef] [PubMed]
49. Brissette, J.L.; Russel, M.; Weiner, L.; Model, P. Phage shock protein, a stress protein of Escherichia coli. *Proc. Natl. Acad. Sci. USA* **1990**, *87*, 862–866. [CrossRef] [PubMed]
50. Fröbel, J.; Rose, P.; Müller, M. Early Contacts between Substrate Proteins and TatA Translocase Component in Twin-arginine Translocation. *J. Biol. Chem.* **2011**, *286*, 43679–43689. [CrossRef] [PubMed]
51. Bageshwar, U.K.; VerPlank, L.; Baker, D.; Dong, W.; Hamsanathan, S.; Whitaker, N.; Sacchettini, J.C.; Musser, S.M. High Throughput Screen for Escherichia coli Twin Arginine Translocation (Tat) Inhibitors. *PLoS ONE* **2016**, *11*, e0149659. [CrossRef] [PubMed]

52. Friman, V.-P.; Hiltunen, T.; Jalasvuori, M.; Lindstedt, C.; Laanto, E.; Örmälä, A.-M.; Laakso, J.; Mappes, J.; Bamford, J.K.H. High Temperature and Bacteriophages Can Indirectly Select for Bacterial Pathogenicity in Environmental Reservoirs. *PLoS ONE* **2011**, *6*, e17651. [CrossRef] [PubMed]
53. Domingues, S.; Harms, K.; Fricke, W.F.; Johnsen, P.J.; da Silva, G.J.; Nielsen, K.M. Natural Transformation Facilitates Transfer of Transposons, Integrons and Gene Cassettes between Bacterial Species. *PLoS Pathog.* **2012**, *8*, e1002837. [CrossRef] [PubMed]
54. Schuster, C.F.; Mechler, L.; Nolle, N.; Krismer, B.; Zelder, M.-E.; Götz, F.; Bertram, R. The MazEF Toxin-Antitoxin System Alters the β-Lactam Susceptibility of Staphylococcus aureus. *PLoS ONE* **2015**, *10*, e0126118. [CrossRef] [PubMed]
55. Magnuson, R.D. Hypothetical functions of toxin-antitoxin systems. *J. Bacteriol.* **2007**, *189*, 6089–6092. [CrossRef] [PubMed]
56. Dy, R.L.; Przybilski, R.; Semeijn, K.; Salmond, G.P.C.; Fineran, P.C. A widespread bacteriophage abortive infection system functions through a Type IV toxin-antitoxin mechanism. *Nucleic Acids Res.* **2014**, *42*, 4590–4605. [CrossRef] [PubMed]
57. Mirzaei, K.M.; Nilsson, A.S. Isolation of Phages for Phage Therapy: A Comparison of Spot Tests and Efficiency of Plating Analyses for Determination of Host Range and Efficacy. *PLoS ONE* **2015**, *10*, e0118557. [CrossRef] [PubMed]
58. Abedon, S.T. Lysis from without. *Bacteriophage* **2011**, *1*, 46–49. [CrossRef] [PubMed]
59. Adams, M.H. *Bacteriophages*; John Wiley and Sons Inc.: New York, NY, USA, 1959.
60. Stuer-Lauridsen, B.; Janzen, T.; Schnabl, J.; Johansen, E. Identification of the host determinant of two prolate-headed phages infecting lactococcus lactis. *Virology* **2003**, *309*, 10–17. [CrossRef]
61. Laemli, U.K. Cleavage of Structural Proteins during the Assembly of the Head of Bacteriophage T4. *Nature* **1970**, *227*, 680–685. [CrossRef]
62. Neuhoff, V.; Arold, N.; Taube, D.; Ehrhardt, W. Improved staining of proteins in polyacrylamide gels including isoelectric focusing gels with clear background at nanogram sensitivity using Coomassie Brilliant Blue G-250 and R-250. *Electrophoresis* **1988**, *9*, 255–262. [CrossRef] [PubMed]

antibiotics

MDPI

Perspective

Silk Route to the Acceptance and Re-Implementation of Bacteriophage Therapy—Part II

Expert round table on acceptance and re-implementation of bacteriophage therapy †, Wilbert Sybesma [1,2,*], Christine Rohde [3], Pavol Bardy [4], Jean-Paul Pirnay [5], Ian Cooper [6], Jonathan Caplin [6], Nina Chanishvili [7], Aidan Coffey [8], Daniel De Vos [5], Amber Hartman Scholz [3], Shawna McCallin [9], Hilke Marie Püschner [3], Roman Pantucek [4], Rustam Aminov [10], Jiří Doškař [4] and D. İpek Kurtböke [11,*]

1 Department of Neuro-Urology, Balgrist University Hospital, University of Zürich, CH-8008 Zürich, Switzerland
2 Nestlé Research Center, Nestec Ltd., Vers-chez-les-Blanc, CH-1000 Lausanne, Switzerland
3 Leibniz Institute DSMZ-German Collection of Microorganisms and Cell Cultures, D-38124 Braunschweig, Germany; chr@dsmz.de (C.R.); Amber.H.Scholz@dsmz.de (A.H.S.); Hilke.Pueschner@dsmz.de (H.M.P.)
4 Department of Experimental Biology, Faculty of Science, Masaryk University, Brno 611 37, Czech Republic; bardy.pavol@mail.muni.cz (P.B.); pantucek@sci.muni.cz (R.P.); doskar@sci.muni.cz (J.D.)
5 Laboratory for Molecular and Cellular Technology, Queen Astrid Military Hospital, B-1120 Brussels, Belgium; jean-paul.pirnay@telenet.be (J.-P.P.); daniel_de_vos@skynet.be (D.D.V.)
6 School of Pharmacy and Biomolecular Sciences and School of Environment & Technology, University of Brighton, Brighton BN2 4GJ, UK; I.Cooper@brighton.ac.uk (I.C.); J.L.Caplin@brighton.ac.uk (J.C.)
7 Eliava Institute of Bacteriophage, Microbiology and Virology, Tbilisi 0160, Georgia; nina.chanishvili@gmail.com
8 Department of Biological Sciences, Cork Institute of Technology, Bishopstown, Cork T12 P928, Ireland; aidan.coffey@cit.ie
9 Department of Fundamental Microbiology, University of Lausanne, CH-1015 Lausanne, Switzerland; shawna.mccallin@unil.ch
10 School of Medicine & Dentistry, University of Aberdeen, Aberdeen AB25 2ZD, UK; rustam.aminov@gmail.com
11 GeneCology Research Centre and the Faculty of Science, Health, Education and Engineering, University of the Sunshine Coast, Maroochydore DC, QLD 4558, Australia
* Correspondence: wilbert.sybesma@gmail.com (W.S.); ikurtbok@usc.edu.au (D.I.K.); Tel.: +41-021-785-8111 (W.S.); +61-07-5430-2819 (D.I.K.); Fax: +41-021-785-8561 (W.S.); +61-07-5430-2881(D.I.K.)
† Phage_therapy@pha.ge; wilbert.sybesma@gmail.com (W.S.); ikurtbok@usc.edu.au (D.I.K.).

Received: 25 March 2018; Accepted: 12 April 2018; Published: 23 April 2018

Abstract: This perspective paper follows up on earlier communications on bacteriophage therapy that we wrote as a multidisciplinary and intercontinental expert-panel when we first met at a bacteriophage conference hosted by the Eliava Institute in Tbilisi, Georgia in 2015. In the context of a society that is confronted with an ever-increasing number of antibiotic-resistant bacteria, we build on the previously made recommendations and specifically address how the Nagoya Protocol might impact the further development of bacteriophage therapy. By reviewing a number of recently conducted case studies with bacteriophages involving patients with bacterial infections that could no longer be successfully treated by regular antibiotic therapy, we again stress the urgency and significance of the development of international guidelines and frameworks that might facilitate the legal and effective application of bacteriophage therapy by physicians and the receiving patients. Additionally, we list and comment on several recently started and ongoing clinical studies, including highly desired double-blind placebo-controlled randomized clinical trials. We conclude with an outlook on how recently developed DNA editing technologies are expected to further control and enhance the efficient application of bacteriophages.

Keywords: antibiotic resistance; bacteriophages; bacteriophage therapy; Nagoya Protocol; CRISPR CAS

1. Introduction

The history of antimicrobial drug discovery includes more than 15 classes of compounds that became a cornerstone in microbial infection control and management and have indisputably saved many lives [1]. Indeed, they have become one of the most successful forms of therapy in clinical medicine. This success, however, is compromised by the emergence and dissemination of antimicrobial resistance, in part due to the widespread (over)use of these compounds in clinical and veterinary medicine and agriculture, thus limiting the efficiency of antibiotics in the control and management of infectious diseases [2]. The extent of the antimicrobial resistance problem in terms of increased morbidity and mortality rates, as well as elevated healthcare costs, has been brought to the public's attention by several national and international health protection agencies, including the Centers for Disease Control and Prevention (CDC), the World Health Organization (WHO), and the European Medicines Agency (EMA) [3–5]. More specifically, WHO resolution 68.7.3 invites international, regional, and national partners to implement the necessary actions in order to contribute to the accomplishment of the five objectives of the global action plan on antimicrobial resistance. If no immediate action is taken, the estimated death toll due to antimicrobial resistance will reach the millions by the year 2050, the cost to the global economy is expected to rise to $100 trillion, and the number of people living in extreme poverty is expected to increase [6].

In view of this alarming situation, we published a first opinion paper as a multidisciplinary expert group on the acceptance and re-implementation of bacteriophage therapy in 2016 [7]. In this present perspective paper, we briefly evaluate the status of the previously-made recommendations for bacteriophage therapy over the short term. In addition, we comment on the consequence of the Nagoya Protocol for bacteriophage therapy and then provide an overview on how limitations of the traditional application of bacteriophage therapy could be overcome by the use of Clustered Regularly Interspaced Short Palindromic Repeats/CRISPR associated systems (CRISPR/Cas) gene-edited bacteriophages in the near future.

1.1. *Factors Impacting the Broad-Scale Application of the Bacteriophages*

Four issues have been identified that are limiting, or even preventing, the application of the bacteriophage therapy in the Western World in the 21st century [7]: (1) Quality and quantity of previously conducted study designs, (2) bacteriophage-cocktail production, composition, and application methods in the context of the current legal framework, (3) Lack of awareness among (para-) medical staff and the general public about the potential use of bacteriophage therapy, and (4) Limitations in intellectual property protection for bacteriophage therapeutic applications.

1.1.1. Quality and Quantity of Previously Conducted Study Set-Ups

Bacteriophage therapy has been used for more than 100 years, mainly in Eastern Europe. However, the number of double-blind, placebo-controlled, randomized clinical trials in different fields of medical applications have been limited. Furthermore, they fall far from providing statistically-relevant conclusions about the efficacy of bacteriophage therapy. As a consequence, health authorities and medical professionals in the Western World have been hesitant to proceed with the bacteriophage therapy. Table 1 provides an overview of recently concluded, or currently running clinical studies with bacteriophages with excerpts from https://clinicaltrials.gov/. Notwithstanding the increase in conducted trial numbers, the number of fully-completed and well-documented trials still remains too low to draw substantial conclusions for the diverse range of medical applications where bacteriophage therapy might be implemented. Of the completed trials, several factors have hampered their ability to conclude on the potential efficacy of phage therapy.

Small patient cohorts and the failure to recruit enough patients have severely limited the conclusions that can be drawn from modern bacteriophage trials. For instance, the recently completed Phagoburn trial, which represented a public investment of 3.85 million euro, enrolled a total of only 27 patients between 11 centers [8,9]. This was far from the pre-calculated 220 patients needed to provide statistically significant results for the study. Reasons cited for the low number of participants were restrictive patient inclusion criteria, a lower incidence of burn wound infections than in previous years, and a shorter recruitment period due to regulatory constraints. Patient enrollment was limited to mono-species infections at the request of the ethical committee, which does not represent the clinical reality of burn-wound infections, and resulted in few eligible patients for the trial. Another trial using bacteriophage for the treatment of pediatric *Escherichia coli* diarrhea also did not reach their estimated patient numbers, because of early trial termination [10]. An in-depth failure analysis of this study revealed that *Streptococcus* sp. may have been a better clinical target than as initially anticipated, *E. coli*.

These two studies, which represent the largest in recent history, highlight several lessons to be learned for future bacteriophage therapy investigations. In addition to individual patient safety, ethical committees should also consider the overarching purpose of clinical trials to produce significant and generalizable results when reviewing clinical trial protocols; the inability to do so is both a detriment to the well-being of society and a waste of trial-eligible patients. Priority should be given to infections with established pathogens, and the test product should reflect the clinical reality to cover indicated pathogens. There is no clinical evidence to suggest a safety concern for targeting multiple pathogens with broad spectrum bacteriophage products, as supported by several recent clinical trials (Table 1), the Polish experience, and the long-standing safe indication for commercial polyvalent bacteriophage cocktails in Russia and Georgia [11–13]. Preliminary data from an ongoing trial on patients suffering from urinary tract infections treated with a broad-spectrum bacteriophage cocktail in Georgia will further assuage safety concerns [14]. Furthermore, in this Randomized Clinical Trials (RCT), one of the inclusion criteria requires the in vitro sensitivity of the identified pathogen(s) to the bacteriophage present in the cocktail, thereby acknowledging the very specific bacteriophage-bacterial host interaction. In order to continuously enhance the spectrum of the cocktail during the study, resistant strains are used for adaptation of Pyo bacteriophage cocktail [15].

Given the time and financial investment required for clinical trials, it might be prudent to exploit information from smaller-scale clinical investigations, as well as the ongoing practice of bacteriophage therapy in Eastern countries. Valuable knowledge to understanding bacteriophage therapy has already been generated by The Polish bacteriophage Therapy Unit at the Hirszfeld Institute, a nonprofit entity, that has accumulated data from years of individual patient reports [16]. Good insights can certainly be derived from the several case reports that have provided more in-depth analysis of clinical samples and clinical parameters, even compared to some of the more formal clinical trials. Furthermore, bacteriophage therapy has been approved under emergency treatment schemes in the USA, Australia, France, and Belgium. Table 2 summarizes a collection of recently reported case studies. A thorough and objective assessment on the cost and benefits of bacteriophage production and therapy applications in Russia, Georgia, and Poland, including production protocols, safety, and efficacy, would reveal underlying strategies developed from decades of empirical bacteriophage use. Proper reporting should be a priority for all uses of clinical bacteriophage therapy, whether it be formal trials or case reports [12].

Table 1. Overview of bacteriophage therapy clinical studies.

Name of Study and Organizations Running the Study	Target Organism(s)	Description and Objectives	Outcome Measures	Additional Comments	Clinical Trials.gov Identifier/ Reference
Bacteriophages for treating urinary tract infections in patients undergoing transurethral resection of the prostate: a randomized, placebo-controlled, double-blind clinical trial. Tzulukidze National Center of Urology, Tbilisi, Georgia; Eliava Institute of Bacteriophages, Microbiology, and Virology in Tbilisi, Georgia; Balgrist University Hospital, Zürich, Switzerland The study is run in The Republic of Georgia.	*Enterococcus* spp., *Streptococcus* spp., *Escherichia coli*, *Proteus* spp., *P. aeruginosa*	Randomized placebo-controlled double-blind clinical trial: Patients planned for transurethral resection of the prostate are screened for UTIs and enrolled if eligible microorganisms in urine culture are $\geq 10^4$ cfu/mL. Patients are randomized in a double-blind fashion to the three study treatment arms of 27 people, each in a 1:1:1 ratio to receive either: (a) bacteriophage (b) placebo solution, or (c) antibiotic treatment according to the antibiotic sensitivity pattern.	Primary: Success of intravesical treatment, defined as normalization of urine culture (no evidence of bacteria, i.e., $<10^4$ colony forming units/mL) after 7 days of treatment. Secondary: Adverse events, in categorization according to the National Cancer Institute Common Terminology Criteria for Adverse Events (CTCAE) version four in grade one to five. Tertiary: Changes in bladder and pain diary assessment of number of voids, number of leakages, post void residual.	The study uses the commercially available Pyo-bacteriophage cocktail as produced by The Eliava institute in Tbilisi. 81 patients are involved. The Pyo-bacteriophage cocktail is subjected to continuous adaptation during the study. The study started in 2016, and is expected to end in 2018.	NCT03140085 [14]
PHAGOPIED - Standard treatment associated with bacteriophage therapy vs. placebo for diabetic foot ulcers infected by *Staphylococcus aureus*. The study is run by the Centre Hospitalier Universitaire de Nīmes, France, with collaboration with Pherecydes Pharma, Romainville, France.	*Staphylococcus aureus*, MSSA and MRSA	This project utilizes anti-*Staphylococcus* bacteriophages, delivered topically, vs. a placebo control. The study uses random allocations in a parallel assignment intervention design. The main objective of this study is to compare the efficacy of standard treatment associated with a topical anti-staphylococcal bacteriophage cocktail versus standard treatment plus placebo for diabetic foot ulcers mono-infected by methicillin-resistant or susceptible *S. aureus* (MRSA or MSSA) as measured by the relative reduction in wound surface area (%) at 12 weeks.	Primary: The relative reduction in wound surface area over 12 weeks. Secondary: Safety effects, local side effects (rash onset or worsening of local inflammatory signs) and general symptoms (vital signs, fever, rash, arthralgia, gastrointestinal symptoms).	First posted online in 2016. This study is not yet recruiting, but 60 patients are expected to join.	NCT02664740
MUCOPHAGES - Bacteriophage effects on *Pseudomonas aeruginosa*. The study is run by the University Hospital Montpellier, France.	*Pseudomonas aeruginosa*	The study is designed to evaluate the efficacy of bacteriophages on *P. aeruginosa* isolates recovered from sputum.	The study utilizes a suspension of ten bacteriophages. These are tested against isolates recovered from cystic fibrosis patients, to determine their ability to infect these strains.	Completed in 2012. No results posted online.	NCT01818206 [17]

Table 1. *Cont.*

Name of Study and Organizations Running the Study	Target Organism(s)	Description and Objectives	Outcome Measures	Additional Comments	Clinical Trials.gov Identifier/ Reference
Experimental bacteriophage therapy of bacterial infections. The study is led by the Polish Academy of Sciences.	*Staphylococcus, Enterococcus, Pseudomonas, Escherichia, Klebsiella, Proteus, Citrobacter, Acinetobacter, Serratia, Morganella, Shigella, Salmonella, Enterobacter, Stenotrophomonas, Burkholderia*	The study uses suspensions of lytic bacteriophages active against clinical isolates of the test species. The program determines to use bacteriophage treatment in a therapeutic role where no other viable treatment is available. For each patient only, specific formulations of single bacteriophage or a bacteriophage mixture that are active against the pathogenic bacterial strain or strains isolated from the patient are used for the treatment (oral, rectal and/or topical application).	The principle focus of the work is to use bacteriophage suspensions to treat the following conditions: bone, upper respiratory, genital and urinary tract infections, as well as post-operative non-healing wounds where antibiotic treatment has not produced positive results.	Start: 2005. Current status unknown. Last update posted in 2013. Number of persons involved has not been stated.	NCT00945087 [11]
PhagoBurn. Phase I/II Clinical Trial. This Project is a European Research & Development (R&D) Project Funded by the European Commission Under the 7th Framework Program for Research and Development Involving seven Clinical Sites in France, Belgium & Switzerland.	*Escherichia coli* & *Pseudomonas aeruginosa*	Evaluation of bacteriophage therapy for the treatment of *Escherichia coli* and *Pseudomonas aeruginosa* wound infections in burned patients A randomized, parallel assignment study assessing tolerance and efficacy of local bacteriophage treatment of wound infections due to *E. coli* or *Ps. aeruginosa* in burned patients.	This study tests the efficacy of *E. coli* and *Ps. aeruginosa* bacteriophage cocktails against silver sulfadiazine to treat wound infections by those bacterial species. Primary: Time necessary for a persistent bacteria reduction of two modes or persistent bacteria eradication relative to D0 adjusted on antibiotic treatment (active on targeted strain) introduced between D1 to D7. Secondary: Assessment of tolerance of treatment over 21 days. Adverse events frequencies will be assessed in each treatment arm. bacteriophage therapy safety profile will be compared to safety profile of standard of care. Incidence and delay of infection reduction with different bacterial species from the targets over a period of seven days. Number of sites cured: The number of infected burns or infected wounds getting a clinical improvement will be described and compared between treatment group over a period of 7 day.	Launched in 2013 and achieved in 2017, PhagoBurn was the world first prospective multicentric, randomized, single-blind and controlled clinical trial of bacteriophage therapy ever performed according to both Good Manufacturing (GMP) and Good Clinical Practices (GCP). Only 27 patients between 11 centers were included, which is far from the pre-calculated 220 patients needed to provide statistically significant results for the study. See main text for more information and lessons learned.	NCT02116010 [8,9]

Table 1. *Cont.*

Name of Study and Organizations Running the Study	Target Organism(s)	Description and Objectives	Outcome Measures	Additional Comments	Clinical Trials.gov Identifier/ Reference
Antibacterial treatment against diarrhea in oral rehydration solution. The study was run by Nestlé, Switzerland in collaboration with Dhaka Hospital of the International Centre for Diarrheal Disease Research, Bangladesh.	*Escherichia. coli* (T4 bacteriophage)	This randomized double-blind, placebo-controlled trial aims to demonstrate the potentials of a new form of therapy for childhood diarrhea, by measuring the effect of oral administered E. coli bacteriophage in children aged 4–60 months of age with proven ETEC and EPEC diarrha.	Primary outcome measures: Assessment of safety, tolerability and efficacy (reduce severity of diarrhea assessed by reduced stool volume and stool frequency) of oral administration of T4 bacteriophages in young children with diarrhea due to ETEC and/or EPEC infections. Time frame: five days. Secondary outcome measures: Clinical assessment, blood tests, morbidity, duration of hospitalization. Time frame: five days.	First posted online 2009; study ended in 2013. Oral coliphages showed a safe gut transit in children, but failed to achieve intestinal amplification and to improve diarrhea outcome, possibly due to insufficient bacteriophage coverage and too low *E. coli* pathogen titers requiring higher oral bacteriophage doses.	NCT00937274 [10]
Existence in the human digestive flora of bacteriophages able to prevent the acquisition of multiresistant Enterobacteria (PHAGO-BMR). The study is led by Assistance Publique-Hôpitaux de Paris, France.	MDR- Enterobacteria	The study plans to recruit 460 people hospitalized in intensive care unit (resuscitation). The choice of this unit is linked to the fact that the monitoring of resistant bacteria is carried out regularly during the hospitalization. On stool samples collected at separate times of the stay (admission and then during the stay), the scientists look for 2 types of bacteria and viruses capable of destroying them.	Primary: Presence or absence of bacteriophages capable of lysing circulating Ec-ESBL/EPC or Kp-ESBE/EPC in resuscitation units in non-carriers having acquired carriers *E. coli* or *K. pneumoniae* producing ESBL or carbapenemases. Secondary: Presence or absence of bacteriophages in patients identified as carriers of Ec-ESBL/EPC or Kp-ESBL/EPC at entry to resuscitation (control population). Isolated bacteriophages will be characterized.	First posted online in 2017. The study is not yet recruiting.	NCT03231267

Table 1. *Cont.*

Name of Study and Organizations Running the Study	Target Organism(s)	Description and Objectives	Outcome Measures	Additional Comments	Clinical Trials.gov Identifier/ Reference
METAKIDS Phages dynamics and influences during human gut microbiome establishment.	Enteric microbial species	This project relies on the ability of Meta3C, a technique developed to identify the bacterial host genomes of the different bacteriophages the investigators will detect thanks to the physical collision these molecules experience. Given the role that human gut bacteriophages may play in shaping the development of host microbiomes, their potential for application is of great interest.	Primary outcome measures: Genomic reconstruction and characterization of the different genomes (phages, bacteria, yeast) present in the human gut during the three first years of life. These outcomes will provide a large catalog of DNA sequences. Characterization of the variation of the different species present in human gut during the three first years of life and characterization of bacteriophages-bacteria interactions. This outcome will provide access to the dynamics of the different species present during this period and possibility to correlate them with environmental variation (dietary, age). Secondary: Characterization of bacteriophages and bacteria variations in response to environmental perturbations during infant gut development. Time frame: two to three weeks.	First posted online in 2017; the study is currently recruiting. Estimated enrollment is 20 persons.	NCT03296631
Evaluate bacteriophage as a useful immunogen in patients with primary immune deficiency diseases (PIDD) The study is run by the University of South Florida, USA.	*Escherihia. coli*	This protocol is designed to ascertain whether the bacteriophage 0X174 neoantigen is safe and effective as an antigen used in the evaluation of primary and secondary immune responses. Bacteriophage 0X174 is given intravenously two billion PFU/Kg of body weight; small blood specimens of 3–5 mL (about 1 teaspoon) are collected after 15 min, 7 days, 14 days, and 28 days.	Primary Outcome Measures: Evidence of capacity of switch from IgM to IgG during 12 weeks of trial. Blood samples are obtained after each immunization of bacteriophages.	Current status unknown. First posted online in 2012. Last update 2012. All patients receive two doses of bacteriophages. Selected patients may receive a tertiary vaccine.	NCT01617122

Table 1. *Cont.*

Name of Study and Organizations Running the Study	Target Organism(s)	Description and Objectives	Outcome Measures	Additional Comments	Clinical Trials.gov Identifier/ Reference
Evaluation and detection of facial *Propionibacterium acnes* bacteria and bacteriophage The study is led by Maccabi Healthcare Services, Israel.	*Propionibacterium acnes*	This multi-center, outpatient study will extract and evaluate the presence of facial *P. acnes* bacteria and bacteriophage strains using pore strips on up to 400 human subjects. An additional *P. acnes* visual detection method (VISIOPOR®PP34N) will be used in this study as per PI decision to explore whether there is a correlation between *P. acnes* bacterial presence and fluorescent signal.	Primary: Detection and analysis of facial *P. acnes* presence. Time frame: Day 0 and week 8. Secondary: Assessing correlation between bacteriophage and *P. acnes* using a. Demographic Questionnaire b. Visual Supportive Methodology (VISIOPOR®PP34N) as per PI decision. Time frame: As above.	First posted online in 2017. Not currently recruiting, but 400 people are estimated to participate.	NCT03009903
Bacteriophages PreforPro cocktails as novel Prebiotics The study is led by Colorado State University, USA. PreforPro is commercialized by Deerland Enzymes, Kennesaw, GA, USA	Enteric bacteria	The bacteriophage Study is a randomized, double-blind, placebo-controlled crossover trial that investigates the utility of four supplemental bacteriophage strains (LH01-Myoviridae, LL5-Siphoviridae, T4D-Myoviridae, and LL12-Myoviridae) to modulate the gut microbiota, and therefore ameliorate common inflammation-related GI distress symptoms (e.g., gas, bloating, diarrhea, constipation, etc.) experienced by healthy individuals. The main goal of this study is to see if consumption of PreforPro, a commercially available prebiotic dietary supplement consisting of a mixture of bacteriophages, improves gut bacteria profiles in individuals relative to a placebo control.	Primary: Microbiota modulation. Time frame: Baseline visit prior to starting treatments, four weeks after starting treatment one, end of two-week washout period, 4-weeks after starting treatment two. Use of 16s rRNA sequencing of stool samples to determine whether the administered interventions resulted in changes to microbial composition. Secondary: Local inflammation Time frame: as above. Inflammation in the bowels will be assessed by use of ELISA test for fecal calprotectin. Systemic Inflammation. Time Frame: As above. Systemic inflammation will be assessed by an ELISA test for CRP and circulating cytokines and immune factors.	The study completed in 2017, but results have not yet been posted online. 43 persons enrolled in the study.	NCT03269617

Table 1. *Cont.*

Name of Study and Organizations Running the Study	Target Organism(s)	Description and Objectives	Outcome Measures	Additional Comments	Clinical Trials.gov Identifier/ Reference
The Use of Bacteriophage Phi X174 to Assess the Immune Competence of HIV-Infected Patients in vivo. The study is run by the National Institute of Allergy and Infectious Diseases, USA.	*Escherichia. coli*	The objective of this study is to evaluate the safety and utility of bacteriophage phi X174 immunization as a tool to assess the immune competence of HIV-infected patients at different stages of disease in vivo, and to assess the impact of viral load levels and therapy-induced changes in viral load levels on the response to immunization with the neo-antigen bacteriophage phi X174.	Primary. Immune parameters (not further published online.	Study started in 1996, and ended in 2000. 52 patients were involved in the study.	NCT00001540 [18]
Randomized and double-blinded placebo-controlled study of topical application of AB-SA01 cocktail to intact skin of healthy adults. The study is run by AmpliPhi bacteriophage Ltd., the US Army, and the Walter Reed Army Institute of Research Clinical Trials Center, USA.	*Staphylococcus aureus*	The study aims to examine the safety of ascending doses of AB-SA01 when topically applied to intact skin of healthy adults. AB-SA01 consists of three bacteriophages (viruses) that target *Staphylococcus aureus* bacteria. The safety of AB-SA01 will be assessed when topically administered once daily to the volar aspect of the forearm at different doses for three consecutive days.	Primary: Occurrence, intensity, and relationship of adverse events (AEs) from first dose through the end of study visit (14 ± 2 days). Change from baseline in clinical laboratory tests. Time frame: Day 0 (pre-dose), Day 3, and Day 14 ± 2 days. Clinical laboratory tests (hematology, chemistry, and urinalysis). Skin Reaction change from Baseline. Time Frame.	First published online in 2016, and the last update was in 2016 as well. 12 persons recruited.	NCT02757755
A prospective, randomized, double-blind controlled study of WPP-201 for the safety and efficacy of treatment of venous leg ulcers. The study was run by Southwest Regional Wound Care Center, USA.	*Pseudomonas aeruginosa*	The study was designed to assess the safety of *Ps. aeruginosa*-specific bacteriophages for the treatment of leg ulcers in human patients. WPP-201 is a pH neutral, polyvalent bacteriophage preparation, which contains 8 bacteriophages lytic for *Ps. aeruginosa*, *S. aureus*, and *E. coli*. The cocktail contains a concentration of approximately 1×10^9 PFU/mL of each of the component monophages.	Primary: Evaluate the safety of the use of WPP-201.	Study started in 2008 and was completed in 2011. 64 patients were involved. This study found no safety concerns with the bacteriophage treatment. Efficacy of the preparation will need to be evaluated in a phase II efficacy study.	NCT00663091 [19]

Table 2. Human bacteriophage therapy related case studies published in peer-reviewed English-language scientific literature over the last ten years.

Case Study Title	Description	Outcomes	Comments	Reference
Refractory *Pseudomonas* Bacteremia in a Two-Year-Old sterilized by bacteriophage therapy	The authors report a complex case that involved a pediatric patient who experienced recalcitrant multidrug-resistant *Pseudomonas aeruginosa* infection complicated by bacteremia/sepsis; antibacterial options were limited because of resistance, allergies, and suboptimal source control.	A cocktail of 2 bacteriophages targeting the infectious organism introduced on 2 separate occasions sterilized the bacteremia.	USA	[20]
Development and use of personalized bacteriophage-based therapeutic cocktails to treat a patient with a disseminated resistant *Acinetobacter baumannii* infection	The authors report on a method used to produce a personalized bacteriophage-based therapeutic treatment for a 68-year-old diabetic patient with necrotizing pancreatitis complicated by an MDR *A. baumannii* infection. Despite multiple antibiotic courses and efforts at percutaneous drainage of a pancreatic pseudocyst, the patient deteriorated over a 4-month period. In the absence of effective antibiotics, two laboratories identified nine different bacteriophages with lytic activity for an *A. baumannii* isolate from the patient.	Administration of bacteriophages intravenously and percutaneously into the abscess cavities was associated with reversal of the patient's downward clinical trajectory, clearance of the *A. baumannii* infection, and a return to health.	USA	[21]
Phage therapy in a 16-year-old boy with Netherton syndrome	The authors report on a 16-year-old male with all the typical manifestations of Netherton Syndrome, including atopic diathesis and ongoing serious staphylococcal infections and allergy to multiple antibiotics whose family sought help at the Eliava bacteriophage Therapy Center when all other treatment options were failing.	Treatment with several antistaphylococcal bacteriophage preparations led to significant improvement within seven days and very substantial changes in his symptoms and quality of life after treatment for six months, including return visits to the Eliava bacteriophage Therapy Center after three and six months of ongoing use of bacteriophage at home	Georgia. (Patient came from France)	[22]
Use of bacteriophages in the treatment of colistin-only-sensitive *Pseudomonas aeruginosa* septicaemia in a patient with acute kidney injury—a case report	A 61-year-old man with gangrene of the peripheral extremities, resulting in the amputation of the lower limbs and the development of large necrotic pressure sores, developed septicaemia with colistin-only-sensitive *P. aeruginosa*. Intravenous colistin therapy was started. Ten days later, the patient developed acute kidney injury and antibiotic therapy was discontinued to prevent further kidney damage. *P. aeruginosa* septicaemia re-emerged and two bacteriophages, which showed in vitro activity against the patient's *P. aeruginosa* isolates, were administered as a 6-h intravenous infusion for ten days.	Immediately upon bacteriophage application, blood cultures turned negative, CRP levels dropped and the fever disappeared. Kidney function recovered after a few days. Hemofiltration was avoided and no unexpected adverse events, clinical abnormalities or changes in laboratory test results that could be related to the application of bacteriophages were observed.	Belgium The patient died four months after bacteriophage therapy of sudden in-hospital refractory cardiac arrest due to blood culture-confirmed *Klebsiella pneumoniae* sepsis.	[23]

Table 2. *Cont.*

Case Study Title	Description	Outcomes	Comments	Reference
Bacteriophage treatment of intransigent diabetic toe ulcers: a case series	The authors present a compassionate-use case series of nine patients with diabetes and poorly perfused toe ulcers containing culture-proven *Staphylococcus aureus* infected bone and soft tissue, who had responded poorly to recommended antibiotic therapy.	All infections responded to the bacteriophage applications and the ulcers healed in an average of seven weeks with infected bone debridement. One ulcer, where vascularity was extremely poor and bone was not removed to preserve hallux function, required 18 weeks of treatment.	USA	[24]
Use of bacteriophages in the treatment of *Pseudomonas aeruginosa* infections	The author reports on bacteriophage treatment of *Pseudomonas aeruginosa* otitis in a pet dog and in a human burn wound patient.	Symptomatic improvement and bacteriophage multiplication were seen in the pet dog and in the human patient.	UK	[25]
Clinical aspects of bacteriophage therapy	The authors present a detailed retrospective analysis of the results of bacteriophage therapy of 153 patients with a wide range of infections resistant to antibiotic therapy admitted for treatment at the bacteriophage therapy unit of the Ludwik Hirszfeld Institute of Immunology and Experimental Therapy, Wrocław, Poland, between January 2008 and December 2010.	Data suggest that bacteriophage therapy provided good clinical results in a significant cohort of patients with otherwise untreatable chronic bacterial infections and is essentially well tolerated.	Poland	[11]
Bacteriophage therapy for refractory *Pseudomonas aeruginosa* urinary tract infection	The authors describe adjunctive bacteriophage therapy for refractory *Pseudomonas aeruginosa* urinary tract infection in the context of bilateral ureteric stents and bladder ulceration, after repeated failure of antibiotics alone.	Combined therapy was well-tolerated, apparently resulting in symptomatic relief and microbiological cure where repeated courses of antibiotics combined with stent removal had failed. Bacteriophage did not persist nor was any antibiotic- or bacteriophage resistant *P. aeruginosa* identified.	Australia Additive effort of combined bacteriophage-antibiotic treatment	[26]
Eradication of *Enterococcus faecalis* by bacteriophage therapy in chronic bacterial prostatitis	The authors report on the treatment of three patients suffering from chronic bacterial prostatitis who were qualified for an experimental bacteriophage therapy protocol managed at the bacteriophage Therapy Unit in Wrocław. The patients had previously been treated unsuccessfully with long-term targeted antibiotics, autovaccines, and laser biostimulation.	Rectal application of bacteriophage lysates targeted against *Enterococcus faecalis* cultured from the prostatic fluid gave encouraging results regarding bacterial eradication, abatement of clinical symptoms of prostatitis, and lack of early disease recurrence.	Poland	[27]
Corneal Infection Therapy with Topical Bacteriophage Administration	A 65-year-old woman suffering from MDR *S. aureus* infection of the cornea as a post-operative complication of a craniotomy. This patient suffered the chronic, persistent infection for years before seeking therapy in Tbilisi, Georgia. A single bacteriophage was administered both topically in the eye and nasal application and intravenous application for four weeks.	The patient's regular physicians published that the patient's ocular and nasal cultures after returning from Georgia were negative at three and six months post-treatment.	Georgia. (Patient came from France)	[28]

Table 2. *Cont.*

Case Study Title	Description	Outcomes	Comments	Reference
The use of a novel biodegradable preparation capable of the sustained release of bacteriophages and ciprofloxacin, in the complex treatment of multi-drug resistant Staphylococcus Aureus-infected local radiation injuries caused by exposure to Sr90.	Authors report the topical use of PhagoBioDerm (phage + ciprofloxacin wound polymer) to treat two wounds infected with *S. aureus* after the failure of conventional treatment. The responsible pathogen was resistant to ciprofloxacin, but sensitive to the bacteriophage.	PhagobioDerm resulted in reduced purulent drainage and symptom amelioration. *S. aureus* was eliminated from the wound	Georgia	[29]
Successful eradication of methicillin-resistant *Staphylococcus aureus* (MRSA) intestinal carrier status in a healthcare worker	Healthcare worker suffered urinary tract infections caused by MRSA that was carried in the GI tract. Authors report that bacteriophage was applied orally	Eradiation of carrier status	Poland	[30]
Phage therapy compassionate use in France in 2017.	Abstract presented by Pherecydes at bacteriophages-sur-Yvette in November 2017 documenting the use of bacteriophage to treat two patients with severe bone and joint infections caused by MDR organisms with topical bacteriophage at a hospital in Lyon.	Symptom amelioration and no reported side effects.	France	[31]
Open-label treatment of RCT with bacteriophages for treating urinary tract infections in patients undergoing transurethral resection of the prostate: a randomized, as mentioned in Table 1.	Prior to the start of the double-blind placebo-controlled trial, nine patients were treated with bacteriophage Pyo cocktail.	In six of nine patients, the titer of the pathogenic bacteria was decreased, varying between 1 log and 7 log (sterile).	Georgia	NCT03140085 [14].

1.1.2. Bacteriophage Production and Application Methods in Context of the Current Legal Framework

Today's legislation and safety requirements for the production and admission of drugs are, for good reasons, heavily controlled. By having strict quality control procedures, it is assured that drugs are effective, safe, and produced with a consistent quality and composition. The production and application of bacteriophage therapy should not be any different. However, the nature of bacteriophages and their effective medical application are not compatible with current production and admission requirements for chemical drugs. The intrinsic strength of bacteriophages relates to their antagonistic evolution with their bacterial hosts. To assure an effective application of bacteriophage therapy, this requires the ability to continuously adjust and adapt the composition of bacteriophage cocktails. Such a flexible and dynamic production system, coupled with the application of an infection-eliminating medication is incompatible with current legislation and safety requirements set for traditional static and chemically produced drugs. Although the use of bacteriophages is already quite old, in fact, their tailor-made use and applications are in line with the growing demand and insights around personalized nutrition and personalized medicine, where DNA, microbiome composition, and personal lifestyle act as leading indicators.

Previously, we referred to a couple of proposals to overcome this incompatibility with today's regulatory frameworks in the Western World, while still assuring safety and efficacy [7]. For instance, the creation of a new EC Directive in Europe concerning bacteriophages and bacteriophage cocktails for human use, or an update of the already existing Medicinal Products Directive 2001/83/EC with a specific amendment for bacteriophages and bacteriophage cocktails, or to register bacteriophages and bacteriophage cocktails under the Council Directive concerning medical devices (93/42/EEC) [32]. A recent breakthrough in this debate has occurred in Belgium, where the national authorities agreed on implementing a pragmatic phage therapy framework that centres on the magistral preparation (compounding pharmacy in the US) of tailor-made phage medicines [33]. There is good reason to believe that this Belgian "magistral phage medicine" framework will be flexible enough to exploit and further explore the specific nature of bacteriophages as co-evolving antibacterials whilst giving precedence to patient safety.

Regarding good manufacturing practices, experts are in agreement that these should be defined in the specific context of bacteriophages as natural entities. The use of whole genome sequencing technologies, together with several additional specific controls [7], can assure the safety of newly identified or adapted bacteriophages and bacteriophage products in general. Rapid sequencing will also allow the safe incorporation of the unique feature of bacteriophage therapy that when in the case of acute infections, new bacteriophages can be isolated within 48 h, or adapted to counteract potential resistance of emerging pathogens. Quality and safety requirements for bacteriophage therapy products have been listed in Table 1 in the previously published paper [7].

Another suggested mechanism that was highlighted by the panel members to ensure the safety of bacteriophage therapy relates to the implementation of a monitoring system. This would function much like that for antibiotic resistance, and should be put into place as soon as bacteriophage therapy has started. The main purpose of the monitoring system is to collect data for prospective analyses, as well as to detect and follow the development of bacterial resistance to bacteriophages.

Several researchers have made proposals for the safe re-implementation of bacteriophages by establishing validated bacteriophage collections in hospitals for compassionate use applications. In this way, as soon as the bacterial pathogen has been identified, which is often practiced already, it could be tested for sensitivity against such a library of bacteriophages [34]. Another suggestion refers to installing dedicated public structures, National Reference Centres for bacteriophage therapy, that can conduct pilot treatments and facilitate production of hospital-based bacteriophage solutions, and application protocols that will ensure adequate product quality, patient safety and monitoring of treatment efficacy [35]. Such a way of working is, in fact, already operational at the bacteriophage therapy centre located at the Hirszfeld Institute of Immunology and Experimental Therapy in Wroclaw, Poland [16].

1.1.3. Lack of Awareness among (Para-) Medical Staff and the Public About Bacteriophage Therapy

For many years, the history and potential of bacteriophage therapy was out of sight for people in the Western World, including medical staff and patients. Following the international increase of antibiotic resistance of pathogens and since the 100th anniversary of bacteriophage therapy in 2017, there has been a clear increase in reporting and communication on bacteriophage therapy in all forms of media in many Western countries, including scientific opinion articles, television programs, and social media initiatives. Also, the increase of medical tourism to Georgia or Poland for bacteriophage treatment illustrates a growing awareness on and demand for bacteriophage therapy.

Despite this increase of public awareness, panel members feel that there is still not enough interface with the medical community. Little focus is given in the curricula used for (para-) medical trainees on phage therapy, including bacteria-phage antagonistic evolution. Expertise on phage therapy could easily be provided through researchers who by reaching out to their medical colleagues operating at local hospitals. In this way, relevant information can be disseminated about the pros and cons of bacteriophage therapy in the context of the emergence and spreading of antimicrobial resistance. The increased awareness could thus lead to more application of bacteriophages for compassionate use as described above. At the same time, hospitals and patients should be aware that bacteriophage therapy will not always be successful and yet unknown safety risks and complications cannot be excluded.

1.1.4. Limitations in Intellectual Property Protection

Bacteriophages are ubiquitous natural organisms that are relatively easy to isolate from the environment and have been in the public domain since the 1920s, and therefore the possibilities for intellectual property (IP) protection are limited. Although this might be seen as a limitation by companies intending to seriously invest in bacteriophage therapy, in fact it also offers the opportunity for local, national, and supra-national state-supported medical care to collectively invest in the development of bacteriophage therapy that is expected to be more cost-effective for treatment of many infectious diseases compared to several of the currently used antibiotics. The expected efficacy and affordability of bacteriophage therapy is also illustrated by initiatives around development of bacteriophage therapy in developing countries. In these situations, people are disproportionately impacted by infectious diseases, leading to a critical disease burden on healthcare budgets, and where standard medical care is already difficult to afford for the majority of the population [36].

We expect IP protection opportunities to exist for bacteriophage production methods, as well as for applications that can enhance the efficiency and quality of bacteriophage therapy, and/or improve shelf-life stability. Opportunities to protect IP will also come from genetically-engineered bacteriophages, as outlined further in this article. In addition, we see options for private/government partnerships with patent pools under supra national governance that should be managed through organizations such as the WHO, CDC, ECDC or UN [37].

2. The Nagoya Protocol and the Implications on Bacteriophage Therapy

In order to increase acceptance and implementation of bacteriophage therapy, it is obvious that rapid and efficient procurement of bacteriophages from the environment with therapeutic potential, and their bacterial hosts, is essential. Bacteriophages, along with all other genetic (biological) resources, are regulated by the Nagoya Protocol (NP). Briefly put, this means that regulations governing the collection of bioresources from natural environments and subsequent benefit-sharing with the country of origin are increasingly important (Box 1).

In 2017, several authors evaluated the impact of the NP on research and international cooperation, the need for best practices for benefit-sharing, and proposed adjustments to the NP to accommodate microbiological research and development (R&D) [2–4]. Since microorganisms are typically ubiquitous and of the same constitution across the world, it was argued that the expectations of the NP as set by lawmakers is scientifically unfeasible. Similar to their bacterial hosts, bacteriophages are also cosmopolitan, and an estimated 10^7 bacteriophage particles might be present in any one milliliter of natural sample. As a result, thousands of new genetic resources can result from a single step of sampling from the environment, but there are thousands of diverse natural environments around the world that are not limited to one specific country or region of interest. New environmental samples are themselves inherently worthless, as the potential for each bacteriophages and bacterial isolates is unknown and intensive research is necessary to purify and evaluate these characteristics. Additionally, high numbers of specific bacterial hosts are required for phage amplification, to determine the activity spectrum, and to increase their efficacy in clinical applications. Taken together, these scientific realities mean that researchers will be likely to search for bacteriophages/hosts where NP restrictions are either not in place, or where the NP highly efficiently organized.

Bacteriophages offer great potential for human and veterinary medicine, but NP regulations could be interpreted as in conflict with the WHO objective (and subsequent G7 and G20 summits) calling for all countries to develop alternative antibacterial strategies in human medicine [38]. Also in the case of serious outbreaks of foodborne infections, requiring urgent response, (e.g., Germany in 2011 with the outbreak of *E. coli* O104:H4 EAHEC that caused more than 50 cases of deaths), the NP might impede progress in the global search for potent therapeutics and straightforward exchange of bacteriophages that could save lives.

For microorganisms in general, and for research and applications of bacteriophages in particular, we share the concerns put forward in the Lactic Acid Bacterial Industrial Platform (LABIP) [39] and Microbiological Research Under the Nagoya Protocol: Facts and Fiction [40] and would welcome the following amendments to the NP: 1. Precise definitions of terms like "utilization" and "research and development", so there is regulatory certainty about what is meant when these words are written in the Nagoya Protocol; 2. Guidelines to consider R&D expenditure around bacteriophages and related investment for basic microbiological research in the terms of agreement on benefit sharing, 3. Simplification of the NP requirements in case of screening activities of a large number of bacteriophages and potential host strains aimed to find a just few candidates with specific characteristics; 4. Research using digital sequence information of bacteriophages and host strains to remain outside the scope of the NP and ABS legislations, as it would be a daunting task to obtain PIC and MAT for all relevant sequences in a database such as GenBank. 5. In case of infection outbreaks, rapid exchange of material should be uncomplicated and governed by a generic international benefit-sharing agreement that is ready for the unexpected.

Box 1. Explanation about the Nagoya Protocol.

The NP is a new international regime that came into effect on 12 October 2014, and has been ratified by more than 100 countries. The NP is the implementing treaty for the Convention on Biological Diversity (CBD, www.cbd.int), which itself has been in force since December 1993, and is intended to harmonize access and benefit-sharing mechanisms for the retrieval of biological resources out of provider countries (often in emerging economies). The purpose of the NP is to achieve the objectives of the CBD: 1. Conservation of biodiversity, 2. Sustainable use of the genetic resources and 3. Balanced and equitable sharing of benefits when genetic resources are used (Access and Benefit Sharing (ABS)). Hereto, before starting any research and development work on biological resources, Prior Informed Consent (PIC) by the 'provider country' is needed, which done according to Mutually Agreed Terms (MAT) to be laid down in a contract describing access to the materials and how benefits will be shared. In practice, benefit sharing can take a variety of forms, including monetary payments, for example with royalties or research funding, but also via non-monetary forms, such as technology transfer or scientific collaborations. Historically, bioresources were considered a shared heritage of humankind, and the NP allows signatories to regulate access to genetic resources to ensure benefit-sharing with such provider country. Generic guidance flowcharts are given by Overmann and Scholz [40] and Smith et al. [41].

Within the EU, member states are varied in the regulation of access to genetic resources, with northern Member States often allowing unrestricted access and southern States considering regulation. The United States neither ratified the CBD nor the NP, but these international agreements do affect U.S. scientists and aligned standard operating procedures are in development [42]. The primary resource for determining whether there are restrictions imposed on a genetic resource is the ABS Clearing-House (ABSCH) (https://absch.cbd.int/), which provides country profiles and, if appropriate, national regulations that, ideally, explain how to access their sovereign genetic resources, including required documents such as PIC and MAT. Due to the current lack of completed country profiles in the ABSCH, it is often difficult to find the practical information needed to be compliant with the NP. The long-term implications for phage research are unclear, but conceivably threatened.

3. The Future of Bacteriophage Therapy by the CRISPR/Cas System

In the era of gene editing, it is relevant to study the potential of genetically-modified bacteriophages as therapeutics. Bacteriophages could be engineered for better efficacy and a broader range of application, and, importantly, would be more attractive for investors due the generation of IP rights (see above). Recent advances in CRISPR/Cas-editing (see Box 2) made bacteriophage DNA editing a hot-topic, since this method can be applied on basically every bacteriophage regardless of its size, host, or properties. Even bacteriophages encoding anti-CRISPR proteins or bacteriophages without a mechanism for repairing CRISPR/Cas-induced breaks can be edited using different CRISPR/Cas types, or by incorporating a repairing protein from another bacteriophage in trans [43]. The strategies based on therapeutic genetically-engineered bacteriophage, which we mentioned previously [7], were constructed almost exclusively for *E. coli*, due to the range of editing strategies already developed for this model bacterium. On the contrary, CRISPR/Cas-editing can be applied also to bacteriophages targeting members of the ESKAPE pathogen group (*Enterococcus faecium*, *Staphylococcus aureus*, *Klebsiella pneumoniae*, *Acinetobacter baumannii*, *Pseudomonas aeruginosa*, and *Enterobacter* species). This has already been proven in *Staphylococcus aureus* and bacteriophage ISP [44], which is member of the same genus *Kayvirus* as a bacteriophage used in the EU-approved therapeutic STAFAL®. Thus, previously proposed applications of genetically-engineered bacteriophages are expected to get relevant clinical significance as summarized in Table 3.

Seven bacteriophages from various hosts have recently been successfully edited using three different types of CRISPR/Cas (Table 4): i. type I-E, prevalent in *E. coli*, ii. type II-A, which requires only one protein Cas 9 and sgRNA for editing ability and is best described so far, and iii. type III-A, which does not require a protospacer adjacent motive and cannot be evaded by simple nucleotide substitution [44].

Table 3. Targets for genetic engineering for bacteriophage therapy applications.

Gene	Modification	Purpose	Advantage to Natural Bacteriophage Therapy	Related Reference
Antimicrobial protein *	Insertion/gene replacement	Phage killing other strains	Product for mixed infections	[7]
Biofilm degrading enzyme *	Insertion/gene replacement	Degrading biofilm	More active against biofilm-producing strains	
Virulence factor	Gene deletion	No virulence transfer	Novel therapeutic bacteriophage	
Baseplate proteins/tail fibers	Gene replacement	Altered host-range	Novel therapeutic bacteriophage for mixed infections	
Receptor-binding protein/structural proteins	Single gene mutations	Broader host-range	More effective, and faster to obtain than by natural selection	[45]
Major capsid protein	Purification tags insertion	More efficient purification	Purer product	[46]
Major capsid protein	Anti-immune tags insertion/single gene mutations	Longer circulation in bloodstream	More effective, and faster to obtain than by natural selection	[47]
Various, e.g., lytic module	Gene knockout	Non-replicative bacteriophage	Replication control of a bacteriophage	[48]
Endotoxin antibody *	Insertion/gene replacement	Endotoxin removal	Safer product	[49]

* Foreign genes.

Table 4. Summary of Clustered Regularly Interspaced Short Palindromic Repeats/CRISPR associated systems (CRISPR/Cas) applications in the editing of bacteriophages.

Bacteriophage	Host	CRISPR/Cas Type Employed	Mutation	Purpose	Gene	Reference
T7	*E. coli*	I-E	Gene deletion	PoC of CRISPR/Cas-editing in bacteriophage	gene 1.7	[43]
2972	*Streptococcus thermophilus*	II-A/Cas9	Gene replacement with methyltransferase/Gene deletion	PoC of native II-A editing, bacteriophage resistant to RM system	orf 33, 39	[50]
ICP1	*Vibrio cholera*	I-E	Gene deletion/Gene replacement with GFP	PoC in *Vibrio* model	*cas* 1, *cas* 2–3	[51]
P2	*Lactococcus lactis*	II-A/Cas9	Deletion/insertion/substitution	PoC of heterologous II-A editing	orf 24, 42, 47, 49	[52]
T4	*E. coli*	II-A/Cas9	Substitutions, deletion	PoC of editing by two gRNA, functional study of gene knock-out	*mcp*, *rnl*B	[53]
Andhra	*Staphylococcus epidermidis*	III-A/Cas10	Substitutions	PoC of native III-A editing	orf 9, 10	[44]
ISP	*Staphylococcus aureus*	III-A/Cas10	Substitutions	PoC of heterologous III-A editing	orf 61	[44]

PoC—proof-of-concept, RM—restriction modification.

Box 2. CRISPR/Cas editing.

CRISPR/Cas (Clustered Regularly Interspaced Short Palindromic Repeats/CRISPR associated systems) was originally identified as a prokaryotic adaptive immune system. It mediates the cleavage of foreign DNA by Cas nuclease if this DNA matches the sequence of spacers in the CRISPR locus. It soon became an important method for eukaryotic genome engineering. However, a limited number of studies have been concerned with its application in bacterial genome engineering due to the poor ability of bacteria to repair CRISPR/Cas-induced breaks [54]. Bacteriophages, on the other hand, have evolved several mechanisms to repair such breaks, as a way to increase the probability of escaping the bacterial immune system. Bacteriophages thus represent an excellent model for optimizing CRISPR/Cas as a prokaryotic genome engineering tool, as well as for various practical applications of modified bacteriophages (for review see Bardy et al. 2016, [55]).

4. Limitations and Concerns for CRISPR/Cas Gene Edited Bacteriophages

4.1. *Efficacy*

One of the unknowns concerning CRISPR/Cas-edited bacteriophages relates to whether the mutations will be preserved and reproduced, or if spontaneous mutants that escape the CRISPR/Cas will arise and, due to increased fitness, will outgrow the recombinant. As a consequence, bacteriophages with many point mutations in inserted foreign genes and tags would emerge, rendering the intended modifications useless. A solution could potentially come through the application of CRISPR/Cas III-A, where there has been no evidence of escape mutants [44]. However, in order to understand the long-term effect of CRISPR/Cas-based selections on the bacteriophage population, more research is necessary. In CRISPR/Cas II-A type, it is possible to utilize several spacers targeting different locations in the gene of interest, which will reduce the probability of escape. This technique, however, becomes more laborious with every additional step and might still prove to be non-efficient.

4.2. *Legislative Hurdles*

CRISPR/Cas-edited bacteriophages with heterologous gene insertions would rank as genetically modified organisms (GMOs), which brings additional hurdles to their approval as therapeutic agents. Their future success will therefore depend on the safety and environmental regulation of GMOs in individual countries. At present, the United States and China appear to have far more amenable policies and prospects for GMO acceptance compared to the EU and Russia. The reproducibility and viability of a genetically modified bacteriophage could also be eliminated by introducing lethal mutations to ensure that the modified bacteriophage would disappear over time, therefore decreasing environmental concerns. Another option is to transform the bacteriophage to a non-replicative delivery vehicle that would mediate killing by an alternative approach (Box 3), which is currently becoming an important focus of research interest in terms of modified bacteriophage therapy. The production of such bacteriophages would require the use of a modified host strain, or presence of a helper bacteriophage to facilitate reproduction. For such applications, the efficacy of such modified constructs in bacteriophage therapy remains to be investigated.

Box 3. CRISPR/Cas-based antibiotics delivered by bacteriophage.

Apart from editing, CRISPR/Cas has been utilized as a weapon for the specific killing of virulent or antibiotic-resistant bacterial strains, by reprogramming CRISPR/Cas to target a gene sequence encoding such a property. A bacteriophage can serve as a delivery vehicle, by binding to the relevant bacterial strain and transferring DNA encoding Cas9 nuclease and sgRNA into the cell. This DNA is packaged into the bacteriophage capsids by introduction of specific bacteriophage packaging sites on both ends of the sequence. As a result, only the targeted strain is eliminated, therefore preserving the rest of the microbiome. In addition, the transfer of the undesired phenotypic traits is abolished among strains infected by the delivery bacteriophage. Many groups are studying this strategy (for review see Fagen et al. 2017 [48]), with several start-ups (Locus in US, Nemesis in UK or Eligo in France) already developing such a bacteriophage for commercial use. Still, classic problems of bacteriophage therapy, such as finding suitable bacteriophages for each individual pathogenic strain or the potential immune reaction of the body to large doses of bacteriophage virions, remain obstacles for application.

4.3. Safety and Environmental Risks

Despite the absence of adverse reports and also the recent progress and resulting knowledge in the area of natural bacteriophage immunogenicity [16], the mechanism of bacteriophage tolerance by the human immune system is not yet entirely understood. Since bacteriophages need to be applied in a relatively large amount or can replicate at the site of the bacterial infection to high titers, their effect on the human immune system requires thorough investigation. Furthermore, especially in the area of GM bacteriophages, there might be unique health and safety risks that require careful evaluation, such as the possible side effects around the application of long-circulating bacteriophages (Table 3). Also, bacteriophages with modified structural proteins, displaying purification tags, or with additional receptor-binding domains may represent novel antigens for the immune system. In the case of using bacteriophages as delivery vehicles, even in the non-lytic killing strategy, cells could be lysed by the patient's immune system and the cell content containing modified proteins or endotoxins thus released into the bloodstream. The (in) stability of modified bacteriophages and the consequences of spontaneous mutations are other safety targets that require investigation.

Given the continuous need for propagating host strains, the probability of survival of bacteriophages, including modified bacteriophages, outside the laboratory or hospital is minimal. However, the risk of recombination between modified and wild-type bacteriophages or horizontal gene transfer among modified bacteriophages and bacteria must be evaluated, particularly in case bacteriophages would carry genes foreign antibacterial products. Theoretically, uptake and incorporation of such sequences by a bacterium may result in a fitness advantage over the strain for which the bacteriophage was originally designed.

5. Discussion and Conclusions

The urgent need for effective alternatives to antimicrobials is self-evident in order to reduce the morbidity and mortality associated with antimicrobial resistance, as well as reducing the healthcare burden on economies. Many proposals have been made on which path bacteriophage therapy should follow in order to be approved. A recurrent conclusion is that this therapy will never become a viable option if we continue assessing this treatment under the same regulatory frameworks as for chemical drugs. In fact, bacteriophage therapy is already safely being applied today, as demonstrated not only from ongoing activities in Eastern Europe, but also from the observation that already for more than a decade several of the successful bacteriophage therapy cases reported in the Western World (Table 2) are led by military hospitals who have the flexibility to invest in therapies which are not yet legally approved.

If bacteriophage therapy is to develop as a means to stop the emergence of antibiotic resistance, courageous decision making is needed that allows the controlled use of bacteriophage. This needs to take into consideration the many years of practice in Russia, Georgia and Poland, and which is in line with many good proposals from scientists and physicians who understand the specific nature of bacteriophages.

At the same time, we should not expect that bacteriophage therapy will always be 100% effective and might occasionally induce side effects, as we will understand better upon the implementation of bacteriophage therapy monitoring systems. But how does the occurrence of potential side effect compare to the opportunity costs of not applying bacteriophage therapy in case of antibiotic resistance? Furthermore, also for antibiotics that in many cases are lifesaving, we accept undesired side effects, for instance when due to the unspecific mode of action of antibiotics beneficial microbes are also affected, which often causes complications such as antibiotic-induced dysbiosis and secondary infections.

A great step forward can be made when more countries will follow the example of the Belgian Ministry of Public Health regarding the set- up of a phage therapy framework that centers on the magistral preparation (compounding pharmacy in the US) of tailor-made phage medicines. Importantly, this Belgian solution assures safety and avoids the application of certain medicinal product requirements that restrain flexible phage therapy approaches, such as compliance to Good Manufacturing Practice [33].

In this paper, we have underlined the call for more RCTs with bacteriophage therapy. In this context, it is also an important responsibility of the researchers involved to properly report the outcome of the studies, as has not always been the case (Table 1). Incomplete reporting on clinical studies does not help to gain confidence on pros and cons of new therapies. Besides the mentioned need for more studies, at the same time we reason that depending on the type of infection, a traditional RCT might not always be the most efficient route to determining bacteriophage efficacy, and it is not only bacteriophage therapy that finds itself in this predicament. An increase in personalized-medicine approaches, particularly for cancer treatments, have necessitated different designs for clinical trials [56]. N-of-1 studies, where a patient is the entire study, has become increasingly frequent to evaluate treatments for rare diseases or pain reducers. Basket designs permit the selection of patients with molecular markers that make them likely to respond to a certain treatment; this is not unlike pre-testing bacterial isolates for bacteriophage sensitivity prior to trial enrollment. The increasing frequency of case-reports for bacteriophage therapy suggests that bacteriophage therapy may be heading in the direction of a personalized treatment, at least for some diseases (Table 2).

With regard to the Nagoya Protocol, we call for pragmatic approaches to obtain Prior Informed Consent (PIC) and Mutually Agreed Terms (MAT) that are proportional to the value of the immediate foreseen benefits. Otherwise, due to the delays that might be encountered in the process, there may not be any significant benefit to share. Since there are, at present, very few culture collections with bacteriophage holdings, new bacteriophages have to be isolated in high numbers to constantly fulfill the WHO resolution 68.7.3 (WHO Assembly October 2016). The proposed amendments and clarifications to NP as mentioned in the section on NP could facilitate that the protocol does not unintendedly retard these innovative developments that are important for humankind in the 21st century. This is especially valid for a rapid and effective application of bacteriophages in the events of emergency since users should ensure legally-complied uses of the genetic resources, including those for collaborative projects.

While it becomes increasingly acknowledged that bacteriophage therapy, as applied already for many years in some countries, has great potential to combat the consequences of increasing antibiotic resistance, we acknowledge that the traditional bacteriophage therapy also has some limitations. In this article, we have described that the use of CRISPR/Cas to specifically edit the genome of natural bacteriophages offers an opportunity for controlled and effective bacteriophage therapy and may facilitate reliable future applications of the bacteriophage therapy. However, today several technical issues and safety concerns of GM bacteriophages still need to be addressed. The first step would be to prove that modified bacteriophages could be prepared in large-scale, without any significant background of undesirable mutants. Next, research should focus on immunogenicity assessment of modified virions and potential horizontal transfer of genes encoding antibacterial products. We expect that in a time frame of five-to-ten years, it will become clear if this approach for bacteriophage therapy is worthwhile. Furthermore, it is foreseen that CRISPR/Cas-edited bacteriophages will revolutionize bacteriophage therapy, if not by direct application of modified bacteriophages, by the enormous possibilities that it will bring to fundamental research on bacteriophage biology.

Although the entire world seems to agree on the clear urgency to implement novel solutions, it is worrying to see that therapy with natural bacteriophages has yet to be seriously considered in the Western World. Notwithstanding some open questions, there are many realistic options and scenarios that would enable a gradual, pragmatic, and responsible re-implementation of bacteriophage therapy in the short term to help avoid needless deaths due to antibiotic-resistant infections. Any further delay could be seen as a supported prolongation of the difficult conditions many patients without alternative treatment options find themselves in today.

Author Contributions: W.S. lead the manuscript preparation. J.-P.P., D.D.V., I.C., and S.M. contributed to the creation of Tables. A.H.S., H.M.P. and C.R. contributed to the Nagoya Protocol paragraph. R.P., P.B. and J.D. wrote the CRISPR/Cas paragraph. D.D.V., N.S., S.M., R.A., J.C. and A.C. provided valuable input on the manuscript. W.S. and D.I.K. constructed and created the final copy.

Conflicts of Interest: The authors declare no conflict of interest.

References

1. Aminov, R. History of antimicrobial drug discovery: Major classes and health impact. *Biochem. Pharmacol.* **2017**, *133*, 4–19. [CrossRef] [PubMed]
2. Aminov, R.I. A brief history of the antibiotic era: Lessons learned and challenges for the future. *Front. Microbiol.* **2010**, *1*, 134. [CrossRef] [PubMed]
3. Centers for Disease Control and Prevention. *Antibiotic Resistance Threats in the United States, 2013*; Centers for Disease Control and Prevention: Atlanta, GA, USA, 2013.
4. European Centre for Disease Prevention and Control; European Food Safety Authority; European Medicines Agency. Ecdc/*efsa*/ema second joint report on the integrated analysis of the consumption of antimicrobial agents and occurrence of antimicrobial resistance in bacteria from humans and food-producing animals. *EFSA J.* **2017**, *15*, e04872. [CrossRef]
5. Organization, W.H. *Global Action Plan on Antimicrobial Resistance*; WHO: Geneva, Switzerland, 2015.
6. O'Neill, J. Tackling Drug–Resistant Infections Globally: Final Report and Recommendations. Available online: http://www.iica.int/en/press/news/tackling-drug-resistant-infections-globally-final-report-and-recommendations (accessed on 18 April 2018).
7. Expert round-table on acceptance and re–implementation of bacteriophage therapy. Silk route to the acceptance and re-implementation of bacteriophage therapy. *Biotechnol. J.* **2016**, *11*, 595–600.
8. Servick, K. Beleaguered phage therapy trial presses on. *Science* **2016**, *352*, 1506. [CrossRef] [PubMed]
9. PhagoBurn. Evaluation of Phage Therapy for the Treatment of *Escherichia coli* and *Pseudomonas aeruginosa* Burn Wound Infections. Available online: http://www.phagoburn.eu/ (accessed on 18 April 2018).
10. Sarker, S.A.; Sultana, S.; Reuteler, G.; Moine, D.; Descombes, P.; Charton, F.; Bourdin, G.; McCallin, S.; Ngom-Bru, C.; Neville, T. Oral phage therapy of acute bacterial diarrhea with two coliphage preparations: A randomized trial in children from bangladesh. *EBioMedicine* **2016**, *4*, 124–137. [CrossRef] [PubMed]
11. Międzybrodzki, R.; Borysowski, J.; Weber-Dąbrowska, B.; Fortuna, W.; Letkiewicz, S.; Szufnarowski, K.; Pawełczyk, Z.; Rogóż, P.; Kłak, M.; Wojtasik, E. Chapter 3–Clinical aspects of phage therapy. In *Advances in Virus Research*; Łobocka, M., Szybalski, W., Eds.; Academic Press: Cambridge, MA, USA, 2012; Volume 83, pp. 73–121.
12. Abedon, S.T.; García, P.; Mullany, P.; Aminov, R. Editorial: Phage therapy: past, present and future. *Front. Microbiol.* **2017**, *8*, 981. [CrossRef] [PubMed]
13. McCallin, S.; Alam Sarker, S.; Barretto, C.; Sultana, S.; Berger, B.; Huq, S.; Krause, L.; Bibiloni, R.; Schmitt, B.; Reuteler, G. Safety analysis of a russian phage cocktail: From metagenomic analysis to oral application in healthy human subjects. *Virology* **2013**, *443*, 187–196. [CrossRef] [PubMed]
14. Leitner, L.; Sybesma, W.; Chanishvili, N.; Goderdzishvili, M.; Chkhotua, A.; Ujmajuridze, A.; Schneider, M.P.; Sartori, A.; Mehnert, U.; Bachmann, L.M. Bacteriophages for treating urinary tract infections in patients undergoing transurethral resection of the prostate: A randomized, placebo-controlled, double-blind clinical trial. *BMC Urol.* **2017**, *17*, 90. [CrossRef] [PubMed]
15. Ujmajuridze, A.; Chanishvili, N.; Goderdzishvili, M.; Leitner, L.; Mehnert, U.; Chkhotua, A.; Kessler, T.; Sybesma, W. Adapted bacteriophages for treating urinary tract infections. **2018**. Submitted for publication.
16. Górski, A.; Międzybrodzki, R.; Weber-Dąbrowska, B.; Fortuna, W.; Letkiewicz, S.; Rogóż, P.; Jończyk-Matysiak, E.; Dąbrowska, K.; Majewska, J.; Borysowski, J. Phage therapy: Combating infections with potential for evolving from merely a treatment for complications to targeting diseases. *Front. Microbiol.* **2016**, *7*, 1515. [CrossRef] [PubMed]
17. Saussereau, E.; Vachier, I.; Chiron, R.; Godbert, B.; Sermet, I.; Dufour, N.; Pirnay, J.P.; De Vos, D.; Carrié, F.; Molinari, N. Effectiveness of bacteriophages in the sputum of cystic fibrosis patients. *Clin. Microbiol. Infect.* **2014**, *20*, O983–O990. [CrossRef] [PubMed]

18. Bernstein, L.J.; Ochs, H.D.; Wedgwood, R.J.; Rubinstein, A. Defective humoral immunity in pediatric acquired immune deficiency syndrome. *J. Pediatr.* **1985**, *107*, 352–357. [CrossRef]
19. Rhoads, D.D.; Wolcott, R.D.; Kuskowski, M.A.; Wolcott, B.M.; Ward, L.S.; Sulakvelidze, A. Bacteriophage therapy of venous leg ulcers in humans: Results of a phase i safety trial. *J. Wound Care* **2009**, *18*, 237–243. [CrossRef] [PubMed]
20. Duplessis, C.; Biswas, B.; Hanisch, B.; Perkins, M.; Henry, M.; Quinones, J.; Wolfe, D.; Estrella, L.; Hamilton, T. Refractory pseudomonas bacteremia in a 2-year-old sterilized by bacteriophage therapy. *J. Pediatr. Infect. Dis. Soc.* **2017**. [CrossRef] [PubMed]
21. Schooley, R.T.; Biswas, B.; Gill, J.J.; Hernandez-Morales, A.; Lancaster, J.; Lessor, L.; Barr, J.J.; Reed, S.L.; Rohwer, F.; Benler, S. Development and use of personalized bacteriophage-based therapeutic cocktails to treat a patient with a disseminated resistant acinetobacter baumannii infection. *Antimicrob. Agents Chemother.* **2017**, *61*, e00954-17. [CrossRef] [PubMed]
22. Zhvania, P.; Hoyle, N.S.; Nadareishvili, L.; Nizharadze, D.; Kutateladze, M. Phage therapy in a 16-year-old boy with netherton syndrome. *Front. Med.* **2017**, *4*, 94. [CrossRef] [PubMed]
23. Jennes, S.; Merabishvili, M.; Soentjens, P.; Pang, K.W.; Rose, T.; Keersebilck, E.; Soete, O.; François, P.-M.; Teodorescu, S.; Verween, G. Use of bacteriophages in the treatment of colistin-only-sensitive pseudomonas aeruginosa septicaemia in a patient with acute kidney injury—A case report. *Crit. Care* **2017**, *21*, 129. [CrossRef] [PubMed]
24. Fish, R.; Kutter, E.; Wheat, G.; Blasdel, B.; Kutateladze, M.; Kuhl, S. Bacteriophage treatment of intransigent diabetic toe ulcers: A case series. *J. Wound Care* **2016**, *25*, S27–S33. [CrossRef] [PubMed]
25. Soothill, J. Use of bacteriophages in the treatment of pseudomonas aeruginosa infections. *Expert Rev. Anti-Infect. Ther.* **2013**, *11*, 909–915. [CrossRef] [PubMed]
26. Khawaldeh, A.; Morales, S.; Dillon, B.; Alavidze, Z.; Ginn, A.N.; Thomas, L.; Chapman, S.J.; Dublanchet, A.; Smithyman, A.; Iredell, J.R. Bacteriophage therapy for refractory pseudomonas aeruginosa urinary tract infection. *J. Med. Microbiol.* **2011**, *60*, 1697–1700. [CrossRef] [PubMed]
27. Letkiewicz, S.; Międzybrodzki, R.; Fortuna, W.; Weber-Dąbrowska, B.; Górski, A. Eradication of enterococcus faecalis by phage therapy in chronic bacterial prostatitis—Case report. *Folia Microbiol.* **2009**, *54*, 457–461. [CrossRef] [PubMed]
28. Fadlallah, A.; Chelala, E.; Legeais, J.-M. Corneal infection therapy with topical bacteriophage administration. *Open Ophthalmol. J.* **2015**, *9*, 167–168. [CrossRef] [PubMed]
29. Jikia, D.; Chkhaidze, N.; Imedashvili, E.; Mgaloblishvili, I.; Tsitlanadze, G.; Katsarava, R.; Glenn Morris, J.; Sulakvelidze, A. The use of a novel biodegradable preparation capable of the sustained release of bacteriophages and ciprofloxacin, in the complex treatment of multidrug-resistant staphylococcus aureus-infected local radiation injuries caused by exposure to sr90. *Clin. Exp. Dermatol.* **2005**, *30*, 23–26. [CrossRef] [PubMed]
30. Leszczyński, P.; Weber-Dabrowska, B.; Kohutnicka, M.; Łuczak, M.; Górecki, A.; Górski, A. Successful eradication of methicillin-resistantstaphylococcus aureus (MRSA) intestinal carrier status in a healthcare worker—Case report. *Folia Microbiol.* **2006**, *51*, 236–238. [CrossRef]
31. Fevre, C.; Ferry, T.; Petitjean, C.; Leboucher, C.; L'hostis, G.; Laurent, F.; Regulski, K. Phage therapy: Compassionate use in france in 2017. In *Phages-sur-Yvette*; Gif-sur-Yvette: Essonne, France, 2017.
32. Verbeken, G.; Pirnay, J.P.; De Vos, D.; Jennes, S.; Zizi, M.; Lavigne, R.; Casteels, M.; Huys, I. Optimizing the european regulatory framework for sustainable bacteriophage therapy in human medicine. *Arch. Immunol. Ther. Exp.* **2012**, *60*, 161–172. [CrossRef] [PubMed]
33. Pirnay, J.-P.; Verbeken, G.; Ceyssens, P.-J.; Huys, I.; De Vos, D.; Ameloot, C.; Fauconnier, A. The magistral phage. *Viruses* **2018**, *10*, 64. [CrossRef] [PubMed]
34. Young, R.; Gill, J.J. Microbiology. Phage therapy redux–What is to be done? *Science* **2015**, *350*, 1163–1164. [CrossRef] [PubMed]
35. Debarbieux, L.; Pirnay, J.-P.; Verbeken, G.; De Vos, D.; Merabishvili, M.; Huys, I.; Patey, O.; Schoonjans, D.; Vaneechoutte, M.; Zizi, M. A bacteriophage journey at the european medicines agency. *FEMS Microbiol. Lett.* **2015**, *363*. [CrossRef] [PubMed]
36. Nagel, T.E.; Chan, B.K.; De Vos, D.; El-Shibiny, A.; Kang'ethe, E.K.; Makumi, A.; Pirnay, J.-P. The developing world urgently needs phages to combat pathogenic bacteria. *Front. Microbiol.* **2016**, *7*, 882. [CrossRef] [PubMed]

37. Van Zimmeren, E.; Vanneste, S.; Matthijs, G.; Vanhaverbeke, W.; Van Overwalle, G. Patent pools and clearinghouses in the life sciences. *Trends Biotechnol.* **2011**, *29*, 569–576. [CrossRef] [PubMed]
38. Inoue, H.; Minghui, R. Antimicrobial resistance: Translating political commitment into national action. *Bull. World Health Organ.* **2017**, *95*. [CrossRef]
39. Johansen, E. Future access and improvement of industrial lactic acid bacteria cultures. *Microb. Cell Fact.* **2017**, *16*, 230. [CrossRef] [PubMed]
40. Overmann, J.; Scholz, A.H. Microbiological research under the nagoya protocol: Facts and fiction. *Trends Microbiol.* **2017**, *25*, 85–88. [CrossRef] [PubMed]
41. Smith, D.; da Silva, M.; Jackson, J.; Lyal, C. Explanation of the nagoya protocol on access and benefit sharing and its implication for microbiology. *Microbiology* **2017**, *163*, 289–296. [CrossRef] [PubMed]
42. McCluskey, K.; Barker, K.B.; Barton, H.A.; Boundy-Mills, K.; Brown, D.R.; Coddington, J.A.; Cook, K.; Desmeth, P.; Geiser, D.; Glaeser, J.A. The U.S. Culture collection network responding to the requirements of the nagoya protocol on access and benefit sharing. *mBio* **2017**, *8*. [CrossRef] [PubMed]
43. Kiro, R.; Shitrit, D.; Qimron, U. Efficient engineering of a bacteriophage genome using the type i-e crispr-cas system. *RNA Biol.* **2014**, *11*, 42–44. [CrossRef] [PubMed]
44. Bari, S.M.N.; Walker, F.C.; Cater, K.; Aslan, B.; Hatoum-Aslan, A. Strategies for editing virulent staphylococcal phages using crispr-cas10. *ACS Synth. Biol.* **2017**, *6*, 2316–2325. [CrossRef] [PubMed]
45. Pouillot, F.; Blois, H.; Iris, F. Genetically engineered virulent phage banks in the detection and control of emergent pathogenic bacteria. *Biosecur. Bioterror.* **2010**, *8*, 155–169. [CrossRef] [PubMed]
46. Oślizło, A.; Miernikiewicz, P.; Piotrowicz, A.; Owczarek, B.; Kopciuch, A.; Figura, G.; Dąbrowska, K. Purification of phage display-modified bacteriophage T4 by affinity chromatography. *BMC Biotechnol.* **2011**, *11*, 59. [CrossRef] [PubMed]
47. Vitiello, C.L.; Merril, C.R.; Adhya, S. An amino acid substitution in a capsid protein enhances phage survival in mouse circulatory system more than a 1000-fold. *Virus Res.* **2005**, *114*, 101–103. [CrossRef] [PubMed]
48. Fagen, J.R.; Collias, D.; Singh, A.K.; Beisel, C.L. Advancing the design and delivery of crispr antimicrobials. *Curr. Opin. Biomed. Eng.* **2017**, *4*, 57–64. [CrossRef]
49. Cross, A. Endotoxin: Back to the future. *Crit. Care Med.* **2016**, *44*, 450–451. [CrossRef] [PubMed]
50. Martel, B.; Moineau, S. Crispr-cas: An efficient tool for genome engineering of virulent bacteriophages. *Nucleic Acids Res.* **2014**, *42*, 9504–9513. [CrossRef] [PubMed]
51. Box, A.M.; McGuffie, M.J.; O'Hara, B.J.; Seed, K.D. Functional analysis of bacteriophage immunity through a type i-e crispr-cas system in vibrio cholerae and its application in bacteriophage genome engineering. *J. Bacteriol.* **2016**, *198*, 578–590. [CrossRef] [PubMed]
52. Lemay, M.-L.; Tremblay, D.M.; Moineau, S. Genome engineering of virulent lactococcal phages using crispr-cas9. *ACS Synth. Biol.* **2017**, *6*, 1351–1358. [CrossRef] [PubMed]
53. Tao, P.; Wu, X.; Tang, W.-C.; Zhu, J.; Rao, V. Engineering of bacteriophage t4 genome using crispr-cas9. *ACS Synth. Biol.* **2017**, *6*, 1952–1961. [CrossRef] [PubMed]
54. Luo, M.L.; Leenay, R.T.; Beisel, C.L. Current and future prospects for crispr-based tools in bacteria. *Biotechnol. Bioeng.* **2016**, *113*, 930–943. [CrossRef] [PubMed]
55. Bardy, P.; Pantucek, R.; Benesik, M.; Doskar, J. Genetically modified bacteriophages in applied microbiology. *J. Appl. Microbiol.* **2016**, *121*, 618–633. [CrossRef] [PubMed]
56. Golan, T.; Milella, M.; Ackerstein, A.; Berger, R. The changing face of clinical trials in the personalized medicine and immuno-oncology era: Report from *The International Congress on Clinical Trials in Oncology & Hemato-Oncology* (Icto 2017). *J. Exp. Clin. Cancer Res.* **2017**, *36*, 192. [PubMed]

antibiotics

MDPI

Opinion

Is Genetic Mobilization Considered When Using Bacteriophages in Antimicrobial Therapy?

Lorena Rodríguez-Rubio, Joan Jofre and Maite Muniesa *

Department of Genetics, Microbiology and Statistics, Faculty of Biology, Av. Diagonal 643, 08028 Barcelona, Spain; lorenarodriguez@ub.edu (L.R.-R.); jjofre@ub.edu (J.J.)

* Correspondence: mmuniesa@ub.edu; Tel.: +34-934-039-386

Academic Editor: Pilar García Suárez

Received: 5 September 2017; Accepted: 4 December 2017; Published: 5 December 2017

Abstract: The emergence of multi-drug resistant bacteria has undermined our capacity to control bacterial infectious diseases. Measures needed to tackle this problem include controlling the spread of antibiotic resistance, designing new antibiotics, and encouraging the use of alternative therapies. Phage therapy seems to be a feasible alternative to antibiotics, although there are still some concerns and legal issues to overcome before it can be implemented on a large scale. Here we highlight some of those concerns, especially those related to the ability of bacteriophages to transport bacterial DNA and, in particular, antibiotic resistance genes.

Keywords: bacteriophages; antimicrobials; lysins; horizontal gene transfer, transduction

1. Revival of Phage Therapy

The World Health Organization (WHO) has identified antibiotic resistance as one of the most challenging problems in public health care on a global scale [1]. The Centers for Disease Control (CDC) estimate that each year around 2 million people in the USA suffer from bacterial infections caused by antibiotic-resistant bacteria, with at least 23,000 resulting in death [2]. In the European Union, antimicrobial resistance is responsible for 25,000 deaths per year and costs around 1.5 billion EUR in healthcare and productivity losses [3]. The report launched by the WHO on 20 September, 2017 [4] shows that as of May 2017, only 8 out of 51 antibiotics in the clinical pipeline belong to new classes, indicating a serious lack of novel antibiotic development. The urgent need for alternative or complementary therapies to control bacterial infections has prompted the rediscovery of phage therapy (the use of bacterial viruses to treat bacterial infections).

As the most abundant entities on Earth, with approximately 10^{31} viral particles [5], bacteriophages, or phages, play a crucial role in the regulation of bacterial populations. After their discovery in the second decade of the 20th century, the capacity of phages to kill pathogenic bacteria led to their therapeutic application against infectious diseases. However, during the early trials of phage therapy (for an extensive review, see [6]), several mistakes were made, mostly attributed to insufficient knowledge about the biological nature of phages. Low titers, preparations contaminated with bacterial antigens, or phages with no infectivity for the bacterial target were used [7]. As a result, the success rate of phage treatment was not constant and after the introduction of antibiotics it was abandoned as unreliable in many parts of the world, with the exception of the Soviet Union and Eastern Europe [8].

The increase of pathogens with multiple resistances to antibiotics has revived interest in phage therapy. Thus, several studies using animal models have been performed against clinically relevant pathogens such as *Pseudomonas aeruginosa*, *Clostridium difficile*, Vancomycin-resistant *Enterococcus faecium*, β-lactamase-producing *Escherichia coli*, *Acinetobacter baumannii*, or *Staphylococcus aureus*, where bacteriophages were used to treat bacteremia or sepsis with a good rate of success in reducing mortality [7]. The phages were mostly administered by intraperitoneal injection, although subcutaneous injection and oral administration were also used in some cases. Clinical trials have also been performed, mostly against antibiotic-resistant *S. aureus* and *P. aeruginosa* (for an extensive review, see [6]). However, these have been primarily focused on safety rather than efficacy [9], since safety concerns are still a major hurdle for the development of phage therapy.

Phages are ubiquitous: both humans and animals carry many different types and are in constant contact with them, so it is reasonable to assume they are not harmful. Although to date no phage products have been approved for human therapy in the USA or EU, several phage cocktails are "Generally Recognized as Safe" (GRAS) products by the US Food and Drug Administration (www.fda.gov) and approved for use in the food industry [10]. On the other hand, the more conservative European Food Safety Authority (EFSA) awarded phages a "Qualified Presumption of Safety" (QPS) in 2007. Recent reports issued by the EFSA on the use of phages in food production [11] indicate that, in the opinion of the EFSA scientific panel of experts, phages have great potential, but further research is advisable for each specific phage application. Most of the reports describing the in vitro efficacy of phages state their usefulness for a variety of antimicrobial applications, although phage–bacteria interactions in an active infection and the involvement of the immune system are difficult to reproduce in vitro [12].

2. Advantages and Potential Disadvantages of Phage Therapy

Many claims have been made about the advantages of phage therapy, notably the highly specific manner in which phages target their host bacterial strain. In addition, phages display great diversity and are relatively easy to isolate. Once the host is infected, phage propagation leads to host lysis at the end of the lytic cycle and the release of virion progeny. The resulting exponential growth in phage numbers amplifies the treatment and the possibilities of success. Phages are also self-limiting, multiplying only as long as host bacteria are present, and they have an inherent low toxicity, since they consist exclusively of proteins and DNA [13]. As clinical trials show, phages are effective against antibiotic-resistant pathogens [6], indicating a lack of cross-resistance with antibiotics.

Opponents of phage therapy always point out the rapid emergence of phage-resistant bacterial mutants and the adverse reaction of the immune system against the phage. However, some strategies to avoid these potential drawbacks have been devised. The most common solution to avoid resistance involves using a phage cocktail, instead of a single phage, since the host is unlikely to become resistant to all the phages simultaneously. Moreover, the emergence of resistant mutants is also a risk of antibiotic treatment, but has never been a reason to discard it. The immune system response could be avoided or minimized by selecting phages with characteristics that are unlikely to trigger an immune response. Interestingly, the immune system does not always thwart phage therapy efficacy. Roach and Leung [14] showed that synergy between the immune system and bacteriophages is essential for the success of phage therapy in the treatment of pulmonary infections caused by *P. aeruginosa*.

3. Bacteriophages and Horizontal Gene Transfer

In our opinion, there is, however, another important issue that is not usually taken into account when selecting phages for therapeutic application. Phages are responsible for a considerable amount of horizontal gene transfer and the evolution of their genomes is characterized by an unusually high degree of horizontal genetic exchange [15]. During the lytic cycle, bacterial rather than phage DNA may be packaged into the phage capsid, producing a transducing particle that upon release from the (donor) host cell can transfer this bacterial DNA to another (recipient) cell [16]. The fact that phages can mobilize bacterial DNA means they can also mobilize and transduce virulence genes [17–19], antibiotic resistance genes [20–24], or genes related to fitness [25,26]. Although transduction was previously thought to be the consequence of errors in the phage packaging machinery and therefore a rare event, occurring approximately once every 10^7–10^9 phage infections [27], recent data suggest this is not the case, as the ratio of transducing particles to lytic phages varies upon prophage induction by different agents and conditions, including antibiotics [28].

By avoiding temperate phages, this problem could certainly be minimized, although not completely resolved. Quite a large number of virulent phages are also capable of mobilizing bacterial sequences, either via generalized transduction events [29–32] or other mechanisms still not completely understood [33]. Phages regulate their own induction and mobilize themselves, but they can also be hijacked by non-self-mobile elements that use phage capsids to spread. Examples can be found in the *Staphylococcus* genera [34–36]. Different parts of the bacterial genome are mobilized, and to date it is not clear if this occurs randomly or is orchestrated to favor the transfer of specific genes. It seems reasonable that the transfer of genes related to the virulence, survival, or fitness of the host strain—such as antibiotic resistance genes (ARGs)—will be favorably selected [19,21,23,37,38]. Moreover, a certain proportion of recombination events (homologous or illegitimate) are responsible for the mosaic structure of phages. These events can take place once phage DNA (even from a virulent phage) reaches the inner cell, and they can occur either between phage and bacterial DNA, or between phage and prophage DNA, as prophage-containing cells are common in most bacterial groups [39,40]. Though quite unpredictable, such events have been observed to take place during generalized transduction [41–43]. The consequences of this recombination could be an additional problem for the inclusion of phages in food or medicines [44], since new recombinant phages could be generated. All these gene transfer mechanisms and recombination events have been reported quite recently, and others will probably appear in the light of new findings resulting from the complete sequencing of bacterial genomes.

4. Bacteriophages as ARG Mobilizing Elements

Despite the growing challenge of antibiotic resistance, we know surprisingly little about how ARGs are transferred between strains, species, and even genera, and even less about how environments and gene-expression levels influence transmission. It seems likely that the use of antibiotics and other antimicrobials increases selective pressure on bacteria that carry ARGs and the vectors that mobilize these genes. From these bacteria, ARGs can be horizontally transferred to other bacteria by mobile genetic elements, most commonly plasmids and transposons, although recently it has been proposed that bacteriophages are also involved [20,45,46]. Bacteriophages basically consist of one nucleic acid molecule (the phage genome) surrounded by a protein coating, the capsid. The packaging of the nucleic acid in a protein capsid confers protection and hence extracellular persistence, which cannot be found in naked DNA or RNA. Since phage-packaged DNA is protected from degradation, and phages can persist in different extracellular environments without losing their infectious capabilities, gene transfer by transduction might be more important than previously thought. Some bacterial genera can produce phage-like elements using information encoded in their own genome. These particles, called gene transfer agents (GTAs), have a bacteriophage-like capsid and although so far they have been reported exclusively in α-proteobacteria [33,47,48], it is reasonable to expect that these mechanisms, or similar ones, could play a role in the spread of bacterial DNA in other bacterial groups. While there is no evidence for the exact origin of GTA genes, there is no doubt they are identical to phage genes.

A recent study has reported that bacteriophages (understood as complete phage particles containing phage DNA) rarely encode ARGs [49], suggesting that instead bacterial DNA is packaged in phage particles. This is supported by our own and other authors' findings that ARGs occur in the bacteriophage DNA fraction of human fecal samples [45], hospital wastewater [50], aquaculture wastewater [51], sludge [52], raw wastewater [53,54], and environmental samples [20]. Providing further evidence, Lekunberri et al. [55] analyzed 33 viromes of different origins and found that while human-associated viromes rarely encode ARGs, the non-human viromes contain a large ARG reservoir, suggesting that bacteriophages or phage particles could play an important role in the spread of resistances in the environment.

It has also become clear that phages play a role in the fine-tuning of all known microbiomes [56]; even their influence on the homeostasis of the microbiota and welfare of the individual cannot be excluded [57–59]. In light of this evidence, we should consider the implications of introducing a cocktail of up to 10^9 PFU/mL of phages into a fine-tuned microbiome, even if they have been confirmed as strictly virulent. The potential impact of this introduction on microbiomes, including gene mobility, is unpredictable. A lack of caution and careful planning when antibiotics first came into use has led to the resistance crisis we are now facing.

5. Phage Lytic Proteins: A Suitable Alternative in Phage Therapy to Avoid the Risk of Genetic Transfer

As phage therapy is one of the most feasible alternatives to antibiotics and due to all the aforementioned concerns, phage lytic enzymes have also been explored as antimicrobials. There are two general classes of phage lytic proteins that mediate the enzymatic cleavage of peptidoglycans (PGs): endolysins and virion-associated peptidoglycan hydrolases (VAPGHs) [60,61]. While VAPGHs degrade PGs in the first stages of phage infection prior to phage DNA injection, endolysins are expressed in the last stages, ensuring the release of the phage progeny via bacterial lysis. The potential of these phage lytic enzymes as antimicrobials and biotechnological tools has been extensively reviewed (see [60–63]), thus here we only point out their applicational advantages over the complete phage particle.

Due to their high specificity and strong activity, bacteriophage lytic enzymes may be as effective as phages while offering additional benefits. Since they are incapable of transferring genetic content, the potential problems of horizontal gene transfer and recombination events are avoided. As enzymes cannot propagate as phages do, their effect is limited to the first doses, allowing better control of a possible influence on microbiota homeostasis. Like phage particles, phage lytic proteins have already proven their efficacy in vitro as well as in animal models [64]. Regarding human applications, Phase I and II clinical trials have been completed by GangaGen Inc. (http://www.gangagen.com, Palo Alto, CA, USA) for the intra-nasal use of an anti-staphylococcal phage protein, and ContraFect (http://www.contrafect.com, Yonkers, NY, USA) is carrying out Phase II trials for the intravenous use of CF-301 to treat *S. aureus* bacteremia (ClinicalTrials.gov Identifier NCT03163446), whose results have not yet been published. For topical application, Staphefekt™ developed by Micreos (http://www.micreos.com, Bilthoven, The Netherlands) is the first endolysin approved for use in humans on intact skin. It is commercialized under the brand Gladskin, which includes products to treat *S. aureus* skin infections (www.gladskin.com, Micreos Human Health, The Hague, The Netherlands).

Phage lysins possess many properties that make them suitable as therapeutics. Their effectivity at low doses would reduce both the immune response and therapy costs, and they display synergistic effects with other antimicrobials [63]. They are highly specific, destroying the target pathogen without affecting commensal microflora. Due to their proteinaceous nature, they are noncorrosive and biodegradable. Several studies have reported the development of antibodies against endolysins upon systemic or mucosal application in animal models [63], but no adverse effects or anaphylaxis were observed and no inactivation by antibodies occurred. Moreover, one of the most interesting features of phage lytic proteins is that no resistance to these enzymes has been reported so far, despite attempts to find it [63,65].

The therapeutic potential of phage lytic proteins has prompted the development of tailor-made antimicrobials based on these enzymes. Their unique modular structure enables domain shuffling, giving rise to antimicrobials with the desired specificity and enhanced activity [64]. Initially, the major disadvantage of phage lytic enzymes as antimicrobials was their inefficacy against Gram-negative bacteria due to the latter's outer membrane barrier. However, this issue seems to have been resolved with the development of Artilysin® engineered enzymes that combine the lytic activity of a phage-derived enzyme with the outer membrane-penetrating activity of an antimicrobial peptide [66,67].

In our opinion, the therapeutic use of phage lytic proteins is more feasible and advisable than that of complete infective phage particles. The limitations to be expected are similar to those of many other therapeutic products: the engineering and production of the enzymes, which require previous optimization.

Acknowledgments: This study was supported by the Generalitat de Catalunya (2014SGR7, the Spanish Ministerio de Innovación y Ciencia (AGL2016-75536-P), the Agencia Estatal de Investigación (AEI) and the European regional fund (ERF) and by the Xarxa de Referència en Biotecnologia (XRB). Lorena Rodríguez-Rubio holds a Juan de la Cierva-Incorporacion grant funded by the Spanish National Plan for Scientific and Technical Research and Innovation 2013-2016.

Author Contributions: L.R.-R., J.J. and M.M. conceived and wrote the paper.

Conflicts of Interest: The authors declare no conflict of interest.

References

1. WHO. *Antibiotic Resistance—A Threat to Global Health Security*; WHO: Geneva, Switzerland, 2013.
2. CDC. *Antibiotic Resistance Threats in the United States*; CDC: Atlanta, GA, USA, 2013.
3. European Commission Antimicrobial Resistance. Available online: https://ec.europa.eu/health/amr/antimicrobial-resistance_en (accessed on 5 December 2017).
4. WHO. *Antibacterial Agents in Clinical Development—An Analysis of the Antibacterial Clinical Development Pipeline, Including Tuberculosis*; WHO: Geneva, Switzerland, 2017.
5. Suttle, C.A. Viruses in the sea. *Nature* **2005**, *437*, 356–361. [CrossRef] [PubMed]
6. Abedon, S.T. Bacteriophage Clinical Use as Antibacterial "Drugs": Utility and Precedent. *Microbiol. Spectr.* **2017**, *5*. [CrossRef] [PubMed]
7. Lin, D.M.; Koskella, B.; Lin, H.C. Phage therapy: An alternative to antibiotics in the age of multi-drug resistance. *World J. Gastrointest. Pharmacol. Ther.* **2017**, *8*, 162. [CrossRef] [PubMed]
8. Abedon, S.T.; Kuhl, S.J.; Blasdel, B.G.; Kutter, E.M. Phage treatment of human infections. *Bacteriophage* **2011**, *1*, 66–85. [CrossRef] [PubMed]
9. Vandenheuvel, D.; Lavigne, R.; Brüssow, H. Bacteriophage Therapy: Advances in Formulation Strategies and Human Clinical Trials. *Annu. Rev. Virol.* **2015**, *2*, 599–618. [CrossRef] [PubMed]
10. *GRAS Notice Inventory—Agency Response Letter GRAS Notice No. GRN 000198*; U.S. Food & Drug: Silver Spring, MD, USA, 2006. Available online: https://www.fda.gov/Food/IngredientsPackagingLabeling/GRAS/NoticeInventory/ucm154675.htm (accessed on 5 December 2017).
11. European Food Safety Authority. The use and mode of action of bacteriophages in food production. *EFSA J.* **2009**, *1076*, 1–26.
12. Reindel, R.; Fiore, C.R. Phage Therapy: Considerations and Challenges for Development. *Clin. Infect. Dis.* **2017**, *64*, 1589–1590. [CrossRef] [PubMed]
13. Stephen, T.; Abedon, A.J.C. Phage Therapy: Emergent Property Pharmacology. *J. Bioanal. Biomed.* **2012**. [CrossRef]
14. Roach, D.R.; Leung, C.Y.; Henry, M.; Morello, E.; Singh, D.; Di Santo, J.P.; Weitz, J.S.; Debarbieux, L. Synergy between the Host Immune System and Bacteriophage Is Essential for Successful Phage Therapy against an Acute Respiratory Pathogen. *Cell Host Microbe* **2017**, *22*, 38–47. [CrossRef] [PubMed]
15. Hatfull, G.F. Bacteriophage genomics. *Curr. Opin. Microbiol.* **2008**, *11*, 447–453. [CrossRef] [PubMed]
16. Zinder, N.D. Bacterial transduction. *J. Cell. Comp. Physiol.* **1995**, *45*, 23–49. [CrossRef]
17. O'Brien, A.; Newland, J.; Miller, S.; Holmes, R.; Smith, H.; Formal, S. Shiga-like toxin-converting phages from Escherichia coli strains that cause hemorrhagic colitis or infantile diarrhea. *Science* **1984**, *226*, 694–696. [CrossRef] [PubMed]
18. Allué-Guardia, A.; García-Aljaro, C.; Muniesa, M. Bacteriophage-encoding cytolethal distending toxin type V gene induced from nonclinical *Escherichia coli* isolates. *Infect. Immun.* **2011**, *79*, 3262–3272. [CrossRef] [PubMed]
19. Penadés, J.R.; Chen, J.; Quiles-Puchalt, N.; Carpena, N.; Novick, R.P. Bacteriophage-mediated spread of bacterial virulence genes. *Curr. Opin. Microbiol.* **2015**, *23*, 171–178. [CrossRef] [PubMed]
20. Colomer-Lluch, M.; Jofre, J.; Muniesa, M. Antibiotic resistance genes in the bacteriophage DNA fraction of environmental samples. *PLoS ONE* **2011**, *6*. [CrossRef] [PubMed]
21. Haaber, J.; Leisner, J.J.; Cohn, M.T.; Catalan-Moreno, A.; Nielsen, J.B.; Westh, H.; Penadés, J.R.; Ingmer, H. Bacterial viruses enable their host to acquire antibiotic resistance genes from neighbouring cells. *Nat. Commun.* **2016**, *7*, 13333. [CrossRef] [PubMed]

22. Muniesa, M.; García, A.; Miró, E.; Mirelis, B.; Prats, G.; Jofre, J.; Navarro, F. Bacteriophages and diffusion of beta-lactamase genes. *Emerg. Infect. Dis.* **2004**, *10*, 1134–1137. [CrossRef] [PubMed]
23. Ross, J.; Topp, E. Abundance of antibiotic resistance genes in bacteriophage following soil fertilization with dairy manure or municipal biosolids, and evidence for potential transduction. *Appl. Environ. Microbiol.* **2015**, *81*, 7905–7913. [CrossRef] [PubMed]
24. Colavecchio, A.; Cadieux, B.; Lo, A.; Goodridge, L.D. Bacteriophages contribute to the spread of antibiotic resistance genes among foodborne pathogens of the *Enterobacteriaceae* family—A review. *Front. Microbiol.* **2017**, *8*. [CrossRef] [PubMed]
25. Lindell, D.; Sullivan, M.B.; Johnson, Z.I.; Tolonen, A.C.; Rohwer, F.; Chisholm, S.W. Transfer of photosynthesis genes to and from *Prochlorococcus* viruses. *Proc. Natl. Acad. Sci. USA* **2004**, *101*, 11013–11018. [CrossRef] [PubMed]
26. Müller, M.G.; Ing, J.Y.; Cheng, M.K.-W.; Flitter, B.A.; Moe, G.R. Identification of a phage-encoded Ig-binding protein from invasive Neisseria meningitidis. *J. Immunol.* **2013**, *191*, 3287–3296. [CrossRef] [PubMed]
27. Bushman, F. *Lateral DNA Transfer. Mechanisms and Consequences*; CSHL Press: New York, NY, USA, 2002.
28. Stanczak-Mrozek, K.I.; Laing, K.G.; Lindsay, J.A. Resistance gene transfer: Induction of transducing phage by sub-inhibitory concentrations of antimicrobials is not correlated to induction of lytic phage. *J. Antimicrob. Chemother.* **2017**, *72*, 1624–1631. [CrossRef] [PubMed]
29. Petty, N.K.; Toribio, A.L.; Goulding, D.; Foulds, I.; Thomson, N.; Dougan, G.; Salmond, G.P.C. A generalized transducing phage for the murine pathogen *Citrobacter rodentium*. *Microbiology* **2007**, *153*, 2984–2988. [CrossRef] [PubMed]
30. Ripp, S.; Ogunseitan, O.A.; Miller, R.V. Transduction of a freshwater microbial community by a new *Pseudomonas aeruginosa* generalized transducing phage, UT1. *Mol. Ecol.* **1994**, *3*, 121–126. [CrossRef] [PubMed]
31. Lee, S.; Kriakov, J.; Vilcheze, C.; Dai, Z.; Hatfull, G.F.; Jacobs, W.R. Bxz1, a new generalized transducing phage for mycobacteria. *FEMS Microbiol. Lett.* **2004**, *241*, 271–276. [CrossRef] [PubMed]
32. Monson, R.; Foulds, I.; Foweraker, J.; Welch, M.; Salmond, G.P.C. The *Pseudomonas aeruginosa* generalized transducing phage φPA3 is a new member of the φKZ-like group of "jumbo" phages, and infects model laboratory strains and clinical isolates from cystic fibrosis patients. *Microbiology* **2011**, *157*, 859–867. [CrossRef] [PubMed]
33. Stanton, T.B. Prophage-like gene transfer agents-novel mechanisms of gene exchange for *Methanococcus, Desulfovibrio, Brachyspira*, and *Rhodobacter* species. *Anaerobe* **2007**, *13*, 43–49. [CrossRef] [PubMed]
34. Quiles-Puchalt, N.; Carpena, N.; Alonso, J.C.; Novick, R.P.; Marina, A.; Penadés, J.R. Staphylococcal pathogenicity island DNA packaging system involving cos-site packaging and phage-encoded HNH endonucleases. *Proc. Natl. Acad. Sci. USA* **2014**, *111*, 6016–6021. [CrossRef] [PubMed]
35. Novick, R.P.; Christie, G.E.; Penadés, J.R. The phage-related chromosomal islands of Gram-positive bacteria. *Nat. Rev. Microbiol.* **2010**, *8*, 541–551. [CrossRef] [PubMed]
36. Frígols, B.; Quiles-Puchalt, N.; Mir-Sanchis, I.; Donderis, J.; Elena, S.F.; Buckling, A.; Novick, R.P.; Marina, A.; Penadés, J.R. Virus Satellites Drive Viral Evolution and Ecology. *PLoS Genet.* **2015**, *11*, e1005609. [CrossRef] [PubMed]
37. Brown-Jaque, M.; Calero-Caceres, W.; Muniesa, M. Transfer of antibiotic-resistance genes via phage-related mobile elements. *Plasmid* **2015**, *79*, 1–7. [CrossRef] [PubMed]
38. Von Wintersdorff, C.J.H.; Penders, J.; Van Niekerk, J.M.; Mills, N.D.; Majumder, S.; Van Alphen, L.B.; Savelkoul, P.H.M.; Wolffs, P.F.G. Dissemination of antimicrobial resistance in microbial ecosystems through horizontal gene transfer. *Front. Microbiol.* **2016**, *7*, 1–10. [CrossRef] [PubMed]
39. Fortier, L.-C.; Sekulovic, O. Importance of prophages to evolution and virulence of bacterial pathogens. *Virulence* **2013**, *4*, 354–365. [CrossRef] [PubMed]
40. Casjens, S. Prophages and bacterial genomics: What have we learned so far? *Mol. Microbiol.* **2003**, *49*, 277–300. [CrossRef] [PubMed]
41. Chiura, H.X. Generalized gene transfer by virus-like particules from marine bacteria. *Aquat. Microb. Ecol.* **1997**, *13*, 75–83. [CrossRef]
42. Thierauf, A.; Perez, G.; Maloy, A.S. Generalized transduction. *Methods Mol. Biol.* **2009**, *501*, 267–286. [PubMed]

43. Beumer, A.; Robinson, J.B. A broad-host-range, generalized transducing phage (SN-T) acquires 16S rRNA genes from different genera of bacteria. *Appl. Environ. Microbiol.* **2005**, *71*, 8301–8304. [CrossRef] [PubMed]
44. Brüssow, H.; Canchaya, C.; Hardt, W.-D. Phages and the evolution of bacterial pathogens: From genomic rearrangements to lysogenic conversion. *Microbiol. Mol. Biol. Rev.* **2004**, *68*, 560–602. [CrossRef] [PubMed]
45. Quirós, P.; Colomer-Lluch, M.; Martínez-Castillo, A.; Miró, E.; Argente, M.; Jofre, J.; Navarro, F.; Muniesa, M. Antibiotic resistance genes in the bacteriophage DNA fraction of human fecal samples. *Antimicrob. Agents Chemother.* **2014**, *58*, 606–609. [CrossRef] [PubMed]
46. Balcazar, J.L. Bacteriophages as Vehicles for Antibiotic Resistance Genes in the Environment. *PLoS Pathog.* **2014**, *10*. [CrossRef] [PubMed]
47. Zhao, Y.; Wang, K.; Budinoff, C.; Buchan, A.; Lang, A.; Jiao, N.; Chen, F. Gene transfer agent (GTA) genes reveal diverse and dynamic *Roseobacter* and *Rhodobacter* populations in the Chesapeake Bay. *ISME J.* **2009**, *3*, 364–373. [CrossRef] [PubMed]
48. Lang, A.S.; Zhaxybayeva, O.; Beatty, J.T. Gene transfer agents: Phage-like elements of genetic exchange. *Nat. Rev. Microbiol.* **2012**, *10*, 472–482. [CrossRef] [PubMed]
49. Enault, F.; Briet, A.; Bouteille, L.; Roux, S.; Sullivan, M.B.; Petit, M.-A. Phages rarely encode antibiotic resistance genes: A cautionary tale for virome analyses. *ISME J.* **2017**, *11*, 237–247. [CrossRef] [PubMed]
50. Subirats, J.; Sànchez-Melsió, A.; Borrego, C.M.; Balcázar, J.L.; Simonet, P. Metagenomic analysis reveals that bacteriophages are reservoirs of antibiotic resistance genes. *Int. J. Antimicrob. Agents* **2016**, *48*, 163–167. [CrossRef] [PubMed]
51. Colombo, S.; Arioli, S.; Guglielmetti, S.; Lunelli, F.; Mora, D. Virome-associated antibiotic-resistance genes in an experimental aquaculture facility. *FEMS Microbiol. Ecol.* **2016**, *92*. [CrossRef] [PubMed]
52. Calero-Cáceres, W.; Melgarejo, A.; Colomer-Lluch, M.; Stoll, C.; Lucena, F.; Jofre, J.; Muniesa, M. Sludge as a potential important source of antibiotic resistance genes in both the bacterial and bacteriophage fractions. *Environ. Sci. Technol.* **2014**, *48*, 7602–7611. [CrossRef] [PubMed]
53. Calero-Cáceres, W.; Muniesa, M. Persistence of naturally occurring antibiotic resistance genes in the bacteria and bacteriophage fractions of wastewater. *Water Res.* **2016**, *95*, 11–18. [CrossRef] [PubMed]
54. Colomer-Lluch, M.; Calero-Cáceres, W.; Jebri, S.; Hmaied, F.; Muniesa, M.; Jofre, J. Antibiotic resistance genes in bacterial and bacteriophage fractions of Tunisian and Spanish wastewaters as markers to compare the antibiotic resistance patterns in each population. *Environ. Int.* **2014**, *73*, 167–175. [CrossRef] [PubMed]
55. Lekunberri, I.; Subirats, J.; Borrego, C.M.; Balcázar, J.L. Exploring the contribution of bacteriophages to antibiotic resistance. *Environ. Pollut.* **2017**, *220*, 981–984. [CrossRef] [PubMed]
56. Virgin, H.W. The virome in mammalian physiology and disease. *Cell* **2014**, *157*, 142–150. [CrossRef] [PubMed]
57. Navarro, F.; Muniesa, M. Phages in the human body. *Front. Microbiol.* **2017**, *8*, 1–7. [CrossRef] [PubMed]
58. Norman, J.M.; Handley, S.A.; Baldridge, M.T.; Droit, L.; Liu, C.Y.; Keller, B.C.; Kambal, A.; Monaco, C.L.; Zhao, G.; Fleshner, P.; et al. Disease-Specific Alterations in the Enteric Virome in Inflammatory Bowel Disease. *Cell* **2015**, *160*, 447–460. [CrossRef] [PubMed]
59. Pérez-Brocal, V.; García-López, R.; Nos, P.; Beltrán, B.; Moret, I.; Moya, A. Metagenomic Analysis of Crohn's Disease Patients Identifies Changes in the Virome and Microbiome Related to Disease Status and Therapy, and Detects Potential Interactions and Biomarkers. *Inflamm. Bowel Dis.* **2015**. [CrossRef] [PubMed]
60. Nelson, D.C.; Schmelcher, M.; Rodriguez-Rubio, L.; Klumpp, J.; Pritchard, D.G.; Dong, S.; Donovan, D.M. Endolysins as Antimicrobials. *Adv. Virus Res.* **2012**, *83*, 299–365. [PubMed]
61. Rodríguez-Rubio, L.; Martínez, B.; Donovan, D.M.; Rodríguez, A.; García, P. Bacteriophage virion-associated peptidoglycan hydrolases: Potential new enzybiotics. *Crit. Rev. Microbiol.* **2013**, *39*, 427–434. [CrossRef] [PubMed]
62. Rodríguez-Rubio, L.; Gutiérrez, D.; Donovan, D.M.; Martínez, B.; Rodríguez, A.; García, P. Phage lytic proteins: Biotechnological applications beyond clinical antimicrobials. *Crit. Rev. Biotechnol.* **2016**, *8551*, 1–11. [CrossRef] [PubMed]
63. Schmelcher, M.; Donovan, D.M.; Loessner, M.J. Bacteriophage endolysins as novel antimicrobials. *Future Microbiol.* **2012**, *7*, 1147–1171. [CrossRef] [PubMed]
64. Roach, D.R.; Donovan, D.M. Antimicrobial bacteriophage-derived proteins and therapeutic applications. *Bacteriophage* **2015**, *5*, e1062590. [CrossRef] [PubMed]

65. Rodriguez-Rubio, L.; Martinez, B.; Rodriguez, A.; Donovan, D.M.; Goetz, F.; Garcia, P. The Phage Lytic Proteins from the *Staphylococcus aureus* Bacteriophage vB_SauS-phiIPLA88 Display Multiple Active Catalytic Domains and Do Not Trigger Staphylococcal Resistance. *PLoS ONE* **2013**, *8*, e64671. [CrossRef] [PubMed]
66. Gerstmans, H.; Rodriguez-Rubio, L.; Lavigne, R.; Briers, Y. From endolysins to Artilysin(R)s: Novel enzyme-based approaches to kill drug-resistant bacteria. *Biochem. Soc. Trans.* **2016**, *44*, 123–128. [CrossRef] [PubMed]
67. Briers, Y.; Walmagh, M.; Van Puyenbroeck, V.; Cornelissen, A.; Cenens, W.; Aertsen, A.; Oliveira, H.; Azeredo, J.; Verween, G.; Pirnay, J.-P.; et al. Engineered endolysin-based "Artilysins" to combat multidrug-resistant gram-negative pathogens. *mBio* **2014**, *5*, e01379-14. [CrossRef] [PubMed]

MDPI
St. Alban-Anlage 66
4052 Basel
Switzerland
Tel. +41 61 683 77 34
Fax +41 61 302 89 18
www.mdpi.com

Antibiotics Editorial Office
E-mail: antibiotics@mdpi.com
www.mdpi.com/journal/antibiotics

www.ingramcontent.com/pod-product-compliance
Lightning Source LLC
LaVergne TN
LVHW070245100526
838202LV00015B/2182
* 9 7 8 3 0 3 9 4 3 4 0 4 6 *